Recent Titles in This Series

138 **Yu. P. Razmyslov,** Identities of algebras and their representations, 1994
137 **F. I. Karpelevich and A. Ya. Kreinin,** Heavy traffic limits for multiphase queues, 1994
136 **Masayoshi Miyanishi,** Algebraic geometry, 1994
135 **Masaru Takeuchi,** Modern spherical functions, 1994
134 **V. V. Prasolov,** Problems and theorems in linear algebra, 1994
133 **P. I. Naumkin and I. A. Shishmarev,** Nonlinear nonlocal equations in the theory of waves, 1994
132 **Hajime Urakawa,** Calculus of variations and harmonic maps, 1993
131 **V. V. Sharko,** Functions on manifolds: Algebraic and topological aspects, 1993
130 **V. V. Vershinin,** Cobordisms and spectral sequences, 1993
129 **Mitsuo Morimoto,** An introduction to Sato's hyperfunctions, 1993
128 **V. P. Orevkov,** Complexity of proofs and their transformations in axiomatic theories, 1993
127 **F. L. Zak,** Tangents and secants of algebraic varieties, 1993
126 **M. L. Agranovskiĭ,** Invariant function spaces on homogeneous manifolds of Lie groups and applications, 1993
125 **Masayoshi Nagata,** Theory of commutative fields, 1993
124 **Masahisa Adachi,** Embeddings and immersions, 1993
123 **M. A. Akivis and B. A. Rosenfeld,** Élie Cartan (1869–1951), 1993
122 **Zhang Guan-Hou,** Theory of entire and meromorphic functions: Deficient and asymptotic values and singular directions, 1993
121 **I. B. Fesenko and S. V. Vostokov,** Local fields and their extensions: A constructive approach, 1993
120 **Takeyuki Hida and Masuyuki Hitsuda,** Gaussian processes, 1993
119 **M. V. Karasev and V. P. Maslov,** Nonlinear Poisson brackets. Geometry and quantization, 1993
118 **Kenkichi Iwasawa,** Algebraic functions, 1993
117 **Boris Zilber,** Uncountably categorical theories, 1993
116 **G. M. Fel′dman,** Arithmetic of probability distributions, and characterization problems on abelian groups, 1993
115 **Nikolai V. Ivanov,** Subgroups of Teichmüller modular groups, 1992
114 **Seizô Itô,** Diffusion equations, 1992
113 **Michail Zhitomirskiĭ,** Typical singularities of differential 1-forms and Pfaffian equations, 1992
112 **S. A. Lomov,** Introduction to the general theory of singular perturbations, 1992
111 **Simon Gindikin,** Tube domains and the Cauchy problem, 1992
110 **B. V. Shabat,** Introduction to complex analysis Part II. Functions of several variables, 1992
109 **Isao Miyadera,** Nonlinear semigroups, 1992
108 **Takeo Yokonuma,** Tensor spaces and exterior algebra, 1992
107 **B. M. Makarov, M. G. Goluzina, A. A. Lodkin, and A. N. Podkorytov,** Selected problems in real analysis, 1992
106 **G.-C. Wen,** Conformal mappings and boundary value problems, 1992
105 **D. R. Yafaev,** Mathematical scattering theory: General theory, 1992
104 **R. L. Dobrushin, R. Kotecký, and S. Shlosman,** Wulff construction: A global shape from local interaction, 1992
103 **A. K. Tsikh,** Multidimensional residues and their applications, 1992
102 **A. M. Il′in,** Matching of asymptotic expansions of solutions of boundary value problems, 1992
101 **Zhang Zhi-fen, Ding Tong-ren, Huang Wen-zao, and Dong Zhen-xi,** Qualitative theory of differential equations, 1992
100 **V. L. Popov,** Groups, generators, syzygies, and orbits in invariant theory, 1992
99 **Norio Shimakura,** Partial differential operators of elliptic type, 1992
98 **V. A. Vassiliev,** Complements of discriminants of smooth maps: Topology and applications, 1992 (revised edition, 1994)

(Continued in the back of this publication)

Identities of Algebras and Their Representations

Translations of

MATHEMATICAL MONOGRAPHS

Volume 138

Identities of Algebras and Their Representations

Yu. P. Razmyslov

American Mathematical Society
Providence, Rhode Island

Ю. П. Размыслов

ТОЖДЕСТВА АЛГЕБР И ИХ ПРЕДСТАВЛЕНИЙ

«НАУКА» МОСКВА, 1989

Translated by A. M. Shtern
Translation edited by Simeon Ivanov

1991 *Mathematics Subject Classification.* Primary 17B99, 16R10;
Secondary 17B30.

ABSTRACT. This book is devoted to the study of an important branch of abstract algebra: analysis of identities in algebras and in representations of algebras. The main technique developed in the book is the method of 2-words and α-functions that was invented by the author in the late seventies. This method, which can be compared to the theory of group characters, enabled the author to solve a number of important problems that remained unsolved for several decades. This is the first monograph which gives a detailed account of recent achievements in the theory of identical relations in algebras and their representations. It will be useful to specialists working in algebra, mathematical logic, algebraic geometry, and other areas of mathematics.

Library of Congress Cataloging-in-Publication Data

Razmyslov, ÎU. P. (ÎUriĭ Pitrimovich)
[Tozhdestva algebr i ikh predstavlenii. English]
Identities of algebras and their representations/Yu. P. Razmyslov.
p. cm. — (Translations of mathematical monographs, ISSN 0065-9282; v. 138)
Includes bibliographical references.
ISBN 0-8218-4608-6 (acid-free)
1. Lie algebras. 2. Representations of algebras. I. Title. II. Series.
QA252.3.R3913 1994
512′.55—dc20 94-20766
CIP

∞ The paper used in this book is acid-free and falls within the guidelines
established to ensure permanence and durability.
♻ Printed on recycled paper.
This volume was typeset using AMS-TeX,
the American Mathematical Society's TeX macro system.

10 9 8 7 6 5 4 3 2 1 98 97 96 95 94

CONTENTS

PREFACE

During the last forty years a new trend in the theory of associative algebras, Lie algebras, and their representations has formed under the influence of mathematical logic and universal algebra—namely, the theory of varieties and identities of associative algebras, Lie algebras, and their representations. The study of two interrelated classes of problems is characteristic for this direction. The first class deals with the problems of finding identities of concrete classical algebras and their representations, while the second class includes the problems of the description of properties of algebras and their representations satisfying nontrivial identical relations, in particular, identities of certain classical algebras. The first class of problems forms a basis of the theory, and it is of interest mainly to specialists in algebra and mathematical logic, whereas the second one is the "applied" part of the theory and this can be used in various fields in mathematics and is of interest to a wide circle of experts in functional analysis, differential and algebraic geometry, superalgebras, and Lie algebras. Presently there are many papers devoted to this subject. Certain monographs have appeared; in particular, "Nauka" (Moscow) recently published a monograph of Yu. A. Bakhturin, "Identical Relations in Lie Algebras", which contains sufficiently complete information on classical problems and methods of the theory of varieties of Lie algebras.

Investigations of the last twenty years in the area of identities of algebras and their representations lead to the creation of a new method—the method of 2-words and $\boldsymbol{\alpha}$-functions. This original method, remotely reminiscent of the character theory in group theory, has arisen at the juncture of universal algebra and representation theory of reductive Lie algebras in the process of my attempts to solve some particular problems about identities that have remained unsolved despite prolonged efforts to apply other methods. The formation of the method was basically completed in 1976. As a result, a number of well-known principal problems of the theory of groups, rings, Lie algebras, and their representations were solved subsequently in a unified way. The results of these investigations have applications in differential and algebraic geometry, in the theory of Lie superalgebras, and in the theory of simple infinite-dimensional Lie algebras.

The possibilities of the method are far from being exhausted. However, an important period appears in its development, in which, continuing our going forward, we are to look back to sum up the applications of the method in the theory of varieties and to give a systematic exposition of contemporary achievements of the theory of identities of algebras and their representations closely related to the method of 2-words and $\boldsymbol{\alpha}$-functions. Understanding of this necessity made me begin writing this book and mainly determined the circle of problems on algebras and their representations treated in it.

There are seven chapters in the book.

Chapter I is introductory; in it we analyse the three classical approaches to the concept of identity of a representation of a Lie algebra, introduce the basic notions of the theory of varieties of representations, and develop the growth theory for varieties of representations of Lie algebras. Theorems included in this chapter show that identities are important invariants of an irreducible representation of a finite-dimensional algebra and that they determine the representation up to an isomorphism.

Chapter II is an introduction to the theory of 2-words and $\boldsymbol{\alpha}$-functions. It forms a conceptual basis for the entire book. Here the main facts of this theory are collected and a general setting of the main problem for the method of 2-words is given. Using the examples of the nonsolvability problem for $(p-2)$th Engel Lie algebras and of the problem on identities of the two-dimensional representation of the simple three-dimensional Lie algebra, we show how the main notions of the method under consideration naturally arise.

The three subsequent chapters are related to the solution of the general problem of the theory of 2-words and $\boldsymbol{\alpha}$-functions posed in Chapter II for three partucular sets of irreducible representations of reductive Lie algebras. Their main aim is to fill the statements discussed in Chapter II with real content and to show how the method of 2-words permits us to obtain results on identities of algebras and their representations. In these chapters we discuss finite basis problems for identities of particular varieties, develop the theory of trace identities and of central polynomials for matrix algebras M_n, for matrix superalgebras, and for irreducible representations of reductive Lie algebras, and give applications of the method of 2-words to varieties of associative nil-algebras, Lie algebras with Engel property, and locally finite groups of prime period.

In Chapter VI we discuss the theory of identities of Lie algebras of Cartan type and give a classification of simple infinite-dimensional Lie algebras for which the identities of Lie algebras of Cartan type are satisfied. These results are closely related to the theory (exposed in Chapter III) of central polynomials of irreducible representations of reductive Lie algebras and also to the existence of a polynomial mapping which permits one to reconstruct the algebra of regular functions of a smooth variety from its Lie algebra of vector fields.

Each of the first six chapters is concluded by comments. We give there precise references to the authors and list the papers in which the results discussed in the chapter have been published; we also mention open problems of importance both to the theory itself and to its applications.

My intent while writing this book was to make it more accessible to a wide circle of readers, in particular, to second year students of physics and mathematics about to choose their specialization. To this end, in Chapter VII of the book I have collected more or less well-known auxiliary facts and theorems of a general nature that are constantly used in the previous chapters, but whose proofs are scattered about various textbooks and monographs. I hope that this makes Chapters I, II, III, and VI completely accessible to the reader acquainted with generalities on linear spaces over fields. Chapter IV requires the knowledge of the classical Young theory of irreducible representations of the symmetric group to the extent of several pages of §28 of the book of Curtis and Reiner [CR]. To be able to read Chapter III, the reader must be acquainted with the theory of semisimple finite-dimensional Lie algebras and their representations over algebraically closed fields of characteristic zero. A vast literature is presently devoted to this topic. I recommend that the reader, in case of difficulties with Chapter III, consult the excellent book [Dix] of J. Dixmier on universal enveloping algebras. Chapters I, III, IV, V, and VI are almost independent of each other, and they can be studied in an arbitrary order after Chapter II. One can learn about the main events of the development of the theory of identities from the short historical review at the end of the book.

I also wanted this book to be a reader's quick guide to the main problems of modern theory of identities of algebras and their representations. So, I shall be glad if the problems treated here help get the reader to those frontiers of the theory where the vast wide space opens up for independent creative work.

CHAPTER I

PRELIMINARY RESULTS

In this chapter we introduce certain concepts, namely: associative-Lie pair, identity of a representation of a Lie algebra, variety of associative-Lie pairs, and variety of representations of Lie algebras. We develop the growth theory in varieties of associative-Lie pairs and give a new construction of the central closure that was first introduced for binary algebras by W. Martindale. The main results of the chapter are the following: the existence theorem for varieties of associative-Lie pairs of arbitrary high growth (see Theorem 2.4), a rank theorem (see §4), and Theorems 5.1–5.4 and their corollaries which imply (in the case of an algebraically closed field) that centrally prime associative-Lie pairs with the same identities are isomorphic.

§1. Associative-Lie pairs, identities of pairs, and varieties of pairs. Connections with varieties of representations of Lie algebras

In this section we describe three existing and presently utilized approaches, based on the notion of identity, to the notion of the variety of representations of Lie algebras.

We denote the ground field by K.

Let A be an arbitrary associative K-algebra. It is well known (e.g., see §52) that in this case the linear space of A can be endowed with the structure of a Lie algebra by introducing on it a new binary multilinear operation $[,]\colon A \otimes_K A \to A$ by means of the formula $[a, b] \rightleftharpoons ab - ba$ for any elements a, b from the space A. The Lie algebra obtained in this way from the associative algebra A will be denoted by $A^{(-)}$.

1.1. The category of associative-Lie pairs. Identities of pairs. A K-algebra A is said to be *enveloping* for a Lie K-algebra L if L is a Lie subalgebra in $A^{(-)}$ and A (considered as an associative K-algebra) is generated by the subspace of L. The pair of objects: a Lie algebra L and its enveloping associative algebra A is called an *associative-Lie pair* (or simply a *pair*) and is denoted by (A, L).

Generators of the pair (A, L) are, by definition, generators of the Lie algebra L.

A pair (A_2, L_2) is said to be a *subpair* of a pair (A_1, L_1) if L_2 is a Lie subalgebra in L_1 and A_2 is an associative subalgebra in A_1.

By a *homomorphism* of a pair (A_1, L_1) into a pair (A_2, L_2) we mean a homomorphism of associative algebras $\zeta\colon A_1 \to A_2$ such that $\zeta(L_1)$ is contained in L_2.

The category whose objects are associative-Lie pairs and morphisms are homomorphisms of pairs is called the *category of associative-Lie pairs* and is denoted by Π_1.

The analog to the Cartesian product is naturally defined in the category of pairs. Let (A_i, L_i), $i \in I$, be a set of pairs, and let $\prod_{i\in I} A_i$, $\prod_{i\in I} L_i$ be the full Cartesian products of associative algebras and Lie algebras, respectively. Obviously, $\prod_{i\in I} L_i$ is a Lie subalgebra in $\left(\prod_{i\in I} A_i\right)^{(-)}$. Denote by L the Lie algebra $\prod_{i\in I} L_i$ and by A the associative subalgebra in $\prod_{i\in I} A_i$ generated by L. The pair (A, L) is called the *full Cartesian product* of the pairs (A_i, L_i), $i \in I$.

Let H be an absolutely free associative algebra with a countable number of free generators $t_1, t_2, t_3, \ldots$. The elements of H are formal K-linear combinations of different words in the alphabet $t_1, t_2, \ldots$ with the natural operation of multiplication and the natural structure of a K-linear space. It is well known (e.g., see §52) that the Lie subalgebra $\mathfrak{M}$ generated in the Lie algebra $H^{(-)}$ by the generators $t_1, t_2, \ldots$ is a free Lie algebra of countable rank with free generators $t_1, t_2, t_3, \ldots$, and the associative K-algebra H is isomorphic to the universal enveloping algebra of the Lie algebra $\mathfrak{M}$. This is equivalent to the following: any mapping ζ of the generators $t_1, t_2, t_3, \ldots$ to any Lie algebra L has an extension to a homomorphism $\zeta\colon \mathfrak{M} \to L$ of Lie algebras and thus for any pair (A, L) it has an extension to a homomorphism $\zeta\colon H \to A$ of associative algebras. Since then $\zeta(\mathfrak{M}) \subset L$, it follows that any mapping ζ of the generators $t_1, t_2, t_3, \ldots$ of the pair $(H, \mathfrak{M})$ has an extension to a homomorphism of pairs $\zeta\colon (H, \mathfrak{M}) \to (A, L)$. This means that the pair $(H, \mathfrak{M})$ is a *free pair* of countable rank.

DEFINITION 1.1. Let $f(t_1, \ldots, t_l)$ be an associative polynomial in a free associative algebra H. The formula $f = 0$ is called an *identity* of the pair (A, L) if for any elements $e_1, \ldots, e_l \in L$ the element $f(e_1, \ldots, e_l)$ is equal to zero in the algebra A.

It is obvious that if $f(t_1, \ldots, t_l) = 0$ is an identity of the pair (A, L), then for any Lie polynomials $g_1, \ldots, g_l$ and words $a, b \in H$ the formula $af(g_1, \ldots, g_l)b = 0$ is also an identity of the pair (A, L). Thus the set $\mathfrak{V}$ of polynomials in H, consisting of all polynomials $f \in H$ such that $f = 0$ is an identity of the pair (A, L), is a two-sided ideal in H invariant under all endomorphisms of the pair $(H, \mathfrak{M})$. This leads to the following definition.

DEFINITION 1.2. An ideal $\mathfrak{V}$ in the free associative algebra H is called an *ideal of weak identities* if for any $f(t_1, \ldots, t_l) \in \mathfrak{V}$ and any $g_1, \ldots, g_l \in \mathfrak{M}$, the polynomial $f(g_1, \ldots, g_l)$ belongs to $\mathfrak{V}$.

It is evident that the set of left-hand sides of all identities of the form $f = 0$ of the pair (A, L) forms an ideal of weak identities in H. Obviously, the converse is also true: if $\mathfrak{V}$ is an ideal of weak identities in H, then all the identities of the pair $(A/\mathfrak{V}, \mathfrak{M}/(\mathfrak{M} \cap \mathfrak{V}))$ are of the form $f = 0$, where f runs over the ideal $\mathfrak{V}$.

For any class of pairs, all the left-hand sides of the identities of the form $f = 0$ that are valid on each pair of this class form an ideal of weak identities, because the intersection of any set of ideals of weak identities is an ideal of weak identities.

DEFINITION 1.3. Let S be a set of polynomials in a free associative algebra H. An ideal of weak identities $\mathfrak{V}$ is said to be *generated* by the set S if the smallest ideal of weak identities $\mathfrak{V}'$ containing S coincides with $\mathfrak{V}$. An identity $f = 0$ is said to be a *consequence* of the identities $g = 0$, where g ranges over the set S, if and only if $f \in \mathfrak{V}'$.

In the ideal of weak identities $\mathfrak{V}(S)$ generated by a subset S in the algebra H, all elements can be defined by induction with the help of the following rules:

a) all polynomials in the set S belong to $\mathfrak{V}(S)$;

b) if $f(t_1, \dots, t_l) \in \mathfrak{V}(S)$ and $g_1, \dots, g_l \in \mathfrak{M}$, $a, b \in H$, then

$$af(g_1, \dots, g_l)b,\ af(g_1, \dots, g_l),\ f(g_1, \dots, g_l)b \in \mathfrak{V}(S);$$

c) if $f_1, \dots, f_k \in \mathfrak{V}(S)$, then any K-linear combination of them belongs to $\mathfrak{V}(S)$.

DEFINITION 1.4. Let S be a set of polynomials in H. The class of all pairs in which all the identities of the form $g = 0$, $g \in S$, are valid is called a *variety of associative-Lie pairs*.

It is easy to see that if an identity $f = 0$ is valid in a certain pair (A, L), then it is also valid in any subpair of this pair and in any homomorphic image of the pair. Moreover, if the identity $f = 0$ is valid in all pairs (A_i, B_i), $i \in I$, then it is valid also in the full Cartesian product of the pairs (A_i, B_i), $i \in I$. This shows that any variety of pairs is closed under the operations of passing to the following objects: a) Cartesian products; b) subpairs; c) homomorphic images. Using the argument that is known in universal algebra as the one due to Birkhoff (e.g., cf. §49), we see that the converse is also true: any class of associative-Lie pairs that is closed under these three operations is a variety.

It follows from Definitions 1.4 and 1.3 that if for all the pairs of a certain variety $\mathfrak{B}$ the identities $g = 0$ are valid for a certain set of polynomials S in a free algebra H, then the identities $f = 0$ are valid for all pairs of the variety $\mathfrak{B}$, where f ranges over the ideal of weak identities $\mathfrak{V}(S)$ generated by the subset S in the algebra H. This shows that any variety $\mathfrak{B}$ of pairs in the algebra H corresponds to a uniquely determined ideal of weak identities $\mathfrak{V}$; a polynomial $f \in H$ belongs to $\mathfrak{V}$ if and only if the identity $f = 0$ is valid in all pairs of the variety $\mathfrak{B}$. It is evident that this correspondence between varieties of associative-Lie pairs and ideals of weak identities in the algebra H is bijective.

DEFINITION 1.5. A pair $(F, \mathfrak{L})$ is called a *relatively free pair* with free generators x_i, where i ranges over a certain set I of indices, if any mapping ζ of the generators x_i to the Lie algebra $\mathfrak{L}$ has an extension to a homomorphism $\zeta\colon (F, \mathfrak{L}) \to (F, \mathfrak{L})$. A pair $(F, \mathfrak{L})$ is called a *free pair* of a certain variety of pairs $\mathfrak{B}$ if $(F, \mathfrak{L})$ is contained in $\mathfrak{B}$ and for all pairs (A, L) from $\mathfrak{B}$ any mapping ζ of the generators x_i to L has an extension to a homomorphism of pairs $\zeta\colon (F, \mathfrak{L}) \to (A, L)$.

DEFINITION 1.6. Let (A, L) be a pair. An ideal of weak identities $\mathfrak{V}$ of the free associative algebra H is said to be an *ideal of identities of the pair* (A, L) provided an identity $f = 0$ is valid in the pair (A, L) if and only if $f \in \mathfrak{V}$. A variety of pairs is called a *variety generated by a pair* (A, L) provided any identity is valid in this variety if and only if it is valid in the pair (A, L); this variety is denoted by $\mathrm{var}(A, L)$.

Birkhoff's argument (e.g., cf. §49) shows that any relatively free pair $(F, \mathfrak{L})$ of the variety $\mathrm{var}(A, L)$ can be obtained as a subpair of the full Cartesian product of a certain set of isomorphic copies of (A, L) and that any pair of $\mathrm{var}(A, L)$ can be obtained as a homomorphic image of a certain subpair of the full Cartesian product of sufficiently many isomorphic copies of the pair (A, L).

1.2. The category of representations of Lie algebras in associative algebras. Let L be a Lie algebra and A an associative algebra. As usual, a homomorphism $\rho\colon L \to A^{(-)}$ of Lie algebras is called a *representation* of the Lie algebra L in the associative algebra A. A representation is said to be *faithful* if the kernel of this homomorphism is equal to zero. Consider a category Π_2 whose objects are arbitrary representations of various Lie algebras L in various algebras A. A *morphism* of representations $\rho_1\colon L_1 \to A_1$ and $\rho_2\colon L_2 \to A_2$ is a pair of homomorphisms $\varphi\colon L_1 \to L_2$ (homomorphism of Lie algebras) and $\psi\colon A_1 \to A_2$ (homomorphism of associative algebras) for which the diagram

$$\begin{array}{ccc} L_1 & \xrightarrow{\rho_1} & A_1 \\ {\scriptstyle\varphi}\downarrow & & \downarrow{\scriptstyle\psi} \\ L_2 & \xrightarrow[\rho_2]{} & A_2 \end{array}$$

is commutative. Let $\rho\colon L \to A$ be a representation. By analogy with what was said above, a formula $f(t_1, \dots, t_l) = 0$, $f \in H$, such that $f(\rho(e_1), \dots, \rho(e_l)) = 0$ in the algebra A for all $e_1, \dots, e_l \in L$ is called an *identity of the representation* ρ. Denote by L_ρ the Lie algebra $\rho(L)$ and by A_ρ the associative subalgebra in A generated by L_ρ. It follows from Definition 1.1 that the pair (A_ρ, L_ρ) and the representation $\rho\colon L \to A$ have the same identities. Evidently, the pairs form a subcategory of the category of representations of Lie algebras in associative algebras, and the assignment of the pair (A_ρ, L_ρ) to a representation $\rho\colon L \to A$ defines a functor from the category of representations Π_2 to the category of pairs Π_1. We denote this functor by O_1.

The category of representations Π_2 of Lie algebras in associative algebras possesses another description from the viewpoint of algebras with two supports and with signature $\{\cdot; [,]; \rho\}$. In fact, any representation $\rho\colon L \to A$ can be considered as an algebra $\{A, L\}$ with two supports and with multilinear operations

$$(1.1) \qquad \cdot\colon A \otimes_K A \to A; \qquad [,]\colon L \otimes_K L \to L; \qquad \rho\colon L \to A;$$

the first of them endows A with the structure of associative algebra, the second one supplies L with the structure of Lie algebra, and the third one satisfies the identity

$$\rho([x, y]) = \rho(x)\rho(y) - \rho(y)\rho(x)$$

for all $x, y \in L$. Under this approach, the morphisms in the category Π_2 are the homomorphisms of algebras with two supports, and the class Π_2 itself of algebras with two supports forms a subvariety of the variety of all algebras with two supports of signature $\{\cdot; [,]; \rho\}$ (the 'arities', ranges and domains of these multilinear operations are indicated in (1.1)). This permits one to use the language of varieties for such algebras with two supports. It is evident that the functor O_1 from the category Π_2 to the category Π_1 commutes with the operations of passing to full Cartesian products, homomorphic images, subalgebras and subpairs. This shows that the functor O_1 maps any subvariety of algebras with two supports in Π_2 into a certain subvariety of the variety of pairs Π_1.

It is possible to construct explicitly the free algebra $\{F, \mathfrak{L}\}$ with two supports and with the set of free generators $\{Y, T\}$ (where $Y \cap T = \varnothing$, $T \subset \mathfrak{L}$, $Y \subset F$) for the variety Π_2 of algebras with two supports. Toward this end, it is sufficient to take for F the free associative algebra with the set of free generators $Y \cup T$, and for $\mathfrak{L}$ the free Lie algebra with the set of free generators T and the natural embedding operation $\rho\colon \mathfrak{L} \to F^{(-)}$ that is the identity on the set T. Let $\{A, L\}$ be any algebra with two supports from Π_2. It is evident that any pair of maps $\varphi\colon T \to L$, $\psi\colon Y \to A$ has a unique extension to a homomorphism $(\psi, \varphi)\colon \{F, \mathfrak{L}\} \to \{A, L\}$, and if T and Y are countable then any subvariety $\mathfrak{B}$ of algebras with two supports in Π_2 is uniquely determined by the ideal $\{\mathfrak{V}_1, \mathfrak{V}_2\}$ in $\{F, \mathfrak{L}\}$, where $\mathfrak{V}_2$ is the ideal in $\mathfrak{L}$ invariant under all endomorphisms of the Lie algebra $\mathfrak{L}$ and $\mathfrak{V}_1$ is the two-sided ideal in F that contains $\rho(\mathfrak{V}_2)$ and is stable under all endomorphisms of the associative algebra F mapping T to $\rho(\mathfrak{L})$. Moreover, any identity $\{f = 0,\ g = 0\}$ with $f \in F$, $g \in \mathfrak{L}$, is valid in the variety $\mathfrak{B}$ if and only if $f \in \mathfrak{V}_1$, $g \in \mathfrak{V}_2$.

Let $T = \{t_1, t_2, t_3, \dots\}$. We identify in a natural way the Lie algebras $\mathfrak{L}$ and $\mathfrak{M}$; moreover, we identify the free associative algebra H (with the free generators $t_1, t_2, t_3, \dots$) with the subalgebra in F generated by T. Then for any ideal of weak identities $\mathfrak{V}$ in H, the subclass of all representations of Lie algebras in associative algebras from Π_2 such that all the identities of the form $\{f = 0,\ 0 = 0\}$ are valid, where f ranges over $\mathfrak{V}$, forms a subvariety of the variety Π_2 of all algebras with two supports for which the ideal of identities has the form $\{\mathfrak{V} + F\mathfrak{V} + \mathfrak{V}F + F\mathfrak{V}F,\ 0\}$. This proves that any subvariety of pairs is obtained from some subvariety of algebras with two supports in Π_2 with the help of the functor O_1.

1.3. The category of representations of Lie algebras in linear spaces. Denote by $\operatorname{End}_K V$ the associative K-algebra of all linear operators in a linear space V. A representation ρ of a Lie algebra L in the associative algebra $\operatorname{End}_K V$ is said to be a representation of the Lie algebra L in the linear space V. For any representation ρ, the space V can be supplied with the structure of a module over the Lie algebra L by setting

$$e \times v \rightleftharpoons \rho(e)v$$

for all $e \in L$, $v \in V$.

Let ρ be a representation of the Lie algebra L in a linear space V. It is possible to give the following definition: an *identity of the representation* ρ is a formula $f(t_1, \dots, t_l) = 0$, $f \in H$, such that $f(\rho(e_1), \dots, \rho(e_l)) = 0$ in $\operatorname{End}_K V$ for all $e_1, \dots, e_l \in L$. This viewpoint on identities of representations of Lie algebras has been adopted by the Riga algebraists [PlV].

Apparently, for a given representation ρ of a Lie algebra L in a space V, the spaces V, L are supplied with the structure of an algebra with two supports of signature $\{[,]; \times\}$ with multilinear operations

$$[,]\colon L \otimes_K L \to L; \qquad \times\colon L \otimes_K V \to V, \tag{1.2}$$

the first of which supplies L with the structure of Lie algebra and the second one supplies V with the structure of a module over the Lie algebra L, i.e., the following identity is valid:

$$[e_1, e_2] \times v = e_1 \times (e_2 \times v) - e_2 \times (e_1 \times v)$$

for all $e_1, e_2 \in L$, $v \in V$.

The converse is also true: any algebra $\langle L, V\rangle$ (L is a Lie algebra, V is an L-module) with two supports defines a representation of the Lie algebra L in the space V. Toward this end, it is sufficient to assign a linear operator $\rho(e)$ such that $\rho(e)\cdot v \rightleftharpoons e \times v$, $v \in V$, to any element e of the Lie algebra L. It is evident that the class of all algebras $\langle L, V\rangle$ with two supports thus obtained from the representations of Lie algebras L in spaces V forms a subvariety Π_3 of the variety of all algebras with two supports of signature $\langle [\,,\,]; \times\rangle$ (the 'arities', ranges and domains of these multilinear operations are described by the scheme in (1.2)).

We may consider the variety Π_3 as a category whose objects are algebras $\langle L, V\rangle$ with two supports (modules over Lie algebras) and morphisms are homomorphisms of these algebras with two supports. This category will be called *the category of representations of Lie algebras in linear spaces* (or the category of modules over Lie K-algebras).

As it was already mentioned, any algebra $\langle L, V\rangle \in \Pi_3$ with two supports defines a representation ρ of the Lie algebra L in the associative algebra $\mathrm{End}_K V$ and thus defines a representation of the algebra L in the smallest subalgebra $(\mathrm{End}_K V)_\rho$ of the algebra $\mathrm{End}_K V$ containing $L_\rho \rightleftharpoons \rho(L)$. Denote by O_3 the correspondence that assigns to any algebra with two supports $\langle L, V\rangle$ from Π_3 the algebra with two supports $\{(\mathrm{End}_K V)_\rho, L\}$ from Π_2 and denote by O_2 the correspondence that assigns to any $\langle L, V\rangle$ from Π_3 the associative-Lie pair $((\mathrm{End}_K V)_\rho, L_\rho)$ from Π_1. It is clear that the representation ρ of the Lie algebra L in the linear space V and the images $O_3(\langle L, V\rangle)$, $O_2(\langle L, V\rangle)$ have the same identities. Unfortunately, unlike the case of $O_1 : \Pi_2 \to \Pi_1$, neither O_3 nor O_2 can be naturally turned into a functor from Π_3 to Π_2 and from Π_3 to Π_1; moreover, generally speaking, for an arbitrary subvariety $\mathfrak{B}$ in Π_3 the class $O_3(\mathfrak{B})$ is not a subvariety in Π_2. However, for any subvariety $\mathfrak{B}$ in Π_3 the class of pairs $O_2(\mathfrak{B})$ forms a subvariety in Π_1. To prove this, it is sufficient to study the structure of identities of algebras $\langle L, V\rangle$ with two supports from Π_3.

For the variety Π_3 of algebras with two supports it is possible to construct explicitly the free algebra $\{\mathfrak{L}, W\}$ with two supports and with the set of free generators $\langle T, Z\rangle$ (where $T \cap Z = \varnothing$, $T \subset \mathfrak{L}$, $Z \subset W$). Toward this end, it is $\mathfrak{L}$ the free Lie algebra with the set of free generators T, and for W the free module with the set of free generators Z over the associative universal enveloping algebra F of the Lie algebra $\mathfrak{L}$. If $T = \{t_1, t_2, t_3, \dots\}$, then it is possible to identify in a natural way the Lie algebra $\mathfrak{L}$ with the Lie algebra $\mathfrak{M}$ and F with H. Let $Z = \{z\}$ be a singleton. Then any subvariety $\mathfrak{B}$ of algebras with two supports in Π_3 is uniquely determined by the ideal $\langle \mathfrak{V}_2, \mathfrak{V}_1 \times z\rangle$ with two supports, where $\mathfrak{V}_2$ is the ideal in the Lie algebra $\mathfrak{M}$ invariant under all endomorphisms of the algebra $\mathfrak{M}$ and $\mathfrak{V}_1$ is the ideal of weak identities in H such that $\mathfrak{V}_2 \times z \subseteq \mathfrak{V}_1 \times z$. This shows that for any ideal of weak identities $\mathfrak{V}$ in H the subclass of representations of Lie algebras in linear spaces (in other words, the subclass of modules over Lie algebras) such that all the identities of the form $f = 0$, where f ranges over $\mathfrak{V}$, are valid for these representations forms a subvariety of the variety of all algebras Π_3 with two supports, and the ideal (with two supports) of identities for this subvariety has the form $\langle 0, \mathfrak{V} \times z\rangle$. This proves that any subvariety of pairs in Π_1 is obtained from a certain subvariety of algebras with two supports in Π_3 with the help of O_2. Moreover, if a subvariety of pairs $\mathfrak{B}$ in Π_1 is given by an ideal $\mathfrak{V}$ of weak identities, then any subvariety of algebras $\mathfrak{B}'$ with two supports in Π_3 such that $O_2(\mathfrak{B}') = \mathfrak{B}$ is determined by the ideal $\langle \mathfrak{V}', \mathfrak{V} \times z\rangle$ with two supports, where $\mathfrak{V}'$

is a Lie ideal of identities contained in $\mathfrak{V}$ (we mean here that $\mathfrak{M}$ is identified with the Lie subalgebra in $H^{(-)}$ spanned by $t_1, t_2, \dots$). It is evident that, conversely, for any subvariety $\mathfrak{B}'$ in Π_3 given by the ideal with two supports $\langle \mathfrak{V}', \mathfrak{V} \times z \rangle$, where $\mathfrak{V}'$ is a Lie ideal of identities contained in the ideal $\mathfrak{V}$ of weak identities of the algebra H and for any variety of pairs $\mathfrak{B}$ given by the ideal $\mathfrak{V}$ the equality $O_2(\mathfrak{B}') = \mathfrak{B}$ is valid.

1.4. Connections with identities of Lie algebras and of associative algebras. All the three approaches discussed above are closely related. They have in common the definition of the notion of identity. The distinction among them lies in the classes of models (objects) in which the validity of these identities is required. We shall see in §5 that the first two approaches can be used not only for associative algebras and Lie algebras, but also for algebras of arbitrary signature Ω, Ω', and in this direction rather substantive results can be obtained (see Theorems 5.2–5.4). The merit of all three approaches is that they all include the study of identities of varieties of associative algebras and varieties of Lie algebras as a particular case. Now we give the corresponding definitions.

DEFINITION 1.7. Let $f(t_1, \dots, t_l)$ be an associative polynomial from the free associative algebra H. A formula $f = 0$ is called an *identity* of the associative algebra A if for any elements $a_1, \dots, a_l \in A$ the element $f(a_1, \dots, a_l)$ is equal to zero in the algebra A.

DEFINITION 1.8. An ideal $\mathfrak{V}$ in the free associative algebra H is called an *ideal of identities* or a *T-ideal* if for any $f(t_1, \dots, t_l) \in \mathfrak{V}$ and any $g_1, \dots, g_l \in H$ the polynomial $f(g_1, \dots, g_l)$ belongs to $\mathfrak{V}$.

It is obvious that if $f(t_1, \dots, t_l) = 0$ is an identity of the associative algebra A, then for all polynomials $g_1, \dots, g_l \in H$ and all words $a, b \in H$ the formula $a f(g_1, \dots, g_l) b = 0$ is also an identity of the associative algebra A. Thus, the set $\mathfrak{V}$ of polynomials in H that consists of all polynomials $f \in H$ such that $f = 0$ is an identity of the algebra A, is a T-ideal. This T-ideal is called the *ideal of identities of the associative algebra A*. It follows immediately from Definitions 1.7 and 1.8 that any T-ideal $\mathfrak{V}$ is the ideal of identities of the quotient algebra $H/\mathfrak{V}$.

DEFINITION 1.9. Let S be a set of polynomials in the free associative algebra H. An ideal of identities $\mathfrak{V}$ is said to be *generated* by the set S if the smallest ideal of identities $\mathfrak{V}'$ containing S coincides with $\mathfrak{V}$. An identity $f = 0$ is said to be a *consequence* of the identities $g = 0$, where g ranges over the set S, if and only if $f \in \mathfrak{V}'$.

DEFINITION 1.10. The identities of an associative algebra A are said to be *finitely based* if the ideal of identities of this algebra is finitely generated. A T-ideal $\mathfrak{V}$ is called a *Specht ideal*, or an *ideal with Specht property*, if it is finitely generated and the T-ideals containing the ideal $\mathfrak{V}$ satisfy the ascending chain condition.

Evidently, any T-ideal is an ideal of weak identities, and the ideal of identities of an arbitrary associative algebra A coincides with the ideal of weak identities of the associative-Lie pair $(A, A^{(-)})$. However, it should be mentioned that, generally speaking, if the identities of an associative algebra A are finitely based, this does not imply that the ideal of weak identities of the pair $(A, A^{(-)})$ is finitely generated. The point is that the consequence rules of identities in associative algebras that follow from Definition 1.9 are much richer than the corresponding consequence rules for identities in associative-Lie pairs (see Definition 1.3). Thus the study of identities of associative algebras is included in the general problem of the study of identities of associative-Lie pairs.

DEFINITION 1.11. Let $g(t_1, \dots, t_l)$ be a Lie polynomial in the free Lie algebra $\mathfrak{M}$. A formula $g = 0$ is called an *identity* of the Lie algebra L if for any elements $v_1, \dots, v_l \in L$ the element $g(v_1, \dots, v_l)$ is equal to zero in the Lie algebra L.

DEFINITION 1.12. An ideal $\mathfrak{W}$ in the free Lie algebra $\mathfrak{M}$ is called an *ideal of identities* or a *verbal ideal* if for any $g(t_1, \dots, t_l) \in \mathfrak{W}$ and any $h_1, \dots, h_l \in \mathfrak{M}$ the Lie polynomial $g(h_1, \dots, h_l)$ belongs to $\mathfrak{W}$.

Let L be an arbitrary Lie K-algebra. It follows immediately from Definitions 1.11 and 1.12 that the set $\mathfrak{W} = \{g \in \mathfrak{M} \mid g = 0 \text{ is an identity of } L\}$ of Lie polynomials is a verbal ideal of the free Lie algebra $\mathfrak{M}$. This ideal is called the *ideal of identities of the Lie algebra L*. Consider the adjoint representation $\operatorname{ad}: L \to \operatorname{End}_K L$. (Recall that by definition we have $\operatorname{ad} v \cdot w = [v, w]$ $(v, w \in L)$.) Put $(\operatorname{Ad} L, \operatorname{ad} L) = ((\operatorname{End}_K L)_{\text{ad}}, \operatorname{ad} L)$. Let $f(t_1, t_2, \dots, t_l) = 0$ be an arbitrary identity of this associative-Lie pair $(\operatorname{Ad} L, \operatorname{ad} L)$. Then it is evident that the Lie polynomial

$$g(t_1, \dots, t_l, t_{l+1}) = f(\operatorname{ad} t, \dots, \operatorname{ad} t_l) \times t_{l+1}$$

is linear with respect to the indeterminate t_{l+1}, and the relation $g = 0$ is an identity of the Lie algebra L. The converse is also true: any identity of the pair $(\operatorname{Ad} L, \operatorname{ad} L)$ can be obtained in this way from Lie polynomials from the ideal of identities $\mathfrak{W}$ of the Lie algebra L that are linear with respect to one of their indeterminates. This shows that if the ideal of identities $\mathfrak{W}$ of the Lie algebra L is generated by polynomials that are linear with respect to one of the indeterminates, then the study of identities of associative-Lie pair $(\operatorname{Ad} L, \operatorname{ad} L)$ is equivalent to the study of identities of the Lie algebra L. However, over fields of characteristic zero any identity is equivalent to a certain multilinear identity, and thus in this case the problem of description of identities in Lie algebras is a particular case of the problem on the identities of associative-Lie pairs.

DEFINITION 1.13. Let S be a set of polynomials in $\mathfrak{M}$. The class of all Lie algebras in which all the Lie identities $g = 0$, $g \in S$, are valid is called a *variety of Lie algebras*.

DEFINITION 1.14. A Lie algebra $\mathfrak{L}$ of a variety $\mathfrak{B}$ is said to be a *relatively free algebra* in this variety if it has generators x_i $(i \in I)$ such that any mapping ζ of the generators x_i to an arbitrary Lie algebra L of the variety $\mathfrak{B}$ has an extension to a homomorphism $\zeta: \mathfrak{L} \to L$. In this case the generators x_i $(i \in I)$ are named *free generators* of the algebra $\mathfrak{L}$.

§2. Complexity of varieties of Lie algebras and their representations

In this section we assign to any proper subvariety $\mathfrak{B}$ of associative-Lie pairs an important characteristic — an entire function of one complex variable describing the structural complexity of this variety.

2.1. Complexity functions for varieties of associative-Lie pairs. Let $(F, \mathfrak{L})$ be a relatively free pair of the variety $\mathfrak{B}$ with a countable set of free generators $X = \{x_1, x_2, \dots\}$. The elements of the algebra F of the form

$$f(x_{i_1}, \dots, x_{i_l}) = \sum_{\sigma \in S_l} \beta_\sigma x_{i_{\sigma(1)}} \cdots x_{i_{\sigma(l)}},$$

where $\beta_\sigma \in K$ and σ ranges over all permutations of the set $1, 2, \dots, l$, are called *multilinear* associative polynomials depending on $x_{i_1}, \dots, x_{i_l}$. For any finite subset λ of the set X, denote by F_λ the linear subspace of multilinear associative polynomials of the K-algebra F that essentially depend on all variables from the set λ. Similarly, let $\mathfrak{L}_\lambda = \mathfrak{L} \cap F_\lambda$ be the linear space of multilinear Lie polynomials of the Lie algebra $\mathfrak{L}$ depending on all variables from λ. It follows from Definition 1.5 of a relatively free pair that the dimensions of the subspaces $F_\lambda, \mathfrak{L}_\lambda$ depend only on the cardinality of the set λ, not on the generators x_i it consists of.

Let the cardinality of the set λ be equal to n. We set

$$e_n \rightleftharpoons \dim_K F_\lambda, \qquad \tilde{e}_n \rightleftharpoons e_n/n!,$$
$$c_n \rightleftharpoons \dim_K \mathfrak{L}_\lambda, \qquad \tilde{c}_n \rightleftharpoons c_n/n!.$$

PROPOSITION 2.1. *For any proper subvariety $\mathfrak{B}$ of the variety of all associative-Lie pairs, the power series*

$$e_{\mathfrak{B}}(z) \rightleftharpoons 1 + \sum_{n=1}^{\infty} \tilde{e}_n z^n, \tag{2.2}$$

$$c_{\mathfrak{B}}(z) \rightleftharpoons \sum_{n=1}^{\infty} \tilde{c}_n z^n, \tag{2.3}$$

define entire finctions of the complex variable z.

(We recall that a function of a complex variable is called an entire function if and only if it can be expanded in a power series that converges at every point of the complex plane (see [Shab]).)

The proof of the proposition is decomposed into a number of lemmas.

Let H be the absolutely free associative K-algebra with the same set of free generators as F, and let H_λ be the space of multilinear polynomials essentially depending on the generators in the set λ. A basis of the space H consists of the words, in the alphabet X, depending only on the generators in the set λ, and any letter from λ appears in such a word exactly once. Introduce a natural total ordering on the alphabet X by setting $x_i < x_{i+1}$ $(i = 1, 2, \dots)$. We order all the words in H_λ lexicographically, i.e., considering a word u greater than a word v if and only if

$$u = a x_i u' \quad \& \quad v = a x_j v' \quad \& \quad x_i > x_j.$$

DEFINITION 2.1. A word $W \in H_\lambda$ is said to be *Lie m-decomposable* if it can be represented in the form

$$w = aw_m w_{m-1} \cdots w_1 b, \tag{2.4}$$

where w_i $(i = 1, \ldots, m)$ are nonvoid words such that

a) the left letter in the word w_i is greater than any other letter in this word $(i = 1, \ldots, m)$;

b) the first letter in the word w_{i+1} is greater than the first letter in the word w_i $(i = 1, \ldots, m-1)$.

The representation (2.4) is called an *m-decomposition* of the word w. If the word w has no m-decompositions, then it is said to be *Lie m-indecomposable*.

LEMMA 2.1. *Let any letter of the word $w = w'x_n w''$ be contained in it exactly once and let x_n be the greatest letter in w. Then w is Lie m-indecomposable if and only if the following conditions are satisfied*:

a) *the word w'' is Lie $(m-1)$-indecomposable*;

b) *the word w' is Lie m-indecomposable.*

PROOF. It follows from Definition 2.1 that if w is Lie m-indecomposable then the conditions a) and b) are satisfied.

Conversely, suppose that the conditions a) and b) are satisfied. Let w be a Lie m-decomposable word and let $w = aw_m w_{m-1} \cdots w_1 b$ be an m-decomposition of it. Then the following three cases may occur.

1. The letter x_n appears in one of the words w_i. Then, from Definition 2.1 it follows that $w_m = x_n w'_m$, and hence the product $w'_m w_{m-1} \cdots w_1 b$ defines an $(m-1)$-decomposition of the word w''.

2. The letter x_n appears in the word a. Then $a = a'x_n a''$, and hence the product $a'' w_m w_{m-1} \cdots w_1 b$ is an m-decomposition of the word w''.

3. The letter x_n appears in the word b. Then $b = b'x_n b''$, and hence the product $aw_m w_{m-1} \cdots w_1 b'$ is an m-decomposition of the word w'.

In all three cases we obtain a contradiction with the conditions a) and b). The lemma is proved.

COROLLARY. *Let $a_m(n)$ be the cardinality of the set of Lie m-indecomposable multilinear words of $x_1, \ldots, x_n$ of length n. Then*

$$a_m(n) = \sum_{i=0}^{n-1} \frac{(n-1)!}{i!(n-1-i)!} a_m(i)\, a_{m-1}(n-1-i). \tag{2.5}$$

(*We set here $a_m(0) = 1$ $(m = 1, 2, \ldots)$.*)

PROOF. Consider in H_λ $(\lambda = \{x_1, \ldots, x_n\})$ the subset P_i of Lie m-indecomposable words of the form $w = w'x_n w''$, where the length of the word w' is equal to i. There are precisely $(n-1)!/(i!(n-1-i)!)$ possibilities of choosing i letters from the set $\{x_1, \ldots, x_{n-1}\}$. By Lemma 2.1, for any such choice of letters we can take any Lie m-indecomposable word of these letters for w', and any Lie $(m-1)$-indecomposable word of the remaining $(n-1-i)$ letters for w''. It follows that the cardinality of the set P_i is equal to $\binom{n-1}{i} a_m(i)\, a_{m-1}(n-1-i)$. This proves (2.5).

LEMMA 2.2. *For any natural number m the power series*

$$q_m(z) \rightleftharpoons \sum_{i=0}^{\infty} \frac{a_m(i)}{i!} z^i \qquad (a_m(0) \rightleftharpoons 1) \tag{2.6}$$

is an entire function of the complex variable z.

PROOF. It follows from (2.5) that

$$q_m(z) q_{m-1}(z) = \frac{d}{dz} q_m(z), \qquad q_m(0) = a_m(0) = 1. \tag{2.7}$$

Since any nonvoid word is 1-decomposable, we have

$$q_1(z) = 1.$$

Integrating the differential equation (2.7) step by step for $m = 2, 3, \dots$ and taking account of the initial condition $q_m(0) = a_m(0) \rightleftharpoons 1$, we obtain an explicit expression for $q_m(z)$ in terms of $q_{m-1}(z)$:

$$q_m(z) = \exp\Big(\int_0^z q_{m-1}(z)\, dz\Big); \tag{2.8}$$

in particular,

$$q_2(z) = e^z, \qquad q_3(z) = e^{e^z - 1}. \tag{2.9}$$

Thus, the complex functions $q_m(z)$ are obtained by (2.8) from the constant function $q_1(z) = 1$ by successive superpositions of integration and exponentiation. Hence the function $q_m(z)$ thus obtained is differentiable at any point of the complex plane and is an entire function (see [Shab]). By (2.7) its Taylor series at the point $z = 0$ must have the form (2.6). The lemma is proved.

LEMMA 2.3. *In any proper subvariety $\mathfrak{B}$ of the variety of all associative-Lie pairs a multilinear identity of the form*

$$0 = x_m x_{m-1} \cdots x_2 x_1 - \sum_{\substack{\sigma \in S_m \\ \sigma \neq 1}} \beta_\sigma x_{\sigma(m)} x_{\sigma(m-1)} \cdots x_{\sigma(1)} \tag{2.10}$$

is satisfied, where $\beta_\sigma \in K$ and the number m depends on the variety $\mathfrak{B}$.

PROOF. The assumptions of the lemma imply that in the variety $\mathfrak{B}$ a nontrivial identity $f(x_1, \dots, x_n) = 0$ takes place. If the polynomial f is multilinear, then we have $f = \sum_{\sigma \in S_l} \beta'_\sigma x_{\sigma(l)} \cdots x_{\sigma(1)}$ and for a certain permutation $\sigma \in S_l$ the coefficient β'_σ is nonzero. Then for $l = m$ the identity $(\beta'_\sigma)^{-1} f\left(x_{\sigma^{-1}(1)}, \dots, x_{\sigma^{-1}(l)}\right) = 0$ is valid in the variety $\mathfrak{B}$ and has the form (2.10).

Suppose that the polynomial f is not multilinear. Then the following two cases are possible.

1. In any monomial with nonzero coefficient in the polynomial f, any participating letter appears precisely once. If $f\big|_{x_1=0}, \dots, f\big|_{x_l=0}$ are zero polynomials, then f has the form $f = \sum_{\sigma\in S_l} \beta'_\sigma x_{\sigma(1)} \cdots x_{\sigma(l)}$ and thus f is a multilinear polynomial. If for a certain $i = 1, 2, \dots, l$ the polynomial $f\big|_{x_i=0}$ is nonzero, then the nontrivial identity $f\big|_{x_i=0} = 0$ takes place in the variety $\mathfrak{B}$ and depends on lesser number of letters. The existence of an identity of the form (2.10) in the variety $\mathfrak{B}$ now follows by induction.

2. There exists a monomial with nonzero coefficient in the polynomial f such that a certain letter appears in this monomial more than once. Let x_l be this letter. Let t be the degree of the polynomial f with respect to x_l. Since $t > 1$, we see that the associative polynomial

$$g(x_1, \dots, x_l, x_{l+1}) \rightleftharpoons f\big|_{x_l=x_l+x_{l+1}} - f - f\big|_{x_l=x_{l+1}}$$

is nonzero in the free associative algebra and its degree with respect to each of the indeterminates x_l, x_{l+1} is strictly less than t. It is evident that the degrees of the polynomials f and g are equal and $g = 0$ is an identity in $\mathfrak{B}$. Continuing this process of diminishing the degrees of the identity with respect to each indeterminate, we reduce the case under consideration to the case 1. The lemma is proved.

LEMMA 2.4. *Suppose that in a variety of associative-Lie pairs $\mathfrak{B}$ an identity of the form* (2.10) *is valid. Then the dimension of the space of multilinear polynomials F_λ, where $\lambda = \{x_1, \dots, x_n\}$ and $(F, \mathfrak{L})$ is a free pair of the variety $\mathfrak{B}$, does not exceed $a_m(n)$.*

PROOF. Let a word w from H_λ be Lie m-decomposable and let

$$w = a w_m \cdots w_1 b$$

be its m-decomposition. Denote by w'_i the left-normed Lie commutator obtained from the word w_i by left-normed positioning of the brackets. (For example, if we have $w_i = y_1 y_2 y_3 \cdots y_t$ $(y_i \in \lambda)$, then $w'_i = [[\dots[[y_1, y_2], y_3], \dots], y_t]$.) Evidently, we have

$$a w'_m \cdots w'_1 b = w + c,$$

where c contains monomials that are lesser than w according to the lexicographic order. Moreover, for any nonidentical permutation σ of the set $\{1, 2, \dots, m\}$ all the monomials arising in the expression $a w'_{\sigma(m)} \cdots w'_{\sigma(1)} b$ are also lesser than w. Hence

$$w = a\, f(w'_1, \dots, w'_m) b + e, \tag{2.11}$$

where $f(x_1, \dots, x_m)$ denotes the polynomial on the right-hand side of (2.10), and all the monomials in the polynomial $e \in H_\lambda$ are lesser than w. Equality (2.11) shows that any Lie m-decomposable monomial in the space F_λ is expressed as a linear combination of Lie m-indecomposable monomials. Therefore we can choose a basis of the space F_λ in the set of Lie m-indecomposable monomials and $\dim_K F_\lambda \le a_m(n)$. The lemma is proved.

PROOF OF PROPOSITION 2.1. It follows from Lemmas 2.3 and 2.4 that all coefficients $\widetilde{e}_n$ of the series (2.2) do not exceed the coefficients $\widetilde{a}_m(n) = a_m(n)/n!$ of the series (2.6). The radius of convergence r of the series (2.2) is defined by

$$1/r = \overline{\lim_{n\to\infty}} \sqrt[n]{\widetilde{e}_n}.$$

Since by Lemma 2.2 the series (2.6) is the Taylor series of an entire function, we see that its radius of convergence is infinite and hence $\overline{\lim}_{n\to\infty} \sqrt[n]{\tilde{a}_m(n)} = 0$. But this implies $\overline{\lim}_{n\to\infty} \sqrt[n]{\tilde{e}_n} = 0$, and the series (2.2) converges for any z in the complex plane. The proposition is proved.

The functions $e_{\mathfrak{B}}(z)$ and $c_{\mathfrak{B}}(z)$ are called the *complexity functions of the variety* $\mathfrak{B}$ *of associative-Lie pairs*, and the pair of functions $(e_{\mathfrak{B}}(z), c_{\mathfrak{B}}(z))$ the *complexity of the variety* $\mathfrak{B}$.

2.2. Varieties of exponential type. We introduce two partial orderings on the set of all entire functions:

$$f(z) = \sum_{n=0}^{\infty} \tilde{f}_n z^n \prec g(z) = \sum_{n=0}^{\infty} \tilde{g}_n z^n \iff \forall n \in \mathbb{N} \quad |\tilde{f}_n| \le |\tilde{g}_n|,$$

$$f(z) \underset{\text{acc}}{\prec} g(z) \iff \exists N \in \mathbb{N} \quad \forall n \in \mathbb{N} \quad (n > N \Rightarrow |\tilde{f}_n| \le |\tilde{g}_n|).$$

It follows from Lemmas 2.3 and 2.4 that for the entire function $e_{\mathfrak{B}}(z)$ of an arbitrary variety $\mathfrak{B}$ (see formula (2.2)) we have the inequality

$$e_{\mathfrak{B}} \prec q_m(z), \tag{2.12}$$

where m is the smallest degree of nontrivial identities that are valid in $\mathfrak{B}$. For many varieties $\mathfrak{B}$, of associative-Lie pairs, naturally arising in the study of identities of representations of Lie algebras, this bound is excessive. It often turns out that

$$e_{\mathfrak{B}} \underset{\text{acc}}{\prec} \exp(Nz) \qquad (N = N(\mathfrak{B}) > 0). \tag{2.13}$$

Varieties $\mathfrak{B}$ such that (2.13) is valid for some $N > 0$ are called *varieties of exponential type*.

An important example of varieties of associative-Lie pairs of exponential type is given by associative algebras that admit a nontrivial identity.

THEOREM 2.1. *Let U be an associative K-algebra and $\mathfrak{g} = U^{(-)}$. Suppose that a nontrivial identity is valid in the pair $(U, \mathfrak{g})$. Then the variety $\mathfrak{B}$ generated by the pair $(U, \mathfrak{g})$ is a variety of exponential type.*

PROOF. It follows from Lemmas 2.3 and 2.4 that in the variety $\mathfrak{B}$ a multilinear identity $f(x_1, \ldots, x_m) = 0$ of the form (2.10) is valid. Since, by the assumption of the theorem, $\mathfrak{g}$ as a vector space coincides with U, we see that for any words $a_1, \ldots, a_n$ in the alphabet X, the identity

$$f(a_1, \ldots, a_n) = 0 \tag{2.14}$$

is valid in the pair $(U, \mathfrak{g})$. Let $a_i = y_i a_i'$, where $y_i \in X$ and $y_m > y_{m-1} > \cdots > y_1$. Then, for any nonidentical permutation σ of the set $\{1, 2, \ldots, m\}$, the word $a_{\sigma(m)} a_{\sigma(m-1)} \cdots a_{\sigma(1)}$ is smaller than the word $a_m a_{m-1} \cdots a_1$, according to the lexicographic order. Therefore, for any words a and b we have the equality

$$a a_m a_{m-1} \cdots a_1 b = a f(a_1, \ldots, a_m) b + e, \tag{2.15}$$

where e is a polynomial whose all monomials are smaller than $a a_m a_{m-1} \cdots a_1 b$.

We denote by $W(X)$ the set of all words in the alphabet X.

DEFINITION 2.2. A word w in the alphabet X is said to be *associatively m-decomposable* if it can be represented in the form

$$w = ay_m a_m y_{m-1} a_{m-1} \cdots y_1 a_1, \tag{2.16}$$

where $y_1, \dots, y_m \in X$, $a, a_m, \dots, a_1 \in W(X)$, and $y_m > y_{m-1} > \cdots > y_1$. A representation of the word w in the form (2.16) is called an *associative m-decomposition* of the word w. If the word w has no associative m-decompositions, we say that the word w is *associatively m-indecomposable*.

Let $(F, \mathfrak{L})$ be a free associative-Lie pair of the variety $\mathfrak{B}$ with the set X of free generators. Identity (2.14) for the variety $\mathfrak{B}$, together with equality (2.15), shows that any associatively m-decomposable monomial w in the algebra F is representable in the form of a linear combination of monomials of the same degree and structure as w, but preceding w in the lexicographic order in $W(X)$. Therefore, we can choose a basis of the space F formed by associatively m-indecomposable monomials. Hence, in order to prove Theorem 2.1, it is sufficient to show that for any $n = 1, 2, \dots$ the number of associatively m-indecomposable words from H_λ ($\lambda = \{x_1, x_2, \dots, x_n\}$) does not exceed N^n for a sufficiently large number $N > 0$.

LEMMA 2.5 (V. N. Latyshev). *Let $a_m^\circ(n)$ be the cardinality of the set of associatively m-indecomposable multilinear words of $x_1, \dots, x_n$ of length n. Then*

$$a_m^\circ(n) \leq (m-1)^{2n}/(m-1)!.$$

PROOF. For any finite disjoint subsets $\lambda_1, \lambda_2, \dots, \lambda_{m-1}$ of letters of the alphabet X, we define inductively a set of multilinear words $W(\lambda_1, \lambda_2, \dots, \lambda_{m-1})$ of letters from $\lambda = \lambda_1 \cup \lambda_2 \cup \cdots \cup \lambda_{m-1}$, considering that a word $w = yw'$ belongs to this set if and only if the letter y is the smallest one in some subset λ_i and

$$w' \in W(\lambda_1, \dots, \lambda_{i-1}, \lambda_i \backslash \{y\}, \lambda_{i+1}, \dots, \lambda_{m-1}).$$

This definition implies that

a) all the words $w \in W(\lambda_1, \dots, \lambda_{m-1})$ are associatively m-indecomposable;
b) for the cardinality of the set $W(\lambda_1, \dots, \lambda_{m-1})$ the following equality is valid:

$$|W(\lambda_1, \dots, \lambda_{m-1})| = \frac{|\lambda|!}{|\lambda_1|! \cdots |\lambda_{m-1}|!}. \tag{2.17}$$

Now we show that any associatively m-indecomposable word w from the collection H_λ belongs to one of sets $W(\lambda_1, \dots, \lambda_{m-1})$. This follows from the Dilworth theorem (see §50). In fact, introduce a partial order on the set λ by setting x_i less than x_j if $x_i < x_j$ and x_i is on the left of the letter x_j in the word w. Then from the definition of associatively m-indecomposable words it follows that any m letters in this word contain a pair of comparable letters (with respect to the order just introduced), and by the Dilworth theorem the set λ can be represented as the union of k disjoint chains $\lambda_1(w), \dots, \lambda_k(w)$, where $k < m$. But then the word w is associatively $(k+1)$-indecomposable and $w \in W(\lambda_1(w), \dots, \lambda_k(w))$.

Hence $a_m^\circ(n) \leq \sum |W(\lambda_1, \dots, \lambda_{m-1})|$, where the sum is taken over all subsets $\lambda_1, \dots, \lambda_{m-1}$ of the set $\lambda = \{x_1, \dots, x_n\}$ such that

$$\lambda_i \cap \lambda_j = \varnothing, \qquad \lambda_1 \cup \cdots \cup \lambda_{m-1} = \lambda, \qquad |\lambda_1| \geq \cdots \geq |\lambda_{m-1}|.$$

This inequality and equality (2.17) imply

$$\begin{aligned} a_m^\circ(n) &\leq \sum_{\substack{n_1+\cdots+n_{m-1}=n \\ n_1 \geq \cdots \geq n_{m-1}}} \left(\frac{n!}{n_1! \cdots n_{m-1}!}\right)^2 \Big/ (m-1)! \\ &\leq \left(\sum_{\substack{n_1+\cdots+n_{m-1}=n \\ n_1 \geq \cdots \geq n_{m-1}}} \frac{n!}{n_1! \cdots n_{m-1}!}\right)^2 \Big/ (m-1)! \\ &\leq \frac{(m-1)^{2n}}{(m-1)!}. \end{aligned}$$

The lemma is proved.

Thus, for the complexity function $e_{\mathfrak{B}}$ of the variety $\mathfrak{B}$, from Lemma 2.5 we obtain the following estimate:

$$e_{\mathfrak{B}} \prec 1 + \sum_{n=1}^{\infty} a_m^\circ(n) z^n/n! \prec \exp((m-1)^2 z).$$

This proves Theorem 2.1.

Theorem 2.1 implies two corollaries.

COROLLARY 1. *Let a nontrivial identity be valid in an associative algebra U over a field K of arbitrary characteristic. Then, for a sufficiently large $l \in \mathbb{N}$ a "rarefied" identity*

$$\sum_{\sigma \in S_l} \beta_\sigma x_{\sigma(1)} y_1 x_{\sigma(2)} y_2 \cdots x_{\sigma(l-1)} y_{l-1} x_{\sigma(l)} = 0 \tag{2.18}$$

holds in U, where not all $\beta_\sigma \in K$ are zero and the sum is taken over all permutations σ of the set $\{1, \dots, l\}$.

PROOF. Let $(F, \mathfrak{L})$ be a free pair of the variety $\operatorname{var}(U, U^{(-)})$ with a countable number of free generators X. It follows from Theorem 2.1 that for a certain N depending on the algebra U and for sufficiently large numbers l we have the inequality

$$\dim_K F_\lambda \leq e^{(2l-1)\log N} \qquad (\lambda = \{x_1, \dots, x_l, y_1, \dots, y_{l-1}\}).$$

On the other hand, by the Stirling formula

$$n! = \frac{1}{\sqrt{2\pi n}} (n^n/e^n) e^{\theta_n/12n} \qquad (0 \leq \theta_n \leq 1),$$

we see that the number of monomials of the form $x_{\sigma(1)} y_1 x_{\sigma(2)} y_2 \cdots x_{\sigma(l-1)} y_{l-1} x_{\sigma(l)}$, where σ ranges over the collection of all permutations of the set $\{1, \dots, l\}$, is equal to $l! > e^{l(\log l - 1)}/\sqrt{2\pi l}$. Therefore, for $l > N^4$ these monomials are linearly dependent in F_λ. Corollary 1 is proved.

COROLLARY 2. *Let V be a simple associative K-algebra over an arbitrary field K and let a nontrivial identity be valid in it. Then V is a finite-dimensional algebra over its centroid C.*

PROOF. Recall that an algebra V is said to be *simple* if $VV = V$ and any of its two-sided ideals coincides either with V or with the zero ideal. By definition, we have

$$C = \{c \in \operatorname{End}_K V \mid c(ab) = (ca)b = a(cb) \quad \text{for any} \quad a, b \in V\}.$$

For any $a \in V$ denote by l_a (respectively, r_a) the linear transformation on V such that $l_a v = av$ ($r_a v = va$). Consider an associative K-subalgebra U in $\operatorname{End}_K V$ generated by all transformations of left and right multiplication l_a and r_a of the algebra V. Then the simplicity of the algebra V is equivalent to the irreducibility of the U-module V, and the centroid C of the algebra V coincides with the centralizer of this module (see §53). By the Schur Lemma, C is a skew field (see §53). Since $VV = V$ and

$$(c_1c_2)v_1v_2 = (c_1v_1)(c_2v_2) = (c_2c_1)v_1v_2,$$

we see that C is a commutative skew field, and hence C is a field.

Since a nontrivial identity is valid in the algebra V, we see that, by Corollary 1, a "rarefied" identity of the form (2.18) is valid in V for some $m \in \mathbb{N}$, and we can assume that $\beta_1 \neq 0$ in it. Let us show that $\dim_C V < m$. Suppose not. Then it is possible to choose elements $v_1, \dots, v_m$ that are linearly independent over C, and from the Jacobson Density Theorem (see §53) it follows that for any element $w_i \in V$ we can choose an element $u_i \in U$ such that

$$u_i \times v_i = w_i, \qquad u_i \times v_j = 0 \quad (j \neq i).$$

Since, by the definition of the algebra U, any element in it is representable in the form $\sum_{a,b \in V} l_a r_b$, we see that identity (2.18) implies the equality

$$\sum_{\sigma \in S_m} \beta_\sigma u_1 \times v_{\sigma(1)} \cdot u_2 \times v_{\sigma(2)} \cdots u_m \times v_{\sigma(m)} = 0.$$

On the other hand, the elements $u_i \in U$ are chosen so that in this sum all the summands corresponding to $\sigma \neq 1$ are zero and the summand with $\sigma = 1$ is equal to $\beta_1 w_1 \cdots w_m$. This means that for any elements $w_i \in V$ the product $w_1 \cdots w_m$ is equal to zero in the algebra V. Hence we have $V^m = 0$. But $V^2 = V$; therefore $V = 0$ and this contradicts to the assumption that $v_1, \dots, v_m$ are linearly independent elements over the field C. The contradiction thus obtained shows that $\dim_C V < m$. Corollary 2 is proved.

Another important class of varieties of exponential type arises in the study of representations of finite-dimensional Lie algebras.

LEMMA 2.6. *Suppose that $(U, \mathfrak{g})$ is an associative-Lie pair, the Lie K-algebra $\mathfrak{g}$ is finite-dimensional, and $\dim_K \mathfrak{g} = m$. Then in the pair $(U, \mathfrak{g})$ all the identities of the form*

$$0 = \sum_{\sigma \in S_{m+1}} \operatorname{sign}(\sigma) x_{\sigma(1)} f_1 x_{\sigma(2)} f_2 \cdots x_{\sigma(m)} f_m x_{\sigma(m+1)}, \tag{2.19}$$

are valid, where $f_1, \dots, f_m$ *are arbitrary associative polynomials,* $\operatorname{sign}(\sigma)$ *is the sign of the permutation* σ, *and the sum is taken over all permutations of the set* $\{1, \dots, m+1\}$.

PROOF. It is sufficient to verify the statement of the lemma for arbitrary multilinear identities of the form (2.19). In this case it is sufficient to check that the right-hand sides of the identities (2.19) vanish in the algebra U for any specialization of the indeterminates of X by the elements of a certain fixed basis $e_1, \dots, e_m$ of the Lie algebra $\mathfrak{g}$. But for any such specialization, a pair of arguments among the $x_1, \dots, x_{m+1}$ will attain equal values, and the skew-symmetry of (2.19) with respect to $x_1, \dots, x_{m+1}$ implies that it becomes a truthful equality in the algebra U. The lemma is proved.

THEOREM 2.2. *Let in a variety* $\mathfrak{B}$ *of associative-Lie pairs over the field* K *all the identities of the form*

$$\sum_{\sigma \in S_m} \beta_\sigma x_{\sigma(1)} g_1 x_{\sigma(2)} g_2 \cdots x_{\sigma(m-1)} g_{m-1} x_{\sigma(m)} = 0, \tag{2.20}$$

be valid for some natural m, *where* β_σ *are fixed elements of the field* K *(not all of them are zero),* $g_1, g_2, \dots, g_{m-1}$ *are arbitrary monomials, and the sum is taken over all permutations of the set* $\{1, \dots, m\}$. *Then* $\mathfrak{B}$ *is a variety of exponential type.*

PROOF. Let $(F, \mathfrak{L})$ be a relatively free pair of the variety $\mathfrak{B}$ with a countable set of free generators X, let $\lambda = \{x_1, \dots, x_n\}$ be a subset of X, and let $y_1 > \cdots > y_m$ in the word $w = a y_1 c_1 y_2 c_2 \cdots c_{m-1} y_m b \in H_\lambda$. Then identity (2.20) for $g_i = c_i$ permits one to express in F the monomial w by monomials from F_λ that are strictly less than w in the lexicographic order. Hence we can choose a basis of the space F_λ in the set of associatively m-indecomposable monomials, and by Lemma 2.5 we have $e_{\mathfrak{B}}(z) \prec \exp((m-1)^2 z)$. The theorem is proved.

COROLLARY. *Any associative-Lie pair* $(U, \mathfrak{g})$, *where the Lie algebra* $\mathfrak{g}$ *is finite-dimensional, generates a variety of exponential type.*

The proof immediately follows from Lemma 2.6 and Theorem 2.2.

A third class of varieties of associative-Lie pairs of exponential type arises in the study of Lie algebras of vector fields acting by derivations on the algebra of all infinitely differentiable functions on a smooth n-dimensional manifold.

Let $\mathcal{E}$ be an arbitrary associative and commutative differential K-algebra with m mutually commuting signature derivations $\partial_1, \dots, \partial_m$ (e.g., see §51). It is immediate that for any $e_1, \dots, e_m \in \mathcal{E}$ a linear transformation $g = g(e_1, \dots, e_m) \in \operatorname{End}_K \mathcal{E}$ such that $g \times e \rightleftharpoons e_1(\partial_1 e) + \cdots + e_m(\partial_m e)$ is a derivation of the K-algebra $\mathcal{E}$. Such derivations are called special derivations of the differential algebra $\mathcal{E}$. They form a Lie subalgebra $\mathfrak{g}$ in $\operatorname{Der}_K \mathcal{E}$. Denote by U the associative subalgebra in $\operatorname{End}_K \mathcal{E}$ generated by $\mathfrak{g}$.

THEOREM 2.3. *Let* $\mathfrak{g}$ *be the Lie algebra of all special derivations of an associative and commutative differential* K*-algebra* $\mathcal{E}$ *with* m *signature derivations and let* U *be the associative subalgebra in* $\operatorname{End}_K \mathcal{E}$ *generated by* $\mathfrak{g}$. *Then the variety* $\mathfrak{B} = \operatorname{var}(U, \mathfrak{g})$ *is a variety of exponential type and its complexity function satisfies the estimate* $e_{\mathfrak{B}}(z) \prec 2^{m-1} \exp(2^{m+1} m z)$.

PROOF. Obviously, any epimorphism of differential algebras $\varphi \colon \mathcal{E} \to \mathcal{E}'$ induces an epimorphism of Lie algebras of special derivations $\mathfrak{g} \to \mathfrak{g}'$ and of pairs $(U, \mathfrak{g}) \to$

$(U', \mathfrak{g}')$. Hence we have $\operatorname{var}(U, \mathfrak{g}) \supset \operatorname{var}(U', \mathfrak{g}')$, and it is sufficient to prove the statement of the theorem for the case where $\mathcal{E}$ is the free differential K-algebra with a countable set of free generators $\{x_i^{(j)} \mid i, j = 1, 2, \dots\}$. In this case a free pair $(F, \mathfrak{L})$ of the variety $\operatorname{var}(U, \mathfrak{g})$ can be constructed in $\operatorname{End}_K \mathcal{E}$ as the K-subpair in $(U, \mathfrak{g})$ freely generated by the set of special derivations

$$X = \left\{ x_i = \sum_{j=1}^{m} x_i^{(j)} \partial_j | i = 1, 2, \dots \right\}.$$

Set $\lambda = \{x_1, \dots, x_n\}$. Then using the relation $e\partial_i = \partial_i e - [\partial_i, e]$ $(e \in \mathcal{E})$ step by step, we can express any element of F_λ in $\operatorname{End}_K \mathcal{E}$ in terms of operators of the form

$$\left(\partial_1^{r_{11}} \cdots \partial_m^{r_{1m}} \left(x_1^{(j_1)}\right)\right) \cdots \left(\partial_1^{r_{n1}} \cdots \partial_m^{r_{nm}} \left(x_n^{(j_n)}\right)\right) \partial_1^{r_{01}} \cdots \partial_m^{r_{0m}},$$

where $j_1, \dots, j_n \in \{1, \dots, m\}$ and $\sum_{i,j} r_{ij} = n$. The number of these operators is equal to $m^n A_{(n+1)m,n}$, where $A_{t,j} = \binom{t+j-1}{j}$ is the number of various monomials of degree j in the algebra of polynomials in t commuting indeterminates. Therefore $\dim_K F_\lambda \leq m^n \cdot 2^{(n+1)m+n-1}$, and the estimate given in the theorem for the complexity function follows from the very definition of the latter (see formula (2.2)). The theorem is proved.

If $\mathcal{E}$ is the algebra of power series in m commuting indeterminates $t_1, \dots, t_m$ over a field of characteristic zero, then the Lie algebra of special derivations $\mathfrak{g} = \mathcal{E}\frac{\partial}{\partial t_1} + \dots + \mathcal{E}\frac{\partial}{\partial t_m}$ coincides with the Lie algebra of all K-derivations of the algebra $\mathcal{E}$. This Lie algebra is called the *Cartan Lie algebra of general type* and is denoted by $\widetilde{W}_m(K)$.

COROLLARY. *The variety $\mathfrak{B}$ of Lie algebras generated by the Cartan Lie algebra of general type $\widetilde{W}_m(K)$ is a variety of exponential type, and for its complexity function $c_{\mathfrak{B}}(z)$ we have $c_{\mathfrak{B}}(z) \prec 2^{m-1} \exp\left(m \cdot 2^{m+1} z\right)$.*

2.3. Existence of varieties of large growth. In the theory of functions of a complex variable, for the study of the rate of growth of the maximum modulus of an entire function $f(z)$ on the disc of radius r as $r \to \infty$, the notion of order of $f(z)$ is introduced. The order is defined with respect to a certain continuously differentiable real function $h(r)$ such that the equation $rh^t(r)h'(r) = \beta$ has a unique solution for any $\beta > 0$, $t > 0$. In this case, the number

$$\rho = \rho(f, h) = \varlimsup_{r \to \infty} \left(\log\log \max_{|z|=r} |f(z)| / \log h(r) \right) \tag{2.21}$$

is called the *order* of the entire function $f(z)$ with respect to the function $h(r)$, and the number

$$\sigma = \sigma(f, h) = \varlimsup_{r \to \infty} \left(\log \max_{|z|=r} |f(z)| / (h(r))^\rho \right) \tag{2.22}$$

is called the *type* or the *growth exponent*. If these limits are infinite, we say that the function $f(z)$ is of *infinite order* (type) with respect to $h(r)$. If $\rho < \infty$ then for the coefficients f_n of the series $f(z) = \sum_{i=0}^{\infty} f_n z^n$ the following asymptotic estimates are valid (see [Evg]):

$$(2.23)\qquad \begin{aligned} |f_n| &< \exp\left((\sigma+\varepsilon)\cdot h(r_n)^{\rho}\right)/(r_n)^n, \\ |f_n| &< \exp\left(h(r_n')^{\rho+\varepsilon}\right)/(r_n')^n \end{aligned} \qquad (\varepsilon > 0),$$

where r_n, r_n' are, respectively, the solutions of the equations

$$(2.24)\qquad \begin{aligned} rh'(r)h(r)^{\rho-1} &= n/(\rho\sigma+\rho\varepsilon), \\ rh'(r)h(r)^{\rho-1+\varepsilon} &= n/(\rho+\varepsilon). \end{aligned}$$

The most useful function for the growth measurement is the function $h(r) = r$. In this case the functions $f(z)$ such that the upper limit (2.21) is finite are said to be the *functions of finite order*. Since for $h(r) = r$ the equations (2.24) possess an explicit solution, we see that the asymptotic formulas (2.23) for functions of finite order are simplified and attain the following form:

$$(2.25)\qquad \begin{aligned} |f_n| &< (\rho\sigma+\rho\varepsilon)^{n/\rho}/(n^{n/\rho}/\exp(n/\rho)), \\ |f_n| &< (\rho+\varepsilon)^{n/(\rho+\varepsilon)}/(n^{n/(\rho+\varepsilon)}/\exp(n/(\rho+\varepsilon))). \end{aligned}$$

It can be shown (see [Levin]) that, conversely, if the inequalities (2.25) are asymptotically valid then the entire function $f(z)$ has the type and the order equal to σ and ρ, respectively. This implies that the complexity functions of varieties considered in the previous subsection are of order one. Moreover, Theorem 2.1 could suggest the idea that the complexity function of any nontrivial variety of associative-Lie pairs is of finite order. Unfortunately, this is not so.

THEOREM 2.4. *Suppose that $q_1(z) = 1$ and the entire functions $q_m(z)$ $(m = 2, 3, \dots)$ are defined by the recurrent formula (2.8). Then for any natural m there exists a variety $\mathfrak{B}$ of associative-Lie pairs such that its complexity functions $e_{\mathfrak{B}}(z)$ and $c_{\mathfrak{B}}(z)$ satisfy the inequality*

$$(2.26)\qquad q_m(z) \prec c_{\mathfrak{B}}(z) \prec e_{\mathfrak{B}}(z).$$

The proof of this theorem is based on the following statement.

PROPOSITION 2.2. *Let $\mathfrak{L}$ be an arbitrary relatively free Lie algebra with a countable set of free generators $X = \{x_1, x_2, \dots\}$, let F be its associative universal enveloping K-algebra, and let the variety $\mathfrak{B}$ be generated by the pair $(F, \mathfrak{L})$. Then the complexity functions $e_{\mathfrak{B}}(z)$, $c_{\mathfrak{B}}(z)$ of this variety are related by*

$$(2.27)\qquad e_{\mathfrak{B}}(z) = \exp(c_{\mathfrak{B}}(z)).$$

THE PROOF of this proposition is based on the following combinatorial formula for the nth derivative of the superposition of functions $f(z)$ and $g(z)$:

$$(2.28)\qquad \frac{d^n}{dz^n} f(g(z)) = \sum_{k=1}^{n} \sum_{\substack{k_1+\dots+k_t=k \\ k_1n_1+\dots+k_tn_t=n \\ n_1<\dots<n_t}} C(k_1,\dots,k_t,n_1,\dots,n_t) f^{(k)}(g(z)) \left(g^{(n_1)}\right)^{k_1} \cdots \left(g^{(n_t)}\right)^{k_t},$$

where $g^{(l)}$ denotes the lth derivative of the function g and

$$C(k_1, \dots, k_t, n_1, \dots, n_t) = \frac{(k_1 n_1 + \dots + k_t n_t)!}{k_1! \cdot (n_1!)^{k_1} \cdots k_t! \cdot (n_t!)^{k_t}}.$$

Now we compute the dimensions of the spaces F_λ, using the Poincaré-Birkhoff-Witt Theorem (see §52). Toward this end, we fix a basis E_μ in any subspace $\mathfrak{L}_\mu$ (μ is an arbitrary finite subset in X). We consider the elements of the set X as ordered naturally: $x_i < x_{i+1}$. We introduce an arbitrary total order in the basis E_μ. For elements $v_1 \in E_{\mu_1}, v_2 \in E_{\mu_2}$ we set $v_1 < v_2$ if either the cardinality of μ_1 is less than the cardinality of μ_2 or $|\mu_1| = |\mu_2|$ and μ_2 contains an element that is greater than all the elements of μ_1. Let $\lambda = \{x_1, \dots, x_n\}$. Then a basis of the space F_λ is constituted by all possible elements of the form $v_1 v_2 \cdots v_p$, where

$$v_1 \le \dots \le v_p, \qquad v_i \in E_{\mu_i}, \qquad \mu_i \cap \mu_j = \varnothing, \qquad \bigcup_{i=1}^{p} \mu_i = \lambda.$$

But in this case we have $|\mu_1| \le |\mu_2| \le \dots \le |\mu_p|$, and if $|\mu_i| = |\mu_{i+1}|$ then there exists an element in μ_{i+1} that is greater than all elements in μ_i. Therefore,

$$\dim_K F_\lambda = \sum_{\mu_1, \dots, \mu_t} \dim_K \mathfrak{L}_{\mu_1} \dim_K \mathfrak{L}_{\mu_2} \cdots \dim_K \mathfrak{L}_{\mu_t},$$

where the sum is taken over all possible partitions $\mu_1, \dots, \mu_t$ of this type of the set λ. Since $\dim_K \mathfrak{L}_{\mu_1} = \dim_K \mathfrak{L}_{\mu_2}$ for $|\lambda_1| = |\lambda_2|$, we see that

$$\text{(2.29)} \qquad \dim_K F_\lambda = \sum_{k=1}^{n} \sum_{\substack{k_1 n_1 + \dots + k_t n_t = n \\ n_1 < \dots < n_t}} B(k_1, \dots, k_t, n_1, \dots, n_t) c_{n_1}^{k_1} \cdots c_{n_t}^{k_t},$$

where c_l denotes the dimension of $\mathfrak{L}_{\{x_1, \dots, x_l\}}$, and $B(k_1, \dots, k_t, n_1, \dots, n_t)$ is the number of all partitions of the set λ into disjoint subsets $\mu_1, \dots, \mu_p$ such that the cardinality of the first k_1 subsets is equal to n_1, the cardinality of the next k_2 subsets is equal to n_2, etc., and, moreover, if $|\mu_i| = |mv\mu_{i+1}|$, then there exists an element in μ_{i+1} that is greater than all elements in μ_i. It is obvious that the number of these partitions coincides with $C(k_1, \dots, k_t, n_1, \dots, n_t)$. Comparing formula (2.29) and equality (2.28) for $f(z) = \exp z$, $g(z) = c_{\mathfrak{B}}(z)$ and $z = 0$, and also recalling relations (2.2) and (2.3) defining the functions $e_{\mathfrak{B}}(z)$ and $c_{\mathfrak{B}}(z)$, we obtain the equality (2.27). Proposition 2.2 is proved.

Let $(F, \mathfrak{L})$ be a relatively free Lie algebra of a variety $\mathfrak{B}$ with a countable set of free generators X. Let V be a left free F-module and let $v_1, v_2, \dots$ be its free generators. Turn the direct sum of spaces $U = F \oplus V$ into an associative algebra by setting

$$(f_1 \oplus w_1)(f_2 \oplus w_2) = f_1 f_2 \oplus f_1 w_2 \qquad (w_1, w_2 \in V,\ f_1, f_2 \in F),$$

and generate in it, by the elements $y_i = x_i \oplus v_i$ ($i = 1, 2, \dots$), an associative subalgebra F' and a Lie subalgebra $\mathfrak{L}'$, respectively.

LEMMA 2.7. *For any relatively free pair $(F, \mathfrak{L})$ of the variety $\mathfrak{B}$, the pair $(F', \mathfrak{L}')$ is a relatively free pair of a variety $\mathfrak{B}_1 \supseteq \mathfrak{B}$ with a set of free generators $\{y_1, y_2, \dots\}$, and the complexity functions of these varieties satisfy the following inequalities*:

$$\int_0^z e_{\mathfrak{B}}(z)\, dz \prec c_{\mathfrak{B}_1}(z) \prec z e_{\mathfrak{L}}(z).$$

PROOF. First we check that for any associative polynomial $g \in H$, the equality $g(y_1, \dots, y_l) = 0$ in the algebra U implies the equality $g(w_1, \dots, w_l) = 0$ for any elements w_i of the Lie algebra $\mathfrak{L} \oplus V$. In fact, since the pair $(F, \mathfrak{L})$ and the F-module V are free, we see that any mapping $x_i \to \mathfrak{L}$, $v_i \to V$ has an extension to an endomorphism of the algebra U. Since any element w_i has the form $w_i' + w_i''$, where $w_i' \in \mathfrak{L}$, $w_i'' \in V$, we see that the extension of the mapping $x_i \mapsto w_i'$, $v_i \mapsto w_i''$ is the endomorphism of the algebra U that maps y_i to w_i. This proves that the pair $(F', \mathfrak{L}')$ is a relatively free pair and that $y_1, y_2, \dots$ are its free generators.

Let $\lambda = \{x_1, \dots, x_n\}$, $\lambda' = \{y_1, \dots, y_{n+1}\}$. Choose in the space F_λ a basis consisting of monomials of the form $x_{i_1} \cdots x_{i_n}$. Since any Lie commutator in $\mathfrak{L}'_{\lambda'}$ is representable in the form

$$[y_{i_1}, [\dots [y_{i_n}, y_{n+1}] \dots]] = \sum_{j=1}^{n+1} g_j v_j \oplus [x_{i_1}, [\dots [x_{i_n}, x_{n+1}] \dots]],$$

where $g_j \in F$ and $g_{n+1} = x_{i_1} \cdots x_{i_n}$, in view of the fact that the F-module V is free, such Lie commutators are linearly independent in $\mathfrak{L}'_{\lambda'}$ and thus we have $\dim_K \mathfrak{L}'_{\lambda'} > \dim_K F_\lambda$. This inequality and formulas (2.1) imply that all coefficients of the power series for the function $(d/dz)c_{\mathfrak{B}_1}(z)$ are not less than the corresponding coefficients of the series for the function $e_{\mathfrak{B}}(z)$, and this is equivalent to the desired statement. The lemma is proved.

THE PROOF OF THEOREM 2.4 is obtained now by successive application of the constructions of Lemma 2.7 and of Proposition 2.2. In fact, if the variety $\mathfrak{B}$ has the complexity function $e_{\mathfrak{B}}(z)$, then we have $c_{\mathfrak{B}'}(z) \succ \int_0^z e_{\mathfrak{B}}(z)\, dz$ for the variety $\mathfrak{B}'$ with the free pair $(F', \mathfrak{L}')$ (see Lemma 2.7), and for the complexity function of the variety $\mathfrak{B}'' = \operatorname{var}(U(\mathfrak{L}'), \mathfrak{L}')$, where $U(\mathfrak{L}')$ is the universal enveloping algebra of the Lie algebra $\mathfrak{L}'$, we have the inequality $\exp\left(\int_0^z e_{\mathfrak{B}}(z)\, dz\right) \prec e_{\mathfrak{B}''}(z)$. Starting from an arbitrary variety of associative-Lie pairs and applying the construction described above repeatedly, we can obtain, on the $(m+1)$th step, a variety whose complexity functions exceed the function $q_m(z)$. Theorem 2.4 is proved.

§3. Central closure for semiprime algebras

In this section, following the paper [Raz11], we shall give, for any semiprime K-algebra A of signature Ω, a construction of a commutative and associative algebra $C = C(A)$ and a semiprime C-algebra $Q = Q(A)$ of the same signature Ω containing A as a K-subalgebra such that $Q = CA$. In the paper [Raz11] the algebras C and Q are called the full central ring of A and the full central ring of quotients of A, respectively. The study of properties of the algebras C and Q shows that in the case of a semiprime algebra A whose signature consists of a single binary multilinear operation, the algebras C, Q can be obtained by another constuction proposed in 1979 by Martindale (see [BM, ErMO]). For this reason, we shall call the algebra $C(A)$ a Martindale centroid, and the algebra $Q(A)$ — a central closure of the algebra A.

3.1. Construction of a central closure. Following §49, by an algebra of signature Ω we shall mean a linear space A over a certain fixed field K such that for any $\omega \in \Omega$ a K-multilinear n-ary $(n = n(\omega))$ operation is defined:

$$\omega\colon \underbrace{A \otimes_K A \otimes_K \cdots \otimes_K A}_{n \text{ times}} \to A.$$

In addition, we shall suppose that among the signature operations there is an operation whose "arity" (n) is ≥ 2.

Obviously, the class of all K-algebras of fixed signature is a variety. Denote by the letter F the absolutely free algebra of countable rank of the same signature Ω with free generators x_i and y_i $(i = 1, 2, \dots)$. The elements of this algebra will be called *polynomials* in the signature Ω. A polynomial $g(x_1, \dots, x_i, \dots, x_l) \in F$ is said to be *linear in the indeterminate* x_i if for any element β of the field K we have the equality $g\big|_{x_i = \beta x_i} = \beta g$; a polynomial g is said to be *multilinear* if it is linear in each of the participating indeterminates. For any algebra A of signature Ω, we can connect an associative K-algebra with unit $D = D(A)$ that is generated in $\operatorname{End}_K A$ by the unit operator and all linear transformations of the form $g(a_1, \dots, \widehat{a_i}, \dots, a_l)$, where $g(x_1, \dots, x_l)$ is an arbitrary polynomial linear in x_i and $a_1, \dots, a_i, \dots, a_l$ are elements of the algebra A whose action on A is defined by the formula

$$g(a_1, \dots, \widehat{a_i}, \dots, a_l) \times a_i \rightleftharpoons g(a_1, \dots, a_i, \dots, a_l)$$

(here a_i is an arbitrary element of the algebra A and the polynomial g and the elements $a_1, \dots, \widehat{a_i}, \dots, a_l$ are fixed). In the algebra A, a K-subspace J is called an *ideal* if J is a D-submodule of the D-module A. Let $J_1, \dots, J_k$ be ideals of the algebra A; the *product* $(J_1 \cdot J_2 \cdot \cdots \cdot J_k)$ *of the ideals* J_i $(i = 1, \dots, k)$ is the ideal generated by all the elements of the form $g(i_1, \dots, i_k, a_1, \dots, a_l)$, where $g(x_1, \dots, x_k, y_1, \dots, y_l)$ is an arbitrary polynomial of signature Ω, multilinear in $x_1, \dots, x_k$; $a_1, \dots, a_l \in A$, $i_1 \in J_1$, $\dots$, $i_k \in J_k$.

REMARK 3.1. This definition of the product of ideals assumes implicitly that in the signature Ω there are operations with "arity" not less than two.

DEFINITION 3.1. An algebra A is said to be *prime* if the equality $(J_1 \cdot J_2) = 0$ implies $J_1 = 0$ or $J_2 = 0$ and to be *semiprime* if the equality $(J \cdot J) = 0$ implies $J = 0$, for any ideals J, J_1, J_2 of the algebra A.

DEFINITION 3.2. A module m over an associative algebra with unit D is said to be *injective* if for any two modules M_1, M_2, any monomorphism $\varphi\colon M_1 \to M_2$, and any homomorphism $\psi\colon M_1 \to M$ there exists a homomorphism $\rho\colon M_2 \to M$ such that $\rho \circ \varphi = \psi$.

It is well known (see §53) that any D-module M can be embedded in an injective module P. It turns out, that among these embeddeings there is a minimal one; the corresponding injective module is said to be an injective hull of the module M.

DEFINITION 3.3. A submodule M of a module P is said to be *large* in P if any nonzero submodule in P has a nonzero intersection with M.

DEFINITION 3.4. A D-module P is called an *injective hull* of a D-module M if the following three conditions are satisfied:

a) P is an injective D-module;
b) M is a submodule in P;
c) M is a large submodule in P.

For any module M there exists an injective hull, and for a given module it is unique up to isomorphism (see §53).

Let A be a K-algebra of signature Ω and let $D = D(A)$ be the related associative subalgebra in $\mathrm{End}_K A$. Let P be an injective hull of the D-module A, let $E = \mathrm{End}_D P$ be the algebra of all endomorphisms of the D-module P, let $Q = EA$ be a D-submodule in P (usually called a *quasi-injective hull* of the D-module A; it is characterized by the two properties: a) A is a large D-submodule in Q; b) any homomorphism of any D-submodule J of the module Q has an extension to an endomorphism of the module Q).

LEMMA 3.1. *If a K-algebra of signature Ω is semiprime, then the restriction ρ of the action of the algebra E to the D-module Q is commutative.*

PROOF. Denote by I_0 the two-sided ideal in the algebra E generated by Lie commutators $[h_1, h_2] = h_1h_2 - h_2h_1$, where h_1, h_2 range over the algebra E. Since $I_0Q = I_0EA = I_0A$, we see that for the proof of the assertion of the lemma it is sufficient to establish the equality $I_0A = 0$. But as a left ideal, the ideal I_0 is generated by the commutators $[h_1, h_2]$. Therefore, it is sufficient to show that for any $h_1, h_2 \in E$, $a \in A$ we have the following equality in P:

$$(h_1h_2 - h_2h_1) \times a = 0. \tag{3.1}$$

Denote by I_i $(i = 1, 2)$ the greatest $D(A)$-submodule in A such that $h_iI_i \subset A$. Obviously, I_i is a large submodule in A. Since the algebra A is semiprime, for any two ideals J_1, J_2 the equality $J_1 \cap J_2 = 0$ is equivalent to the relation $(J_1 \cdot J_2) = 0$ (in fact, the inclusion $J_1 \cap J_2 \supset (J_1 \cdot J_2)$ shows that the first equality implies the second one; conversely, if $(J_1 \cdot J_2) = 0$ then, because $J_1 \cap J_2$ is contained in J_1 and J_2, we see that $((J_1 \cap J_2) \cdot (J_1 \cap J_2)) = 0$ and the equality $J_1 \cap J_2 = 0$ follows from the definition of semiprimeness). This implies that for any nonzero ideal J of the algebra A we have $0 \neq (I_1 \cdot (I_2 \cdot J)) \subseteq (I_1 \cdot I_2) \cap J$, and thus $(I_1 \cdot I_2)$ is also a large submodule of the D-module A.

It is obvious that $(h_1h_2 - h_2h_1) \times (I_1 \cdot I_2) = 0$. Suppose that for certain elements $h_1, h_2 \in E$, $a \in A$ the element b on the left-hand side of the equality (3.1) is nonzero. Then from the property c) of the injective hull (see the Definition 3.4) it follows that the ideal $J = D \times b \cap A$ of the algebra A is nonzero. Moreover, $(J \cdot (I_1 \cdot I_2)) \subseteq ((h_1h_2 - h_2h_1) \times D \times a \cdot (I_1 \cdot I_2)) = (D \times a \cdot [h_1, h_2] \times (I_1 \cdot I_2)) = 0$. However, as is was mentioned above, this implies $J \cap (I_1 \cdot I_2) = 0$, and this contradicts the fact that $(I_1 \cdot I_2)$ is a large submodule of A. The contradiction thus obtained proves the equality (3.1) for any $a \in A$, $h_1, h_2 \in E$. The lemma is proved.

Since $Q = EA$ by definition and by Lemma 3.1 the algebra $C = E/\mathrm{Ker}\,\rho$ is commutative, we can extend all operations of the signature Ω by C-linearity from the algebra A to Q, and endow Q with the structure of a C-algebra of signature Ω. The algebra C will be called a *Martindale centroid* of the (semiprime) algebra A, and the C-algebra $Q(A)$ will be called a *central closure* of the algebra A.

3.2. Simplest properties of the central closure.

PROPOSITION 3.1. *If A is a semiprime K-algebra of signature Ω, then the centroid $C(A)$ and the central closure $Q(A)$ are semiprime algebras that are characterized by the following properties*:

1. *C is a commutative K-algebra with unit and $Q = CA$.*
2. *The intersection with A of an arbitrary nonzero D-submodule in Q is a nonzero ideal of the algebra A.*
3. *For any D-homomorphism $\varkappa$ of a nonzero D-submodule J of the D-module Q into Q, there exists an element $c \in C$ such that $cj = \varkappa(j)$, where j is an arbitrary element of the module J. Moreover, if J is a large D-submodule in Q, then the element c is uniquely defined by the homomorphism $\varkappa$.*

PROOF. Suppose that the algebra Q is not semiprime. Then we have $(I \cdot I) = 0$ for some nonzero ideal I in the algebra Q. The ideal I is a D-submodule in Q, hence $J = I \cap A$ is a nonzero ideal in the algebra A (because A is a large submodule in its injective hull P and hence also in Q) and we have $(J \cdot J) = 0$. This contradicts the semiprimeness of the algebra A. The contradiction thus obtained proves the semiprimeness of the algebra Q.

Suppose that the K-algebra C is not semiprime. Then $c^2 = 0$ for some nonzero element c in C. It is obvious that $J = cQ$ is a nonzero ideal in Q such that $(J \cdot J) = 0$. This contradicts to the semiprimeness of the algebra Q.

The first property of the algebras C and Q is evident. The second one reflects the fact that A is a large D-module in the injective hull P, and, hence, in Q. The definition of injective modules implies that for an arbitrary homomorphism of D-modules $\varkappa\colon J \to Q$ there exists an endomorphism $h \in E = \mathrm{End}_D P$ of the injective module P whose restriction to J coincides with $\varkappa$. However, h maps Q to Q. Hence the restriction of h to Q is the desired element of the algebra $C = E/\mathrm{Ker}\,\rho$.

To complete the proof, it remains to check that if for an element c in C and a large module J we have $cJ = 0$, then $c = 0$. Consider two ideals in Q: $I_1 = cQ$ and $I_2 = \{ q \mid q \in Q \ \& \ c \cdot q = 0 \}$. Obviously, we have $(I_1 \cdot I_2) = 0$. By the semiprimeness of Q, this implies $I_1 \cap I_2 = 0$; however, $I_2 \supset J$ and hence it is a large submodule in Q. Therefore we have $I_1 = 0$, i.e., $cQ = 0$, and this means that $c = 0$. The proposition is proved.

PROPOSITION 3.2. *If A is a prime K-algebra of signature Ω, then the centroid $C(A)$ is a field, $Q(A)$ is a prime K-algebra, and the C-algebra Q is characterized by the following properties*:

1) *$Q = CA$;*
2) *any nonzero D-submodule in Q has a nonzero intersection with A;*
3) *any partial D-endomorphism $\varkappa$ of a nonzero ideal J of the algebra A into A determines a unique endomorphism $c \in C$ whose restriction to J coincides with $\varkappa$.*

PROOF. Properties 1)–3) follow immediately from Proposition 3.1, because in a prime algebra A any nonzero ideal of the algebra A is a large D-submodule in A.

Since A is a large D-submodule in Q, we see that for any nonzero ideals I_1, I_2 in Q the ideals $J_i = I_i \cap A$ $(i = 1, 2)$ in A are nonzero. Thus $(I_1 \cdot I_2) \neq 0$, for otherwise we would have the equality $(J_1 \cdot J_2) = 0$, and this contradicts to the primeness of the algebra A.

It remains to show that C is a field. Let c be an arbitrary element of C. Define an endomorphism $\varkappa$ of the D-module Q by setting $\varkappa(q) \rightleftharpoons cq$, where q ranges over Q. Two cases are possible: either $\operatorname{Ker} \varkappa \neq 0$ or $\operatorname{Ker} \varkappa = 0$. In the first case $\operatorname{Ker} \varkappa$ is a large D-submodule in Q (because the property 2) implies $\operatorname{Ker} \varkappa \cap A \neq 0$ and the primeness of A implies now that $\operatorname{Ker} \varkappa \cap A$ is a large submodule in Q). Therefore, by the property 3), we would have $c = 0$. In the second case $c \neq 0$ and the modules Q and cQ are isomorphic. By property 3) there exists an element $c' \in C$ such that $c'(\varkappa(q)) = q$ for any $q \in Q$. Recalling the definition of the homomorphism $\varkappa$, we obtain $c'c = 1$ in the algebra C. The proposition is proved.

3.3. Centrally closed prime algebras, sufficient closedness conditions. The following lemmas permit the computation of the Martindale centroid in the simplest cases.

LEMMA 3.2. *Suppose that A is an associative K-algebra and A has a faithful irreducible representation in a K-space V. Then there exists a natural K-monomorphism δ of the Martindale centroid $C(A)$ into the centralizer $\operatorname{End}_A V$ of the A-module V.*

PROOF. The algebra A is prime, hence the centroid $C(A)$ is well defined and it is a field. Let $c \in C$; then there exists a two-sided ideal I in A such that $cI \subset A$ and $c(aib) = a(ci)b$ for any $a, b \in A$, $i \in I$. Since the representation is faithful and irreducible, any element in V is representable in the form $v = r_1 v_1 + \cdots + r_l v_l$, where $r_i \in I$, $v_1, \ldots, v_l \in V$. Define a mapping δ by setting $\delta(c)v \rightleftharpoons (cr_1)v_1 + \cdots + (cr_l)v_l$. If $v = 0$, then for any element $i \in I$ we have

$$i \sum_{j=1}^{l} (cr_j)v_j = (ci)v = 0,$$

and hence $\delta(c)v = 0$ by the faithfulness and irreducibility of V. This argument shows that the mapping δ is well defined and does not depend on the choice of the ideal I with the property $cI \subset A$. From the definition of δ it follows that $\delta(c)$ is contained in the centralizer $\operatorname{End}_A V$ of the A-module V and that δ is a nonzero K-homomorphism. The lemma is proved.

LEMMA 3.3. *Let A be a simple K-algebra of signature Ω and let $D(A)$ be the related associative subalgebra in $\operatorname{End}_K A$. Then the Martindale centroid $C(A)$ of the algebra A coincides with the centroid $\operatorname{End}_D A$ of the algebra A.*

PROOF. Since A is a simple algebra, A is the only nontrivial ideal in A and $(A \cdot A) \neq 0$. Hence, A is a prime algebra. As in the construction of a central closure, suppose that P denotes the injective hull of the D-module A and E denotes the algebra of endomorphisms End_P of the D-module P. Since the D-module A is irreducible, for any $h \in E$ the D-submodule hA is either zero or coincides with A (because A is a large submodule in P). This shows that $Q = EA$ coincides with A. Thus, by property 3) of central closure of a prime algebra A (Proposition 3.2), the Martindale centroid C must coincide with the centroid $\mathrm{End}_D A$ of the algebra A. The lemma is proved.

The next lemma gives sufficient conditions for the centralizer of an irreducible module over an associative algebra to be equal to $K \cdot \mathbf{1}$.

LEMMA 3.4. *Let A be a primitive associative algebra over an algebraically closed field K, let V be a faithful irreducible A-module, and let the dimension of the algebra A over K be less than the cardinality of the field K. Then the centralizer* $\mathrm{End}_A V$ *of the A-module V coincides with $K \cdot \mathbf{1}$.*

PROOF. By the Schur Lemma, $\mathrm{End}_A V$ is a skew field. Hence for any $c \in \mathrm{End}_A V$ and any $\beta \in K \cdot \mathbf{1}$, with the possible exception of $c = \beta$, the elements $c - \beta$ are invetrible. Since the A-module V is isomorphic to a quotient module of the module A with respect to a maximal left ideal, we see that $\dim_K V \leq \dim_K A <$ the cardinality of the field K. Hence there are linearly dependent elements among the ones of the form $(c-\beta)^{-1}v$, where $0 \neq v \in V$ and β ranges over the field K, and thus for some $\beta_1, \dots, \beta_m \in K$ ($m \geq 2$, $\beta_1 \neq 0$), $\varepsilon_1, \dots, \varepsilon_m \in K$ we have

$$\sum_{i=1}^{m} \beta_i (c - \varepsilon_i)^{-1} v = 0.$$

This equality and the uniqueness of decomposition of a rational function in elementary fractions imply that for a nonzero polynomial $f(x)$ we have the equality $f(c)v = 0$. Decomposing the polynomial f to linear factors, we see from the last relation that for a certain root $\beta \in K$ of the polynomial f there exists a nonzero vector $w \in V$ such that $(c-\beta)w = 0$. However, since V is irreducible and $c, \beta \cdot \mathbf{1} \in \mathrm{End}_A V$, we now see that $0 = A(c-\beta)w = (c-\beta)Aw = (c-\beta)V$. Hence $c = \beta \cdot \mathbf{1}$. The lemma is proved.

DEFINITION 3.5. A prime K-algebra of signature Ω is said to be *centrally prime* (*centrally closed*) if its Martindale centroid $C(A)$ coincides with the ground field K, in particular, $Q(A)$ coincides with A.

REMARK 3.2. The Lemmas 3.2–3.4 show that if the cardinality of the algebraically closed field K is greater than the dimension of a simple algebra A of signature Ω then the algebra A is centrally prime. In particular, a simple finite-dimensional algebra A of signature Ω over an algebraically closed field is centrally prime. These results are valid also for a primitive associative K-algebra.

For irreducible representations of finite-dimensional Lie K-algebras the following lemma is known.

LEMMA 3.5. *The centralizer of any irreducible representation of a finite-dimensional Lie algebra over an algebraically closed field K coincides with $K \cdot \mathbf{1}$.*

The proof is given in §54.

REMARK 3.3. This lemma, together with Lemma 3.2, shows that if $\rho \colon \mathfrak{g} \to \mathrm{End}_K V$ is a faithful irreducible representation of a finite-dimensional algebra $\mathfrak{g}$ in a vector space V (possibly infinite-dimensional), then the associative subalgebra in A generated in $\mathrm{End}_A V$ by the Lie algebra $\rho(\mathfrak{g}) \simeq \mathfrak{g}$ is centrally prime if the ground field K is algebraically closed.

LEMMA 3.6. *A finite-dimensional K-algebra A of signature Ω is prime if and only if in the algebra A there exists a unique nonzero minimal ideal I such that $(I \cdot I) \neq 0$. Moreover, if the field K is algebraically closed, then a prime finite-dimensional algebra A is centrally prime.*

PROOF. The first assertion of the lemma is obvious. Let a prime algebra A be finite-dimensional over an algebraically closed field K. Then the related associative algebra $D(A)$ is also finite-dimensional over K. But then the injective hull P of the D-module A is also finite-dimensional over K, and thus $C(A)$ and $Q(A)$ are finite-dimensional over K. From Proposition 3.2 it follows that $C(A)$ is a field. Thus $C(A)$ must coincide with the ground field K. The lemma is proved.

REMARK 3.4. A Lie algebra is said to be semisimple if it has no commutative ideals. Lemma 3.6 shows that any semisimple finite-dimensional Lie algebra over an algebraically closed field is prime and centrally prime if it contains a unique minimal ideal.

§4. Capelli identities and the rank theorem

We continue to denote by A a K-algebra of signature Ω (where among the signature operations there are ones with 'arity' not less than two) and by F an absolutely free K-algebra of the same signature Ω with free generators x_i, y_i $(i = 1, 2, \dots)$.

DEFINITION 4.1. A polynomial $d_k(x_1, \dots, x_k, y_1, \dots, y_l)$ in the algebra F that it is linear and skew-symmetric with respect to $x_1, \dots, x_k$ is called a *Capelli polynomial* of order k. Let V be an arbitrary K-subspace in A; we say that all *Capelli identities of order k* are valid on V if for any Capelli polynomial of order k and any elements $v_1, \dots, v_k \in V$, $a_1, \dots, a_l \in A$ the equality

$$d_k(v_1, \dots, v_k, a_1, \dots, a_l) = 0$$

is valid in A.

DEFINITION 4.2. The least number k such that all *Capelli identities of order k* are valid on V is called the *rank* of the K-linear subspace V with respect to the algebra A. We denote this number by $\operatorname{rank}(A, V)$.

THEOREM 4.1. *Let V be a K-subspace in a prime K-algebra A of signature Ω. If* $\operatorname{rank}(A, V) < \infty$, *then in the central closure $Q(A)$ of the algebra A we have*

$$\dim_{C(A)} C(A)V = \operatorname{rank}(A, V) - 1.$$

PROOF. Let $m = \dim_C CV$, $n = \operatorname{rank}(A, V) - 1$. It is obvious that $m \geq n$. From the definition of the rank it follows that there exist elements $v_1, \dots, v_n \in V$, $b_i \in A$ and a polynomial $d_n(x_1, \dots, x_n, y_1, \dots, y_l)$ of signature Ω that is skew-symmetric and multilinear with respect to the first n indeterminates such that

$$a = d_n(v_1, \dots, v_n, b_1, \dots, b_l) \neq 0.$$

For any $v \in V$ and $i = 1, \dots, n$ we set $a_i = d_n(v_1, \dots, v, \dots, v_n, b_1, \dots, b_l)$ (we substitute v instead of x_i). Applying a Capelli identity of order $n + 1$ to the elements

$v_1, \dots, v_n$ in the expression for a and to the element v in the expression for a_i, we obtain

$$\begin{aligned} & g\,(w \times a_i, \dots, u \times a) \\ &= \sum_{j=1}^{n} g\left(w \times d_n\Big|_{\substack{x_1=v_1,\dots,x_i=v_j,\dots,x_n=v_n \\ y_1=b_1,\dots,y_l=b_l}}, \dots, u \times d_n\Big|_{\substack{x_1=v_1,\dots,x_j=v,\dots,x_n=v_n \\ y_1=b_1,\dots,y_l=b_l}}\right) \\ &= g\left(w \times d_n\Big|_{\substack{x_1=v_1,\dots,x_i=v_i,\dots,x_n=v_n \\ y_1=b_1,\dots,y_l=b_l}}, \dots, u \times d_n\Big|_{\substack{x_1=v_1,\dots,x_i=v,\dots,x_n=v_n \\ y_1=b_1,\dots,y_l=b_l}}\right) \\ &= g\,(w \times a, \dots, u \times a_i), \end{aligned} \tag{4.1}$$

where w, u are any elements of the associative algebra $D(A)$, g is a polynomial of signature Ω that is multilinear with respect to two chosen indeterminates. Equality (4.1) implies that if $w \times a = 0$ for a certain $w \in D$, then the product of the ideals $D \times a$ and $D \times w \times a_i$ is zero. Since $a \neq 0$ and A is prime, we see now that the equality $w \times a = 0$ implies $w \times a_i = 0$. Hence there exists a homomorphism of the D-module $D \times a$ into $D \times a_i$, which defines a unique element $c_i(v) \in C(A)$ such that $a_i = c_i(v)a$ (see Proposition 3.2). It remains to show that the element $b = v - \sum_{i=1}^{n} c_i(v) \cdot v_i$ in $Q(A)$ is zero. Suppose $b \neq 0$. Then there exists $w \in D$ such that $0 \neq w \times b \in A$ (see property 2) of central closure in Proposition 3.2). However, for any polynomial g that is multilinear with respect to a certain pair of indeterminates and for any $u_1, u_2 \in D$ we have

$$\begin{aligned} & g\,(\dots, u_2 \times w \times b, \dots, u_1 \times a, \dots) \\ &\qquad = g\,(\dots, u_2 \times w \times v, \dots, u_1 \times a, \dots) \\ &\qquad\quad - \sum_{i=1}^{n} g\,(\dots, u_2 \times w \times v_i, \dots, u \times a_i, \dots). \end{aligned} \tag{4.2}$$

The right-hand side of equality (4.2) is skew symmetric with respect to the elements $v_1, \dots, v_n, v$; moreover, if we replace formally these elements by indeterminates $x_1, \dots, x_n, x_{n+1}$ then the polynomial thus obtained will be multilinear with respect to these indeterminates. Hence, the right-hand side of (4.2) is zero. This implies that the product of the ideals $D \times a$, $D \times w \times b$ is zero. Since $a \neq 0$, we see that $w \times b = 0$ because A is prime. The contradicition thus obtained proves the theorem.

The following two propositions follow easily from the rank theorem (Theorem 4.1) and play a useful role in the proof of results of the next section.

PROPOSITION 4.1. *Let A_1 be a prime K-subalgebra of a prime K-algebra A_2 of signature Ω. Moreover, let A_2 be an algebra over a commutative K-algebra $\mathcal{E}$ (with unit), and let A_2 be equal to $\mathcal{E}A_1$. Then there exist K-monomorphisms $\delta\colon C(A_1) \to C(A_2)$, $\rho\colon Q(A_1) \to Q(A_2)$, associated with the embedding of A_1 in A_2, such that the $C(A_2)$-algebras $C(A_2) \otimes_{\delta C(A_1)} \rho Q(A_1)$, $Q(A_2)$ of signature Ω are isomorphic.*

PROOF. Let $c \in C(A_1)$. Then there exists a nonzero ideal I of the algebra A_1 such that $cI \subset A_1$. Denote by J the ideal $\mathcal{E}I$ of A_2. Any element of the ideal J has the form

$$v = \sum_{i=1}^{n} f_i v_i \qquad (f_i \in \mathcal{E},\ v_i \in I).$$

Set $\delta(c)\cdot v \rightleftharpoons f_1(cv_1)+\cdots+f_n(cv_n)$. If $v=0$, then $(D(A_2)\times\delta(c)v\cdot J)=\mathcal{E}(D(A_1)\times\delta(c)v\cdot I)=\mathcal{E}(D(A_1)v\cdot cI)=0$, and this implies $\delta(c)v=0$, because the algebra A_2 is prime. This argument shows that $\delta(c)$ is a well defined homomorphism of the $D(A_2)$-module J into the $D(A_2)$-module A_2, and hence $\delta(c)$ defines uniquely an element of $C(A_2)$; moreover, this element does not depend on the choice of the ideal I (see property 3) of central closure in Proposition 3.2). Thus, δ is a well defined nonzero mapping of the field $C(A_2)$ into the field $C(A_1)$. Obviously, δ is a K-homomorphism.

An arbitrary element of the algebra $Q(A_1)$ has the form $w=\sum_{i=1}^n c_iv_i$, $c_i\in C(A_1)$, $v_i\in A_1$. Set $\rho(w)\rightleftharpoons\sum_{i=1}^n\delta(c_i)v_i$. If $w=0$, then we choose a nonzero ideal J of the algebra A_1 such that $c_iJ\subset A_1$ $(i=1,\dots,n)$. Then

$$\begin{aligned}(C(A_2)D(A_2)\rho(w)\cdot C(A_2)J)&\\ &=C(A_2)\mathcal{E}(D(A_1)\rho(w)\cdot J)\subseteq C(A_2)(D(A_1)w\cdot J)=0\end{aligned}$$

(the latter inclusion takes place, since, by the definition of δ, the elements $\delta(c_i)$ and c_i have the same action on J for each $i=1,2,\dots,n$). Since the algebra $Q(A_2)$ is prime, this implies $\rho(w)=0$. Thus, ρ is a well defined K-linear mapping of $Q(A_1)$ into $Q(A_2)$. Obviously, ρ is a K-homomorphism that fixes the elements of the subalgebra A_1. Let V be an arbitrary finite-dimensional K-subspace in the algebra A_1. Since $\mathcal{E}A_1=A_2$, we see that $\operatorname{rank}(A_2,V)=\operatorname{rank}(A_1,V)$. This implies that ρ is a monomorphism and that there exists an isomorphism $Q(A_2)\simeq C(A_2)\otimes_{\delta C(A_1)}\rho Q(A_1)$. The proposition is proved.

PROPOSITION 4.2. *Let a K-algebra A_1 be centrally prime (i.e., $C(A_1)=K$ and $Q(A_1)=A_1$). Then, for any extension $\widehat{K}$ of the field K, the K-algebra $A_2=\widehat{K}\otimes_K A_1$ is prime, and $C(A_2)=\widehat{K}$ ($Q(A_2)=A_2$).*

PROOF. Let $b_1,\dots,b_n$ be arbitrary elements of the algebra A_2 that are linearly independent over $\widehat{K}$. Let $e_1,\dots,e_k$ be linearly independent over K elements of A_1 such that the elements $b_1,\dots,b_n$ are expressed as linear combinations of $e_1,\dots,e_k$ with coefficients in $\widehat{K}$. Obviously, we have $k\ge n$ and we can choose $k-n$ elements among the e_i such that $b_1,\dots,b_n,e_{i_1},\dots,e_{i_{k-n}}$ are linearly independent over $\widehat{K}$. Since A_1 is centrally prime, from the rank theorem (Theorem 4.1) it follows that there exist a polynomial $d_k(x_1,\dots,x_k,y_1,\dots,y_l)$ that is multilinear and skew-symmetric with respect to $x_1,\dots,x_k$ and $a_1,\dots,a_l\in A_1$ such that

$$\begin{aligned}(4.3)\quad 0\ne d_k(e_1,\dots,e_k,a_1,\dots,a_l)&\\ &=\beta d_k(b_1,\dots,b_n,e_{i_1},\dots,e_{i_{k-n}},a_1,\dots,a_l),\qquad \beta\in\widehat{K}.\end{aligned}$$

From equality (4.3) it follows that, first, any nonzero $D(A_1)$-submodule in A_2 contains a nonzero element of the form βa, where $\beta\in\widehat{K}$, $a\in A_1$, and thus, since A_1 is centrally prime, we see that A_2 is centrally prime; next, the rank of the elements $b_1,\dots,b_n$ in the algebra A_2 is equal to $n+1$ and, by the rank theorem, any elements of the algebra A_2 linearly independent over $\widehat{K}$ must also be linearly independent over $C(A_2)$ in $Q(A_2)$. Thus, $C(A_2)=\widehat{K}$, and the proposition is proved.

§5. Isomorphism of centrally prime associative-Lie pairs with the same identities over an algebraically closed field

In this section we shall prove the following

THEOREM 5.1. *Let the ground field K be algebraically closed. Let ρ_1, ρ_2 be faithful irreducible representations of finite-dimensional Lie K-algebras $\mathfrak{g}_1$ and $\mathfrak{g}_2$ in (not necessarily finite-dimensional) vector spaces V_1 and V_2 over K, respectively. Denote by A_i $(i = 1, 2)$ the associative subalgebra in $\operatorname{End}_K V_i$ generated by the Lie algebra $\rho(\mathfrak{g}_i) \simeq \mathfrak{g}_i$. In this case, if the identities of the associative-Lie pairs $(A_1, \mathfrak{g}_1)$, $(A_2, \mathfrak{g}_2)$ are the same, then these pairs are isomorphic.*

COROLLARY 1. *Over an algebraically closed field, any finite-dimensional simple Lie algebra is uniquely defined by its identities up to an isomorphism.*

COROLLARY 2. *Semisimple finite-dimensional Lie algebras over an algebraically closed field, containing a unique minimal ideal, possess the same identities if and only if they are isomorphic.*

In our opinion, this theorem and its Corollary 1 are of fundamental character. It provides a convenient language for the description of simple Lie algebras and explains why it is reasonable to study identities of Lie algebras (identities of representations of Lie algebras) and to investigate Lie algebras (representations of Lie algebras) from the viewpoint of identity relations.

The class of Lie algebras and their representations plays an important role in algebra and in mathematics. Therefore, we distinguish specially our Theorem 5.1 as a separate assertion. However, actually, under a correct understanding of the notion of "representation", this result has a natural generalization in the class of universal algebras of arbitrary signature Ω with multilinear operations.

As we have seen already in §1, there are three closely related approaches to the study of representations of Lie algebras from the viewpoint of identity relations. Obviously, the first two of them, related to the category of associative-Lie pairs and the category of representations of Lie algebras in associative algebras, are applicable to algebras of arbitrary signature. In fact, there is a natural way to endow any K-algebra of the variety of algebras of signature Ω with the structure of a "poorer" signature Ω'. Toward this end, it is sufficient to fix once and for all a certain set of multilinear polynomials $g_\omega(t_1, \dots, t_l)$ in an absolutely free algebra H of signature Ω and take them as basic operations in the signature Ω'. In this case, any K-algebra A of signature Ω is a K-algebra with signature Ω' with respect to these operations, and a representation of an algebra L of signature Ω' in the algebra A is a homomorphism $\rho\colon L \to A$ of algebras of signature Ω'. If we suppose that ρ is a faithful representation ($\operatorname{Ker} \rho = 0$), then we can identify the algebra L of signature Ω' with the vector subspace $\rho(L)$ of the algebra A, and the operations $\omega \in \Omega'$ with multilinear polynomials g_ω in the free algebra H of signature Ω. If the K-subspace L in the K-algebra A is stable with respect to Ω' and spans the whole algebra A relative to all operations from Ω, then (A, L) is called a *pair of signature* (Ω, Ω').

If for a polynomial $f(t_1, \dots, t_l)$ in an absolutely free algebra H of signature Ω and for any elements $v_1, \dots, v_l \in L$ the equality $f(v_1, \dots, v_l) = 0$ takes place in the algebra A, we say that in the pair (A, L) the identity $f = 0$ holds.

If it is possible to choose a system of generators x_i, $i \in M$, in the algebra L such that any mapping of x_i into L has an extension to an endomorphism of the algebra A, then the pair is called a (relatively) free pair of signature (Ω, Ω') with free generators x_i, $i \in M$. As in §1 for associative-Lie pairs, by a homomorphism φ of a pair (A_1, L_1) into a pair (A_2, L_2) we mean a homomorphism φ of the algebra A_1 into the algebra A_2 (both algebras of signature Ω) such that $\varphi(L_1)$ is contained in L_2. By analogy with associative-Lie pairs in §1, the notions of product of pairs, variety of pairs, and also the Birkhoff theorem on free algebras (i.e., the assertion on the existence, in an arbitrary variety $\mathfrak{B}$ of pairs of signature (Ω, Ω'), of a relatively free pair $(F, \mathfrak{L})$ with an arbitrary set of free generators x_i, $i \in M$, such that any mapping of the generators x_i into L, where (A, L) is an arbitrary pair in the variety $\mathfrak{B}$, extends to a homomorphism of $(F, \mathfrak{L})$ into (A, L)), are carried over to the pairs of signature (Ω, Ω') in a natural way.

Under this approach to the notion of a pair of signature (Ω, Ω'), the following assertions are natural generalizations of Theorem 5.1.

THEOREM 5.2. *Let $(A_1, \mathfrak{g}_1)$ and $(A_2, \mathfrak{g}_2)$ be pairs of signature (Ω, Ω') over an infinite field K, let the K-algebras A_1 and A_2 be centrally prime and let the algebras $\mathfrak{g}_1$ and $\mathfrak{g}_2$ be finite-dimensional over K. The pairs $(A_1, \mathfrak{g}_1)$ and $(A_2, \mathfrak{g}_2)$ have the same identities if and only if there exists a K-linear automorphism σ of an algebraically closed field $\overline{K}$ of infinite transcendence degree over K such that the pairs*

$$(\overline{K} \otimes_K A_1, \overline{K} \otimes_K \mathfrak{g}_1), \qquad (\overline{K} \otimes_K A_2, \overline{K} \otimes_K \mathfrak{g}_2)$$

are K-isomorphic, and this isomorphism is σ-semilinear.

THEOREM 5.3. *Let $(A_1, \mathfrak{g}_1)$ and $(A_2, \mathfrak{g}_2)$ be pairs of signature (Ω, Ω') over an algebraically closed field K of arbitrary characteristic, where the algebras $\mathfrak{g}_1$ and $\mathfrak{g}_2$ of signature Ω' are finite-dimensional and the algebras A_1 and A_2 of signature Ω are centrally prime. The pairs $(A_1, \mathfrak{g}_1)$ and $(A_2, \mathfrak{g}_2)$ have the same identities if and only if there exist K-homomorphisms $\psi_1 \colon A_1 \to A_2$, $\psi_2 \colon A_2 \to A_1$ such that $\psi_1(\mathfrak{g}_1) = \mathfrak{g}_2$, $\psi_2(\mathfrak{g}_2) = \mathfrak{g}_1$.*

COROLLARY 1. *Simple finite-dimensional algebras of the same signature Ω over an algebraically closed field and with the same identities are isomorphic.*

COROLLARY 2. *Prime finite-dimensional algebras of the same signature Ω over an algebraically closed field possess the same identities if and only if they are isomorphic.*

We shall derive Theorem 5.2 from the following assertion, of independent interest in itself.

THEOREM 5.4. *Let $(A, \mathfrak{g})$ be a pair of signature (Ω, Ω') over an infinite field K, let the K-algebra A be centrally prime, and let $\mathfrak{g}$ be finite-dimensional over K. Denote by $(F, \mathfrak{L})$ a relatively free pair of the variety generated by the pair $(A, \mathfrak{g})$ with a countable set of free generators and by K_0 a field of rational functions over K that has a countable transcendence degree. Then there exists an embedding δ of the Martindale centroid $C(F)$ of the prime K-algebra F of signature Ω in the field K_0 such that the K_0-pairs*

$$(K_0 \otimes_K A, K_0 \otimes_K \mathfrak{g}), \qquad (K_0 \otimes_{\delta C(F)} Q(F), K_0 \otimes_{\delta C(F)} C(F)\mathfrak{L})$$

are isomorphic in the signature (Ω, Ω').

REMARK 5.1. As we have already seen in §3, it is implicitly assumed in the definition of a prime algebra that some of the signature operations are of 'arity' greater than one. Hence we are to assume in Theorems 5.2–5.4 and their consequences that in the signature Ω there exist operations with 'arity' greater than one. We do not impose any assumptions on the signature Ω'; e.g., we can suppose that Ω' is the empty set.

PROOF OF THEOREM 5.4. Choose a basis $e_1, \dots, e_m$ in the K-space $\mathfrak{g}$. Let $x_i^{(j)}$ $(j = 1, \dots, m)$ be algebraically independent over K generators of the field of rational functions K_0. Denote by (R, L) the K_0-pair $(K_0 \otimes_K A,\ K_0 \otimes_K \mathfrak{g})$ of signature (Ω, Ω'). Let $(F, \mathfrak{L})$ be the K-subpair of signature (Ω, Ω') generated in the K_0-pair (R, L) by the elements

$$x_i \rightleftharpoons \sum_{j=1}^{m} x_i^{(j)} e_j \qquad (i = 1, 2, \dots).$$

If the ground field is infinite, then we have the following

LEMMA 5.1. *The pair $(F, \mathfrak{L})$ is a relatively free pair of the variety of pairs* $\operatorname{var}(A, \mathfrak{g})$, *generated by the pair $(A, \mathfrak{g})$, with free generators $x_1, x_2, \dots$.*

PROOF. Since the field K is infinite, any identity $f(t_1, \dots, t_m) = 0$ that is valid in the pair $(A, \mathfrak{g})$ is also valid in the K_0-pair (R, L), and therefore also in the K-pair $(F, \mathfrak{L})$. This shows that $(F, \mathfrak{L}) \in \operatorname{var}(A, \mathfrak{g})$. Hence it remains to show that $(F, \mathfrak{L})$ is a relatively free pair, i.e., that any mapping φ of the generators $x_1, x_2, \dots$ to elements of the algebra $\mathfrak{g}$ of signature Ω' is extendable to a homomorphism of the pair $(F, \mathfrak{L})$ into the pair $(A, \mathfrak{g})$. Let $\varphi(x_i) = v_i$ $(i = 1, 2, \dots)$. Since $v_i = \sum_{j=1}^{m} \beta_i^{(j)} e_j$ for some elements $\beta_i^{(j)}$ of the field K and the mapping $x_i^{(j)} \mapsto \beta_i^{(j)}$ is extendable to a homomorphism $\varkappa$ of the algebra of polynomials $K\left[x_i^{(j)} \mid i = 1, 2, \dots;\ j = 1, \dots, m\right]$ into the field K, we see that $\varkappa$ can be extended to a homomorphism $\varkappa^*$ of the algebra $\widetilde{A} = K\left[x_i^{(j)}\right] \otimes_K A$ into A by setting

$$\varkappa^*\left(\sum_{j=1}^{m} p_j \otimes e_j\right) \rightleftharpoons \sum_{j=1}^{m} \varkappa(p_j) \otimes e_j \qquad \left(p_j \in K\left[x_i^{(j)}\right]\right).$$

Obviously, $\varkappa^*(x_i) = v_i$ and the restriction of the homomorphism $\varkappa^*$ to the K-subalgebra F of the algebra $\widetilde{A}$ of signature Ω is an extension of the mapping φ to a homomorphism of the pair $(F, \mathfrak{L})$ into the pair $(A, \mathfrak{g})$. The lemma is proved.

Since the K-algebra A of signature Ω is centrally prime, Proposition 4.2 implies that the K_0-algebra R is centrally prime, i.e., $C(R) = K_0$, $Q(R) = R$. Obviously, $R = K_0 F$ (for this, it is sufficient that the number of free generators x_i be not less than the dimension of the algebra $\mathfrak{g}$ over K). This implies that the algebra F is prime. Since the K_0-algebra R is centrally prime, we obtain, by applying Proposition 4.1 to the K-algebras $R \supset F$ of signature Ω, that there exist monomorphisms $\delta\colon C(F) \to K_0$, $\rho\colon Q(F) \to R$ defining a K_0-isomorphism of the pairs (R, L) and $(K_0 \otimes_{\delta C(F)} Q(F),\ K_0 \otimes_{\delta C(F)} C(F)\mathfrak{L})$ of signature (Ω, Ω'). The equality $L = K_0 \mathfrak{L}$ implies that the image of L under this isomorphism contains $\mathfrak{L}$. Theorem 5.4 is proved.

PROOF OF THEOREM 5.2. Let the K-pairs $(A_1, \mathfrak{g}_1)$ and $(A_2, \mathfrak{g}_2)$ of signature (Ω, Ω') satisfy the conditions of Theorem 5.2 and have the same identities. In an algebraically closed field $\overline{K}$ of infinite transcendence degree over K, we choose a subfield of rational functions K_0 of countable transcendence degree such that the transcendence degree of $\overline{K}$ over K is infinite. Since the pairs $(A_1, \mathfrak{g}_1)$ and $(A_2, \mathfrak{g}_2)$ have the same identities, we see that relatively free pairs with countable number of generators of the varieties $\operatorname{var}(A_1, \mathfrak{g}_1)$ and $\operatorname{var}(A_2, \mathfrak{g}_2)$ are isomorphic. Denote this free pair by $(F, \mathfrak{L})$. Choosing bases in $\mathfrak{g}_1$ and $\mathfrak{g}_2$ and using the subfield K_0 as in the proof of Theorem 5.4, we can construct monomorphisms $\delta_i \colon C(F) \to \overline{K}$, $\rho_i \colon Q(F) \to \overline{K} \otimes_K A_i$ that give a $\overline{K}$-isomorhism of pairs

$$(\overline{K} \otimes_K A_i,\ \overline{K} \otimes_K \mathfrak{g}_i), \qquad (\overline{K} \otimes_{\delta_i C(F)} Q(F),\ \overline{K} \otimes_{\delta_i C(F)} C(F)\mathfrak{L}).$$

The construction of δ_i implies that actually the subfield $\delta C(f)$ lies in K_0, and $\overline{K}$ has an infinite transcendence degree over this subfield. It is known (see [Lan]) that in this case there exists a K-automorphism σ of the field $\overline{K}$ such that the image $\delta_1 C(f)$ coincides with $\delta_2 C(f)$ and we have $\sigma\delta_1 = \delta_2$. Choose elements $a_1, a_2, \dots$ in the algebra F so that they form a basis of the $C(F)$-algebra $Q(F)$, and the first m elements form a basis in $C(F)\mathfrak{L}$. Define a K-linear mapping ψ of the algebra $\overline{K} \otimes_K A_1$ onto the algebra $\overline{K} \otimes_K A_2$ by the formula

$$\psi\left(\sum_{i=1}^{n} f_i \rho_1(a_i)\right) \rightleftharpoons \sum_{i=1}^{n} \sigma(f_i)\rho_2(a_i) \qquad (f_i \in \overline{K}).$$

Obviously, ψ is a σ-semilinear mapping and ψ is a K-isomorphism of the K-pairs $(\overline{K} \otimes_K A_1,\ \overline{K} \otimes_K \mathfrak{g}_1)$, $(\overline{K} \otimes_K A_2,\ \overline{K} \otimes_K \mathfrak{g}_2)$. Theorem 5.2 is proved.

The proof of Theorem 5.3 is an immediate consequence of the following lemma.

LEMMA 5.2. *Let $\mathfrak{g}_1$ and $\mathfrak{g}_2$ be finite-dimensional vector spaces over an algebraically closed field K that generate certain K-algebras A_1, A_2 of signature Ω, respectively. Suppose that for a certain extension $\overline{K}$ of the field K and an automorphism σ of the field $\overline{K}$ over K, the K-algebras $\overline{K} \otimes_K A_1$ and $\overline{K} \otimes_K A_2$ are σ-semi-isomorphic and the spaces $\overline{K} \otimes_K \mathfrak{g}_1$ and $\overline{K} \otimes_K \mathfrak{g}_2$ are mapped to each other under this isomorphism. Then there exist homomorphisms $\varphi_1 \colon A_1 \to A_2$ and $\varphi_2 \colon A_2 \to A_1$ such that $\varphi_1(\mathfrak{g}_1) = \mathfrak{g}_2$ and $\varphi_2(\mathfrak{g}_2) = \mathfrak{g}_1$.*

PROOF. Denote by φ the σ-semilinear automorphism of the $\overline{K}$-algebra $\overline{K} \otimes_K A_1$ into $\overline{K} \otimes_K A_2$. Let $e_1, \dots, e_m$ be a basis of the K-space $\mathfrak{g}_1$ and let $v_1, \dots, v_k$ be a basis of the K-space $\mathfrak{g}_2$. By assumption of the lemma, $\varphi(\overline{K} \otimes_K \mathfrak{g}_1) = \overline{K} \otimes_K \mathfrak{g}_2$; hence $k = m$, because the isomorphism φ is semilinear. Then we have

$$\varphi(e_i) = \sum_{j=1}^{m} f_i^{(j)} v_j \qquad (i = 1, \dots, m)$$

for some $f_i^{(j)} \in \overline{K}$, and the determinant f of the matrix with elements $f_i^{(j)}$ $(i, j = 1, \dots, m)$ is not zero in the field $\overline{K}$. Denote by $\mathcal{E}$ the commutative K-subalgebra in $\overline{K}$

generated by the elements $f_i^{(j)}$ $(i, j = 1, \dots, m)$ and f^{-1}. The algebra $\mathcal{E}$ is a finitely generated K-algebra and contains the elements f, f^{-1} $(f \cdot f^{-1} = 1)$. Since the field K is algebraically closed by assumption, we have, from one of the forms of the Hilbert Nullstellensatz (see §51), that there exists a K-homomorphism $\varkappa$ of the algebra $\mathcal{E}$ onto the field K. This homomorphism has a unique extension to a K-homomorphism $\varkappa^*$ of the algebra $\mathcal{E} \otimes_K A_2$ onto A_2 such that $\varkappa^*(g \otimes b) = \varkappa(g) \otimes b$, where $b \in A_2$, $g \in \mathcal{E}$. Then the mapping $\varphi_1 = \varkappa^* \circ \varphi\big|_{A_1}$ is a K-homomorphism of the algebra A_1 into the algebra A_2, and we have $\varphi_1(\mathfrak{g}_1) \subset \mathfrak{g}_2$. Since the element $\varkappa(f)$ is invertible, we see that $\varphi_1(e_1), \dots, \varphi_1(e_m)$ are linearly independent over K elements of $\mathfrak{g}_2$, and thus $\varphi_1(\mathfrak{g}_1) = \mathfrak{g}_2$. However, the subspace $\mathfrak{g}_2$ generates the algebra A_2; hence we have $\varphi_1(A_1) = A_2$. A homomorphism $\varphi_2\colon A_2 \to A_1$ such that $\varphi_2(\mathfrak{g}_2) = \mathfrak{g}_1$ is constructed analogously. The lemma is proved.

Applying Theorem 5.2 and Lemma 5.2, in succession, to the pairs $(A_1, \mathfrak{g}_1)$ and $(A_2, \mathfrak{g}_2)$ of signature (Ω, Ω'), we obtain the proof of Theorem 5.3.

Proof of the Corollaries of Theorem 5.3. Suppose the finite-dimensional simple algebras $\mathfrak{g}_1$ and $\mathfrak{g}_2$ of signature Ω over an algebraically closed field K have the same identities. As noted in Remark 3.2, the simple algebras $\mathfrak{g}_1$ and $\mathfrak{g}_2$ are centrally prime. We set $A_1 = \mathfrak{g}_1$ and $A_2 = \mathfrak{g}_2$. Consider formally the pairs $(A_1, \mathfrak{g}_1)$ and $(A_2, \mathfrak{g}_2)$ of signature (Ω, Ω). Obviously, these pairs have the same identities of pairs. Therefore, we can apply Theorem 5.3 to them and obtain the existence of epimorphisms $\varphi_1\colon \mathfrak{g}_1 \to \mathfrak{g}_2$ and $\varphi_2\colon \mathfrak{g}_2 \to \mathfrak{g}_1$ of algebras of signature Ω. Since the algebras $\mathfrak{g}_1$ and $\mathfrak{g}_2$ are finite-dimensional, this implies their isomorphism. The Corollaries are proved.

Proof of Theorem 5.1. Let ρ_i, V_i, $\mathfrak{g}_i$, A_i $(i = 1, 2)$ have the same meaning as in the statement of Theorem 5.1. Since the ground field K is algebraically closed, the Lie algebra $\mathfrak{g}_i$ is finite-dimensional, and V_i is a faithful irreducible $\mathfrak{g}_i$-module, by Remark 3.3 the associative subalgebra A_i is centrally prime. Applying Theorem 5.3 to the associative-Lie pairs $(A_i, \mathfrak{g}_i)$, $i = 1, 2$, having the same identities, we see that there exist epimorphisms $\varphi_1\colon A_1 \to A_2$ and $\varphi_2\colon A_2 \to A_1$ such that $\varphi_1(\mathfrak{g}_1) = \mathfrak{g}_2$ and $\varphi_2(\mathfrak{g}_2) = \mathfrak{g}_1$. Since the universal enveloping associative K-algebra of a finite-dimensional Lie K-algebra is Noetherian (see §54), we see that the enveloping algebra A_i of the algebra $\mathfrak{g}_i$ $(i = 1, 2)$ is Noetherian. Therefore, the epimorphism $\varphi_2 \circ \varphi_1$ of the algebra A_1 must be an automorphism. This proves that the kernel of the epimorphism φ_1 is zero, and hence φ_1 is an isomorphism of the pairs $(A_1, \mathfrak{g}_1)$ and $(A_2, \mathfrak{g}_2)$. Theorem 5.1 is proved.

Corollary 1 of Theorem 5.1 is a particular case of Corollary 1 of Theorem 5.3. Corollary 2 of Theorem 5.1 follows from Remark 3.4 and Corollary 2 of Theorem 5.3.

Comments

The use of the entire function $e_{\mathfrak{B}}(z)$ in §2 in the study of the rate of growth (as $n \to \infty$) of the dimensions of spaces of multilinear polynomials F_λ in n indeterminates of a variety $\mathfrak{B}$, as well as Theorem 2.4, are seen in the literature, apparently, for the first time. In connection with Proposition 2.1 the following asymptotic result of A. N. Grishkov is worth noticing: $\dim_K F_\lambda \leq |\lambda|!/N^\lambda$ $(N = 1, 2, \dots)$ (oral communication). Theorem 2.3 is due to S. P. Mishchenko [Mi, Bak]. Theorem 2.1 and its Corollary 1 were first proved by A. Regev [Reg, AR]. The proof of Theorem 2.1 given in §2 is due to V. N. Latyshev [Lat1]. Its Corollary 2 is close in spirit to the paper of I. Kaplansky [Kap1]. The Corollary of Theorem 2.2 and Theorem 2.3 show that exponential type varieties contain the most interesting classes of associative-Lie pairs. Therefore, for the general theory of these varieties and its applications (see Chapter 6), it is important to solve the following problem.

PROBLEM 1.1. Is it true that an arbitrary variety of associative-Lie pairs of exponential type can be embedded in a variety of exponential type defined by a finite number of identities? Find such finite systems of identities.

Various applications of the central closure and Martindale centroid are contained in the papers of W. E. Baxter, W. S. Martindale, J. S. Erickson [BMa, ErMO], and V. K. Kharchenko [Kha1, Kha2]. In connection with central closure of algebras, the following problem deserves attention.

PROBLEM 1.2. Does there exist a finitely generated associative primitive algebra A over a countable algebraically closed field K, the Martindale centroid of which does not coincide with $K \cdot 1$ (for which $Q(A) \neq A$)?

Proposition 4.2 was first published by J. Matzuk [Mat]. Theorems 4.1 and 5.1–5.4 and their corollaries were published in author's papers [Raz13, Raz16, KuR]. An important particular case of Theorem 5.1 for faithful irreducible finite-dimensional representations of semisimple splittable Lie algebras over a field of characteristic zero was first obtained by A. Kh. Kushkuleĭ, who derived it from a finite-dimensional version of Razmyslov's Theorem 5.2 with the help of Chevalley's basis [KuR]. In connection with Corollary 1 of Theorem 5.1, the following problem arises.

PROBLEM 1.3. Let the ground field K be algebraically closed and of characteristic zero. Is it true that for any two simple finite-dimensional Lie K-algebras $\mathfrak{g}_1$ and $\mathfrak{g}_2$ the inclusion $\operatorname{var}\mathfrak{g}_1 \subseteq \operatorname{var}\mathfrak{g}_2$ holds if and only if $\mathfrak{g}_1$ can be embedded in $\mathfrak{g}_2$?

In spite of the fact that over fields of positive characteristic, I know that the answer to this problem is negative, it would be important for the general theory of simple Lie algebras over such fields to find necessary and sufficient conditions for the inclusion $\operatorname{var}\mathfrak{g}_1 \subseteq \operatorname{var}\mathfrak{g}_2$ (where $\mathfrak{g}_1$ and $\mathfrak{g}_2$ are simple and finite-dimensional).

CHAPTER II

CHARACTERS AND α-FUNCTIONS ON 2-WORDS AND VARIETIES OF REPRESENTATIONS OF LIE ALGEBRAS DISTINGUISHED BY THEM

In this chapter we state the main notions and problems of the theory of 2-words and α-functions, presently the most powerful method of investigation of identities of Lie algebras belonging to varieties of exponential type and and of representations of these Lie algebras. The substantive aspect of this method will be uncovered in the following three chapters. In this chapter we restrict ourselves to a conceptual consideration of the theory of 2-words and α-functions, outline some directions of possible further investigations, and give all formal definitions that are necessary in the subsequent chapters. However, first we consider results preceding the creation of the new method of studying identical relations.

§6. An important example

A Lie K-algebra L over a field of an arbitrary characteristic is said to be *nth Engel* if the identity

$$[x, \underbrace{y, y, \dots, y}_{n \text{ times}}] = 0 \tag{6.1}$$

is valid in L. At present it is known that any pth Engel Lie algebra over a field of characteristic $p > 0$ is locally nilpotent (see [Kos2]). This means that for any natural number k there exists a number $N(p, k)$ such that in any pth Engel Lie algebra over a field of characteristic p with k generators the following nilpotence identity is valid:

$$[x_1, x_2, \dots, x_N] = 0. \tag{6.2}$$

In 1959, A. I. Kostrikin (to whom this result is due) posed the problem on nilpotency of pth Engel Lie algebras with arbitrary number of generators. In 1985, E. I. Zel′manov (see [Kos4]) succeeded in obtaining the positive answer to this problem for any nth Engel Lie algebras over a field of characteristic zero. A negative answer to the problem of A. I. Kostrikin for any $p > 3$ was obtained in 1969 (see [Raz2]).

In this section we discuss the original proof of that fact and further modifications of this proof in order to show how the main notions of the method of 2-words and α-functions have arisen intrinsically.

6.1. The solvability problem for 3rd Engel Lie algebras over fields of characteristic 5. Let us define inductively Lie polynomials r_k in 2^k indeterminates by setting

$$r_1 = [x_1, x_2], \qquad r_{k+1} = \left[r_k, r_k\big|_{x_1 = x_{2^k+1}, \dots, x_{2^k} = x_{2^{k+1}}}\right].$$

A Lie algebra in which $r_k = 0$ is an identity for some k is said to be a *solvable Lie algebra of class* k, and the identity itself is called the *solvability identity*. Obviously, the nilpotence identity (6.2) implies the solvability identity $r_k = 0$ for $k > \log_2 N$. For pth Engel Lie algebras the converse is also true.

THEOREM 6.1 (Higgins). *Any pth Engel Lie algebra over a field of characteristic p satisfying a solvability identity* $r_k = 0$ *is nilpotent and identity* (6.2) *holds in it for* $N > \frac{(p-1)^k-1}{p-2} + 1$ (see [Higg1, Bak]).

This result implies that the negative solution of the nilpotency problem for pth Engel Lie algebras is equivalent to the negative solution of the solvability problem for these algebras.

THEOREM 6.2. *The variety of* 3rd *Engel Lie algebras over a field of characteristic* 5 *is nonsolvable* (*nonnilponent*).

In order to prove this assertion, it is necessary to indicate a 3rd Engel Lie algebra M with a countable number of generators for which no identity of the form (6.2) is valid in M. In order to construct such an algebra M, we shall elaborate on the properties of Lie algebras that follow from the 3rd Engel identity

$$[x, y, y, y] = 0. \tag{6.3}$$

Let us substitute in it an element $y + \beta x$ ($\beta \in K$) instead of the variable y. Then

$$([z, y, y, y]) + \beta([z, y, y, x] + [z, y, x, y] + [z, x, y, y]) \\ + \beta^2([z, x, x, y] + [z, x, y, x] + [z, y, x, x]) + \beta^3([z, x, x, x]) = 0.$$

Considering these equations for different $\beta \in K$ as a linear system of equations with respect to the expressions in parentheses and noticing that the determinant of this system is the Vandermonde determinant, we conclude that any expression at β^k ($k = 0, 1, \dots$) is also equal to zero. (This discussion is of general nature and it is applicable to any identity $f = 0$, thus obtaining "new" identities from a given one. This process is called a (partial) *linearization of the identity* $f = 0$ *with respect to the variable* y. It guarantees the creation of "new" identities if the Vandermonde determinant is nonzero, and the latter condition can easily be satisfied if the cardinality of the field is greater than the degree of the polynomial f with respect to the variable y.) Thus, the following identity is valid in M:

$$[z, x, y, y] + [z, y, x, y] + [z, y, y, x] = 0. \tag{6.4}$$

The Jacobi identity permits us to write any Lie commutator containing the variable x in the form of a linear combination of left-normed commutators, in each of which the letter x appears in the first position from the left. Hence, we can rewrite identity (6.4) in the form

$$-3[x, z, y, y] + 3[x, y, z, y] - [x, y, y, z] = 0.$$

Replacing x by z and z by x, we obtain a "new" identity

$$3[z, x, y, y] - 3[z, y, x, y] + [z, y, y, x] = 0.$$

From now on, side by side with the Lie algebra L, it is convenient to consider an associative-Lie pair (A, L) corresponding to the adjoint representation of the Lie algebra M. By definition, L is the linear subspace in $\operatorname{End}_K M$ spanned by all transformations $\operatorname{ad} g$ $(g \in M)$ that map $v \in M$ to $[g, v]$ and A is the associative subalgebra in $\operatorname{End}_K M$ generated by L. It is clear that the properties of the pair (A, L) are closely related to those of the Lie algebra M: first, M is isomorphic to the quotient algebra of the algebra M with respect to its center; second, the associative algebra A is nilpotent (i.e., $A^k = 0$ for some $k \in \mathbb{N}$) if and only if the Lie algebra M is nilpotent; third, any identity $f(x_1, \dots, x_l) = 0$ of the pair (A, L) gives the Lie identity $f(\operatorname{ad} x_1, \dots, \operatorname{ad} x_l) \times z = 0$ of the Lie algebra M, and any identity of the pair (A, L) is obtained in this way from a suitable Lie identity of M; fourth, if M is a relatively free Lie algebra with the free generators $y_1, y_2, \dots$, then the pair (A, L) is relatively free and $x_1 = \operatorname{ad} y_1$, $x_2 = \operatorname{ad} y_2$, $\dots$ are its free generators. Thus, the identities (6.3–6.5) of the Lie algebra M give us the following identities of the pair (A, L):

(6.6)
$$y^3 = 0, \qquad xy^2 + yxy + y^2x = 0,$$
$$xy^2 - 3yxy + 3y^2x = 0,$$

which are equivalent, for $p \neq 2$, to the following identities:

(6.7)
$$yxy = \tfrac{1}{2}y^2x, \qquad xy^2 = -\tfrac{3}{2}y^2x.$$

These identities show that

$$y^2x^2 = 2yxyx = yx^2y = -\tfrac{2}{3}y^2x^2.$$

Therefore, if the characteristic p of the ground field K is not equal to 5, then in the pair (A, L) we have the following identity:

(6.8)
$$x^2y^2 = 0.$$

LEMMA 6.1. *In an arbitrary associative-Lie pair over a field of characteristic $p \neq 2$ the identity*

(6.9)
$$x_1^2x_2^2 \cdots x_l^2 = 0$$

implies the nilpotence identity $x_1 \cdots x_N = 0$ for $N \geq 3l$.

THE PROOF of this assertion will be given by induction on the number l. The base of induction is $l = 1$. In this case identity (6.9) has the form $x_1^2 = 0$ and its linearization gives the identity $xy + yx = 0$. For any elements a, b denote by $a \circ b$ the element $ab + ba$. A direct verification shows that

(6.10)
$$4v_1v_2v_3 = -([v_1, v_2] \circ v_3) + 3(v_1 \circ [v_2, v_3]) + ([v_1, v_3] \circ v_2)$$
$$+ 2v_1(v_2 \circ v_3) - 2v_2(v_1 \circ v_3) + 2v_3(v_1 \circ v_2).$$

Therefore, over a field of characteristic $\neq 2$, the product of any three Lie elements is equal to zero in any pair with the identity $x_1^2 = 0$. The base of induction is proved.

Since for any Lie elements u, v we have

$$u \circ v = (u+v)^2 - u^2 - v^2,$$

formula (6.10) shows that, replacing in identity (6.9) the variable x_1 by the elements of the Lie algebra in y_1, y_2, y_3 and multiplying them from the left by the elements of the same kind, we can express the identity

$$y_1 y_2 y_3 x_2^2 \cdots x_l^2 = 0$$

as a linear combination of the former identities. Using similar substitutions, it is possible to derive from this identity the following one

$$y_{11} y_{12} y_{13} y_{21} y_{22} y_{23} x_3^2 \cdots x_l^2 = 0.$$

Continuing this process of replacing the square of a variable by the product of three variables, we obtain the identity $y_1 y_2 \cdots y_{3l} = 0$. This proves the assertion of the lemma.

From this lemma and (6.8) it follows that over fields whose characteristic is different from 2 and 5, the algebra A in the pair (A, L) is nilpotent and, therefore, any 3rd Engel Lie algebra over such fields is nilpotent.

In the case of fields of characteristic 5 we can rewrite identities (6.7) in the following form:

$$y^2 x = x y^2, \qquad y x y = \tfrac{1}{2} y^2 x. \tag{6.11}$$

Denote by $(F, \mathfrak{L})$ the relatively free pair of the variety given by identities (6.11) with the countable set of free generators $X = \{x_1, x_2, \dots\}$.

LEMMA 6.2. *In the pair $(F, \mathfrak{L})$, the Lie algebra $\mathfrak{L}$ is 3rd Engel, and the associative algebra F is nilpotent if and only if any 3rd Engel Lie algebra over a field of characteristic 5 is nilpotent.*

PROOF. Identities (6.11) successively imply

$$\begin{aligned} &(1 - \tfrac{1}{2}) y^3 = 0, \\ &[z, y, y, y] = z y^3 - 3 y z y^2 + 3 y^2 z y - y^3 z = 4(z y^3 - y^3 z) = 0. \end{aligned} \tag{6.12}$$

This proves that the algebra $\mathfrak{L}$ is 3rd Engel.

Suppose that the algebra F is nilpotent. Then for any 3rd Engel Lie algebra M, the pair (A, L), constructed according to the adjoint representation of M, satisfies identities (6.11), and the algebra A is to be nilpotent. However, then the Lie algebras $A^{(-)}$ and L are nilpotent, and the nilpotency class of the Lie algebra M is greater by 1 than the one of the algebra L, because L is isomorphic to the quotient algebra of the algebra M by its center.

Suppose that an arbitrary 3rd Engel Lie algebra is nilpotent. Then the 3rd Engel Lie algebra $\mathfrak{L}$ is also nilpotent. By the definition of nilpotency, identity (6.2) is to hold in this algebra for some sufficiently large N. In particular, the identity

$$(\operatorname{ad} x_n)^2 \cdots (\operatorname{ad} x_1)^2 z = 0 \qquad (n > [(N+1)/2])$$

is to hold in $\mathfrak{L}$. Since in the algebra F the Lie element $(\operatorname{ad} y)^2 z = [z, y, y]$ can be rewritten in the form $2 \cdot (1 - \frac{1}{2}) z y^2$ with the help of identities (6.11), we see that identity (6.13), modulo identities (6.11), is equivalent to the identity

$$z x_1^2 \cdots x_n^2 = 0,$$

which implies that the algebra F is nilpotent, by Lemma 6.1. The lemma is proved.

LEMMA 6.3. *Any monomial $u = x_{i_1} \cdots x_{i_n}$ in free generators x_{i_j} of the pair $(F, \mathfrak{L})$ can be written in the algebra F as a linear combination of monomials of the same degree and the same composition that have the form*

$$x_{j_1}^2 \cdots x_{j_s}^2 x_{k_1} \cdots x_{k_t} \qquad (s, t \geq 0,\ 2s + t = n), \tag{6.14}$$

where $j_1, \dots, j_s$; $k_1, \dots, k_t$ are mutually distinct.

PROOF. The first of identities (6.11) means that the square of any element in $\mathfrak{L}$ belongs to the center of the algebra F. Moreover, since the linearization of this identity has the form

$$(y \circ z) x = x (y \circ z) \qquad (y \circ z \rightleftharpoons yz + zy), \tag{6.15}$$

we see that the element $v_1 \circ v_2$ also belongs to the center of the Lie algebra F for any elements $v_1, v_2 \in \mathfrak{L}$. Let u contain a subword $x_i a x_i$, where a is nonempty and does not contain x_i. If $a = x_j a'$ then, since $x_i \circ x_j$ and x_i^2 belong to the center, we see that

$$x_i x_j a' x_i = -x_j x_i a' x_i + a'(x_i \circ x_j) x_i = -x_j x_i a' x_i + (1 + \tfrac{1}{2}) x_i^2 a' x_j$$

(the last equality follows from the second identity in (6.11)). This immediately implies that if $a = x_{q_1} \cdots x_{q_l}$ then

$$x_i a x_i = x_i^2 \left[-\frac{3}{2} \sum_{t=1}^{l} (-1)^t x_{q_1} \cdots \widehat{x}_{q_t} \cdots x_{q_l} x_{q_t} + (-1)^l a \right]. \tag{6.16}$$

This formula implies directly that it is possible to represent u by monomials of the form (6.14). We only note that formula (6.16) and identity (6.12) imply that any monomial containing a generator x_i three times is equal to zero in the algebra F. The lemma is proved.

DEFINITION 6.1. A word u in an alphabet Y is said to be a 2-*word* if any letter that occurs in u has exactly two occurrences in u. Linear combinations of 2-words over the ground field are said to be 2-*elements*.

PROPOSITION 6.1. *In the algebra F, any 2-element u in the alphabet X can be reduced to the form*

$$\alpha x_{j_1}^2 x_{j_2}^2 \cdots x_{j_l}^2 \qquad (j_1 < j_2 < \cdots < j_l), \tag{6.17}$$

where α is a scalar factor and $x_{j_1}, \dots, x_{j_l}$ are all the letters occurring in u. Moreover, a free 3rd Engel Lie algebra with a countable number of free generators is not nilpotent *over a field of characteristic* 5 *if and only if the element* α *is uniquely determined by the* 2*-element* u, *i.e., it is independent of the way of reduction of* u *in the algebra* F *to the form* (6.17).

PROOF. The existence of the representation (6.17) for u follows from Lemma 6.3.

If a 2-element u could be expressed in the form (6.17) in two different ways, then it would mean that identity (6.9) were valid in the pair $(F, \mathfrak{L})$ and then, by Lemma 6.1, the algebra F would be nilpotent. Hence, by Lemma 6.2, the free 3rd Engel Lie algebra would also be nilpotent.

On the other hand, if any 2-element is reducible to a unique form (6.17), then this means, in particular, that all the elements of the form $x_1^2 \cdots x_l^2$ $(l = 1, 2, \ldots)$ are nonzero in the algebra F. But then the associative algebra F is nonnilpotent and, therefore, so is the free 3rd Engel Lie algebra of countable rank as well, by Lemma 6.2. The proposition is proved.

Proposition 6.1 accomplishes the first part in the study of the solvability problem for 3rd Engel Lie algebras over a field of characteristic 5. This shows that the existence of nonsolvable 3rd Engel Lie algebra, which is still a conjecture, is related to the existence of a certain function $\alpha(u)$ on 2-words (a linear function on 2-elements). Moreover, it follows from our consideration that if this function α exists, then we are able to evaluate it with the help of formula (6.16) (see the proof of Lemma 6.3).

Now we pass to the second part of the investigation. We show that in the pair $(F, \mathfrak{L})$ the representation (6.17) is uniquely determined by the 2-element u, and the assignment of an element of the field $\alpha(u)$ to u is a mapping. We divide the proof of this fact into three steps.

1ST STEP. We consider an associative algebra A_0 over a field of characteristic 5; this algebra is given by generators $\overline{x}_1, \overline{x}_2, \ldots$ and defining relations

$$\overline{x}_i^2 \overline{x}_j = \overline{x}_j \overline{x}_i^2, \qquad \overline{x}_i^3 = 0 \qquad (i, j = 1, 2, \ldots), \tag{6.18}$$

$$\overline{x}_i a \overline{x}_i = \overline{x}_i^2 \left(\gamma \sum_{j=1}^{l} (-1)^j \overline{x}_{i_1} \cdots \widehat{\overline{x}}_{i_j} \cdots \overline{x}_{i_l} \overline{x}_{i_j} + (-1)^l a \right), \tag{6.19}$$

where $a = \overline{x}_{i_1} \cdots \overline{x}_{i_l}$ are arbitrary monomials in $\overline{x}_1, \overline{x}_2, \ldots$ and $\gamma = -3/2$. It follows from the definition of the pair $(F, \mathfrak{L})$ that in F all these relations with respect to x_i are satisfied (see the identities (6.11), (6.12), (6.16)). Hence, the algebra F is the image of the algebra A_0 under the homomorphism transforming $\overline{x}_i$ into x_i.

Just as in Lemma 6.3, using the relations (6.18) and (6.19), we can represent any word $u = \overline{x}_{i_1} \overline{x}_{i_2} \cdots \overline{x}_{i_n} \in A_0$ in the form of a linear combination of words of the same length and the same composition having the form

$$\overline{x}_{j_1}^2 \cdots \overline{x}_{j_s}^2 \overline{x}_{k_1} \cdots \overline{x}_{k_t} \qquad (j_1 < j_2 < \cdots < j_s), \tag{6.20}$$

were $j_1, \ldots, j_s, k_1, \ldots, k_t$ are pairwise distinct. This form of recording of the word u will be said to be *regular*. In particular, if u contains some generator $\overline{x}_i$ three or more times, then $u = 0$.

LEMMA 6.4. *In the algebra A_0, any element is uniquely representable in the form of linear combination of regularly written words.*

PROOF. Denote by π_i the operation that replaces the entry of the subword $\overline{x}_i a \overline{x}_i$ in u by the sum

$$\overline{x}_i^2 \left[\gamma \sum_{j=1}^{l} (-1)^j \overline{x}_{i_1} \cdots \widehat{\overline{x}}_{i_j} \cdots \overline{x}_{i_l} \overline{x}_{i_j} + (-1)^l a \right],$$

if $a = \overline{x}_{i_1} \cdots \overline{x}_{i_l}$. Denote the expression in square brackets by $\pi(a)$. The operation π_i can be extended linearly to any linear combination. In order to prove the lemma, it is sufficient to verify that the operations π_i and π_j commute and agree with the transformations of the form $\overline{x}_t \overline{x}_k^2 = \overline{x}_k^2 \overline{x}_t$. The last assertion follows immediately from the formula

$$\pi_i(\overline{x}_i y_1 \cdots y_s \overline{x}_j^2 z_1 \cdots z_t \overline{x}_i) = \overline{x}_j^2 \pi_i(\overline{x}_i y_1 \cdots y_s z_1 \cdots z_t \overline{x}_i),$$

which can be checked directly (for convenience, here some of the generators $\overline{x}_k$ of the algebra A_0 are denoted by new letters y and z).

The proof of the statement that the operators π_i and π_j commute will be given by induction. The following two cases are possible:

A) $u = \overline{x}_i a \overline{x}_j b \overline{x}_i c \overline{x}_j$;
B) $u = \overline{x}_i a \overline{x}_j b \overline{x}_j c \overline{x}_i$,

where $a = x_1 x_2 \cdots x_s$, $b = y_1 y_2 \cdots y_t$, $c = z_1 z_2 \cdots z_k$.

Consider the case A). We apply induction on s, using the recurrent relation

$$\pi(x_1 v) = -\gamma v x_1 - x_1 \pi(v).$$

If $s \geq 1$, then $a = x_1 a'$ and

$$\begin{aligned}
\pi_i(u) &= \overline{x}_i^2 \{-\gamma a' \overline{x}_j b x_1 c \overline{x}_j\} - x_1 \pi_i(\overline{x}_i a' \overline{x}_j b \overline{x}_i c \overline{x}_j), \\
\pi_j \pi_i(u) &= -\gamma \overline{x}_i^2 \overline{x}_j^2 a' \pi(b x_1 c) - x_1 \pi_j \pi_i(\overline{x}_i a' \overline{x}_j b \overline{x}_i c \overline{x}_j) \\
&= -\gamma \overline{x}_i^2 \overline{x}_j^2 a' \pi(b x_1 c) - x_1 \overline{x}_j^2 \pi_i(\overline{x}_i a' \pi(b \overline{x}_i c)).
\end{aligned}$$

Simultaneously, we have

$$\pi_i \pi_j(u) = \overline{x}_j^2 \pi_i(\overline{x}_i a \pi(b \overline{x}_i c)).$$

Setting $\pi(b \overline{x}_i c) = \sum_l u_l \overline{x}_i^2 v_l$, we obtain now

$$\begin{aligned}
\pi_i\Big(\sum_l \overline{x}_i a u_l \overline{x}_i v_l\Big) &= -\gamma \overline{x}_i^2 \sum_l a' u_l x_1 v_l - x_1 \pi_i\Big(\sum_l \overline{x}_i a' u_l \overline{x}_i v_l\Big) \\
&= -\gamma \overline{x}_i^2 a' \pi(b x_1 c) - x_1 \pi_i(\overline{x}_i a' \pi(b \overline{x}_i c)).
\end{aligned}$$

This implies $\pi_i \pi_j(u) = \pi_j \pi_i(u)$.

Suppose now that $s = 0$; we proceed by induction on t. If $t \geq 1$, then $b = y_1 b'$ and

$$\pi_i(\overline{x}_i\overline{x}_j b\overline{x}_i c\overline{x}_j) = -\gamma\overline{x}_i^2 b\overline{x}_j c\overline{x}_j + \gamma\overline{x}_i^2\overline{x}_j b' y_1 c\overline{x}_j + \overline{x}_j y_1 \pi_i(\overline{x}_i b'\overline{x}_i c\overline{x}_j),$$

$$\begin{aligned} \pi_j\pi_i(u) &= -\gamma\overline{x}_i^2\overline{x}_j^2 b\pi(c) \\ &\quad + \gamma\overline{x}_i^2\overline{x}_j^2\pi(b'y_1c) - \gamma\overline{x}_i^2\overline{x}_j^2\pi(b')cy_1 - y_1\overline{x}_i^2\overline{x}_j^2\pi(\pi(b')c), \\ \pi_j(u) &= -\gamma\overline{x}_j^2\overline{x}_i b'\overline{x}_i c y_1 + \overline{x}_i y_1 \pi_j(\overline{x}_j b'\overline{x}_i c\overline{x}_j), \\ \pi_i\pi_j(u) &= \overline{x}_j^2\Big(-\gamma\overline{x}_i^2\pi(b')cy_1 - \pi_i\Big(\sum_l \overline{x}_i y_1 u_l \overline{x}_i v_l\Big)\Big) \\ &= \overline{x}_j^2\Big(-\gamma\overline{x}_i^2\pi(b')cy_1 + y_1\pi_i\Big(\sum_l \overline{x}_i u_l\overline{x}_i v_l\Big) + \gamma\sum_l \overline{x}_i^2 u_i y_1 v_i\Big) \\ &= -\gamma\overline{x}_i^2\overline{x}_j^2\pi(b')cy_1 + y_1\pi_i\pi_j(\overline{x}_i\overline{x}_j b'\overline{x}_i c\overline{x}_j) + \gamma\overline{x}_i^2\pi_j(\overline{x}_j b' y_1 c\overline{x}_j). \end{aligned}$$

Comparing the expressions just obtained, we see that we are to verify the equality

$$y_1\pi_i\pi_j(\overline{x}_i\overline{x}_j b'\overline{x}_i c\overline{x}_j) = -y_1\overline{x}_i^2\overline{x}_j^2\pi(\pi(b')c) - \gamma\overline{x}_i^2\overline{x}_j^2 b\pi(c) = y_1\pi_j\pi_i(\overline{x}_i\overline{x}_j b'\overline{x}_i c\overline{x}_j),$$

which is valid by the induction assumption.

Finally, if $s = 0$, $t = 0$, $u = \overline{x}_i\overline{x}_j\overline{x}_i c\overline{x}_j$, then

$$\pi_i\pi_j(u) = \pi_i\{\overline{x}_j^2(-\gamma\overline{x}_i c\overline{x}_i - \overline{x}_i^2\pi(c))\} = (-1-\gamma)\overline{x}_i^2\overline{x}_j^2\pi(c) = \pi_j\pi_i(u).$$

The case B) is treated similarly.

If $s \geq 1$, then $a = x_1 a'$ and

$$\begin{aligned} \pi_i(u) &= \overline{x}_i^2\{-\gamma a'\overline{x}_j b x_j c\overline{x}_1\} - x_1\pi_i(\overline{x}_i a'\overline{x}_j b\overline{x}_j c\overline{x}_i), \\ \pi_j\pi_i(u) &= -\gamma\overline{x}_i^2\overline{x}_j^2 a'\pi(b)cx_1 - x_1\pi_j\pi_i(\overline{x}_i a'\overline{x}_j b\overline{x}_j c\overline{x}_i) \\ &= -\gamma\overline{x}_i^2\overline{x}_j^2 a'\pi(b)cx_1 - x_1\overline{x}_j^2\pi_i(\overline{x}_i a'\pi(b)c\overline{x}_i). \end{aligned}$$

Simultaneously, we have

$$\pi_j\pi_i(u) = \overline{x}_j^2\pi_i(\overline{x}_i a\pi(b)c\overline{x}_i) = -\gamma\overline{x}_i^2\overline{x}_j^2 a'\pi(b)cx_1 - x_1\overline{x}_j^2\pi_i(\overline{x}_i a'\pi(b)c\overline{x}_i).$$

This implies $\pi_i\pi_j(u) = \pi_j\pi_i(u)$.

Let $s = 0$; we use the case A):

$$\begin{aligned} \pi_j\pi_i(u) &= \pi_j\pi_i(\overline{x}_i\overline{x}_j b\overline{x}_j c\overline{x}_i) \\ &= -\gamma\overline{x}_i^2\pi_j(b\overline{x}_j c\overline{x}_j) - \pi_j(\overline{x}_j\pi_i(\overline{x}_i b\overline{x}_j c\overline{x}_i)) \\ &= -\gamma\overline{x}_i^2\overline{x}_j^2 b\pi(c) - \pi_j\pi_i(\overline{x}_j\overline{x}_i b\overline{x}_j c\overline{x}_i) \\ &= -\gamma\overline{x}_i^2\overline{x}_j^2 b\pi(c) - \pi_i\pi_j(\overline{x}_j\overline{x}_i b\overline{x}_j c\overline{x}_i) \\ &= -\gamma\overline{x}_i^2\overline{x}_j^2 b\pi(c) + \gamma\overline{x}_i^2\overline{x}_j^2 b\pi(c) + \pi_i(\overline{x}_i\pi_j(\overline{x}_j b\overline{x}_j c\overline{x}_i)) \\ &= \pi_i\pi_j(\overline{x}_i\overline{x}_j b\overline{x}_j c\overline{x}_i). \end{aligned}$$

Finally, if $s = 0$, $t = 0$, $u = \overline{x}_i\overline{x}_j^2 c\overline{x}_i$, then, as it was mentioned above,

$$\pi_i(\overline{x}_i\overline{x}_j^2 c\overline{x}_i) = \overline{x}_j^2\pi_i(\overline{x}_i c\overline{x}_i)$$

and this means that $\pi_i\pi_j(u) = \pi_j\pi_i(u)$. The lemma is proved.

COROLLARY. *In the algebra A_0 any 2-word $u = u(\overline{x}_{j_1}^2 \cdots \overline{x}_{j_l}^2)$ can be uniquely represented in the form*

$$u = \alpha(u)\overline{x}_{j_1}^2 \cdots \overline{x}_{j_l}^2 \quad (j_1 < j_2 < \cdots < j_l), \tag{6.21}$$

where $\alpha(u)$ is a scalar factor.

Lemma 6.4 and its Corollary show that the desired function α on 2-words is well defined by the transformations π_i $(i = 1, 2, \ldots)$. It remains to verify that this definition agrees with the identical relations (6.11) defining the pair $(F, \mathfrak{L})$. This verification is the subject of the following two steps.

2ND STEP. Let us consider the ideal I_0 generated in the algebra A_0 by the elements

$$[(\overline{x}_i \circ \overline{x}_j), \overline{x}_k], \tag{6.22}$$

$$[([\overline{x}_i, \overline{x}_t] \circ \overline{x}_j), \overline{x}_k]. \tag{6.23}$$

Let B_0 be the subalgebra in A_0 generated by the 2-words in the generators $\overline{x}_1, \overline{x}_2, \ldots$. Denote by B_1 the image of B_0 under the natural homomorphism $A_0 \to A_1 = A_0/I_0$.

LEMMA 6.5. *The algebra B_1 is isomorphic to B_0 for $\gamma = -2/3$.*

PROOF. By the theorem on homomorphisms (see §49), it is sufficient to prove that $B_0 \cap I_0 = 0$. First of all, we note that if some of the lower indices in the expressions (6.22) and (6.23) are equal, then the result of their reduction to the regular form, via formulas (6.18) and (6.19), is equal to zero. Let us consider the most cumbersome case

$$\begin{aligned}[([\overline{x}_i, \overline{x}_j] \circ \overline{x}_t), \overline{x}_i] = {}& \{\overline{x}_i(\overline{x}_t \circ \overline{x}_j)\overline{x}_i - \overline{x}_i^2(\overline{x}_t \circ \overline{x}_j)\} \\ &- \{\overline{x}_j\overline{x}_i\overline{x}_t\overline{x}_i - \overline{x}_i\overline{x}_j\overline{x}_i\overline{x}_t\} + \{\overline{x}_t\overline{x}_i\overline{x}_j\overline{x}_i - \overline{x}_i\overline{x}_t\overline{x}_i\overline{x}_j\}.\end{aligned}$$

Applying the transformation π_i to every pair of braces, we obtain zero.

Therefore, it is sufficient to take for generators of the ideal I_0 only elements (6.22) and (6.23) with pairwise distinct indices. Then, taking into account the form of the basis in the algebra A_0 (see Lemma 6.4), we note that 2-elements in I_0 can be obtained when multiplying elements (6.22) or (6.23) from the left or from the right by one of the generators taking part in these expressions, and these elements generate $I_0 \cap B_0$. Furthermore, it can be checked directly that the following relations hold (we take into account that $\gamma = -2/3$):

$$\begin{aligned}
\overline{x}_k[\overline{x}_i, \overline{x}_j]\overline{x}_k &= (2\gamma + 1)\overline{x}_k^2[\overline{x}_i, \overline{x}_j] = -(1+\gamma)\overline{x}_k^2[\overline{x}_i, \overline{x}_j], \\
([\overline{x}_i, \overline{x}_t] \circ \overline{x}_j) &= (\overline{x}_i \circ [\overline{x}_t, \overline{x}_j]) + (\overline{x}_i \circ \overline{x}_j)\overline{x}_t - \overline{x}_t(\overline{x}_i \circ \overline{x}_j), \\
\overline{x}_j[([\overline{x}_i, \overline{x}_t] \circ \overline{x}_j), \overline{x}_k] &= -\gamma\overline{x}_j^2[\overline{x}_i, \overline{x}_j]\overline{x}_k \\
&\quad + \gamma\overline{x}_j^2([\overline{x}_i, \overline{x}_t] \circ \overline{x}_k) - \gamma\overline{x}_j^2\overline{x}_k[\overline{x}_i, \overline{x}_t] = 0, \\
\overline{x}_i[([\overline{x}_i, \overline{x}_t] \circ \overline{x}_j), \overline{x}_k] &= \overline{x}_i[[(\overline{x}_i \circ \overline{x}_j), \overline{x}_t], \overline{x}_k] = \gamma[(\overline{x}_k \circ \overline{x}_j), \overline{x}_i]\overline{x}_i^2, \\
\overline{x}_k[([\overline{x}_i, \overline{x}_t] \circ \overline{x}_j), \overline{x}_k] &= \gamma\overline{x}_k^2\{[(\overline{x}_i \circ \overline{x}_j), \overline{x}_t] - [(\overline{x}_t \circ \overline{x}_j), \overline{x}_i]\},
\end{aligned} \tag{6.24}$$

and similar relations for the right multiplication of elements (6.22) and (6.23) by $\overline{x}_i$, $\overline{x}_j$, and $\overline{x}_k$. Hence, the 2-elements in I_0 are generated by the elements of the form (6.22). However, any of elements (6.22) passes to zero in the algebra A_0 under the left and right multiplication by $\overline{x}_i$, $\overline{x}_j$, and $\overline{x}_k$. For instance,

$$\begin{aligned}[(\overline{x}_i \circ \overline{x}_j), \overline{x}_k]\overline{x}_i &= \pi_i(\overline{x}_i\overline{x}_j\overline{x}_k\overline{x}_i + \overline{x}_j\overline{x}_i\overline{x}_k\overline{x}_i - \overline{x}_k(\overline{x}_i \circ \overline{x}_j)\overline{x}_i) \\ &= \overline{x}_i^2\{-\gamma\overline{x}_k\overline{x}_j - \overline{x}_j\pi(\overline{x}_k) + \overline{x}_j\pi(\overline{x}_k) + \gamma\overline{x}_k\overline{x}_j\} = 0.\end{aligned}$$

Hence, $I_0 \cap B_0 = 0$. The lemma is proved.

Denote by L_1 the Lie subalgebra in A_1 generated by the images of the elements $\overline{x}_1$, $\overline{x}_2, \dots$, under the canonical homomorphism $A_0 \to A_1$.

LEMMA 6.6. *In the pair* (A_1, L_1) *the first of the identities* (6.11) *is valid.*

PROOF. First we show that in the pair (A_1, L_1) identity (6.15) is valid.

Since the element (6.22) belongs to the ideal I_0, we see that the following relations are valid in the algebra A_1:

$$(\overline{x}_i \circ \overline{x}_j)\overline{x}_k = \overline{x}_k(\overline{x}_i \circ \overline{x}_j), \tag{6.25}$$

and the elements $\overline{x}_i \circ \overline{x}_j$ belong to the center C_1 of the algebra A_1. The identity $[a, b, c] = (a \circ (b \circ c)) - (b \circ (a \circ c))$ is valid in any associative algebra, hence, in the algebra A_1 we have the relation

$$[\overline{x}_i, \overline{x}_j, \overline{x}_k] = 2 \cdot (\overline{x}_j \circ \overline{x}_k)\overline{x}_i - 2(\overline{x}_i \circ \overline{x}_k)\overline{x}_j, \tag{6.26}$$

and this implies that L_1 is contained in the left module over C_1 generated by $\overline{x}_i$ and $[\overline{x}_i, \overline{x}_j]$. Since identity (6.15) is multilinear, we see that its verification in the pair (A_1, L_1) is equivalent to the proof of the equality

$$(v_i \circ v_j)\overline{x}_k = \overline{x}_k(v_i \circ v_j), \tag{6.27}$$

where v_i, v_j are commutator monomials in the generators $\overline{x}_t$ of length not greater than 2. If v_i, v_j are commutators of length 1, then (6.27) follows from (6.25); if one of these commutators is of length 2 and the other one is of length one, then (6.27) is satisfied, because the elements (6.23) are contained in the ideal I_0 of the algebra A_0. In the case $v_i = [\overline{x}_i, \overline{x}_t]$, $v_j = [\overline{x}_j, \overline{x}_s]$ the relation

$$v_i \circ v_j = -([\overline{x}_i, \overline{x}_t, \overline{x}_s] \circ \overline{x}_j) + [([\overline{x}_i, \overline{x}_t] \circ \overline{x}_j), \overline{x}_s)$$

is valid, and now (6.26), together with the fact that the elements (6.22) and (6.23) are contained in the ideal I_0, implies formula (6.27).

Thus, identity (6.15) is valid in the pair (A_1, L_1). Substituting $z = y$ in this identity and using the fact that the characteristic of the ground field differs from two, we see that identity (6.11) is also valid in (A_1, L_1). The lemma is proved.

Since over a field of characteristic 5 the parameter $\gamma = -3/2$ taking part in the defining relations (6.19) of the algebra A_0, coincides with $-2/3$, it follows from Lemmas 6.5 and 6.6 that the function α on 2-elements, defined by the algebra A_0, agrees with the first identity in (6.11) for the pair $(F, \mathfrak{L})$, and, side by side with (6.18) and (6.19), we can use identity (6.15) in computing α and in reducing a 2-element to a regular form.

3RD STEP. In the algebra A_1, let us consider the ideal I_1 generated by the elements of the form

$$v_1v_2v_3 + v_2v_1v_3 + v_1v_3v_2 + v_3v_1v_2 + v_2v_3v_1 + v_3v_2v_1, \tag{6.28}$$

denoted by $\mathrm{Sym}(v_1, v_2, v_3)$, where v_1, v_2, v_3 are commutators of arbitrary length in the generators $\overline{x}_1, \overline{x}_2, \dots$, taken $\mod I_0$. Denote by A_2 the quotient algebra A_1/I_1 and denote by L_2 the image of the Lie algebra L_1 under the canonical homomorphism $A_1 \to A_2$. Obviously, the pair (A_2, L_2) is canonically isomorphic to the pair $(F, \mathfrak{L})$ (see Lemma 6.2).

Since relations (6.26) are satisfied in the algebra A_1, we see that any element of the Lie algebra L_1 can be expressed as a linear combination of commutators of length 1 and 2 with coefficients from the center C_1 of the algebra A_1. Hence, the ideal I_1 is generated by the elements (6.28), where v_1, v_2, v_3 are commutators of length less than 3. Denote by B_2 the image of B_1 under the natural homomorphism $A_1 \to A_2$.

LEMMA 6.7. *If the characteristic of the ground field is equal to* 5, *then the algebra* B_2 *is isomorphic to* B_1.

PROOF. Taking the commutator of (6.28) with $\overline{x}_t$, we obtain

$$\begin{aligned}[\mathrm{Sym}(v_1, v_2, v_3), \overline{x}_t] = {} & \mathrm{Sym}([v_1, \overline{x}_t], v_2, v_3) \\ & + \mathrm{Sym}(v_1, [v_2, \overline{x}_t], v_3) + \mathrm{Sym}(v_1, v_2, [v_3, \overline{x}_t]),\end{aligned} \tag{6.29}$$

whence it follows that the ideal I_1 is generated by the elements (6.28), where v_3 is a commutator of length 1. Moreover, if $v_3 = \overline{x}_k = v_2$ in relation (6.29), then

$$2\,\mathrm{Sym}(v_1, [\overline{x}_k, \overline{x}_t], \overline{x}_k) = [\mathrm{Sym}(v_1, \overline{x}_k, \overline{x}_k), \overline{x}_t] - \mathrm{Sym}([v_1, \overline{x}_t], \overline{x}_k, \overline{x}_k).$$

Hence, the elements (6.28), where the commutator v_1 is of length less than 3, $v_3 = \overline{x}_k$, and $v_2 = [\overline{x}_k, \overline{x}_t]$, belong to the ideal generated by the elements (6.28), where $v_2 = v_3 = \overline{x}_k$. However, for any $v \in L_1$ we have

$$\mathrm{Sym}(v, \overline{x}_k, \overline{x}_k) = 2(v\overline{x}_k^2 + \overline{x}_k v \overline{x}_k + \overline{x}_k^2 v) = 2(1 - (1+\gamma) + 1)\overline{x}_k^2 v = 0$$

in the algebra A_1 over a field of characteristic 5, because $\gamma = -\frac{3}{2}$ and $1 - \gamma = \frac{5}{2}$. These equalities imply that $I_1 \cap B_1$ is generated by 2-elements obtained from elements (6.28), where the commutators depend on pairwise distinct generators and the commutator v_3 is of length one.

Applying once more (6.24), as well as (6.27) and (6.19), we obtain

$$\begin{aligned}\mathrm{Sym}(v_1, v_2, \overline{x}_k)\overline{x}_k = {} & (v_1 \circ v_2)\overline{x}_k^2 + \{v_1\overline{x}_k v_2 \overline{x}_k + v_2 \overline{x}_k v_1 \overline{x}_k\} \\ & + \overline{x}_k^2(v_1 \circ v_2) = (2 - (1+\gamma))\overline{x}_k^2(v_1 \circ v_2) = 0 \pmod 5.\end{aligned} \tag{6.30}$$

Multiplying relation (6.29), where we set $v_3 = \overline{x}_k$, by $\overline{x}_k$ and taking into account (6.30) and also (6.25) and the relation $(\overline{x}_t \circ \overline{x}_k)\overline{x}_k = -\gamma\overline{x}_k^2\overline{x}_t$, we obtain

$$\mathrm{Sym}(v_1, v_2, \overline{x}_k)(\overline{x}_t \circ \overline{x}_k) = -\gamma\overline{x}_k^2\,\mathrm{Sym}(v_1, v_2, \overline{x}_t) = \mathrm{Sym}(v_1, v_2, [\overline{x}_k, \overline{x}_t])\overline{x}_k,$$

i.e., the 2-elements in the ideal I_1 are generated by the elements (6.28), where v_1, v_2, v_3 are commutators of length 1. Taking into account the form of the basis of the algebra A_0 (see Lemma 6.4), we note that 2-elements in I_1 that generate $I_1 \cap B_1$ as an ideal in B_1 are obtained from these elements (6.28) by left and right multiplication by generators that enter in these expressions. But from (6.30) it follows that $\mathrm{Sym}(\overline{x}_{i_1}, \overline{x}_{i_2}, \overline{x}_{i_3})\overline{x}_{i_j} = 0 \pmod 5$, where $j \in \{1, 2, 3\}$. Similarly it is proved that also $\overline{x}_{i_j}\,\mathrm{Sym}(\overline{x}_{i_1}, \overline{x}_{i_2}, \overline{x}_{i_3}) = 0$. Hence, $I_1 \cap B_1 = 0$, and Lemma 6.7 is proved.

PROOF OF THEOREM 6.2. Since the pairs $(F, \mathfrak{L})$ and (A_2, L_2) are isomorphic, Lemma 6.7 shows that any 2-word possesses a unique representation of the form (6.17) in the algebra F. Hence, by Proposition 6.1, the free 3rd Engel Lie algebra with a countable number of generators is nonnilpotent over fields of characteristic 5. The nonsolvability of this algebra follows from the Higgins theorem. Theorem 6.2 is proved.

6.2. Existence of nonsolvable $(p-2)$th Engel Lie algebras over fields of characteristic $p \geq 5$. Let us consider once more the proof of Theorem 6.2 and let us try to single out the parts that are relatively self-standing and independent of the specificity of the problem. First of all, it is striking that the first two steps of this proof are hardly dependent on the characteristic of the ground field. Indeed, the proof of Lemma 6.4 shows that if in the definition of A_0 we mean that this algebra is given over a field K of arbitrary characteristic and if we set γ, in the defining relations (6.19), equal to an arbitrary element of this field, then the assertion of Lemma 6.4 and its Corollary remain valid in this case as well. Hence, any K-algebra A_0 of this kind determines, via equality (6.21), a function α on 2-words with values in the field K. The second step of the proof of Theorem 6.2, which contains Lemmas 6.5 and 6.6, is also valid for arbitrary fields whose characteristic differs from three, provided we set $\gamma = -2/3$ from the very beginning. In this case, the first identity of (6.11), which is equivalent to (6.15), is satisfied in the pair (A_1, L_1), and this identity permits one to simplify the process of computation of the function α and of reduction of a 2-word to a regular form via transformations π_i, using more general transformations

$$\overline{x}_1 v_1 \cdots v_l \overline{x}_i = \overline{x}_i^2 \left\{ -\frac{2}{3} \sum_{j=1}^{l} (-1)^j v_1 \cdots \widehat{v}_j \cdots v_l v_j + (-1)^l v_1 \cdots v_l \right\}, \tag{6.31}$$

where $v_1, \dots, v_l \in L_1$ (see (6.24) and the proof of Lemma 6.3). The third step in the proof of Theorem 6.2 is very specific, however, this still contains an independent detail: in this step a "homogeneous" ideal I_1 of the algebra A_1 such that $I_1 \cap B_1 = 0$ is studied. The homogeneity of an ideal in the algebra A_1 is understood here in the following sense. Let λ, μ be any finite disjoint subsets of the set $\overline{X} = \{\overline{x}_1, \overline{x}_2, \dots\}$. Denote by ${}_\mu T_\lambda$ the linear space in the algebra A_1 generated by monomials of the algebra A_1 that essentially depend on all $\overline{x}_i \in \lambda \cup \mu$, whereas the degree of any of these monomials with respect to any of indeterminates in μ equals two, and the degree with respect to any of indeterminates in λ equals one. If λ or μ is empty, then we shall use the notation ${}_\mu T$ or T_λ, respectively. It is clear from the definition of the algebra A_1 and the ideal I_1 that $A = \sum^{\cdot}_{\mu,\lambda} {}_\mu T_\lambda$, $I_1 = \sum^{\cdot}_{\mu,\lambda}(I \cap {}_\mu T_\lambda)$, where the direct sums are taken over all disjoint pairs of subsets λ, μ of the set $\overline{X}$. Ideals I such that $I = \sum^{\cdot}_{\mu,\lambda}(I \cap_\mu T_\lambda)$ are called homogeneous ideals of the algebra A_1. It is obvious that for any two homogeneous ideals I', I'' of the algebra A_1 having zero intersections with the subalgebra B_1, the sum $I' + I''$ is also a homogeneous ideal and $(I' + I'') \cap B_1 = 0$. Therefore, in the algebra A_1 there exists the largest homogeneous ideal that has zero intersection with B_1. Denote this ideal by I_1.

Since for fields of characteristic 5 the pair $(A_2, L_2) \rightleftharpoons (A_1/I_1, L_1/L_1 \cap I_1)$ gives us a solution of the problem of A. I. Kostrikin, there naturally arises a desire to study the identities of the pair (A_2, L_2) over fields of arbitrary characteristic and to determine whether the nth Engel identity (6.1) is valid in this pair. However, before studying the identity of this pair, it is necessary to learn to recognize, given any representation of an element in terms of the generators $\overline{x}_i$ of A_2, whether this element is equal to zero in the algebra A_2. From the definition of the algebra A_1 it follows that this is equivalent to the analogous determination of whether an element f of the algebra A_1 belongs to the ideal I_1. Since the ideal I_1 is homogeneous, we may suppose that the element f belongs to some subspace ${}_\mu T_\lambda$. It turns out that it is possible to do that relying only on the function α, which can be calculated effectively, using the transformations π_i and their generalizations (see (6.21)).

Define a bilinear pairing $\mathfrak{b}\colon {}_\mu T_\lambda \otimes_K {}_\nu T_\lambda \to K$ by setting $\mathfrak{b}(u, v) = \alpha(uv)$ $(u \in {}_\mu T_\lambda$, $v \in {}_\nu T_\lambda)$.

Lemma 6.8. *An element f of ${}_\mu T_\lambda$ belongs to the ideal I_1 if and only if $\mathfrak{b}(f, g) = 0$ for any $g \in T_\lambda$.*

Proof. Let $u = \overline{x}_j u'$ be a 2-word; then all generators except for $\overline{x}_j$ occur twice in u'. Hence, $u' = \overline{x}_j \overline{x}_{i_1}^2 \cdots \overline{x}_{i_t}^2$ in the algebra A_0, and $\overline{x}_j u' = u' \overline{x}_j$. This means that the value of a 2-word is invariant under any cyclic permutation, i.e.,

$$\alpha(ab) = \alpha(ba), \qquad ab = ba \qquad (ab \text{ is a 2-word}). \tag{6.32}$$

Let $f \in {}_\mu T_\lambda \cap I_1$. Since $I_1 \cap B_1 = 0$, we see that for any $g \in T_\lambda$ we have $fg \in B_1 \cap I_1$ and, therefore, $fg = 0$ in the algebra A_1. This means that $0 = \alpha(fg) = \mathfrak{b}(f, g)$.

Let $f \in {}_\mu T_\lambda$ and $\mathfrak{b}(f, g) = 0$ for any $g \in T_\lambda$. Denote by $I(f)$ the two-sided ideal of the algebra A_1 generated by f. As a linear space, $I(f) \cap B_1$ is spanned by the elements of the form cfd, where c and d are monomials in the generators $\overline{x}_i$ and these monomials satisfy $dc \in {}_\nu T_\lambda$ $(\nu \cap \mu = 0)$. However, by Lemma 6.4,

$$dc = \overline{x}_{i_1}^2 \cdots \overline{x}_{i_k}^2 g \qquad (g \in T_\lambda,\ \overline{x}_{i_j} \in \nu,\ j = 1, \dots, k).$$

Then from relation (6.32) we obtain $\alpha(cfd) = \alpha(fdc) = \alpha(fg) = 0$. Hence, $cfd = 0$ in the algebra A_1 and $I(f) \cap B_1 = 0$. Since the ideal $I(f)$ is homogeneous and I_1 is the largest homogeneous ideal of the algebra A_1 such that $I_1 \cap B_1 = 0$, we see that $I_1 \supset I(f)$ and $f \in I_1$. The lemma is proved.

Denote by $\operatorname{Ann}_{T_\lambda}\mathfrak{b}$ the left annihilator of the bilinear form $\mathfrak{b}\colon T_\lambda \otimes_K T_\lambda \to K$. Since the form $\mathfrak{b}$ is symmetric (see (6.32)), we see that this annihilator coincides with its right annihilator.

LEMMA 6.9. *Let $\lambda = \{\overline{x}_1, \dots, \overline{x}_l\}$, and let $f(x_1, \dots, x_l)$ be an arbitrary multilinear polynomial in an absolutely free associative algebra. Then the identity $f = 0$ holds in the pair (A_2, L_2) if and only if $f(\overline{x}_1, \dots, \overline{x}_l) \in \operatorname{Ann}_{T_\lambda}\mathfrak{b}$.*

PROOF. The necessity of the assumption of the lemma follows from the definition of the pair (A_2, L_2) and from Lemma 6.8.

Let $f(\overline{x}_1, \dots, \overline{x}_l) \in \operatorname{Ann}_{T_\lambda}\mathfrak{b}$. We shall show that $f' = f(v_1, \dots, v_l)$ belongs to I_1 in this case for any left-normed Lie commutators $v_1, \dots, v_l$ of the generators $\overline{x}_t$ of the algebra A_1. Let $v_1, \dots, v_l \in {}_\mu T_\lambda$; then we have $f' \in {}_\mu T_\nu$, and, by Lemma 6.8, we are to verify that $\mathfrak{b}(f', g) = 0$ for any $g \in T_\nu$. First we make this verification for the case $(\mu \cup \nu) \cap \lambda = \varnothing$. Toward this end, we use the fact that the elements $(\overline{x}_i \circ v_i)$ belong to the center of the algebra A_1, together with the relation

$$(\overline{x}_i \circ v_i)\overline{x}_i = -\gamma \overline{x}_i^2 v_i \qquad (\gamma = -\tfrac{2}{3}) \tag{6.33}$$

(see (6.31), and also (6.24), (6.26), (6.15)). Then we have

$$\alpha(f'g) = \alpha(f(\overline{x}_1, \dots, \overline{x}_l)\{(\overline{x}_1 \circ v_1) \cdots (\overline{x}_l \circ v_l) g\}). \tag{6.34}$$

However, the element in the braces belongs to ${}_{\mu\cup\nu}T_\lambda$ and it can be reduced in the algebra A_0 to the form $\left(\prod_j \overline{x}_j^2\right)g'$, where $\overline{x}_j$ ranges over $\mu \cup \nu$ and $g' \in T_\lambda$. Therefore, the right-hand side of (6.34) is equal to the element $\alpha(f(\overline{x}_1, \dots, \overline{x}_l)g)$ of the field, which is equal to zero by the original assumption. Thus, $f(v_1, \dots, v_l) = 0$ in the algebra A_2 for any commutators $v_1, \dots, v_l$ whose notation does not contain the generators $\overline{x}_1, \dots, \overline{x}_l$. However, then this is true also for arbitrary $v_1, \dots, v_l$, since any mapping of the set $\overline{X}$ of generators of the algebra A_2 into $\overline{X}$ can be extended to an endomorphism of A_2. By the multilinearity of the polynomial f with respect to all of its variables, this implies that $f = 0$ is an identity of the pair (A_2, L_2). The lemma is proved.

Lemmas 6.8 and 6.9 show that the bilinear form $\mathfrak{b}$ and the function α give us a sufficiently effective algorithm for verification of validity of a concrete multilinear idenity in the pair (A_2, L_2). We apply this algorithm to prove the following theorem.

THEOREM 6.3. *The variety of $(p-2)$th Engel Lie algebras over a field of positive characteristic p is nonsolvable (nonnilpotent) for $p \geq 5$.*

The proof of this theorem follows from the next statement.

PROPOSITION 6.2. *If the ground field K is of characteristic p ($p > 0$ and $p \neq 3$), then in the associative-Lie pair (A_2, L_2) the identities*

$$y^2 x = xy^2, \qquad y^{p-2} = 0 \tag{6.35}$$

are satisfied. Moreover, in the Lie algebra L_2, the $(p-2)$th Engel identity is satisfied, and this algebra is nonsolvable for $p \geq 5$.

PROOF. The validity of the first identity in (6.35) in the pair (A_2, L_2) for $p \neq 3$ follows from the definition of the pair and from Lemma 6.6. Applying the total linearization of the second identity in (6.35), we see that the identity $\mathrm{Sym}_{p-2} = 0$ is its consequence. Substituting in the identity $\mathrm{Sym}_{p-2} = 0$ a variable y instead of all variables x_i, we see that the converse is also true: the identity $y^{p-2} = 0$ is a consequence of the identity $\mathrm{Sym}_{p-2} = 0$. In order to verify that the identity $\mathrm{Sym}_{p-2} = 0$ is valid in the pair (A_2, L_2), we apply Lemma 6.9. Taking into account that the polynomial Sym is symmetric with respect to all its variables, we find that we have to verify only the equality

$$\alpha(\mathrm{Sym}_{p-2}(\overline{x}_1, \dots, \overline{x}_{p-2})\overline{x}_{p-2}\cdots\overline{x}_1) = 0. \tag{6.36}$$

However, in the algebra A_1 we have the relation

$$\begin{aligned}
&\mathrm{Sym}_{p-2}(\overline{x}_1, \dots, \overline{x}_{p-2})\overline{x}_{p-2}\\
&\quad= \frac{p-3}{2}\sum_{j=1}^{p-3}\mathrm{Sym}_{p-4}\left(\overline{x}_1, \dots, \widehat{\overline{x}_j}, \dots, \overline{x}_{p-3}\right)(\overline{x}_j \circ \overline{x}_{p-2})\overline{x}_{p-2}\\
&\qquad+ \mathrm{Sym}_{p-3}(\overline{x}_1, \dots, \overline{x}_{p-3})\overline{x}_{p-2}^2\\
&\quad= \left(\frac{p-3}{2}(-\gamma) + 1\right)\overline{x}_{p-2}^2\,\mathrm{Sym}_{p-3}(\overline{x}_1, \dots, \overline{x}_{p-3})\\
&\quad= \left(\frac{p-3}{2}\cdot\frac{2}{3} + 1\right)\overline{x}_{p-2}^2\,\mathrm{Sym}_{p-3}(\overline{x}_1, \dots, \overline{x}_{p-3}) = 0 \pmod p.
\end{aligned}$$

This proves (6.36) and (6.35).

A partial linearization of the identity $y^{p-2} = 0$ leads to $\sum_{i=0}^{p-3} y^i x y^{p-3-i} = 0$; this, with the help of (6.15), implies $yxy^{p-4} = \beta xy^{p-3}$. However, in this case we have

$$\begin{aligned}
(\mathrm{ad}\, y)^{p-2}x &= \sum_{i=0}^{p-2}(-1)^i\binom{p-2}{i}y^{p-2-i}xy^i\\
&= \beta_1 xy^{p-2} + \beta_2 yxy^{p-3} = (\beta_1 + \beta\beta_2)xy^{p-2} = 0.
\end{aligned}$$

This proves that in the Lie algebra L_2 the $(p-2)$th Engel identity is satisfied.

Finally, in order to verify that L_2 is not nilpotent, we use the relation

$$[v, \overline{x}_i]\overline{x}_i = (2+\gamma)\overline{x}_i^2 v \qquad (\gamma = -\tfrac{2}{3}), \tag{6.37}$$

which is equivalent to the relation (6.33). Then

$$\begin{aligned}
&\alpha([\overline{x}_1, \dots, \overline{x}_{N-1}, \overline{x}_N]\overline{x}_N\overline{x}_{N-1}\cdots\overline{x}_1)\\
&\quad= (2+\gamma)\alpha([\overline{x}_1, \dots, \overline{x}_{N-1}]\overline{x}_{N-1}\cdots\overline{x}_1) = (2+\gamma)^{N-1}\alpha(\overline{x}_1, \overline{x}_1) = (4/3)^{N-1},
\end{aligned}$$

and, therefore, by Lemma 6.9, identity (6.2) is not satisfied in L_2 for any N ($p \neq 2, 3$). The nonsolvability of the Lie algebra L_2 now follows from Theorem 6.1. Proposition 6.2 and Theorem 6.3 are proved.

6.3. Identities of the two-dimensional irreducible representation of the simple three-dimensional Lie algebra. Let us consider once more the proof of Theorems 6.2 and 6.3. In spite of the attained success, the proof of Theorem 6.3 leaves a feeling of inadequacy. First, it is carried out by analogy with the proof of Theorem 6.2 and thus by no means reveals intrinsic reasons of $(p-2)$th Engel Lie algebras for the existence of precisely this kind of proof. Second, the function α on 2-words, which plays the main role in this proof, was constructed and applied rather artificially, and in the first place this applies to the defining relations (6.19) and Lemma 6.9. Although these assertions are valid, they do not explain, in any way, why the function α, introduced in this way, has any relationship with identities of associative-Lie pairs and Lie algebras. And the connection between the value $-2/3$ of the parameter γ and the identities

$$y^2x = xy^2, \qquad (y \circ z)x = x(y \circ z) \tag{6.38}$$

(see Lemmas 6.6 and 6.7) seems to be completely out of the blue.

In this subsection, we add some new observations to the results of the preceding investigation. As a whole, these observations do not facilitate the understanding of the situation under consideration, but they reveal some new relationships (and partially explain the older ones) between the function α, which has been already constructed, and the identities of associative-Lie pairs.

First of all, we shall try to find objects that satisfy the identities (6.38). Toward this end, we consider the algebra M_2 of second order matrices over an arbitrary algebraically closed field K. Denote the matrices

$$\begin{pmatrix} 1 & 0 \\ 0 & 1 \end{pmatrix}, \begin{pmatrix} \sqrt{-1} & 0 \\ 0 & -\sqrt{-1} \end{pmatrix}, \begin{pmatrix} 0 & \sqrt{-1} \\ \sqrt{-1} & 0 \end{pmatrix}, \begin{pmatrix} 0 & -1 \\ 1 & 0 \end{pmatrix}$$

by the symbols e_0, e_1, e_2, e_3, respectively. It can be immediately checked that

$$\begin{aligned} &e_1e_2 = -e_2e_1 = e_3, &&e_2e_3 = -e_3e_2 = e_1, &&e_3e_1 = -e_1e_3 = e_2, \\ &e_1^2 = e_2^2 = e_3^2 = -e_0, &&e_0e_i = e_ie_0 = e_i \quad (i = 0, 1, 2, 3). \end{aligned} \tag{6.39}$$

These are the well-known relations of the quaternion algebra. They imply

$$(\beta_1e_1 + \beta_2e_2 + \beta_3e_3)^2 = -(\beta_1^2 + \beta_2^2 + \beta_3^2)e_0. \tag{6.40}$$

Since the K-subspace spanned by e_1, e_2, e_3 is a Lie algebra in $M_2^{(-)}$ coinciding with the well-known three-dimensional Lie algebra $\mathfrak{g} = \mathrm{sl}(2, K)$ of second-order matrices with zero trace, we see that the relation (6.40) implies the following assertion.

LEMMA 6.10. *The identity* (6.38) *is valid in the pair* $(M_2, \mathfrak{g})$.

Let $\overline{\mathcal{F}}$ be a commutative algebra given by the generators $\overline{x}_i^{(1)}$, $\overline{x}_i^{(2)}$, $\overline{x}_i^{(3)}$ and by the defining relations

$$\overline{x}_i^{(k)}\overline{x}_i^{(l)} = \delta_{kl}\left(\overline{x}_i^{(1)}\right)^2, \tag{6.41}$$

where δ_{kl} is the Kronecker symbol. Note that (6.41) implies

$$(6.42)\qquad \begin{gathered}\left(\overline{x}_i^{(1)}\right)^3 = \left(\overline{x}_i^{(1)}\right)^2\left(\overline{x}_i^{(1)}\right) = \left(\overline{x}_i^{(2)}\right)^2\left(\overline{x}_i^{(1)}\right) = 0,\\ \overline{x}_i^{(j)}\overline{x}_i^{(k)}\overline{x}_i^{(l)} = 0 \qquad (i = 1, 2, \dots)\end{gathered}$$

for any $j, k, l \in \{1, 2, 3\}$.

Let A and L be the associative and the Lie algebra, respectively, generated by the elements

$$\overline{x}_i \rightleftharpoons \sqrt{-1}(\overline{x}_i^{(1)} \otimes e_1 + \overline{x}_i^{(2)} \otimes e_2 + \overline{x}_i^{(3)} \otimes e_3) \qquad (i = 1, 2, \dots)$$

in $\overline{M}_2 \rightleftharpoons \overline{\mathcal{F}} \otimes_K M_2$. It follows from relations (6.39) and (6.40) that

$$(6.43)\qquad \overline{x}_i^2 = 3\left(\overline{x}_i^{(1)}\right)^2, \qquad \overline{x}_i e_j \overline{x}_i = -\left(\overline{x}_i^{(1)}\right)^2 \otimes e_j \quad (j = 1, 2, 3).$$

LEMMA 6.11. *Identities* (6.38) *and the relations*

$$(6.44)\qquad \begin{gathered}\overline{x}_i^3 = 0, \qquad \overline{x}_i v \overline{x}_i = -\tfrac{1}{3}\overline{x}_i^2 v \quad (v \in L),\\ (\overline{x}_i \circ v)\overline{x}_i = -\gamma \overline{x}_i^2 v \quad (\gamma = -\tfrac{2}{3})\end{gathered}$$

hold in the pair (A, L).

PROOF. It follows from the definition of the Lie algebra L that any element of it can be represented in the form $\sum_{i=1}^3 c_i \otimes e_i$, where $c_i \in \overline{\mathcal{F}}$. Since the second identity in (6.38) and relations (6.44) are linear with respect to v, we see that it is sufficient to verify them on the basis of the Lie algebra $\mathfrak{g}$. However, in this case the assertion of the lemma follows from (6.43) and from Lemma 6.10. The first relation in (6.44) follows from the equality (6.42). The lemma is proved.

COROLLARY. *In the algebra A, any 2-word is reducible to a unique regular form* (6.17).

The proof of Lemma 6.3 shows that in order to reduce a 2-word u to the form $\alpha\overline{x}_{i_1}^2 \cdots \overline{x}_{i_t}^2$, it is sufficient to use only a particular form of relations (6.38) and (6.42). The uniqueness of this representation follows from the first equality in (6.43) and from the definition of the algebra A. Moreover, it is clear that different regular representations of 2-words are linearly independent in the algebra $\overline{M}_2$. The corollary is proved.

We sum up our first observations. The discovered model (A, L) in second order matrices for the identity (6.38) permits one to simplify substantially the proof of Theorem 6.3. Indeed, Lemma 6.3 and its Corollary show that the pair (A, L) has all the properties of the pair (A_1, L_1) used in the proof of Lemmas 6.8 and 6.9. This offers us the possibility of replacing the purely technical (in their proofs) Lemmas 6.4–6.6 in the proof of Theorem 6.3 by model considerations of Lemma 6.11 and its Corollary, of replacing the pair (A_1, L_1) in Lemmas 6.8 and 6.9 and in Proposition 6.2 by the pair (A, L), and of obtaining the following refinements of Proposition 6.2 and Theorem 6.3.

PROPOSITION 6.3. *The variety* $\mathrm{var}(A_2, L_2)$ *is contained in the variety* $\mathrm{var}(M_2, \mathfrak{g})$.

THEOREM 6.4. *If the characteristic of the ground field K is equal to p and if $p \geq 5$, then there exists a nonsolvable $(p-2)$th Engel Lie algebra such that all the identities of the three-dimensional simple Lie algebra $\mathfrak{g} = \mathrm{sl}(2, K)$ are satisfied in it.*

The variety $\mathrm{var}(A_2, L_2)$ of associative-Lie pairs determined by the function α on 2-words provides a sufficiently nontrivial example, over fields of characteristic $p \geq 5$, of a variety in which the $(p-2)$th Engel identity is satisfied. Now we find out, what variety is given by the function α over fields of characteristic zero.

Let the ground field K be the complex field. Denote by $\boldsymbol{\sigma}$ the anti-isomorphism of the matrix algebra M_2 that is the composition of matrix transposition and of complex conjugation. We extend $\boldsymbol{\sigma}$ to an anti-isomorphism of the algebra $\overline{M}_2 = \overline{\mathcal{F}} \otimes_K M_2$ in such a way that $\boldsymbol{\sigma}(\overline{x}_i^{(j)} a) = \overline{x}_i^{(j)} \boldsymbol{\sigma}(a)$, where $a \in \overline{M}_2$, $j = 1, 2, 3$, $i = 1, 2, \ldots$. Then $\boldsymbol{\sigma}(e_i) = -e_i$ $(i = 1, 2, 3)$ and $\boldsymbol{\sigma}(\overline{x}_i) = \overline{x}_i$ $(i = 1, 2, \ldots)$. Hence, $\boldsymbol{\sigma}(A) = A$, $\boldsymbol{\sigma}(L) = L$ and

$$\boldsymbol{\sigma}(\beta \overline{x}_{i_1} \cdots \overline{x}_{i_k}) = \overline{\beta} \overline{x}_{i_k} \cdots \overline{x}_{i_1} \quad (\beta \in K). \tag{6.45}$$

Define the subspaces ${}_\mu T_\lambda$ and the bilinear pairing $\mathfrak{b}\colon {}_\mu T_\lambda \otimes {}_\nu T_\lambda \to K$ for the algebra A in the same way as it was done above for the algebra A_1.

LEMMA 6.12. *Let $f(x_1, \ldots, x_l)$ be an arbitrary multilinear associative polynomial and let $\overline{f}(\overline{x}_1, \ldots, \overline{x}_l) \in A$. Then*

$$\mathfrak{b}(\overline{f}, \boldsymbol{\sigma}(\overline{f})) = \alpha(\overline{f} \boldsymbol{\sigma}(\overline{f})) \geq 0, \tag{6.46}$$

and the equality is attained if and only if in the pair $(M_2, \mathfrak{g})$ (in the pair (A, L)) the identity $f = 0$ is satisfied.

PROOF. Since $\overline{f}\boldsymbol{\sigma}(\overline{f}))$ is a 2-element, we see by the Corollary of Lemma 6.11 that

$$\overline{f}\boldsymbol{\sigma}(\overline{f}) = \alpha(\overline{f}\boldsymbol{\sigma}(\overline{f}))\overline{x}_1^2 \cdots \overline{x}_l^2 = 3^l \alpha(\overline{f}\boldsymbol{\sigma}(\overline{f}))\left(\overline{x}_1^{(1)}\right)^2 \cdots \left(\overline{x}_l^{(1)}\right)^2 e_0 \tag{6.47}$$

(see (6.43)), and $\overline{f}\boldsymbol{\sigma}(\overline{f})$ belongs to the center of the algebra $\overline{M}_2$. On the other hand, the elements of the matrix $\overline{f}$ have the form

$$a_{ij} = \sum_{r_1, \ldots, r_l} \beta^{ij}_{r_1, \ldots, r_l} \overline{x}_1^{(r_1)} \ldots \overline{x}_l^{(r_l)} \qquad (i, j = 1, 2),$$

where $\beta^{ij}_{r_1, \ldots, r_l}$ are some complex numbers. Using the equality

$$\left(\overline{x}_1^{(r_1)} \cdots \overline{x}_l^{(r_l)}\right)\left(\overline{x}_1^{(q_1)} \cdots \overline{x}_l^{(q_l)}\right) = \delta^{r_1, \ldots, r_l}_{q_1, \ldots, q_l}\left(\overline{x}_1^{(1)}\right)^2 \cdots \left(\overline{x}_l^{(1)}\right)^2$$

(where $\delta^{r_1, \ldots, r_l}_{q_1, \ldots, q_l}$ is the Kronecker symbol), which follows from (6.40), we obtain

$$\mathrm{tr}\left(\overline{f}\boldsymbol{\sigma}(\overline{f})\right) = \left(\overline{x}_1^{(1)}\right)^2 \cdots \left(\overline{x}_l^{(1)}\right)^2 \sum_{r_1, \ldots, r_l, i, j} |\beta^{i,j}_{r_1, \ldots, r_l}|^2.$$

Comparing this equality with (6.47), we conclude that $\alpha(\overline{f}\sigma(\overline{f})) \geq 0$, and that equality is attained if and only if $\overline{f} = 0$ in the algebra $\overline{M}_2$. However,

$$\overline{f} = \sum_{r_1,\dots,r_l} \overline{x}_1^{(r_1)} \cdots \overline{x}_l^{(r_l)} \otimes f\left(e_{r_1}, \dots, e_{r_l}\right) \qquad (r_i = 1, 2, 3).$$

This relation implies $\overline{f} = 0$ in $\overline{M}_2$ only if all the values $f\left(e_{r_1}, \dots, e_{r_l}\right)$ of the polynomial f on the basis e_1, e_2, e_3 of the Lie algebra $\mathfrak{g}$ vanish. Since the polynomial $f(x_1, \dots, x_l)$ is multilinear, this means that $f = 0$ is an identity of the pair $(M_2, \mathfrak{g})$. The lemma is proved.

Lemma 6.12 shows that, over fields of characteristic zero, the function α contains in itself all the information on the identities of the two-dimensional representation of the simple three-dimensional Lie algebra $\mathfrak{g} = \mathrm{sl}(2, K)$. Now we apply this function to find a basis of identities of this representation.

Let g be a multihomogeneous associative polynomial in variables $x_1, x_2, \dots$ such that the element $\overline{g} = g(\overline{x}_1, \overline{x}_2, \dots)$ belongs to some subspace ${}_\mu T_\lambda$ of the algebra A. It has been mentioned already above, that relations (6.44) and identities (6.38) permit one to find a multilinear polynomial h such that $h \in T_\lambda$ and the equality $\overline{g} = \prod_{\overline{x}_i \in \mu} \overline{x}_i^2 \overline{h}$ is satisfied in the algebra A. It is clear that the polynomial h can be chosen in different ways, but if $\overline{g} = \prod_{\overline{x}_i \in \mu} \overline{x}_i^2 \overline{h}'$ is another representation of this kind, then $h - h'$ belongs to $\mathrm{Ann}_{T_\lambda} \mathfrak{b}$, and, according to Lemma 6.12, $h = h'$ is an identity of the pair $(M_2, \mathfrak{g})$. This means that, modulo the ideal of weak identities of the pair $(M_2, \mathfrak{g})$, the polynomial h is uniquely determined by g. We write, by definition, $h = \pi(g)$ (π can be regarded also as a mapping in a free associative algebra from the subspace of multihomogeneous polynomials of type $\langle \mu, \lambda \rangle$ into the subspace of multilinear polynomials of type λ if we use Lemma 6.4, but this viewpoint is unnecessary).

Let $\lambda = \{x_1, \dots, x_l\}$, $x_j \in \lambda$, and $\lambda_j = \lambda \setminus \{x_j\}$. Denote by $f(x_1, \dots, x_l)$ an arbitrary multilinear associative polynomial and denote by $s(f)$ the polynomial $f(x_{s(1)}, \dots, x_{s(l)})$, where $s \in S_l$ is a permutation of the set $\{1, 2, \dots, l\}$. Everywhere below, denote by τ_{ij} $(i \neq j)$ the permutation that transposes the indices i and j and acts identically on other indices. It is clear that if $f = 0$ is an identity of the pair $(M_2, \mathfrak{g})$, then $\overline{f} \in \mathrm{Ann}_{T_\lambda} \mathfrak{b}$ and $\overline{\pi(f|_{x_i=[x_i,x_j]})} \in \mathrm{Ann}_{T_{\lambda_j}} \mathfrak{b}$. Hence, by Lemma 6.12, $g_{ij} = 0$ is an identity of the pair $(M_2, \mathfrak{g})$, where $g_{ij} = \pi(f|_{x_i=[x_i,x_j]})$. However, it follows from relations (6.26), (6.37), and (6.43), that

$$\begin{aligned} \overline{(g_{ij}|_{x_i=[x_i,x_j]})} &= \overline{\pi(f|_{x_i=[x_i,x_t],x_j=x_t})|_{x_i=[x_i,x_j]}} = \overline{\pi(f|_{x_i=[x_i,x_j,x_t],x_j=x_t})} \\ &= \overline{2\pi(f|_{x_j=x_t}(x_j \circ x_t) - f|_{x_i=x_j,x_j=x_t}(x_i \circ x_t))} = \tfrac{4}{3}(\overline{f} - \tau_{ij}\overline{f}). \end{aligned}$$

By Lemma 6.12, this means that for any polynomial f in the pair $(M_2, \mathfrak{g})$ the multilinear identity

$$\tfrac{4}{3}(f - \tau_{ij} f) = g_{ij}|_{x_i=[x_i,x_j]} \tag{6.48}$$

is satisfied. Note that in the proof of the identity (6.48) we do not use the fact that $f = 0$ is an identity of the pair $(M_2, \mathfrak{g})$; therefore, an identity of this kind exists for any multilinear polynomial f.

LEMMA 6.13. *Let the ground field K be of characteristic zero. Then the system of identities* (6.48), *where f ranges over all multilinear polynomials, forms a basis of the identities of the pair* $(M_2, \mathfrak{g})$.

PROOF. Since, over a field of characteristic zero, any identity $f = 0$ is equivalent to a multilinear identity that is obtained as a result of the total linearization of the identity $f = 0$, we see that it is sufficient to verify that any multilinear identity $f(x_1, \dots, x_l) = 0$ of the pair $(M_2, \mathfrak{g})$ is a consequence of identities (6.48). We shall carry out this verification by induction with respect to the degree l of the polynomial f. The base of induction $l = 1$ is valid, because identities of degree 1 fail in the pair $(M_2, \mathfrak{g})$. Now we prove the possibility of the induction step. Let $f = 0$ be an identity of the pair $(M_2, \mathfrak{g})$. Then, as already mentioned above, $g_{ij} = 0$ is also an identity of the pair $(M_2, \mathfrak{g})$. Since the degree of g_{ij} is one less than that of the polynomial f, we see that the identity $g_{ij} = 0$ is a consequence of identities (6.48). But this means that the latter imply also the identities $0 = g_{ij}\big|_{x_i=[x_i,x_j]}$ and $f - \tau_{ij} f = 0$ for any $i, j \in \{1, 2, \dots, l\}$. Moreover, the identities $f = s(f)$ also follow from (6.48) for any permutation s. Adding them up, we obtain another consequence of identities (6.48):

$$l!f = \sum_{s \in S_l} s(f) = \beta \operatorname{Sym}_l(x_1, \dots, x_l) \qquad (\beta \in K). \tag{6.49}$$

Setting $x_1 = \cdots = x_l = \begin{pmatrix} 1 & 0 \\ 0 & -1 \end{pmatrix}$ in this identity, we obtain the relation $0 = l!\beta \begin{pmatrix} 1 & 0 \\ 0 & (-1)^l \end{pmatrix}$. This implies $\beta = 0$, and the consequence (6.49) of identities (6.48) obtains the form $l!f = 0$. The lemma is proved.

THEOREM 6.5. *If the characteristic of the field K is equal to zero, then all the identities of the pair $(M_2, \mathfrak{g})$ are equivalent to identity* (6.38):

$$[(y \circ z), x] = 0.$$

PROOF. The proof of Lemma 6.13 shows us that the construction of a finite basis of identities for the pair $(M_2, \mathfrak{g})$ requires a more detailed study of the structure of the polynomial g_{ij} in identity (6.48).

LEMMA 6.14. *Let $(F, \mathfrak{L})$ be a relatively free associative-Lie pair with the set $X = \{x_1, x_2, \dots\}$ of free generators, defined by the identity* (6.38), *and let* $\operatorname{char} K \neq 2$. *Then any element of the algebra F can be represented in the form of a linear combination of elements of the following two types*:

1*st type:*

$$(x_{i_1} \circ x_{j_1}) \cdots (x_{i_k} \circ x_{j_k}), \quad (x_{i_1} \circ x_{j_1}) \cdots (x_{i_k} \circ x_{j_k})([x_i, x_j] \circ x_t);$$

2*nd type:*

$$(x_{i_1} \circ x_{j_1}) \cdots (x_{i_k} \circ x_{j_k})[x_i, x_j], \quad (x_{i_1} \circ x_{j_1}) \cdots (x_{i_k} \circ x_{j_k}) x_i.$$

PROOF. We use the corollary (6.26) of the identity (6.38) and decompose the Lie commutator of weight 4 in two ways. Then

$$[[x,y],v,z] = 2(v\circ z)[x,y] - 2([x,y]\circ z)v = 2(y\circ v)[x,z] - 2(x\circ v)[y,z],$$
$$2([x,y]\circ z)v = 2(v\circ z)[x,y] - 2(y\circ v)[x,z] + 2(x\circ v)[y,z].$$

The last identity and (6.38) imply

$$\begin{aligned} x_1x_2\cdots x_l &= \tfrac{1}{2}(x_1\circ x_2)x_3\cdots x_l + \tfrac{1}{2}[x_1,x_2]x_3\cdots x_l, \\ [x_1,x_2]x_3\cdots x_l &= \tfrac{1}{2}([x_1,x_2]\circ x_3)x_4\cdots x_l + \tfrac{1}{2}[x_1,x_2,x_3]x_4\cdots x_l \\ &= \tfrac{1}{2}\{(x_3\circ x_4)[x_1,x_2] - (x_2\circ x_4)[x_1,x_3] + (x_1\circ x_4)[x_2,x_3]\}x_5\cdots x_l \\ &\quad + (x_2\circ x_3)x_1x_4\cdots x_l - (x_1\circ x_3)x_2x_4\cdots x_l. \end{aligned}$$

These equalities, together with inductive arguments, show that any monomial can be expressed as a linear combination of elements of the 1st and of the 2nd type. The lemma is proved.

LEMMA 6.15. *For any multilinear polynomial f, we can choose a polynomial g_{ij} such that the identity* (6.48) *of the pair* $(M_2, \mathfrak{g})$ *is a consequence of the identity* (6.38).

PROOF. By Lemma 6.14, it is sufficient to prove the assertion of the lemma for the polynomials f that are elements of the 1st or of the 2nd type in the generators $x_1,\dots,x_l$. Moreover, it follows from the identity

$$([x,y]\circ z) = (x\circ[y,z]), \tag{6.50}$$

which is a consequence of the identity (6.38), and also from the fact that the element $(x_t\circ v)$, where v is a commutator of weight 1 or 2, commutes with any x_k, that it is sufficient to consider only the following cases:

1) $f=(x_i\circ x_j)$; 2) $f=[x_i,x_j]$; 3) $f=(x_i\circ x_t)x_j$.

Let us consider each of them.

1) $\pi(f|_{x_i=[x_i,x_j]}) = \pi([x_i,x_j]\circ x_j) = 0 = \frac{4}{3}(f-\tau_{ij}(f))$. Hence, $g_{ij}=0$.

2) $\pi(f|_{x_i=[x_i,x_j]}) = \pi([x_i,x_j,x_j]) = 2\pi(x_j^2x_i - x_jx_ix_j) = \frac{8}{3}x_i$ (see (6.43)). However, $\frac{4}{3}(f-\tau_{ij}(f)) = \frac{8}{3}[x_i,x_j]$; hence, $g_{ij}=\frac{8}{3}x_i$.

3) Using (6.50) and (6.44), we obtain

$$\pi(f|_{x_i=[x_i,x_j]}) = \pi(([x_i,x_j]\circ x_t)x_j) = \pi((x_j\circ[x_t,x_i])x_j) = \tfrac{2}{3}[x_t,x_i].$$

By setting $g_{ij}=\frac{2}{3}[x_t,x_i]$, from the corollary (6.26) of the identity (6.38) we get:

$$g_{ij}|_{x_i=[x_i,x_j]} = \tfrac{2}{3}[x_j,x_i,x_t] = \tfrac{4}{3}\{(x_i\circ x_t)x_j - (x_j\circ x_t)x_i\} = \tfrac{4}{3}(f-\tau_{ij}(f)).$$

The lemma is proved.

Lemma 6.15 shows that we can choose the system of identities (6.48), which forms a basis of identities of the pair $(M_2,\mathfrak{g})$ by Lemma 6.13, in such a way that it becomes equivalent to identity (6.38). This proves the assertion of Theorem 6.5.

§7. Characters on 2-words and $\boldsymbol{\alpha}$-functions

We fix a certain alphabet $\overline{X}$ with a countable number of letters $\overline{x}_1, \overline{x}_2, \dots$.

DEFINITION 7.1. A word u in the alphabet $\overline{X}$ is called a *2-word* if any letter $\overline{x}_i \in \overline{X}$ that occurs in the notation of this word has exactly two entries in u. Formal linear combinations of 2-words with coefficients from the ground field K are called *2-elements*.

Let $H(\overline{X})$ be the free associative K-algebra with the set $\overline{X}$ of free generators. Obviously, under the natural identification of 2-elements with elements of the algebra $H(\overline{X})$, the set of 2-elements forms a subspace B in $H(\overline{X})$.

DEFINITION 7.2. Let $\mathcal{E}$ be a commutative K-algebra. A function α, defined on 2-words and taking its values in the algebra $\mathcal{E}$, is called a *character* on 2-words, if the following two properties are valid for this function.

1. For any 2-word u of the form $a\overline{x}_i\overline{x}_j b\overline{x}_i c$, where a, b, and c are words in the alphabet $\overline{X}$, some of them possibly empty, the following relation is satisfied:

$$\alpha(a\overline{x}_i\overline{x}_j b\overline{x}_i c) + \alpha(a\overline{x}_i b\overline{x}_i\overline{x}_j c) = \alpha(a\overline{x}_j\overline{x}_i b\overline{x}_i c) + \alpha(a\overline{x}_i b\overline{x}_j\overline{x}_i c).$$

2. For any permutation δ of the set $\{1, 2, \dots\}$ and any 2-word u the following relation is satisfied:

$$\alpha\big(u(\overline{x}_{i_1}, \dots, \overline{x}_{i_l})\big) = \alpha\big(u(\overline{x}_{\delta(i_1)}, \dots, \overline{x}_{\delta(i_l)})\big).$$

A character α on 2-words is called a *multiplicative character*, or an $\boldsymbol{\alpha}$*-function*, if it satisfies the following property.

3. For any 2-word $w = uv$, where u and v are 2-words, the following relation is satisfied in $\mathcal{E}$:

$$\alpha(w) = \alpha(u)\alpha(v).$$

REMARK 7.1. We extend the $\boldsymbol{\alpha}$-function α by K-linearity from the set of 2-words to the whole linear space of 2-elements B. Denote the extended function by the same letter α. Then properties 1, 2, and 3 for the linear $\boldsymbol{\alpha}$-function α are equivalent to the following ones.

1. For any 2-element u of the form $a\overline{x}_i\overline{x}_j b\overline{x}_i c$ in $\mathcal{E}$ the following relation is satisfied:

$$\alpha(a[\overline{x}_i, \overline{x}_j]b\overline{x}_i c) = \alpha(a\overline{x}_i b[\overline{x}_j, \overline{x}_i]c).$$

2. For any permutation δ of the set $\{1, 2, \dots\}$ and any 2-element u in $\mathcal{E}$ the following relation is satisfied:

$$\alpha\big(u(\overline{x}_{i_1}, \dots, \overline{x}_{i_l})\big) = \alpha\big(u(\overline{x}_{\delta(i_1)}, \dots, \overline{x}_{\delta(i_l)})\big).$$

3. For any 2-element $w = uv$, where u and v are 2-elements, the following relation is satisfied in $\mathcal{E}$:

$$\alpha(w) = \alpha(u)\alpha(v).$$

Consider the category whose objects are diagrams $\alpha\colon B \to \mathcal{E}$, where $\mathcal{E}$ are commutative algebras and α are multiplicative characters on 2-words, and morphisms are homomorphisms $\varkappa$ of commutative algebras for which the following diagram is commutative:

$$\begin{array}{ccc} B & \xrightarrow{\alpha_1} & \mathcal{E}_1 \\ \| & & \downarrow{\scriptstyle\varkappa} \\ B & \xrightarrow[\alpha_2]{} & \mathcal{E}_2 \end{array}$$

We denote this category by the word ALPHA.

LEMMA 7.1. *In the category ALPHA there exist products of objects and a universal object, i.e., a diagram* $\alpha_0\colon B \to \mathcal{E}_0$ *such that for any other diagram* $\alpha\colon B \to \mathcal{E}$ *there exists a unique homomorphism* $\varkappa\colon \mathcal{E}_0 \to \mathcal{E}$ *for which the diagram*

$$\begin{array}{ccc} B & \xrightarrow{\alpha_0} & \mathcal{E}_0 \\ \| & & \downarrow{\scriptstyle\varkappa} \\ B & \xrightarrow[\alpha]{} & \mathcal{E} \end{array} \tag{7.1}$$

is commutative.

PROOF. Let $\{\alpha_s\colon B \to B_s,\ s \in S\}$ be a set of diagrams in the category ALPHA. Denote by the letter $\mathcal{E}$ the full Cartesian product $\prod_{s\in S} \mathcal{E}_s$ of commutative algebras $\mathcal{E}_s$. Obviously, the diagram $\alpha\colon B \to \mathcal{E}$, where $\alpha(b) \rightleftharpoons (\dots, \alpha_s(b), \dots)$ (the sth coordinate of the element $\alpha(b)$ is $\alpha_s(b) \in \mathcal{E}_s$), is an object of the category ALPHA and this object is the product of the objects $\alpha_s\colon B \to \mathcal{E}_s$ of this category.

Denote by $\mathcal{E}_0$ the commutative algebra given by the formal generators $\alpha_0(u)$, where u ranges over all 2-words in the alphabet $\overline{X}$, and defining relations mentioned in the properties 1, 2, and 3 of the definition of $\boldsymbol{\alpha}$-function. Obviously, the linear function $\alpha_0\colon B \to \mathcal{E}_0$ that transforms any 2-word u into $\alpha_0(u) \in \mathcal{E}_0$, is an $\boldsymbol{\alpha}$-function. Moreover, for any diagram $\alpha\colon B \to \mathcal{E}$ in the category ALPHA, the mapping that transforms $\alpha_0(u)$ into $\alpha(u) \in \mathcal{E}$ can be uniquely extended to a homomorphism of commutative algebras $\varkappa\colon \mathcal{E}_0 \to \mathcal{E}$ such that the diagram (7.1) is commutative. This shows that $\alpha_0\colon B \to \mathcal{E}_0$ is a universal object in the category ALPHA. The lemma is proved.

The algebra $\mathcal{E}_0$ will be called the *universal algebra of coefficients* for the $\boldsymbol{\alpha}$-function, and the $\boldsymbol{\alpha}$-function α_0 will be called the *universal* $\boldsymbol{\alpha}$*-function*.

§8. The variety of pairs $\overline{\mathfrak{B}}_\alpha$ defined by the character α

Suppose that $(H, \mathfrak{M})$ is an absolutely free associative-Lie pair (H is a free associative algebra, $\mathfrak{M}$ is a free Lie algebra) in the variety of all associative-Lie pairs Π_1 with an infinite set of free generators $\{t_1, t_2, \dots\}$. Let μ and λ be finite disjoint subsets of the set of generators $\{t_1, t_2, \dots\}$ and let ${}_\mu H_\lambda$ be the linear subspace in H that consists of multihomogeneous polynomials of degree 1 with respect to each variable in λ, of degree 2 with respect to each variable in μ, and of degree 0 with respect to all other variables.

Let a character α with values in some commutative algebra $\mathcal{E}$ be given on 2-words in the alphabet $\overline{X} = \{\overline{x}_1, \overline{x}_2, \dots\}$. This means that its extension, by K-linearity, to the subspace B of 2-elements satisfies the conditions 1 and 2 of Remark 7.1. With the help of the character α, we define a K-bilinear pairing $\mathfrak{b}\colon {}_\nu H_\lambda \otimes {}_\mu H_\lambda \to \mathcal{E}$ by putting

$$\begin{aligned}(8.1)\quad &\mathfrak{b}(h(t_1,\dots,t_l,z_1,\dots,z_k),g(t_1,\dots,t_l,y_1,\dots,y_m))\\ &\qquad \rightleftharpoons \alpha\big(h(\overline{x}_1,\dots,\overline{x}_l,\overline{x}_{i_1},\dots,\overline{x}_{i_k})g(\overline{x}_1,\dots,\overline{x}_l,\overline{x}_{j_1},\dots,\overline{x}_{j_m})\big),\end{aligned}$$

where $h\in {}_\nu H_\lambda$, $g\in {}_\mu H_\lambda$, $t_i\in\lambda$, $z_i\in\nu$, $y_j\in\mu$, and all the letters in the collection $\overline{x}_1,\dots,\overline{x}_l,\overline{x}_{i_1},\dots,\overline{x}_{i_k},\overline{x}_{j_1},\dots,\overline{x}_{j_m}$ are pairwise distinct. Property 2 of the character α guarantees that this definition is well posed. Property 1 shows that if $h=f\big|_{x_1=[x_1,v_1],\dots,x_l=[x_l,v_l]}$, where $x_1,\dots,x_l\in\lambda$, $v_1,\dots,v_l\in\nu\cup\lambda$, then in $\mathcal{E}$ we have the relation

$$(8.2)\qquad \mathfrak{b}(f\big|_{x_1=[x_1,v_1],\dots,x_l=[x_l,v_l]},g)=\mathfrak{b}(f,g\big|_{x_1=[x_1,v_1],\dots,x_l=[x_l,v_l]}).$$

We set, by definition, $\operatorname{Ann}_{H_\lambda}\mathfrak{b}\rightleftharpoons\{f\in H_\lambda\mid \mathfrak{b}(f,g)=0$ for any polynomial $g\in {}_\mu H_\lambda$ and any finite subset μ of the set $\{t_1,t_2,\dots\}\}$.

PROPOSITION 8.1. *Suppose that a multilinear polynomial f of the algebra H belongs to the subspace H_λ and that $V(f)$ is the ideal of weak identities generated in H by this polynomial. If in this case we have $f\in\operatorname{Ann}_{H_\lambda}\mathfrak{b}$, then for any finite subset ν of the set of generators $\{t_1,t_2,\dots\}$ the inclusion $V(f)\cup H_\nu\subseteq\operatorname{Ann}_{H_\nu}\mathfrak{b}$ holds.*

PROOF. To be definite, suppose that $\lambda=\{t_1,t_2,\dots,t_l\}$. Then the set $V(f)\cap H_\nu$ is generated over K by arbitrary elements of the form $f\big|_{t_1=v_1,\dots,t_l=v_l}v$, where $v_1,\dots,v_l$ are left-normed multilinear commutators belonging to $\mathfrak{M}$, v is a multilinear polynomial belonging to H, and all $v_1,\dots,v_l,v$ depend on different generators $t_i\in\nu$. Therefore, it is sufficient to verify only for these elements that they belong to $\operatorname{Ann}_{H_\nu}\mathfrak{b}$. However, for any multilinear polynomial h we have the relation

$$h\big|_{t_i=v_i}=\{h\big|_{t_i=v_i'}\}\big|_{y_m=[y_m,y_n]},$$

where $v_i=[y_m,y_n,y_k,\dots,y_r]$, $v_i'=[y_m,y_k,\dots,y_r]$, and $y_m,y_n,y_k,\dots,y_r$ denote various generators t_j, different from those in the notation of h. It follows from this relation, together with Property 2 of the character on 2-words α and inductive arguments, that it suffices to prove that the inclusions

$$p(t_1,\dots,t_q)\in\operatorname{Ann}_{H_\mu}\mathfrak{b}\qquad(\mu=\{t_1,\dots,t_q\})$$

imply the inclusions

a) $p(t_1,\dots,t_q)t_{q+1}\in\operatorname{Ann}_{H_{\mu\cup\{t_{q+1}\}}}\mathfrak{b}$,

b) $p\big|_{t_i=[t_i,t_{q+1}]}\in\operatorname{Ann}_{H_{\mu\cup\{t_{q+1}\}}}\mathfrak{b}$.

The first inclusion immediately follows from the definition of $\mathfrak{b}$ and $\operatorname{Ann}_{H_\mu}\mathfrak{b}$. In the second case, applying Property 1 of the character α (see formula (8.2)), we obtain the equality

$$\mathfrak{b}(p\big|_{t_i=[t_i,t_{q+1}]},g)=\mathfrak{b}(p,g\big|_{t_i=[t_{q+1},t_i]})$$

for any $g\in {}_\nu H_{\mu\cup\{t_{q+1}\}}$.

The polynomial $g\big|_{t_i=[t_{q+1},t_i]}$ belongs to ${}_{\nu\cup\{t_{q+1}\}}H_\mu$, and, hence, the right-hand side of this relation is equal to zero in the algebra $\mathcal{E}$, because $p\in\operatorname{Ann}_{H_\mu}\mathfrak{b}$. The proposition is proved.

COROLLARY. *Suppose that the ideal of weak identities* $\mathfrak{V}_\alpha$ *of the algebra* H *is generated by all polynomials in* $\operatorname{Ann}_{H_\lambda} \mathfrak{b}$, *where* λ *ranges over all finite subsets of the set of generators* $\{t_1, t_2, \dots\}$. *Then for any finite set of generators* λ *we have the equality* $\mathfrak{V}_\alpha \cap H_\lambda = \operatorname{Ann}_{H_\lambda} \mathfrak{b}$.

In the case of zero characteristic of the ground field K, any ideal of weak identities is uniquely determined by its multilinear polynomials. Therefore, in this case the equalities $\mathfrak{V} \cap H_\lambda = \operatorname{Ann}_{H_\lambda} \mathfrak{b}$ (for any finite subset λ and for a fixed ideal of weak identities $\mathfrak{V}$) immediately give the equality $\mathfrak{V} = \mathfrak{V}_\alpha$. The variety of associative-Lie pairs determined by the ideal of weak identities $\mathfrak{V}_\alpha$ is called the variety (of pairs) defined by the character α and is denoted by $\mathfrak{B}_\alpha$.

For fields of positive characteristic, denote by $\overline{\mathfrak{V}}_\alpha$ the largest ideal of weak identities such that $\overline{\mathfrak{V}}_\alpha \cap H_\lambda = \operatorname{Ann}_{H_\lambda} \mathfrak{b}$ for any finite set λ. Obviously, for any ideal of weak identities $\mathfrak{V}$ such that $\mathfrak{V} \cap H_\lambda = \operatorname{Ann}_{H_\lambda} \mathfrak{b}$, where λ ranges over all finite subsets of the set $\{t_1, t_2, \dots\}$, the inclusions $\mathfrak{V}_\alpha \subset \mathfrak{V} \subset \overline{\mathfrak{V}}_\alpha$ take place. We denote by $\mathfrak{B}_\alpha$ the variety of pairs determined by the ideal $\overline{\mathfrak{V}}_\alpha$.

§9. The construction of $\boldsymbol{\alpha}$-functions for any representation of a finite-dimensional Lie algebra possessing a nondegenerate invariant symmetric bilinear form

Let $\mathfrak{g}$ be a finite-dimensional Lie algebra over a field K, and let a symmetric invariant nondegenerate bilinear form $(\ ,\ )\colon \mathfrak{g} \otimes_K \mathfrak{g} \to K$ be given on $\mathfrak{g}$. Let ρ be a representation of the Lie algebra $\mathfrak{g}$ in a linear space V, let U be an associative K-subalgebra in $\operatorname{End}_K V$ generated by operators $\rho(g)$, and let $e_1, \dots, e_m$ be an orthonormal basis of the Lie algebra $\mathfrak{g}$ in accordance with the form $(\ ,\ )$.

Denote by $\overline{\mathcal{F}}$ the commutative K-algebra with unit given by the generators $\overline{x}_i^{(1)}, \dots, \overline{x}_i^{(m)}$, where $i = 1, 2, \dots$, and by the defining relations

$$(9.1)\qquad \begin{aligned} \overline{x}_i^{(j)} \overline{x}_i^{(t)} &= \delta \big(\overline{x}_i^{(1)}\big)^2, \\ \overline{x}_i^{(j)} \overline{x}_i^{(k)} \overline{x}_i^{(l)} &= 0 \end{aligned} \qquad (i = 1, 2, 3, \dots).$$

Let $\overline{\mathfrak{g}} \rightleftharpoons \overline{\mathcal{F}} \otimes_K \mathfrak{g}$, $\overline{U} \rightleftharpoons \mathcal{F} \otimes_K U$, and $\overline{x}_i = \sum_{j=1}^m \overline{x}_i^{(j)} \otimes \rho(e_j)$. In the associative algebra $\overline{U}$, let us consider the associative subalgebra $\overline{F}$ generated by the elements $\overline{x}_i$ $(i = 1, 2, \dots)$, and denote by $\overline{B}$ the subalgebra of the algebra $\overline{F}$ generated by all monomials such that in some notation of these monomials any generator $\overline{x}_j$ taking part in this notation enters in it exactly twice. Let $\overline{\lambda}$ and $\overline{\mu}$ be any finite disjoint subsets of the set of generators $\{\overline{x}_1, \overline{x}_2, \dots\}$, and let ${}_{\overline{\mu}}\overline{F}_{\overline{\lambda}}$ be the linear subspace in the algebra $\overline{F}$ that consists of multihomogeneous polynomials depending essentially only on $\overline{x}_i \in \overline{\lambda} \cup \overline{\mu}$, whereas the degree of any monomial of this polynomial with respect to any generator $\overline{x}_i$ in $\overline{\mu}$ is equal to two, and the degree of this monomial with respect to any generator in $\overline{\lambda}$ is equal to one. If $\overline{\lambda}$ or $\overline{\mu}$ is empty, we shall use the notation ${}_{\overline{\mu}}\overline{F}$ and $\overline{F}_{\overline{\lambda}}$, respectively. It follows from the definition of the algebra $\overline{F}$ that $\overline{F} = \sum_{\overline{\lambda}, \overline{\mu}} {}_{\overline{\mu}}\overline{F}_{\overline{\lambda}}$, where the direct sum is taken over all disjoint pairs of finite subsets $\overline{\mu}, \overline{\lambda}, \overline{\mu} \cup \overline{\lambda} \neq \varnothing$.

LEMMA 9.1. *Let* $u(t_1, \dots, t_l)$ *be a multihomogeneous polynomial from the free associative algebra* H *having degree two with respect to each free variable* $t_1, \dots, t_l$. *Then in the algebra* $\overline{F}$ *we have* $\overline{u} \rightleftharpoons u(\overline{x}_{i_1}, \dots, \overline{x}_{i_l}) \in {}_{\overline{\mu}}\overline{F}$, *where* $\overline{\mu} = \{\overline{x}_{i_1}, \dots, \overline{x}_{i_l}\}$, *and this element satisfies the equality*

$$\overline{u} = \left(\overline{x}_{i_1}^{(1)}\right)^2 \cdots \left(\overline{x}_{i_l}^{(1)}\right)^2 u' \tag{9.2}$$

in the algebra $\overline{U}$, *where* u' *belongs to the center of the algebra* U. *Moreover, the element* u' *does not depend on the choice either of the orthonormal basis* $e_1, \dots, e_m$ *of the Lie algebra* $\mathfrak{g}$, *or of the subset* $\overline{\mu}$.

PROOF. It is sufficient to prove the assertion of the lemma for the case in which $u(t_1, \dots, t_l)$ is a monomial. We prove first that for any elements $a \in \overline{U}$, $h \in \mathfrak{g}$ we have the equality

$$[\overline{x}_i, h]a\overline{x}_i = \overline{x}_i a[h, \overline{x}_i] \tag{9.3}$$

in $\overline{U}$.

Since the bilinear form $([x, h], y) = (x, [h, y])$, $x, y, h \in \mathfrak{g}$, is invariant, we see that the linear transformation $\operatorname{ad} h$ of the space $\mathfrak{g}$ is written in an orthonormal basis $e_1, \dots, e_m$ by a skew-symmetric matrix, and hence, $\operatorname{ad} h(e_i) = c_i^{(1)} e_1 + c_i^{(2)} e_2 + \dots + c_i^{(m)} e_m$, where $-c_i^{(j)} = c_j^{(i)}$. This implies

$$\begin{aligned}
[\overline{x}_i, h]a\overline{x}_i &= \left(\overline{x}_i^{(1)}\right)^2 \sum_{j=1}^{m} \rho([e_j, h])a\rho(e_j) \\
&= \left(\overline{x}_i^{(1)}\right)^2 \sum_{j=1}^{m}\sum_{t=1}^{m} (-c_j^{(t)})\rho(e_t)a\rho(e_j) \\
&= \left(\overline{x}_i^{(1)}\right)^2 \sum_{j=1}^{m}\sum_{t=1}^{m} \rho(e_t)ac_t^{(j)}\rho(e_j) = \overline{x}_i a[h, \overline{x}_i].
\end{aligned}$$

This proves equality (9.3).

We distinguish in the monomial $\overline{u}$ the two entries of each variable $\overline{x}_{i_j} \in \overline{\mu}$: $\overline{u} = a_j \overline{x}_{i_j} b_j \overline{x}_{i_j} c_j$. Then for any element $h \in \mathfrak{g}$ the equality

$$[\overline{u}, h] = \sum_{j=1}^{l} \left\{a_j \left[\overline{x}_{i_j}, h\right] b_j \overline{x}_{i_j} c_j + a_j \overline{x}_{i_j} b_j \left[\overline{x}_{i_j}, h\right] c_j\right\}$$

is valid in $\overline{U}$. The expression in braces is equal to zero in $\overline{U}$ by formula (9.3). This proves that $\overline{u}$ belongs to the center of the algebra $\overline{U}$. The defining relations (9.1) of the algebra $\overline{\mathcal{F}}$ show that the element $\overline{u}$ is representable in the form (9.2) and the element u' is contained in the center of the algebra U and is independent of the choice of the set μ.

Let $h_i = \sum_{k=1}^{m} c_i^{(k)} e_k$ $(i = 1, 2, \dots, m)$ be another orthonormal basis of the Lie algebra $\mathfrak{g}$ in accordance with the form $(\,,\,)$. Then, as it is well known, we have

$e_i = \sum_{k=1}^{m} c_k^{(i)} h_k$. Therefore, we have the equality

$$\begin{aligned}\sum_{j=1}^{m} \rho(h_j)a\rho(h_j) &= \sum_{j=1}^{m}\sum_{k=1}^{m} c_j^{(k)}\rho(e_k)a\rho(h_j) \\ &= \sum_{k=1}^{m} \rho(e_k)a\left(\sum_{j=1}^{m} c_j^{(k)}\rho(h_j)\right) = \sum_{k=1}^{m}\rho(e_k)a\rho(e_k)\end{aligned}$$

in $\overline{U}$. This equality, together with the defining relations (9.1) of the algebra $\overline{\mathcal{F}}$, proves that the element $\overline{u}$ is independent of the choice of an orthonormal basis in the Lie algebra $\mathfrak{g}$. The lemma is proved.

COROLLARY. *The mapping α that assigns to every 2-word $\overline{u}$ in the alphabet $\overline{X} = \{\overline{x}_1, \overline{x}_2, \dots\}$, by formula (9.2), the element u' of the center $\mathcal{E}$ of the algebra U (constructed with respect to a representation ρ, in a linear space V, of the finite-dimensional Lie algebra $\mathfrak{g}$ that posesses an invariant nondegenerate symmetric bilinear form), is an $\boldsymbol{\alpha}$-function.*

PROOF. The first property of an $\boldsymbol{\alpha}$-function is satisfied for the function α by formula (9.3). The second and the third ones follow from equality (9.2). The corollary is proved.

§10. The correspondence between the ideals of weak identities and ideals of the commutative algebra defined by a multiplicative character on 2-words

We fix a K-linear $\boldsymbol{\alpha}$-function on the K-space B of 2-elements that takes its values in a commutative K-algebra $\mathcal{E}$. Let $\mathfrak{b}$ be the bilinear pairing (defined in §8 by the function α) of K-subspaces ${}_{\nu}H_{\lambda}$ and ${}_{\mu}H_{\lambda}$ of the free associative algebra H, with values in the algebra $\mathcal{E}$, and let $\mathfrak{V}_{\alpha}$ and $\overline{\mathfrak{V}}_{\alpha}$ be the ideals of weak identities given by the function α.

Let I be an ideal of the algebra $\mathcal{E}$. Let ξ be a canonical homomorphism of the algebra $\mathcal{E}$ onto the quotient algebra $\mathcal{E}/I$. Obviously, the K-linear function $\alpha' \rightleftharpoons \xi \circ \alpha \colon B \to \mathcal{E}/I$ is an $\boldsymbol{\alpha}$-function, and we can consruct the corresponding bilinear pairing $\mathfrak{b}' \colon {}_{\nu}H_{\lambda} \otimes_K {}_{\mu}H_{\lambda} \to \mathcal{E}/I$ and the ideal of weak identities $\overline{\mathfrak{V}}_{\alpha'}$ of the free associative algebra H given by the $\boldsymbol{\alpha}$-function α'. It is clear that for any finite subset λ we have

$$\begin{aligned}\overline{\mathfrak{V}}_{\alpha'} \cap H_{\lambda} &= \operatorname{Ann}_{H_{\lambda}} \mathfrak{b}' \\ &= \{f \in H_{\lambda} \mid \mathfrak{b}(f, g) \in I \text{ for any } g \in {}_{\mu}H_{\lambda} \text{ and for any finite} \\ &\qquad \text{subset } \mu \text{ of the set } \{t_1, t_2, \dots\} \text{ such that } \mu \cap \lambda = \varnothing\}.\end{aligned} \tag{10.1}$$

Denote by Ψ_{α} the mapping that puts in correspondence to any ideal I of the algebra $\mathcal{E}$ the ideal of weak identities $\overline{\mathfrak{V}}_{\alpha'}$ of the free associative algebra H.

Relation (10.1) suggests an idea of assigning to any ideal of weak identities $\mathfrak{V}$ in H the ideal I of the commutative algebra $\mathcal{E}$ generated by the sets $\mathfrak{b}((\mathfrak{V} \cap H_{\lambda}), {}_{\mu}H_{\lambda})$, where μ and λ range over all disjoint finite subsets of the set $\{t_1, t_2, \dots\}$.

Denote by Φ_{α} the mapping that assigns to any ideal of weak identities $\mathfrak{V}$ the smallest ideal I of the algebra $\mathcal{E}$ containing all the sets $\mathfrak{b}((\mathfrak{V} \cap H_{\alpha})$ and ${}_{\mu}H_{\lambda})$, where μ and λ $(\mu \cap \lambda = \varnothing)$ are arbitrary finite sets. We write

$$\theta_{\alpha} \rightleftharpoons \Psi_{\alpha} \circ \Phi_{\alpha}, \qquad \eta_{\alpha} \rightleftharpoons \Phi_{\alpha} \circ \Psi_{\alpha}.$$

PROPOSITION 10.1. *Suppose that the ground field K is infinite. For any ideals of weak identities $\mathfrak{V}_1$, $\mathfrak{V}_2$, and $\mathfrak{V}$ of the free associative algebra H of countable rank and for any ideals I_1, I_2, and I of the algebra $\mathcal{E}$ the following relations are satisfied:*

$$\Phi_\alpha(\mathfrak{V}_1+\mathfrak{V}_2)=\Phi_\alpha(\mathfrak{V}_1)+\Phi_\alpha(\mathfrak{V}_2), \quad \Psi_\alpha(I_1\cap I_2)=\Psi_\alpha(I_1)\cap\Psi_\alpha(I_2); \tag{10.2}$$

$$\theta_\alpha(\mathfrak{V})\supseteq\mathfrak{V}, \qquad \eta_\alpha(I)\subseteq I; \tag{10.3}$$

$$\Phi_\alpha\Psi_\alpha\Phi_\alpha=\Phi_\alpha, \qquad \Psi_\alpha\Phi_\alpha\Psi_\alpha=\Psi_\alpha; \tag{10.4}$$

$$(\theta_\alpha)^2=\theta_\alpha, \qquad (\eta_\alpha)^2=\eta_\alpha. \tag{10.5}$$

PROOF. Equalities (10.2), (10.3) follow immediately from the definition of the mappings Φ_α and Ψ_α. The first equality of (10.2) implies that if $\mathfrak{V}_1\subseteq\mathfrak{V}_2$, then $\Phi_\alpha(\mathfrak{V}_1)\subseteq\Phi_\alpha(\mathfrak{V}_2)$, and the second equality of (10.2) implies that if $I_1\subseteq I_2$ then $\Psi_\alpha(I_1)\subseteq\Psi_\alpha(I_2)$. This proves that the mappings Φ_α and Ψ_α are monotone.

For any ideal of weak identities $\mathfrak{V}$, the inclusions (10.3) imply $\theta_\alpha(\mathfrak{V})\supseteq\mathfrak{V}$, so that $\Phi_\alpha\Psi_\alpha\Phi_\alpha(\mathfrak{V})=\Phi_\alpha\theta_\alpha(\mathfrak{V})\supseteq\Phi_\alpha(\mathfrak{V})$ by the monotonicity of Φ_α. On the other hand, we have $\Phi_\alpha\Psi_\alpha\Phi_\alpha(\mathfrak{V})=\eta_\alpha\Phi_\alpha(\mathfrak{V})\subseteq\Phi_\alpha(\mathfrak{V})$, by the inclusions (10.3). This proves the first of the equalities (10.4). Similarly, using the monotonicity of the mapping Ψ_α and the inclusions (10.3), we prove the second equality in (10.4). Relations (10.5) follow immediately from (10.4). The proposition is proved.

Proposition 10.1 shows that the mapping θ_α defines a closure operation on the set of all ideals of weak identities of the free associative algebra H and that all θ_α-closed ideals of weak identities have the form $\Psi_\alpha(I)$, where I ranges over all ideals of the algebra $\mathcal{E}$. In the same way, the mapping η_α defines a closure operation on the set of all ideals of $\mathcal{E}$, and all η_α-closed ideals have the form $\Phi_\alpha(\mathfrak{V})$, where $\mathfrak{V}$ ranges over all ideals of weak identities of the algebra H.

PROPOSITION 10.2. *The collection Θ_α of all θ_α-closed ideals of weak identities forms a subsemilattice of the modular lattice of all ideals of weak identities of the free associative algebra H with respect to the intersection operation $\cap$, and the ideals $\mathfrak{V}_\alpha$ and H are the minimal and the maximal elements of this subsemilattice, respectively. The collection Ξ_α of all η_α-closed ideals forms a subsemilattice of the modular lattice of all ideals of the algebra $\mathcal{E}$ with respect to the operation $+$ ("union") in the set of ideals. The semilattices Θ_α and Ξ_α are lattices, and the mutually inverse mappings Φ_α and Ψ_α determine an isomorphism of these lattices.*

PROOF. The equalities (10.4) show that $\Phi_\alpha\colon\Theta_\alpha\to\Xi_\alpha$ and $\Psi_\alpha\colon\Xi_\alpha\to\Theta_\alpha$ are mutually inverse mappings. Relations (10.2) imply that the mappings Φ_α and Ψ_α are monotone on the sets of all ideals of weak identities in H and of all ideals in $\mathcal{E}$, respectively. Moreover, they imply that the infimum of two θ_α-closed ideals of weak identities is a θ_α-closed ideal of weak identities, and the supremum of two η_α-closed ideals is a η_α-closed ideal. This proves that Θ_α and Ξ_α are semilattices with respect to the operations $\cap$ and $+$, respectively. Since the mappings Φ_α and Ψ_α are monotone and mutually inverse on Θ_α and Ξ_α, respectively, we see that Θ_α and Ξ_α are lattices. The proposition is proved.

REMARK 10.1. It is easy to see that in Proposition 10.1 relations (10.2) can be replaced by more general ones:

$$\Phi_\alpha\Big(\sum_{i\in S}\mathfrak{V}_i\Big) = \sum_{i\in S}\Phi_\alpha(\mathfrak{V}_i),$$
$$\Psi_\alpha\Big(\bigcap\nolimits_{i\in S} I_i\Big) = \bigcap\nolimits_{i\in S}\Psi_\alpha(I_i),$$

where $\{\mathfrak{V}_i \mid i \in S\}$ is an arbitrary subset of the set of ideals of weak identities of the algebra H and $\{I_i \mid i \in S\}$ is an arbitrary subset of the set of ideals of the algebra $\mathcal{E}$. Hence, Θ_α is a full subsemilattice of the full semilattice of all ideals of weak identities with respect to the intersection operation and Ξ_α is a full subsemilattice of the full semilattice of all ideals of the algebra $\mathcal{E}$ with respect to the addition operation.

Let $\varkappa\colon \mathcal{E} \to \mathcal{E}'$ be a homomorphism of commutative K-algebras. For any ideal J of the algebra $\mathcal{E}'$, denote by the symbol $\varkappa^{-1}(J)$ the ideal of the algebra $\mathcal{E}$ that is the preimage of the ideal J, and, for any ideal I of the algebra $\mathcal{E}$, denote by the symbol $\overline{\varkappa}(I)$ the least ideal of the algebra $\mathcal{E}'$ containing $\varkappa(I)$. Obviously, $\varkappa^{-1}$ and $\overline{\varkappa}$ are monotone mappings on the sets of ideals and satisfy the following relations:

$$\varkappa^{-1}\overline{\varkappa}\varkappa^{-1} = \varkappa^{-1}, \qquad \overline{\varkappa}\varkappa^{-1}\overline{\varkappa} = \overline{\varkappa}, \tag{10.6}$$
$$\overline{\varkappa}\varkappa^{-1}(J) \subseteq J, \qquad \varkappa^{-1}\overline{\varkappa}(I) \supseteq I, \tag{10.7}$$
$$\overline{\varkappa}\Big(\sum_{i\in S} I_i\Big) = \sum_{i\in S}\overline{\varkappa}(I_i), \tag{10.8}$$

where J is any ideal of the algebra $\mathcal{E}'$ and I and I_i are arbitrary ideals of the algebra $\mathcal{E}$.

PROPOSITION 10.3. *Let $\varkappa\colon \mathcal{E} \to \mathcal{E}'$ be a homomorphism of commutative K-algebras and let $\alpha\colon B \to \mathcal{E}$ and $\alpha'\colon B \to \mathcal{E}'$ be $\boldsymbol{\alpha}$-functions such that $\varkappa \circ \alpha = \alpha'$. Then*

$$\Phi_{\alpha'} = \overline{\varkappa}\Phi_\alpha, \qquad \Psi_{\alpha'} = \Psi_\alpha\varkappa^{-1}, \tag{10.9}$$
$$\theta_\alpha \circ \Psi_{\alpha'} = \Psi_{\alpha'}, \qquad \Phi_{\alpha'} \circ \theta_\alpha = \Phi_{\alpha'}, \tag{10.10}$$
$$\eta_\alpha\varkappa^{-1}(J) \subseteq \varkappa^{-1}\eta_{\alpha'}(J), \qquad \eta_{\alpha'}\overline{\varkappa}(I) \supseteq \overline{\varkappa}\eta_\alpha(I). \tag{10.11}$$

PROOF. Equalities (10.9) follow from the definition of Φ_α, $\Phi_{\alpha'}$, Ψ_α, $\Psi_{\alpha'}$, $\overline{\varkappa}$, and $\varkappa^{-1}$. Equalities (10.10) follow from the definition of θ_α and from (10.9) and (10.4). Since $\eta_{\alpha'} = \overline{\varkappa}\eta_\alpha\varkappa^{-1}$, we see that relations (10.11) follow from the second relation of (10.7).

The first equality of (10.10) means that any $\theta_{\alpha'}$-closed ideal of weak identities is θ_α-closed and, therefore, the semilattice $\Theta_{\alpha'}$ is contained in Θ_α as a full subsemilattice with respect to the operation $\cap$. Denote by j this embedding of $\Theta_{\alpha'}$ into Θ_α. The first relation of (10.9) shows that the ideal $\overline{\varkappa}(I)$ is $\eta_{\alpha'}$-closed in $\mathcal{E}'$ for any η_α-closed ideal I in $\mathcal{E}$. Together with equality (10.8), this implies that $\overline{\varkappa}\colon \Xi_\alpha \to \Xi_{\alpha'}$ is an epimorphism of full semilattices with respect to the operation $+$.

PROPOSITION 10.4. *The diagrams*

$$\begin{array}{ccc} \Theta_\alpha & \xrightarrow{\Phi_\alpha} & \Xi_\alpha \\ {\scriptstyle j}\uparrow & & \uparrow{\scriptstyle \eta_\alpha \varkappa^{-1}} \\ \Theta_{\alpha'} & \xrightarrow{\Phi_{\alpha'}} & \Xi_{\alpha'} \end{array} \qquad \begin{array}{ccc} \Theta_\alpha & \xleftarrow{\Psi_\alpha} & \Xi_\alpha \\ {\scriptstyle j}\uparrow & & \uparrow{\scriptstyle \eta_\alpha \varkappa^{-1}} \\ \Theta_{\alpha'} & \xleftarrow{\Phi_{\alpha'}} & \Xi_{\alpha'} \end{array}$$

$$\begin{array}{ccc} \Theta_\alpha & \xrightarrow{\Phi_\alpha} & \Xi_\alpha \\ {\scriptstyle \theta_{\alpha'}}\downarrow & & \downarrow{\scriptstyle \bar{\varkappa}} \\ \Theta_{\alpha'} & \xrightarrow{\Phi_{\alpha'}} & \Xi_{\alpha'} \end{array} \qquad \begin{array}{ccc} \Theta_\alpha & \xleftarrow{\Psi_\alpha} & \Xi_\alpha \\ {\scriptstyle \theta_{\alpha'}}\downarrow & & \downarrow{\scriptstyle \bar{\varkappa}} \\ \Theta_{\alpha'} & \xleftarrow{\Phi_{\alpha'}} & \Xi_{\alpha'} \end{array}$$

are commutative.

PROOF. Each element of $\Theta_{\alpha'}$ has the form $\Psi_{\alpha'}(J)$, where J is an ideal in $\mathcal{E}'$, and any element of $\Xi_{\alpha'}$ has the form $\Phi_{\alpha'}(\mathfrak{V})$. Therefore, the commutativity of all the diagrams under consideration follows from equalities (10.9) and from equalities (10.4) for the functions α and α'. The proposition is proved.

§11. A general approach and the setting of the problem of studying varieties of pairs by the method of 2-words

In this section we shall advance the main principle in the study of varieties of associative-Lie pairs (in particular, of varieties of Lie algebras and varieties of associative algebras) by the method of 2-words and $\boldsymbol{\alpha}$-functions that will be significant in our further investigations. Roughly speaking, the principle is this: *in order to obtain an important information on varieties of pairs and identical relations, one should successfully select a commutative K-algebra $\mathcal{E}$ and an efficiently computable $\boldsymbol{\alpha}$-function $\alpha\colon B \to \mathcal{E}$ on 2-elements and try to find for this $\boldsymbol{\alpha}$-function a reasonable description of the lattice of θ_α-closed ideals of weak identities (η_α-closed ideals of the algebra $\mathcal{E}$).*

As it was mentioned in §7, in the category of all $\boldsymbol{\alpha}$-functions there is a universal object $\alpha_0\colon B \to \mathcal{E}$, i.e., a universal $\boldsymbol{\alpha}$-function. Therefore, by Remark 10.1 and Proposition 10.4, the problem of describing all lattices Θ_α, where α ranges over all $\boldsymbol{\alpha}$-functions, is a special case of the problem of describing all full subsemilattices of the full semilattice Θ_{α_0} with respect to the operation of intersection of ideals of weak identities.

Apparently, the problem of describing the semilattice Θ_{α_0} and all the more of all of its subsemilattices hardly has a reasonable solution. Therefore, the use of the principle indicated above is first of all in finding a commutative K-algebra $\mathcal{E}$ and an $\boldsymbol{\alpha}$-function $\alpha\colon B \to \mathcal{E}$ for which such a solution is possible.

The results of §9 give us the simplest models of $\boldsymbol{\alpha}$-functions. Considerations of algebraic geometry suggest a natural way of constructing a class of commutative algebras $\mathcal{E}$ and $\boldsymbol{\alpha}$-functions $\alpha\colon B \to \mathcal{E}$ where, in the first place, we are to search for $\boldsymbol{\alpha}$-functions α for which the corresponding full subsemilattice Ξ_α of η_α-closed ideals of the algebra $\mathcal{E}$ admits an explicit description.

Let the ground field K be algebraically closed and of characteristic zero. Let S be a set and let for any $s \in S$ a reductive finite-dimensional Lie K-algebra $\mathfrak{g}_s$, a K-bilinear symmetric nondegenerate invariant form $(\ ,\ )_s\colon \mathfrak{g}_s \otimes_K \mathfrak{g}_s \to K$, and a faithful irreducible representation $\rho_s : \mathfrak{g}_s \to \mathrm{End}_K V_s$ of the Lie algebra $\mathfrak{g}_s$ in a linear space V_s be given. Denote by U_s the associative K-subalgebra in $\mathrm{End}_K V_s$ generated by $\rho_s(\mathfrak{g}_s) = \mathfrak{g}_s$. By Lemma 3.5 and Remark 3.3, the center $\mathcal{E}_s$ of the algebra U_s either coincides with the field $K \cdot \mathbf{1}$ or equals zero. Therefore, the construction carried out in §9 defines for any representation ρ_s its $\boldsymbol{\alpha}$-function $\alpha_s : B \to \mathcal{E}_s = K \cdot \mathbf{1}$ with values in the ground field K. By Lemma 7.1, in the category ALPHA there exists the product $\alpha : B \to \prod_{s \in S} \mathcal{E}_s$ of the objects $\alpha_s : B \to \mathcal{E}_s$. Denote by the letter $\mathcal{E}$ the commutative subalgebra in the Cartesian product $\prod_{s \in S} \mathcal{E}_s$ generated by the elements $\alpha(u)$, where u ranges over B. The algebra $\mathcal{E}$ and the $\boldsymbol{\alpha}$-function α are uniquely defined by the set of irreducible representations ρ_s, where $s \in S$.

Under this approach to the construction of an $\boldsymbol{\alpha}$-function, the study of varieties of pairs with the help of the principle stated above includes necessarily three subproblems:

1. To describe explicitly the algebra $\mathcal{E}$ for a given set of irreducible representations ρ_s $(s \in S)$ of reductive Lie algebras $\mathfrak{g}_s$ with a nondegenerate invariant symmetric form given on $\mathfrak{g}_s$.

2. To find an efficient way of calculating an $\boldsymbol{\alpha}$-function $\alpha : B \to \mathcal{E}$ determined by this set of representations ρ_s, $s \in S$.

3. To describe explicitly the full subsemilattice Θ_α of θ_α-closed ideals of weak identities with respect to the intersection operation (or, what is the same, to describe the full subsemilattice Ξ_α of η_α-closed ideals of the algebra $\mathcal{E}$ with respect to the operation of addition of ideals).

REMARK 11.1. By an explicit description of the algebra $\mathcal{E}$ we mean an identification of the algebra $\mathcal{E}$ with a certain well-known commutative algebra or at least the description of $\mathcal{E}$ via generators and defining relations.

REMARK 11.2. Apparently, in connection with the problem of whether the identities of concrete varieties of associative-Lie pairs and their subvarieties are finitely based, the solution of the third subproblem on the description of the lattices Θ_α and Ξ_α may be considered satisfactory only when it includes the answer to the following question: are there infinite strictly increasing totally ordered chains of elements of the lattice θ_α?

As we shall see in the subsequent chapters, a substantial information on weak identities of varieties of pairs can be learned if a constructive structure of the closure operation η_α on the set of all ideals of the algebra $\mathcal{E}$ or a reasonable construction of the closure operation θ_α on the set of ideals of weak identities is discovered. Therefore, it is natural to add to the problems 1, 2, 3 on the set of irreducible representations ρ_s, $s \in S$, the following problems:

4. To give an efficient description of the closure operations η_α and θ_α.

5. To describe efficiently the mappings Ψ_α and Φ_α.

The five problems stated for a given set of faithful irreducible representations ρ_s, $s \in S$, of reductive finite-dimensional Lie algebras, with an invariant nondegenerate form $(\, , \,)$, represent a precise formulation of the $\boldsymbol{\alpha}$-function problem in the method of 2-words.

Since this problem is not a self-objective and it is considered as a problem of study of identities of associative-Lie pairs and varieties of pairs, whereas the list of well-known nontrivial pairs is continually increasing in the course of the natural process of studying mathematical objects, we see that it is desirable to include the following as a subproblem:

6. For any θ_α-closed ideal of weak identities $\mathfrak{V}$, find a pair from the list of already known associative-Lie pairs whose ideal of weak identities coincides with $\mathfrak{V}$.

Comments

I proved Theorem 6.2 in December 1969; S. Bachmuth, H. J. Mochizuki, and D. W. Walkup announced a similar result in 1970 ([BMW]); Theorem 6.3 and Proposition 6.2 were published for the first time in my paper in 1971 ([Raz1]); in the same year a complete account of the result of S. Bachmuth and H. J. Mochizuki on nonsolvability of 3rd Engel groups with the identity $x^5 = 1$ has appeared (see [BM]). The paper [Raz2], where an example was constructed of a locally finite nonsolvable group of simple period $p \geq 5$ with the solvable word problem, generating an almost Cross variety of groups, served as a foot-bridge to the use of the function α in the proof of Theorem 6.5 (published in the paper [Raz1]). Substantially, the proof of Theorem 6.4 was given in the paper [Raz3], where the full matrix algebra of second order was called on to simplify the original proof of Theorem 6.2. Theorem 6.4 suggests an idea to modify the problem of A. I. Kostrikin [Kos2] on the solvability of Engel Lie algebras in the following way.

PROBLEM 2.1. Whether it is true that in any $(p-1)$th Engel (nth Engel, where $n > p$) Lie algebra over a field of characteristic $p > 0$ all the identities of some finite-dimensional Lie algebra are satisfied. (Note that the positive solution of this problem does not contradict to Theorem 6.4 and implies the statement on local nilpotence of an nth Engel Lie algebra ([Raz6]).)

The approach to the study of identities pf associative-Lie pairs described in §§7–11 was stated for the first time in the author's report in 1978 (see [Raz8]). Of course, in this approach there is an indefiniteness in the choice of the set S of irreducible representations of semisimple reductive finite-dimensional Lie algebras in order to construct the corresponding α-function. The examples considered in the subsequent three chapters show that when choosing the set S, it is worthy to aim for the corresponding $\boldsymbol{\alpha}$-function $\alpha: B \to \mathcal{E}$ to take its values in a sufficiently simple algebra $\mathcal{E}$ and to have a clear algorithm for its computation. Besides the sets S considered in Chapters 3–5, I know another two sets of representations satisfying these two conditions. However, in these two cases I have no idea to approach the problem of describing θ_α-closed and η_α-closed ideals. The confidence that these two problems can have a nice solution and this solution will lead to results that have significant applications stimulates me to pose the following two problems.

PROBLEM 2.2. Find a solution for all the six problems of §11 for the $\boldsymbol{\alpha}$-function constructed with respect to the set of representations of skew-symmetric matrices $\mathrm{so}(n, K)$ in n-dimensional spaces V_n $(n = 2, 3, \dots)$ over an algebraically closed field K of characteristic zero.

At present, it is well known that for the $\boldsymbol{\alpha}$-function constructed for the set of representations of $\mathrm{sp}(n, K) = \{g \in \mathrm{End}_K V_{2n}, g^* = -g\}$, where $*$ denotes the symplectic involution in $2n$-dimensional spaces $(n = 1, 2, \dots)$ the answer to these six problems will be the same as in the case $\mathrm{so}(n, K)$ $(n = 2, 3, \dots)$.

PROBLEM 2.3. Find a solution for the six problems of §11 for the $\boldsymbol{\alpha}$-function constructed with respect to the set of adjoint representations of "all" simple finite-dimensional Lie algebras over an algebraically closed field K of characteristic zero.

CHAPTER III

$\boldsymbol{\alpha}$-FUNCTIONS RELATED TO THE KILLING FORM AND TO IRREDUCIBLE REPRESENTATIONS OF SEMISIMPLE LIE ALGEBRAS. CENTRAL POLYNOMIALS OF IRREDUCIBLE REPRESENTATIONS OF REDUCTIVE LIE ALGEBRAS

§12. Statement of the main results of the chapter

In this chapter we consider the most special case of the problem posed in §11 of the previous chapter, i.e., we suppose that the algebraically closed field K is of characteristic zero, that the set S is a singleton consisting of a faithful irreducible representation $\rho\colon \mathfrak{g} \to \mathrm{End}_K V$ of the semisimple finite-dimensional Lie K-algebra $\mathfrak{g}$ in a linear (possibly infinite-dimensional) K-space V; for a nondegenerate symmetric bilinear form on $\mathfrak{g}$ we take the Killing form $\mathfrak{k}\colon \mathfrak{g} \otimes_K \mathfrak{g} \to K$, and the $\boldsymbol{\alpha}$-function α is constructed with respect to the representation ρ in the same way as in §9. Denote by the letter U the associative subalgebra in $\mathrm{End}_K V$ generated by the Lie algebra $\rho(\mathfrak{g}) \simeq \mathfrak{g}$. We shall concentrate our main efforts on obtaining necessary and sufficient conditions for the validity of the equality $\overline{\mathfrak{B}}_\alpha = \mathrm{var}(U, \mathfrak{g})$ and thus to giving an answer to Problem 6 of §11. A curious and surprising consequence of this investigation is Theorem 12.2 given below, which gives, from the viewpoint of differential geometry, a completely constructive method of recovering a smooth n-dimensional manifold from the Lie algebra of its vector fields.

In the case under consideration the $\boldsymbol{\alpha}$-function α take its values in the center of the algebra U, which is either zero or coincides with the field $K \cdot \mathbf{1} \simeq K$ by Lemma 3.5 and Remark 3.3.

If the center of the algebra U is zero then $\mathcal{E} = 0$ and for any ideal of weak identities $\mathfrak{V}$ we have the equality $\theta_\alpha(\mathfrak{V}) = H = \overline{\mathfrak{V}}_\alpha$, and therefore all the problems of §11 have the trivial answer.

In this chapter it will be shown that if the center of the algebra U is nonzero, then there are nonzero values of the $\boldsymbol{\alpha}$-function α (this follows from Theorem 16.1). Hence, $\alpha(B) = \mathcal{E} = K \cdot \mathbf{1}$. This gives a description of the algebra $\mathcal{E}$. Since the function α is not a zero function, we see that in $\mathcal{E}$ there exist exactly two η_α-closed ideals: $\{0\}$ and $\mathcal{E}$; hence, $\overline{\mathfrak{V}}_\alpha$ and H are all θ_α-closed ideals of weak identities in H. This gives a complete description of the lattices Θ_α and Ξ_α, of the closure operations θ_α and η_α, and of the mapping Ψ_α. Within the limits of the method of 2-words, it remains to clarify a nonformal question: for what pair $(U_1, \mathfrak{g}_1)$ its ideal of identities coincides with the ideal $\overline{\mathfrak{V}}_\alpha$.

It follows from the construction of the function α that if an identity $f = 0$ is satisfied in the pair $(U, \mathfrak{g})$ then $f \in \overline{\mathfrak{V}}_\alpha$. This shows that $\mathrm{var}(U, \mathfrak{g}) \supseteq \mathfrak{B}_\alpha$. The following considerations explain why this inclusion is strict in the case of a nonsimple algebra U. Let U be not simple and let I be a nonzero maximal two-sided ideal of the algebra U (the ideal I exists by Zorn's lemma, because U is an algebra with unit). Then the simple quotient algebra $U_1 = U/I$ contains a unit and possesses a faithful irreducible representation in some linear space V. Hence,

$$\rho_1 : \mathfrak{g}_1 = \mathfrak{g}/(\mathfrak{g} \cap I) \hookrightarrow U_1 \to \operatorname{End}_K V_1$$

is a faithful irreducible representation of the semisimple Lie algebra $\mathfrak{g}_1$. The ideal I has zero intersection with the center of the algebra U. Therefore, the canonical homomorphism $\varkappa: U \to U_1$ induces an isomorphism $\varkappa: \mathcal{E} \to \mathcal{E}'$ of the centers of the algebras U and U_1. Since $\mathfrak{g} = \mathfrak{g}_1 \oplus \mathfrak{g} \cap I$ and the restriction of the Killing form of the Lie algebra $\mathfrak{g}$ to $\mathfrak{g}_1$ coincides with the Killing form of the Lie algebra $\mathfrak{g}_1$, we see that the $\boldsymbol{\alpha}$-function α_1, constructed with respect to the irreducible representation ρ_1 in the same way as in §9, satisfies the equalities $\varkappa\alpha = \alpha_1$ and $\varkappa^{-1}\alpha_1 = \alpha$. Hence, $\overline{\mathfrak{V}}_{\alpha_1} = \overline{\mathfrak{V}}_{\alpha}$ and the following obvious inclusions hold:

$$\operatorname{var}(U, \mathfrak{g}) \supset \operatorname{var}(U_1, \mathfrak{g}_1) \supseteq \overline{\mathfrak{B}}_{\alpha_1} = \overline{\mathfrak{B}}_{\alpha}.$$

The first of these inclusions is strict, because in the contrary case the pairs $(U, \mathfrak{g})$ and $(U, \mathfrak{g}_1)$ would have the same identities and would be isomorphic by Theorem 5.1; however, by the fact that the enveloping algebra of a finite-dimensional Lie algebra is Noetherian, the algebra U cannot be isomorphic to its proper quotient algebra $U_1 = U/I$.

This discussion shows that the conditions that the algebra U_1 is simple and the center of this algebra is nonzero are necessary for the validity of the equality $\operatorname{var}(U_1, \mathfrak{g}_1) = \mathfrak{B}_{\alpha_1}$. In fact, these conditions are also sufficient (see Theorem 16.1 of the present chapter). Therefore, the ideal of weak identities $\mathfrak{V}_\alpha$ constructed with respect to the pair $(U, \mathfrak{g})$, where $\mathfrak{g}$ is a semisimple Lie algebra and U is its primitive associative enveloping algebra with nonzero center, coincides with the ideal of identities of the pair $(U_1, \mathfrak{g}_1)$. This gives an answer to Problem 6 of §11. Since, by Theorem 5.1, nonisomorphic pairs of the form $(U, \mathfrak{g})$, where U is a simple associative K-algebra, have different identities, we see that the equality $\operatorname{var}(U, \mathfrak{g}) = \mathfrak{B}_\alpha$ implies also that these nonisomorphic pairs $(U, \mathfrak{g})$ determine substantially different $\boldsymbol{\alpha}$-functions $\alpha : B \to K$.

The discussion above shows that the key moment in the study of necessary and sufficient conditions for the validity of the equality $\operatorname{var}(U, \mathfrak{g}) = \mathfrak{B}_\alpha$ is the proof of this equality in the following special case: U is a simple associative enveloping K-algebra, with a nonzero center, of a semisimple Lie algebra $\mathfrak{g}$. This proof is quite nontrivial and gives an important information about the identities of the pair $(U, \mathfrak{g})$. This proof is based upon the existence of so-called central polynomials for the pair $(U, \mathfrak{g})$ and exhibits the explicit form of identities distinguishing nonisomorphic pairs $(U, \mathfrak{g})$.

DEFINITION 12.1. An associative polynomial $f(t_1, \dots, t_l) \in H$ is called a *central polynomial* of the pair $(A, \mathfrak{b})$ if for any $h_1, \dots, h_l \in \mathfrak{h}$ the value $f(h_1, \dots, h_l)$ belongs to the center of the algebra A and the relation $f = 0$ is not an identity for the pair $(A, \mathfrak{h})$.

Denote by $R_{m,k}$ the set of multilinear polynomials of degree km in the algebra H in each of which all the variables can be decomposed into k disjoint subsets of cardinality

m so that this polynomial is skew-symmetric with respect to the variables belonging to the same subset. For instance, the polynomial

$$\sum_{\sigma\in S_3}\sum_{\delta\in S_3}\operatorname{sign}(\sigma\delta)x_{\sigma(1)}y_{\delta(1)}y_{\delta(2)}x_{\sigma(2)}y_{\delta(3)}x_{\sigma(3)},$$

where the sum is taken over all permutations σ and δ of the set of three elements, and sign is the sign of the permutation, belongs to $R_{3,2}$, and the polynomial $[x_1, x_2][y_1, y_2][z_1, z_2]$ belongs to $R_{2,3}$.

As we shall see in §§16–20, for a simple algebra U the equality $\operatorname{var}(U, \mathfrak{g}) = \mathfrak{B}_\alpha$ is essentially equivalent to the following theorem, which does not mention the $\boldsymbol{\alpha}$-function α explicitly.

THEOREM 12.1. *Let a semisimple finite-dimensional Lie algebra $\mathfrak{g}$ over an algebraically closed field of characteristic zero have the dimension m. Let its enveloping algebra U be simple and let its center be nonzero. Then, for some positive integer k, there exists a central polynomial of the pair $(U, \mathfrak{g})$ in the set of polynomials $R_{m,k}$.*

COROLLARY. *Let $\rho\colon \mathfrak{g} \to \operatorname{End}_K V$ be a faithful irreducible representation of the semisimple m-dimensional Lie algebra $\mathfrak{g}$ over the algebraically closed field K of characteristic zero in a linear space V and let the associative subalgebra U be generated in $\operatorname{End}_K V$ by multiplications by elements of $\mathfrak{g}$. If the center of the algebra U is nonzero and the largest proper ideal in U has the zero intersection with $\rho(\mathfrak{g}) \simeq \mathfrak{g}$, then in the pair $(U, \mathfrak{g})$ there is a central polynomial belonging to the set $R_{m,k}$.*

REMARK 12.1. Theorem 12.1 and its Corollary are valid not only for semisimple, but also for reductive Lie algebras $\mathfrak{g}$ (see §20).

REMARK 12.2. By the Duflo theorem [Duf], any maximal primitive ideal of the universal enveloping algebra $U(\mathfrak{g})$ of a reductive Lie algebra $\mathfrak{g}$ coincides with the annihilator of some quotient of a Verma module. Therefore, the condition imposed in the corollary on the largest proper ideal is equivalent to the algebra U having a faithful irreducible representation in a quotient module of a Verma module such that its highest weight Λ is not identically zero on the Cartan subalgebra of any minimal ideal of the algebra $\mathfrak{g}$.

The most important consequence of Theorem 12.1 is the following.

THEOREM 12.2. *Let W_n be the Lie algebra of all derivations of the algebra of commutative polynomials $\mathcal{F}$ in n algebraically independent indeterminates over a field K of characteristic zero. The algebra W_n can be regarded as a left $\mathcal{F}$-module, and the algebra $\mathcal{F}$ can be naturally identified with a subalgebra in $\operatorname{End}_K W_n$. Then for $m = 2n + n^2$ there exists an associative polynomial $f \in R_{m,k}$ such that for any $w_1, \dots, w_l \in W_n$ $(l = mk)$ the value $f(\operatorname{ad} w_1, \dots, \operatorname{ad} w_l)$ belongs to $\mathcal{F}$ and the linear mapping*

$$f\circ\operatorname{ad}\colon \underbrace{W_n\otimes\cdots\otimes W_n}_{l \text{ times}} \to \mathcal{F}$$

is an epimorphism.

This theorem indicates a sufficiently constructive way of recovering a commutative algebra of infinitely differentiable functions on a smooth n-dimensional manifold from the Lie algebra of all its vector fields.

In the chapter, the material is disposed as follows. In §13 we gather the information we need on the centers of associative enveloping algebras of a semisimple Lie algebra. In §14 we prove that under the assumptions of Theorem 13.1 there exists a central polynomial for the pair $(U, \mathfrak{g})$ (Theorem 14.1). This result is weaker than Theorem 12.1, because the central polynomial constructed in Theorem 14.1 can be outside the set $R_{m,k}$. In the course of the proof of Theorem 14.1 an algebra with three supports $\{K, \mathfrak{g}, U\}$ arises implicitly, with the new operation given by the Killing form $\mathfrak{k}\colon \mathfrak{g} \otimes_K \mathfrak{g} \to K$. In §15 a precise definition of this algebra with three supports is given and a construction of the free algebra with three supports $\{\mathcal{F}_0, \mathfrak{L}_1, F_1\}$ in the variety $\mathrm{var}\{K, \mathfrak{g}, U\}$ is presented. Section 16 is devoted to the construction, for an irreducible representation of the semisimple Lie algebra $\mathfrak{g}$, of an auxiliary algebra with three supports $\{\overline{\mathcal{F}}_0, \overline{\mathfrak{L}}_1, \overline{F}_1\}$ and of a pair $(\overline{F}, \overline{\mathfrak{L}})$ that permit one to extend the $\boldsymbol{\alpha}$-function constructed in §9 from 2-words to generalized 2-elements, preserving its main properties. In the same section, we give the statement of Theorem 16.1, where it is asserted that in the case of a simple associative enveloping algebra U of the semisimple Lie algebra $\mathfrak{g}$ the ideal of weak identities $\mathfrak{V}_\alpha$, consructed with respect to the $\boldsymbol{\alpha}$-function defined by the pair $(U, \mathfrak{g})$, coincides with the ideal of identities of this pair. The proof of this theorem for a finite-dimensional algebra U is given in §17. In §18 we show how we can prove Theorem 12.1 using the results of §17, and in §19 we achieve the proof of Theorem 16.1 for the case of an infinite-dimensional simple enveloping algebra U with nonzero center. In §20 we discuss some consequences of Theorems 12.1 and 16.1; in particular, we show the form of identities distinguishing nonisomorphic pairs $(U, \mathfrak{g})$ and prove Theorem 12.1 for reductive Lie algebras $\mathfrak{g}$. Section 21 is devoted completely to the proof of Theorem 12.2 that uses only Theorem 12.1 in the case of the adjoint representation of the Lie algebra $\mathrm{sl}(n+1, K)$ and requires the knowledge of the most elementary facts about the algebra W_n.

In this chapter the following notation is used:

$(H, \mathfrak{M})$ denotes the absolutely free pair with free generators $t_1, t_2, \dots$;

$\mathfrak{K}_m$ denotes the ideal of weak Capelli identities of order m, i.e., the least ideal of weak identities containing all polynomials of the form $d(t_1, \dots, t_m, t_{m+1}, \dots, t_{m+l})$ that are skew-symmetric and multilinear with respect to $t_1, \dots, t_m$ (m is fixed, $l = 0, 1, 2, \dots$).

For any polynomial $f(t_1, \dots, t_l)$ and any elements $v_1, \dots, v_l$ of an algebra A, we denote by $f|_{t_1=v_1,\dots,t_l=v_l}$ the element of the algebra A equal to $f(v_1, \dots, v_l)$.

§13. Some remarks on the enveloping algebras of semisimple Lie algebras

In this section we collect auxiliary results, which will be used in the proof of Theorem 12.1.

The proof of the following three assertions can be found in the book of J. Dixmier [Dix].

For any finite-dimensional Lie algebra $\mathfrak{g}$, denote by $\mathfrak{g}^*$ the $\mathfrak{g}$-module dual to the adjoint $\mathfrak{g}$-module $\mathfrak{g}$ and denote by $S(\mathfrak{g}^*)$ the commutative algebra of polynomial functions on $\mathfrak{g}$ (in one vector variable) with the natural action of the Lie algebra $\mathfrak{g}$ on $S(\mathfrak{g}^*)$ by derivations. An associative subalgebra in $S(\mathfrak{g}^*)$ consisting exactly of the polynomial functions on which $\mathfrak{g}$ acts trivially is called the *subalgebra of invariants* in $S(\mathfrak{g}^*)$ with respect to $\mathfrak{g}$.

PROPOSITION 13.1. *Let $\mathfrak{g}$ be a finite-dimensional semisimple Lie algebra over a field K of characteristic zero. Then the subalgebra of all invariants of the algebra $S(\mathfrak{g}^*)$ with* *respect to* $\mathfrak{g}$ *is spanned* (*as a linear space*) *by the functions* $g \to \operatorname{tr}(\rho(g)^m)$ (*which are polynomials on* $\mathfrak{g}$), *where* ρ *ranges over all finite-dimensional representations of the Lie algebra* $\mathfrak{g}$ *and* $m = 1, 2, \ldots$.

Denote by $\operatorname{Sym}_l(t_1, \ldots, t_l)$ the multilinear associative polynomial of degree l that is symmetric with respect to all its variables. The following assertion is an easy consequence of Proposition 13.1.

PROPOSITION 13.2. *Let* $\mathfrak{g}$ *be a finite-dimensional semisimple Lie algebra over a field* K *of characteristic zero, let* $e_1, \ldots, e_n$ *be its basis, and let* $e'_1, \ldots, e'_n$ *be the dual basis with respect to the Killing form. Then the center of the universal enveloping algebra* $U(\mathfrak{g})$ *of the Lie algebra* $\mathfrak{g}$ *is spanned* (*as a linear space*) *by elements of the form*

$$f_{l,\rho} = \sum_{i_1,\ldots,i_l} \operatorname{tr}\left(\operatorname{Sym}_l|_{t_1=\rho(e_{i_1}),\ldots,t_l=\rho(e_{i_l})}\right) e'_{i_1} \cdots e'_{i_l},$$

where ρ *is an arbitrary finite-dimensional representation of the Lie algebra* $\mathfrak{g}$, $l = 1, 2, 3, \ldots$, *and the sum is taken over all collections of indices* $i_1, \ldots, i_l \in \{1, 2, \ldots, m\}$. *Moreover, the elements* $f_{l,\rho}$ *are independent of the choice of the basis* $e_1, \ldots, e_m$.

PROPOSITION 13.3. *Let* $\mathfrak{g}$ *be a finite-dimensional semisimple Lie algebra over an arbitrary field* K *of characteristic zero and let* $U(\mathfrak{g})$ *be its universal enveloping algebra. Then for any ideal* I *of the algebra* $U(\mathfrak{g})$ *the center of the quotient algebra* $U = U(\mathfrak{g})/I$ *coincides with the image of the center of* $U(\mathfrak{g})$ *under the canonical homomorphism* $\varphi \colon U(\mathfrak{g}) \to U$.

PROPOSITION 13.4. *Let* $\mathfrak{g}_1$ *be a semisimple Lie algebra over an arbitrary field* K_1 *of characteristic zero, let* $\mathfrak{g}_2 = K_2 \otimes_{K_1} \mathfrak{g}_1$ *for some extension* K_2 *of the field* K_1, *let* U_2 *be an enveloping* K_2*-algebra of the Lie algebra* $\mathfrak{g}_2$, *and let* U_1 *be the* K_1*-subalgebra generated in* U_2 *by the elements of* $\mathfrak{g}_1$. *Suppose that the equality* $K_2 I_1 = U_2$ *holds for some* K_1*-ideal* I_1 *of the algebra* U_1. *Then we have* $\mathfrak{Z} \cap I_1 \neq 0$ *if the center* $\mathfrak{Z}$ *of the algebra* U_2 *is nonzero.*

PROOF. Let us consider U_1 and U_2 as $\mathfrak{g}_1$-modules with respect to the adjoint action $g \times u = gu - ug$, where $g \in \mathfrak{g}_1$, $u \in U_2$. Since the universal enveloping algebra of an arbitrary finite-dimensional Lie algebra is the sum of finite-dimensional modules (with respect to the adjoint action), and any finite-dimensional module over a semisimple Lie algebra of characteristic zero is decomposed into a direct sum of irreducible submodules (see [Ser]), we see that U_1 is decomposed into the direct sum of irreducible $\mathfrak{g}_1$-submodules. Obviously, I_1 is a $\mathfrak{g}_1$-submodule and, hence, I_1 is a direct sum of irreducible finite-dimensional $\mathfrak{g}_1$-submodules. However, in this case, by the equality $K_2 \cdot I_1 = U_2$, we see that the $\mathfrak{g}_1$-module U_2 is the direct sum of irreducible modules, each of which is isomorphic to some module in the direct decomposition of the $\mathfrak{g}_1$-module I_1. Therefore, if the center of the algebra U_2 is nonzero, then, by the equality $\mathfrak{g}_2 = K_2 \otimes_{K_1} \mathfrak{g}_1$, in I_1 there exist one-dimensional (over K_1) $\mathfrak{g}_1$-submodules that must belong to the center of the algebra U_2. The proposition is proved.

PROPOSITION 13.5. *Let* $\mathfrak{g}$ *be a semisimple Lie algebra over an algebraically closed field* K *of characteristic zero and let* U *be an associative enveloping* K*-algebra of* $\mathfrak{g}$. *Denote by* K_2 *and* K_1 *some extensions of the field* K ($K_2 \supset K_1 \supset K$). *Let* $\mathfrak{g}_2 = K_2 \otimes_K \mathfrak{g}$

and $U_2 = K_2 \otimes_K U$, *let* $\mathfrak{g}_1$ *be some Lie* K_1*-subalgebra of* $\mathfrak{g}_2$, *let* U_1 *be the* K_1*-subalgebra in* U_2 *generated by* $\mathfrak{g}_1$, *and let* $\dim_{K_1} \mathfrak{g}_1 = \dim_K \mathfrak{g}$ *and* $\mathfrak{g}_2 = K_2\mathfrak{g}_1$. *Then the center of the* K_1*-algebra* U_1 *is contained in* $K_1 \otimes_K \mathfrak{Z}$, *where* $\mathfrak{Z}$ *is the center of the algebra* U.

PROOF. It follows from the equalities $\dim_{K_1} \mathfrak{g}_1 = \dim_{K_2} \mathfrak{g}_2$ ($= \dim_K \mathfrak{g}$) and $\mathfrak{g}_2 = K_2\mathfrak{g}_1$ that $\mathfrak{g}_2 \simeq K_2 \otimes_K \mathfrak{g}_1$ and, hence, the algebra $\mathfrak{g}_1$ is semisimple and any bases of the algebra $\mathfrak{g}_1$ that are dual with respect to its Killing form are dual with respect to the Killing form of the Lie algebra $\mathfrak{g}_2$. Therefore, it follows from Propositions 13.2 and 13.3 that any element of the center of the K_1-algebra U_1 is representable as a linear combination of elements $f_{l,\rho}$ with coefficients in the field K_1, where ρ_1 ranges over all K_1-representations of the Lie algebra $\mathfrak{g}_1$ in finite-dimensional spaces V_1. However, in this case, for the algebraic closure $\overline{K}_2$ of the field K_2, the product $V_2 = \overline{K}_2 \otimes_{K_1} V_1$ is a finite-dimensional $\overline{K}_2$-representation ρ_2 of the semisimple Lie algebra $\mathfrak{g}_2$ and, hence, due to the structure of finite-dimensional irreducible representations of a semisimple Lie algebra over an algebraically closed field, there exists a finite-dimensional K-representation ρ of the Lie algebra $\mathfrak{g}$ in the space V such that $\overline{V}_2 \simeq \overline{K}_2 \otimes_K V$. Proposition 13.2 implies that the elements generating the center are independent of the choice of dual bases; therefore, in the algebra $\overline{K}_2 \otimes_{K_2} U_2$ the following equalities are satisfied:

$$f_{l,\rho_1} = f_{l,\rho_2} = f_{l,\rho}.$$

The elements on the right-hand side of this equality belong to the center $\mathfrak{Z}$ of the algebra U. The proposition is proved.

§14. Existence of central polynomials in simple enveloping algebras

Denote by $\mathfrak{K}_m$ the ideal of weak Capelli identities of order m (see §12).

In this section we shall prove the following

THEOREM 14.1. *Let* $\mathfrak{g}$ *be a finite-dimensional semisimple Lie algebra over an algebraically closed field* K *of characteristic zero, let* U *be its simple associative enveloping* K*-algebra with nonzero center, and let* $\dim_K \mathfrak{g} = m$. *Let* $(F, \mathfrak{L})$ *be a free pair of the variety* $\operatorname{var}(U, \mathfrak{g})$ *with an infinite number of free generators* $x_1, x_2, \dots$. *Then for any nonzero ideal* I *of the* K*-algebra* F, *for the pair* $(U, \mathfrak{g})$ *there exists a central polynomial belonging to* $I\mathfrak{K}_m$.

The proof of this theorem is divided into several steps. Denote by $\mathcal{F}_2$ the rational function field generated by the elements $x_i^{(j)}$ ($j = 1, \dots, m; i = 1, 2, \dots$) that are algebraically independent over K and denote by $\mathcal{F}$ the K-subalgebra with unit generated by the same elements in $\mathcal{F}_2$. Let $U_2 = \mathcal{F}_2 \otimes_K U$ and $\mathfrak{g}_2 = \mathcal{F}_2 \otimes_K \mathfrak{g}$. We choose a basis $e_1, \dots, e_m$ in the Lie algebra $\mathfrak{g}$ and generate in U_2 and $\mathfrak{g}_2$, by elements $x_i = \sum_{j=1}^m x_i^{(j)} e_j$, an associative and a Lie subalgebra and denote them by F and $\mathfrak{L}$, respectively.

LEMMA 14.1. *For an arbitrary associative polynomial* $f(t_1, \dots, t_l)$, *the equality* $f(x_1, \dots, x_l) = 0$ *in the algebra* U *implies* $f(v_1, \dots, v_l) = 0$ *in the algebra* U *for any elements* $v_1, \dots, v_l \in \mathfrak{g}$ *and, hence, the pair* $(F, \mathfrak{L})$ *is a free pair of the variety* $\operatorname{var}(U, \mathfrak{g})$ *with free generators* $x_1, x_2, \dots$.

PROOF. An arbitrary element of the Lie algebra $\mathfrak{g}$ can be represented as a K-linear combination of the basis elements $e_1, \dots, e_m$. Therefore, the assertion of the lemma is obtained by specialization of the variables $x_i^{(j)}$ in the field K. The lemma is proved.

In the sequel we assume that the free pair $(F, \mathfrak{L})$ belongs to the pair $(U_2, \mathfrak{g}_2)$.

Since the Lie algebra $\mathfrak{g}_2$ is finite-dimensional over the field $\mathcal{F}_2$, we see that for any elements $u, v \in \mathfrak{g}_2$ the function $\operatorname{tr}(\operatorname{ad} u \operatorname{ad} v)$ with values in the field $\mathcal{F}_2$ can be defined.

LEMMA 14.2. *For any elements $u, v \in \mathfrak{L}$ the element $\operatorname{tr}(\operatorname{ad} u \operatorname{ad} v)$ belongs to the polynomial algebra $\mathcal{F}$ with generators $x_i^{(j)}$ $(j = 1, \dots, m;\ i = 1, 2, \dots)$ and for any polynomial $d(t_1, \dots, t_m, t_{m+1}, \dots, t_{m+l})$ that is multilinear and skew-symmetric with respect to $t_1, \dots, t_m$ the following relation is valid in the associative algebra U_2:*

$$\operatorname{tr}(\operatorname{ad} u \operatorname{ad} v) d(x_1, \dots, x_m, y_1, \dots, y_l) = \sum_{i=1}^{m} d\big|_{x_i = [x_i, u, v]}. \tag{14.1}$$

PROOF. The first assertion of the lemma follows from the observation that the basis $e_1, \dots, e_m$ of the Lie K-algebra $\mathfrak{g}$ is a basis of the Lie $\mathcal{F}_2$-algebra $\mathfrak{g}_2$.

In order to prove the equality (14.1), we note that the elements $x_1, \dots, x_m$ form a basis of the Lie algebra $\mathfrak{g}_2$ over the field $\mathcal{F}_2$. This follows from the fact that the variables $x_i^{(j)}$ $(i, j = 1, \dots, m)$ are algebraically independent over the field K and the determinant of the matrix formed by these elements is nonzero. Therefore, for any elements $u, v \in \mathfrak{g}_2$ there exist elements $c_i^{(k)}$ of the field $\mathcal{F}_2$ such that the equality $\sum_{k=1}^{m} c_i^{(k)} x_k = [x_i, u, v]$ is satisfied in $\mathfrak{g}_2$. However, then we have the relation

$$\sum_{i=1}^{m} d\big|_{x_i = [x_i, u, v]} = \sum_{i=1}^{m} \sum_{k=1}^{m} c_i^{(k)} d\big|_{x_i = x_k},$$

because the polynomial d is multilinear with respect to $x_1, \dots, x_m$, and, because the polynomial d is skew-symmetric with respect to $x_1, \dots, x_m$, only one summand in the inner sum can be nonzero for $i = k$. Hence, the right-hand side of this equality is equal to

$$\sum_{i=1}^{m} c_i^{(i)} d = \operatorname{tr}(\operatorname{ad} u \operatorname{ad} v) d.$$

The lemma is proved.

Denote by $\mathcal{F}_0$ the K-subalgebra in $\mathcal{F}$ generated by the elements $\operatorname{tr}(\operatorname{ad} u \operatorname{ad} v)$, where u and v range over the Lie K-algebra $\mathfrak{L}$, denote by $\mathcal{F}_1$ its field of quotients belonging to $\mathcal{F}_2$, and denote by $\mathfrak{g}_1$ the Lie $\mathcal{F}_1$-subalgebra $\mathcal{F}_1 \cdot \mathfrak{L}$ in the Lie algebra $\mathfrak{g}_2$.

COROLLARY. $\mathcal{F}_0 \mathfrak{K}_m \subset \mathfrak{K}_m$.

LEMMA 14.3. *The Lie $\mathcal{F}_1$-algebra $\mathfrak{g}_1$ is semisimple and its dimension over $\mathcal{F}_1$ equals $m = \dim_K \mathfrak{g}$. Moreover, $\mathcal{F}_2 \mathfrak{g}_1 = \mathfrak{g}_2$.*

PROOF. Let $x_1, \dots, x_m, y_1, \dots, y_m$ denote different free generators of the pair $(F, \mathfrak{L})$, and let $\mathfrak{k}$ be the square matrix of order m for which on the intersection of the ith row with jth column the element $\mathrm{tr}(\mathrm{ad}\, x_i\, \mathrm{ad}\, y_j) \in \mathcal{F}_0 \subset \mathcal{F}_1$ is posed. Denote by $c(x_1, \dots, x_m, y_1, \dots, y_m)$ the determinant of the matrix $\mathfrak{k}$. Obviously, the expression c is multilinear and skew-symmetric both with respect to the variables $x_1, \dots, x_m$ and to the variables $y_1, \dots, y_m$. Therefore, for any element $v \in \mathfrak{L}$ the expression $cv - \sum_{i=1}^{m} c\big|_{x_i=v} x_i$ is skew-symmetric and multilinear with respect to $x_1, \dots, x_m, v$ and, hence, it is equal to zero in the Lie algebra $\mathfrak{g}_2$, because $\dim_K \mathfrak{g} = m$. However, in this case for any $v \in \mathfrak{g}_1$ we have the equality

$$c \cdot v = \sum_{i=1}^{m} c\big|_{x_i=v} x_i$$

in the Lie algebra $\mathfrak{g}_2$.

The Lie algebra $\mathfrak{g}_2$ is semisimple and, as we saw in the proof of Lemma 14.2, the elements $x_1, \dots, x_m$ form a basis of it over the field $\mathcal{F}_2$. Therefore, by the Cartan criterion for semisimplicity of a Lie algebra (see [Dix]), the element

$$c(x_1, \dots, x_m, x_1, \dots, x_m)$$

is nonzero. However, then $c(x_1, \dots, x_m, y_1, \dots, y_m)$ is also nonzero, and we can divide both sides of equality (14.2) by it. This proves that any element of the $\mathcal{F}_1$-algebra $\mathfrak{g}_1$ is representable in the form of a linear combination of the elements $x_1, \dots, x_m$ with coefficients in the field $\mathcal{F}_1$, i.e., we have $\dim_{\mathcal{F}_1} \mathfrak{g}_1 \le m$. Since $x_1, \dots, x_m$ are linearly independent in $\mathfrak{g}_2$ over $\mathcal{F}_2 \supset \mathcal{F}_1$, we see that

$$\dim_{\mathcal{F}_1} \mathfrak{g}_1 = m \quad \text{and} \quad \mathcal{F}_2\mathfrak{g}_1 = \mathfrak{g}_2.$$

The semisimplicity of the Lie algebra $\mathfrak{g}_1$ is now obvious. The lemma is proved.

Denote by U_1 the associative $\mathcal{F}_1$-subalgebra $\mathcal{F}_1 F$ in the algebra U_2 that is $\mathcal{F}_1$-enveloping for the Lie $\mathcal{F}_1$-algebra $\mathfrak{g}_1$ in U_2.

COROLLARY. *The center of the algebra U_1 coincides with $\mathcal{F}_1$.*

The proof immediately follows from Propositions 13.5 and 13.3, because the center of the algebra U equals K by Lemma 3.5.

LEMMA 14.4. *Let I be an arbitrary nonzero ideal of the K-algebra F. Then among the elements of $\mathcal{F}_1 I$ there exists a nonzero central element of the algebra U_2.*

PROOF. Obviously, $I_1 = \mathcal{F}_1 I$ is an ideal in U_1. It follows from Lemma 14.3 that $\mathfrak{g}_1$ is semisimple in $\mathfrak{g}_2 = \mathcal{F}_2 \otimes_{\mathcal{F}_1} \mathfrak{g}_1$. Since the ground field K is algebraically closed, it follows from Lemma 3.5 that the algebra U is centrally prime. Hence, the $\mathcal{F}_2$-algebra U_2 is prime. Since $\mathcal{F}_2 I_1$ is a nonzero ideal in U_2, we see that $U_2 = \mathcal{F}_2 I_1$, and for U_2, $\mathfrak{g}_2$, U_1, and $\mathfrak{g}_1$ the assumptions of Proposition 13.4 are satisfied. Now the assertion of the lemma follows from this proposition.

LEMMA 14.5. *For any nonzero ideal I of the K-algebra F the product of ideals $I\mathfrak{K}_m$ is nonzero.*

PROOF. First we show that the ideal of weak Capelli identities $\mathfrak{K}_m$ is nonzero. Suppose that $\mathfrak{K}_m = 0$ in the algebra F. Then all the Capelli identities of order

$m = \dim_K \mathfrak{g}$ would be valid in the pair $(U, \mathfrak{g})$. This means that $\operatorname{rank}(U, \mathfrak{g}) \leq m$ (see §4). Since the algebra U is simple and contains the unit, we see that it possesses a faithful irreducible representation ρ in some linear space V. However, in this case ρ is a faithful irreducible representation of the Lie algebra $\mathfrak{g}$ and, by Remark 3.3 and the algebraic closedness of the ground field K, the algebra U is centrally prime. Hence, by the rank theorem (see Theorem 4.1), we have $\dim_K \mathfrak{g} = \operatorname{rank}(U, \mathfrak{g}) - 1$, which contradicts the fact that the Lie algebra $\mathfrak{g}$ is m-dimensional. The contradiction $m < m$ thus obtained proves that $\mathfrak{K}_m \neq 0$.

As it was shown in the proof of Lemma 14.4, U_2 is a simple $\mathcal{F}_2$-algebra. Obviously, $\mathcal{F}_2 I$ and $\mathcal{F}_2 \mathfrak{K}_m$ are nonzero ideals in U_2. Therefore, $\mathcal{F}_2 I \mathfrak{K}_m \neq 0$ and, hence, $I \mathfrak{K}_m \neq 0$. The lemma is proved.

PROOF OF THEOREM 14.1. By Lemma 14.5, $I\mathfrak{K}_m \neq 0$ for any ideal I in F. It follows from Lemma 14.4 that among the elements of $\mathcal{F}_1(I\mathfrak{K}_m)$ there exists a nonzero central element of the algebra U_2. However, in this case Corollary of Lemma 14.2 implies that this central element is representable in the form $f^{-1}v$ for some $v \in I\mathfrak{K}_m$ and $f \in \mathcal{F}_0$. Therefore, $v \neq 0$ and v is a central polynomial for the pair $(U, \mathfrak{g})$. The theorem is proved.

REMARK 14.1. As we can see from Corollary of Lemma 14.3, a central polynomial $c \in I\mathfrak{K}_m$ can be represented in the algebra U_2 as a fraction f_1/f_2, where $f_1, f_2 \in \mathcal{F}_0$. However, in this case, by Corollary of Lemma 14.2, there exists a nonzero polynomial $g \in I\mathfrak{K}_m$ that is equal in the algebra U_2 to the elements $f_2 c$ and f_1. If I is a nonzero ideal of weak identities, then, linearizing the identity $g = f_1$, we find that in the ideal $I\mathfrak{K}_m$ there exists a multilinear central polynomial, that is equal in U_2 to a multilinear expression in the algebra $\mathcal{F}_0$.

§15. Varieties of algebras with three supports var$(K, \mathfrak{g}, U)$

In §14, in the course of the proof of Theorem 14.1, we implicitly extended the signature of operations by introducing the new binary operation $\operatorname{tr}(\operatorname{ad} u \operatorname{ad} v)$ defined on the Lie algebra $\mathfrak{g}$ and taking its values in the ground field K. The main relation (14.1), which played a crucial role in the proof of Theorem 14.1, was in fact an identity for another algebra. Therefore, before proceeding further, it is useful to introduce this new algebra explicitly.

Let $\mathfrak{g}$ be a semisimple Lie algebra over an algebraically closed field K of characteristic zero and let ρ be an arbitrary representation of $\mathfrak{g}$ (possibly infinite-dimensional) in a linear space V. Denote by U the associative subalgebra in $\operatorname{End}_K V$ generated by the elements of $\rho(\mathfrak{g})$ and denote by $\mathfrak{k}(x, y)$ the bilinear Killing form of the algebra $\mathfrak{g}$.

Then the three K-linear spaces K, $\mathfrak{g}$, and U can be considered as an algebra with three supports, where, besides the operations which endow K, $\mathfrak{g}$, and U with the structures of commutative K-algebra, Lie K-algebra $\mathfrak{g}$ and associative K-algebra U, respectively, there are multilinear operations

$$\rho\colon \mathfrak{g} \to U, \qquad \mathfrak{k}\colon \mathfrak{g} \otimes_K \mathfrak{g} \to K, \qquad K \otimes_K \mathfrak{g} \to \mathfrak{g}, \qquad K \otimes_K U = U,$$

satisfying the relations

$$\begin{gathered} \rho([x, y]) = \rho(x)\rho(y) - \rho(y)\rho(x), \\ \mathfrak{k}(x, y) = \mathfrak{k}(y, x), \qquad \mathfrak{k}([x, y], z) = \mathfrak{k}(x, [y, z]). \end{gathered} \tag{15.1}$$

If we consider the class of all algebras with three supports $(\mathcal{E}, L, A)$, where $\mathcal{E}$ is a commutative K-algebra, L is a Lie $\mathcal{E}$-algebra, and A is an associative $\mathcal{E}$-algebra, with additional $\mathcal{E}$-multilinear operations ρ and $\mathfrak{k}$ satisfying the identities (15.1), then it is obvious what the notions "homomorphism", "subalgebra", "Cartesian product", and "ideal" (see §49) do mean for algebras with three supports. Therefore, we can discuss varieties of algebras with three supports and the description of free algebras with three supports of these varieties. It is worthy to notice, however, that there are free generators of three types: $X_1 \subset \mathcal{E}$, $X_2 \subset L$, $X_3 \subset A$ in relatively free algebras with three supports, and it is possible that in some cases in a relatively free algebra with three supports some of the X_i $(i = 1, 2, 3)$ may be empty. In this chapter, relatively free algebras with three supports $(\mathcal{F}_0, \mathfrak{L}_1, F_1)$ of the variety $\operatorname{var}(K, \mathfrak{g}, U)$ will have free generators of the form $(\varnothing, X, \varnothing)$, i.e., free generators are taken in the Lie algebra $\mathfrak{L}_1$ and there are no free generators in the commutative algebra and in the associative algebra. In this chapter algebras with three supports mainly play an auxiliary role, and we introduce them here because the technique of several supports has been well developed (see §49, [BL, Higg1]) and it is useful for the subsequent discussion and for proving our results. A substantial role will be played by these algebras in Chapter 5 of the book in the study of identities of representations of the simple three-dimensional Lie algebra.

Let $\dim_K \mathfrak{g} = m$. Denote by the letter $\mathcal{F}$ the K-algebra of polynomials with unit and with free generators $x_i^{(j)}$, $j = 1, \dots, m$, $i = 1, 2, \dots$. Then the triple $(\mathcal{F}, \widetilde{\mathfrak{g}}, \widetilde{U})$, where $\widetilde{\mathfrak{g}} = \mathcal{F} \otimes_K \mathfrak{g}$, $\widetilde{U} = \mathcal{F} \otimes_K U$ with operations ρ and $\mathfrak{k}$ that are extended by $\mathcal{F}$-linearity, is an algebra with three supports and belongs to the variety $\operatorname{var}(K, \mathfrak{g}, U)$ generated by the algebra with three supports $(K, \mathfrak{g}, U)$.

Let X be the set of elements $x_i = \sum_{j=1}^{m} x_i^{(j)} e_j$ $(i = 1, 2, \dots)$, where $e_1, \dots, e_m$ is a basis of the Lie algebra $\mathfrak{g}$. As in §14, the elements x_i generate in $\widetilde{\mathfrak{g}}$ a free Lie K-algebra (we denote it by $\mathfrak{L}$) in the variety $\operatorname{var} \mathfrak{g}$, and the elements $\rho(x_i)$ generate in $\widetilde{U}$ a K-subalgebra F.

Denote by $\mathcal{F}_0$ the commutative K-subalgebra in $\mathcal{F}$ generated by the elements $\mathfrak{k}(u, v)$, where u and v range over all the Lie algebra $\mathfrak{L}$. Obviously, the subalgebra $\mathcal{F}_0$ depends on the choice of the basis $e_1, \dots, e_m$, but it does not depend on the choice of the enveloping algebra U. However, for any choice of the basis, all these subalgebras $\mathcal{F}_0$ are isomorphic.

Let $F_1 = \mathcal{F}_0 F + F$ and $\mathfrak{L}_1 = \mathcal{F}_0 \mathfrak{L} + \mathfrak{L}$. As in §14, the following assertion is valid.

PROPOSITION 15.1. *The pair* $(F, \mathfrak{L})$ *is free in the variety of pairs* $\operatorname{var}(U, \mathfrak{g})$. *The algebra with three supports* $(\mathcal{F}_0, \mathfrak{L}_1, F_1)$ *is relatively free in the variety of algebras with three supports* $\operatorname{var}(K, \mathfrak{g}, U)$ *and possesses free generators* $\{\varnothing, X, \varnothing\}$.

The proof is obvious.

In the sequel we assume that the representation ρ is faithful and that the algebra U has a nontrivial center; therefore, we shall identify $\mathfrak{g}$ with the Lie subalgebra $\rho(\mathfrak{g})$ in the algebra $U^{(-)}$ and we assume that the field K is contained in the algebra U. Moreover, we shall omit ρ and $\mathfrak{k}$ in the notation of elements of the algebras $\mathcal{F}_0$, $\mathfrak{L}_1$, and F_1, and we shall write $(u_1, v_1) \cdots (u_k, v_k) w_1 \cdots w_l$ instead of $\mathfrak{k}(u_1, v_1) \cdots \mathfrak{k}(u_k, v_k) \rho(w_1) \cdots \rho(w_l)$, where $u_i, v_j, w_t \in \mathfrak{L}$. The elements of the algebra F_1 will be called *polynomials*.

§16. An auxiliary algebra with three supports $(\overline{\mathcal{F}}_0, \overline{\mathfrak{L}}_1, \overline{F}_1)$ and the extension of an $\boldsymbol{\alpha}$-function α to the space of generalized 2-elements

In this section we extend the results of §9 to a more general situation.

LEMMA 16.1. *Let a multilinear mapping $f(t_1, \dots, t_l)$ taking values in a linear space V be given on a semisimple Lie algebra $\mathfrak{g}$ of dimension m. Then for any elements $g_3, \dots, g_l \in \mathfrak{g}$ the element $\sum_{j=1}^m f(e_j, e'_j, g_3, \dots, g_l)$ in V does not depend on the choice of dual bases $e_1, \dots, e_m; e'_1, \dots, e'_m$ of the Lie algebra $\mathfrak{g}$ with respect to the Killing form. Moreover, for any element $h \in \mathfrak{g}$ we have the equality*

$$
\sum_{j=1}^{m} f([e_j, h], e'_j, g_3, \dots, g_l) = \sum_{j=1}^{m} f(e_j, [h, e'_j], g_3, \dots, g_l) \tag{16.1}
$$

in the space V.

PROOF. Let $h_i = \sum_{k=1}^m c_i^{(k)} e_k$ $(i = 1, 2, \dots, m)$ be another basis of the Lie algebra $\mathfrak{g}$. Then, as is is well known, $e'_i = \sum_{k=1}^m c_k^{(i)} h_k$. Therefore, in V we have the equalities

$$
\sum_{j=1}^{m} f(h_j, h'_j, g_3, \dots, g_l) = \sum_{j=1}^{m} \sum_{k=1}^{m} c_j^{(k)} f(e_k, h'_j, g_3, \dots, g_l) = \sum_{k=1}^{m} f(e_k, e'_k, g_3, \dots, g_l),
$$

and this proves the first part of the lemma.

By the invariance of the Killing form we have

$$
[e_i, h] = \sum_{k=1}^{m} c_i^{(k)} e_k, \qquad [e'_i, h] = -\sum_{k=1}^{m} c_k^{(i)} e'_k \quad (i = 1, 2, \dots, m).
$$

Therefore,

$$
\begin{aligned}
\sum_{i=1}^{m} f([e_i, h], e'_i, g_3, \dots, g_l) &= \sum_{i=1}^{m} \sum_{k=1}^{m} c_i^{(k)} f(e_k, e'_i, g_3, \dots, g_l) \\
&= -\sum_{k=1}^{m} f(e_k, [e'_k, h], g_3, \dots, g_l).
\end{aligned}
$$

The lemma is proved.

Denote by $\overline{\mathcal{F}}$ the commutative K-algebra with unit and with the generators $\overline{x}_i^{(j)}$ $(j = 1, \dots, m; i = 1, 2, \dots)$ and the defining relations

$$
\overline{x}_i^{(j)} \overline{x}_i^{(k)} = \delta_{jk} \left(\overline{x}_i^{(1)}\right)^2, \qquad \overline{x}_i^{(j)} \overline{x}_i^{(k)} \overline{x}_i^{(l)} = 0. \tag{16.2}
$$

Let $\overline{\mathfrak{g}} = \overline{\mathcal{F}} \otimes_K \mathfrak{g}$ and $\overline{U} = \overline{\mathcal{F}} \otimes_K U$, and let $\overline{x}_i = \sum_{j=1}^m \overline{x}_i^{(j)} e_j$ and $\overline{x}'_i = \sum_{j=1}^m \overline{x}_i^{(j)} e'_j$.

REMARK 16.1. Let f, $\mathfrak{g}$, and V denote the same objects as in Lemma 16.1. We extend the mapping f by $\overline{\mathcal{F}}$-linearity to the mapping

$$
f\colon \underbrace{\overline{\mathfrak{g}} \otimes \overline{\mathfrak{g}} \otimes \cdots \otimes \overline{\mathfrak{g}}}_{l \text{ times}} \to \overline{V} = \overline{\mathcal{F}} \otimes_K V.
$$

Then, by the defining relations (16.2), the first assertion of Lemma 16.1 is equivalent the element $f(\overline{x}_i, \overline{x}'_i, g_3, \dots, g_l)$ of the space $\overline{V}$ being independent of the choice of dual bases $\{e_i\}$, $\{e'_i\}$ $(i = 1, \dots, m)$, and the equality (16.2) gets the form

$$f([\overline{x}_i, h], \overline{x}'_i, g_3, \dots, g_l) = f(\overline{x}_i, [h, \overline{x}'_i], g_3, \dots, g_l). \tag{16.3}$$

Now let the ground field K be algebraically closed and let $e_1, \dots, e_m$ be an orthonormal basis of the semisimple Lie algebra $\mathfrak{g}$ with respect to the Killing form. Let $(\mathcal{F}_0, \mathfrak{L}_1, F_1)$, $(F, \mathfrak{L})$, and $(\mathcal{F}, \widetilde{\mathfrak{g}}, \widetilde{U})$ denote the same objects as in §15. Since $\mathcal{F}$ is a free commutative K-algebra, we see that the mapping $\varphi\colon x_i^{(j)} \to \overline{x}_i^{(j)}$ $(j = 1, \dots, m;$ $j = 1, 2, \dots)$ is uniquely extendable to an epimorphism of the algebra $\mathcal{F}$ onto the algebra $\overline{\mathcal{F}}$. In its turn, this epimorphism can be uniquely extended to a K-homomorphism of the algebra $\widetilde{U}$ onto $\overline{U}$ acting identically on the subalgebra U. We denote this homomorphism also by the letter φ. Denote the images of the algebras $\mathcal{F}_0$, $\mathfrak{L}_1$, F_1, F, and $\mathfrak{L}$ under the homomorphism φ by $\overline{\mathcal{F}}_0$, $\overline{\mathfrak{L}}_1$, $\overline{F}_1$, $\overline{F}$, and $\overline{\mathfrak{L}}$, respectively, and denote the elements $\varphi(x_i)$ by $\overline{x}_i$. Obviously, the algebras with three supports $(\overline{\mathcal{F}}_0, \overline{\mathfrak{L}}_1, \overline{F}_1)$ and $(\overline{\mathcal{F}}, \overline{\mathfrak{g}}, \overline{U})$, where the operations ρ and $\mathfrak{k}$ are extended from the algebra $(K, \mathfrak{g}, U)$ by $\overline{\mathcal{F}}$-linearity, belong to the variety $\operatorname{var}(K, \mathfrak{g}, V)$, and the pairs $(\overline{F}, \overline{\mathfrak{L}})$ and $(\overline{U}, \overline{\mathfrak{g}})$ belong to the variety of pairs $\operatorname{var}(U, \mathfrak{g})$. Moreover, the elements $\overline{x}_1, \overline{x}_2, \dots \in \overline{\mathfrak{L}} \subset \overline{\mathfrak{L}}_1$ are the generators both of the algebra with three supports $(\overline{\mathcal{F}}_0, \overline{\mathfrak{L}}_1, \overline{F}_1)$ and of the pair $(\overline{F}, \overline{\mathfrak{L}})$.

Let λ and μ be any finite disjoint subsets of the set of generators $\overline{x}_1, \overline{x}_2, \dots$ and let ${}_\mu T_\lambda$ be the linear space of the algebra $\overline{F}_1$ consisting of multihomogeneous polynomials that depend essentially only on the generators $\overline{x}_i \in \lambda \cup \mu$, therewith the degree of each monomial of this polynomial with respect to a variable $\overline{x}_i$ from μ equals two and the degree of this monomial with respect to each variable from λ equals one. If λ or μ is empty, then we shall use the notation ${}_\mu T$ or T_λ, respectively. It follows from the definition of the algebra $\overline{F}_1$ that $\overline{F}_1 = \sum_{\lambda,\mu} {}_\mu T_\lambda$, where the direct sum is taken over all disjoint pairs of subsets μ and λ, $\mu \cup \lambda \neq \varnothing$.

The next lemma generalizes Lemma 9.1 of Chapter 2.

LEMMA 16.2. *Let $u(x_{i_1}, \dots, x_{i_k})$ be a multihomogeneous polynomial in the algebra F_1 of degree two with respect to each variable x_{i_j} $(j = 1, \dots, k)$. Then in the algebra $\overline{F}_1$ we have $\varphi(u) \in {}_\mu T$, where $\mu = \{\overline{x}_i^{(1)}, \dots, \overline{x}_i^{(k)}\}$, and for this element we have the equality*

$$u(\overline{x}_{i_1}, \dots, \overline{x}_{i_k}) = \left(\overline{x}_{i_1}^{(1)}\right)^2 \cdots \left(\overline{x}_{i_k}^{(1)}\right)^2 u', \tag{16.4}$$

where u' belongs to the center of the algebra U. Moreover, the element u' does not depend on the choice of the orthonormal basis $e_1, \dots, e_m$ of the Lie algebra $\mathfrak{g}$.

PROOF. An arbitrary element of the algebra F_1 can be represented as a linear combination of elements of the form

$$(u_1, v_1) \dots (u_p, v_p) u_0, \tag{16.5}$$

where u_0 is a nonempty word in generators x_i of the Lie algebra $\mathfrak{L}_1$ and u_i and v_j are some commutators in variables x_i of the Lie algebra $\mathfrak{L}$. Therefore, it is sufficient to prove the assertion of the lemma for multihomogeneous polynomials of the form

(16.5) only. The case $p = 0$ follows from Lemma 9.1. The general case is discussed similarly with the use of the equality

$$([u_i, h], v_i) + (u_i, [v_i, h]) = \operatorname{tr}([\operatorname{ad} u_i \operatorname{ad} v_i, \operatorname{ad} h]) = 0$$

expressing the invariance of the Killing form. Indeed, applying this equality and Remark 16.1 for any element $h \in \mathfrak{L}_1$, we have

$$\begin{aligned}[\overline{u}, h] &= (\overline{u}_1, \overline{v}_1) \cdots (\overline{u}_p, \overline{v}_p)[u_0, h] \\ &= \sum_{j=1}^{p} \operatorname{tr}(\operatorname{ad} \overline{u}_1 \cdot \operatorname{ad} \overline{v}_1) \cdots \operatorname{tr}([\operatorname{ad} \overline{u}_j \cdot \operatorname{ad} \overline{v}_j, \operatorname{ad} h]) \cdots \operatorname{tr}(\operatorname{ad} \overline{u}_p \cdot \operatorname{ad} \overline{v}_p) u_0 = 0,\end{aligned}$$

where, for any $w \in F_1$, we have denoted by $\overline{w}$ the element $\varphi(w)$ of the algebra $\overline{F}_1$. Thus, the element $\varphi(u)$ commutes with any element of the Lie algebra $\mathfrak{L}_1$ and, in particular, with the variable $\overline{x}_j \notin \mu$. The defining relations (16.2) of the algebra $\overline{\mathcal{F}}$ show that in this case the element $\overline{u}$ can be represented in the form (16.4) and it necessarily commutes in $\overline{U}$ with any element $h \in \mathfrak{g}$.

The independence of the element u' of the choice of orthonormal basis in the Lie algebra $\mathfrak{g}$ follows from Remark 16.1 and Lemma 16.1. The lemma is proved.

REMARK 16.2. For any permutation s of the set $\{1, \dots, l\}$ and for any finite-dimensional K-representation ρ of a semisimple Lie algebra $\mathfrak{g}$, we introduce a K-multilinear mapping on the Lie K-algebra $\mathfrak{g}$, taking its values in the universal enveloping algebra $U(\mathfrak{g})$, by the formula

$$f(t_1, \dots, t_l, y_1, \dots, y_l) \rightleftharpoons \operatorname{tr}(\rho(t_{s(1)}, \dots, t_{s(l)}))\, y_1 \cdots y_l.$$

Lemma 16.1 and the discussion used in the proof of Lemma 16.2 show that

$$f\big|_{t_1 = \overline{x}_1, y_1 = \overline{x}_1, \dots, t_l = \overline{x}_l, y_l = \overline{x}_l} = (\overline{x}_1^{(1)})^2 \cdots (\overline{x}_l^{(1)})^2 f_{s,\rho},$$

where the element $f_{s,\rho}$ belongs to the center of the algebra $U(\mathfrak{g})$ and is independent of the choice of the dual bases in $\mathfrak{g}$. Since the element $\sum_{s \in S_l} f_{s,\rho}$ coincides with the element $f_{l,\rho}$ of Proposition 13.2, Lemma 16.1 implies that the element $f_{l,\rho}$ is independent of the choice of basis.

Let ρ be a faithful irreducible representation of a semisimple Lie algebra $\mathfrak{g}$ over an algebraically closed field K in a linear space V such that the center of the associative algebra U generated in $\operatorname{End}_K V$ by the Lie algebra $\mathfrak{g}$ is nonzero. Then, by Lemma 3.5, the center of the algebra U coincides with the field $K \cdot \mathbf{1}$. Hence, for any multihomogeneous polynomial $u \in {}_{\mu}T$, equality (16.4) determines the element u' of the ground field K uniquely, whereas the mapping $u \mapsto u'$ is linear. This permits us to define a function $\widetilde{\alpha}$ for any multihomogeneous element u of the form (16.5) in the letters $\overline{x}_i$ such that any letter enters exactly twice in u, setting by definition $\widetilde{\alpha}(u) \rightleftharpoons u'$. We shall call formal linear combinations of these multihomogeneous elements *generalized 2-elements* and denote by $\widetilde{B}$ the space of generalized 2-elements. Obviously, the space B of 2-elements is contained in $\widetilde{B}$ and the K-linear function $\widetilde{\alpha}\colon \widetilde{B} \to K$ is an extension of the $\boldsymbol{\alpha}$-function α constructed in §9 for the pair $(U, \mathfrak{g})$.

The following lemma extends the assertion of the Corollary of Lemma 9.1 to the function $\widetilde{\alpha}$.

LEMMA 16.3. *Any irreducible representation* $\rho\colon \mathfrak{g} \to \operatorname{End}_K V$ *of a semisimple Lie algebra* $\mathfrak{g}$ *over an algebraically closed field* K *of characteristic zero determines, by formula* (16.4), *a linear function* $\widetilde{\alpha}\colon \widetilde{B} \to K$ *with the following properties:*

1. *For any generalized* 2*-element* $u([\overline{x}_i, \overline{x}_j], \overline{x}_i, \dots)$
$$\widetilde{\alpha}(u([\overline{x}_i, \overline{x}_j], \overline{x}_i, \dots)) = \widetilde{\alpha}(u(\overline{x}_i, [\overline{x}_j, \overline{x}_i], \dots)).$$
2. *For any permutation* σ *of the set* $\{1, 2, \dots\}$ *and for any generalized* 2*-element* u
$$\widetilde{\alpha}(u(\overline{x}_{i_1}, \dots, \overline{x}_{i_l})) = \widetilde{\alpha}(u(\overline{x}_{\sigma(i_1)}, \dots, \overline{x}_{\sigma(i_l)})).$$
3. *For any generalized* 2*-element* $w = uv$, *where* u *and* v *are generalized* 2*-elements,*
$$\widetilde{\alpha}(w) = \widetilde{\alpha}(u)\widetilde{\alpha}(v).$$

The proof of Properties 1–3 follows immediately from Lemma 16.2 and Remark 16.1.

REMARK 16.3. The defining relations (16.2), Lemma 16.1, and Remark 16.1 show that for any multihomogeneous polynomial $u(x_{i_1}, \dots, x_{i_k})$ in the algebra $\mathcal{F}_0$ of degree two with respect to each variable we have the equality

$$u(\overline{x}_{i_1}, \dots, \overline{x}_{i_l}) = (\overline{x}_{i_1}^{(1)})^2 \cdots (\overline{x}_{i_k}^{(1)})^2 \cdot u' \tag{16.6}$$

in $\overline{\mathcal{F}}$, where $u' \in K$ and u' does not depend on the choice of orthonormal basis in the Lie algebra $\mathfrak{g}$ with respect to the Killing form. Moreover, since the elements of the subalgebra $\mathcal{F}_0$ of the algebra $\mathcal{F}$ depend on the Killing form only, for a fixed basis $e_1, \dots, e_m$ of the Lie algebra $\mathfrak{g}$, we see that these elements of the algebra $\mathcal{F}_0$ in $\mathcal{F}$ do not depend on the choice of faithful representation of the Lie algebra $\mathfrak{g}$. Therefore, the scalar coefficient u' in formula (16.6) does not depend either on the representation of the Lie algebra $\mathfrak{g}$ or on the choice of orthonormal basis in $\mathfrak{g}$. We shall denote the element u' also by $\widetilde{\alpha}(u)$.

Now we are technically completely prepared for the proof of the following result.

THEOREM 16.1. *Let* $\mathfrak{g}$ *be a semisimple Lie algebra over an algebraically closed field* K *of characteristic zero and let* U *be its simple enveloping algebra with nonzero center. Let the* **α***-function* $\alpha\colon B \to K$ *be constructed by the pair* $(U, \mathfrak{g})$ (*see* §9 *of Chapter* 2, *formula* (9.2)). *Then the ideal of identities of the pair* $(U, \mathfrak{g})$ *coincides with the ideal of weak identities* $\mathfrak{V}_\alpha$.

If $(H, \mathfrak{M})$ denotes the absolutely free associative-Lie pair with an infinite number of free generators $t_1, t_2, \dots$, λ and μ denote finite disjoint subsets of generators of the set $\{t_1, t_2, \dots\}$, and ${}_\mu H_\lambda$ is the linear space in H consisting of multihomogeneous polynomials of degree 1 with respect to each variable from λ, of degree 2 with respect to each variable from μ, and of degree 0 with respect to any other variable, we recall that then the ideal $\mathfrak{V}_\alpha$ is characterized by the following property: for any finite subset λ of the set $\{t_1, t_2, \dots\}$ we have

$$\mathfrak{V}_\alpha \cap H_\lambda = \operatorname{Ann}_{H_\lambda} \mathfrak{b} = \{f \in H_\lambda \mid \mathfrak{b}(f, g) \rightleftharpoons \alpha(\varphi(fg)) = 0$$
$$\text{for any } g \in {}_\mu H_\lambda \text{ and for any subset } \mu, \mu \cap \lambda = \varnothing\},$$

where φ is the mapping of ${}_\nu H$ into B that produces the formal replacement of the variables $t_1, t_2, \dots$ by the letters $\overline{x}_1, \overline{x}_2, \dots$, respectively.

The proof of Theorem 16.1 will be carried out simultaneously with the proof of Theorem 12.1.

§17. For finite-dimensional U, the identities of the pair $(U, \mathfrak{g})$ are determined by its α-function

In this section we assume that the ground field K is algebraically closed and of characteristic zero, V is a faithful finite-dimensional $\mathfrak{g}$-module, and U is a simple associative enveloping K-algebra of the Lie algebra $\mathfrak{g}$, coinciding with $\mathrm{End}_K V$.

Let $(\widetilde{U}, \widetilde{\mathfrak{g}})$, $(\mathcal{F}_0, \mathfrak{L}_1, F_1)$, $(F, \mathfrak{L})$, $(\overline{U}, \overline{\mathfrak{g}})$, $(\overline{\mathcal{F}}_0, \overline{\mathfrak{L}}_1, \overline{F}_1)$, $(\overline{F}, \overline{\mathfrak{L}})$, and φ denote the same objects as in §§15, 16.

We choose in K a certain maximal formally real subfield R. It is well known (see [VdW]), that $K = R \oplus \mathbf{i}R$, where $\mathbf{i}^2 = -1$, and there exists a unique order on R; moreover, positive elements of R are exactly all nonzero sums of squares of elements of R. For any $\beta = a + b\mathbf{i} \in K$, where $a, b \in R$, we write $\overline{\beta} = a - b\mathbf{i}$. The automorphism of second order of the field K mapping the element $\beta \in K$ to $\overline{\beta}$ will be called the complex conjugation (by analogy with the case when K is the complex field). In the structural theory of finite-dimensional Lie algebras over an algebraically closed field of characteristic zero it is shown [Zhe] that in a semisimple Lie algebra $\mathfrak{g}$ there exists an R-subalgebra $\mathfrak{h}$ such that $\mathfrak{g} = \mathfrak{h} \oplus \mathbf{i}\mathfrak{h}$ and the Killing form on $\mathfrak{h}$ is negative definite. (If K is the complex field and R is the subfield of the reals, then the subalgebra $\mathfrak{h}$ is usually called the compact form of the Lie algebra $\mathfrak{g}$.) From the structure of finite-dimensional irreducible modules over a semisimple Lie algebra $\mathfrak{g}$ it follows that we can choose a basis in a K-space V such that all the elements of the Lie algebra $\mathfrak{h}$ are represented in this basis by skew-Hermitian matrices. This means that there exists an involution $\boldsymbol{\sigma}$ in the algebra U (U can be considered as a full matrix algebra over K) that is the composition of matrix transposition and of complex conjugation such that $\boldsymbol{\sigma}(\beta h) = -\overline{\beta} h$, where $\beta \in K$ and $h \in \mathfrak{h}$.

In the construction of the algebra $(\overline{\mathcal{F}}_0, \overline{\mathfrak{L}}_1, \overline{F}_1)$ with three supports and of the pair $(\overline{F}, \overline{\mathfrak{L}})$, we choose an orthonormal basis in the Lie algebra $\mathfrak{g}$ arbitrarily. We recall that, by Lemma 16.2, the construction of the function $\widetilde{\alpha}$ for the pair $(U, \mathfrak{g})$ does not depend on the choice of orthonormal basis with respect to the Killing form. Thus, in order to prove Theorem 16.1 for the case in which the algebra U is finite-dimensional, we choose this basis in the R-space $\mathbf{i}\mathfrak{h}$. Then for any $h_1, \dots, h_k \in \mathbf{i}\mathfrak{h}$ we have $\boldsymbol{\sigma}(h_1 \cdots h_k) = h_k \cdots h_1$. We extend the involution $\boldsymbol{\sigma}$ from the algebra U to the algebra $\overline{U}$ by setting $\boldsymbol{\sigma}(\overline{x}_j^{(i)} u) = \overline{x}_j^{(i)} \boldsymbol{\sigma}(u)$ for any generator $\overline{x}_j^{(i)}$ of the algebra $\overline{\mathcal{F}}$ and for any element $u \in \overline{U}$. Then for any $\overline{x}_i$ ($i = 1, 2, \dots$) we have $\boldsymbol{\sigma}(\overline{x}_i) = \overline{x}_i$, and the following equalities are valid:

$$\boldsymbol{\sigma}(\beta \overline{x}_{i_1} \cdots \overline{x}_{i_k}) = \overline{\beta} \overline{x}_{i_k} \cdots \overline{x}_{i_1}, \qquad \boldsymbol{\sigma}((u, v)) = (\boldsymbol{\sigma}(u), \boldsymbol{\sigma}(v)),$$

where β and $\overline{\beta}$ are complex conjugate elements of the field K and u and v are arbitrary elements of the Lie algebra $\overline{\mathfrak{g}}$. These relations show that the associative K-algebras $\overline{F}_1$, $\overline{\mathcal{F}}_0$, and $\overline{F}$ are stable under the involution $\boldsymbol{\sigma}$. In the same way, the involution $\boldsymbol{\sigma}$ can be extended from the algebra U to the algebra $\overline{U}$.

THEOREM 17.1. *Let $\mathfrak{g}$ be a semisimple finite-dimensional Lie algebra over an algebraically closed field K of characteristic zero, let U be a simple finite-dimensional associative enveloping algebra of it, and let the function $\widetilde{\alpha}$ be constructed for the pair $(U, \mathfrak{g})$ in the same way as in* §16. *Then for any multilinear associative polynomial $f(t_1, \dots, t_l)$*

we have $\widetilde{\alpha}(\varphi(f)\sigma\varphi(f)) \in R$, *and this element satisfies the inequality*

$$\widetilde{\alpha}(f(\overline{x}_1, \dots, \overline{x}_l)\sigma f(\overline{x}_1, \dots, \overline{x}_l)) \geq 0.$$

Moreover, equality is attained if and only if $f = 0$ *is an identity for the pair* $(U, \mathfrak{g})$.

PROOF. For an arbitrary multilinear polynomial $f(t_1, \dots, t_l)$ we have the following relation in the algebra $\overline{U}$:

$$f(\overline{x}_1, \dots, \overline{x}_l) = \sum_{j_1, \dots, j_l} \overline{x}_1^{(j_1)} \cdots \overline{x}_l^{(j_l)} f(e_{j_1}, \dots, e_{j_l}),$$

where $e_1, \dots, e_m$ is the basis of the Lie algebra $\mathfrak{g}$ and the sum is extended over all collections $\{j_1, \dots, j_l\}$ ($j_k \in \{1, \dots, m\}$). This equality, together with Proposition 15.1 and defining relations (16.2) of the algebra $\overline{\mathcal{F}}$, implies that the relation $f(t_1, \dots, t_l)$ is an identity for the pair $(U, \mathfrak{g})$ if and only if $f(\overline{x}_1, \dots, \overline{x}_l) = 0$ in the algebra $\overline{F}$.

The element $u = f(\overline{x}_1, \dots, \overline{x}_l)$ belongs to $\overline{U}$ and it is a matrix of order $\dim_K V$ with an element of the commutative K-algebra $\overline{\mathcal{F}}$ of the form

$$\sum_{i_1, \dots, i_l} \beta^{ij}_{i_1, \dots, i_l} \overline{x}_1^{(i_1)} \cdots \overline{x}_l^{(i_l)}$$

at the intersection of the ith row and the jth column, where the sum is taken over all collections of indices $i_1, \dots, i_l$ in the set $\{1, \dots, m\}$ and $\beta^{ij}_{i_1, \dots, i_l}$ are some elements of the field K. Then from the definition of the function $\widetilde{\alpha}$ it follows that

$$\operatorname{tr}(u\sigma(u)) = \dim_K V \, (\overline{x}_1^{(1)})^2 \cdots (\overline{x}_l^{(1)})^2 \widetilde{\alpha}(u\sigma(u)).$$

By the defining relations (16.2) of the algebra $\overline{\mathcal{F}}$, the left-hand side of this relation equals

$$(\overline{x}_1^{(1)})^2 \cdots (\overline{x}_l^{(1)})^2 \sum_{i, j, i_1, \dots, i_l} \beta^{ij}_{i_1, \dots, i_l} \overline{\beta}^{ij}_{i_1, \dots, i_l},$$

where the sum is taken over all collections of indices $i_1, \dots, i_l \in \{1, \dots, m\}$ and $i, j \in \{1, 2, \dots, \dim_K V\}$. Thus,

$$\dim_K V \, \overline{\alpha}(u\sigma(u)) = \sum_{\substack{i_1, \dots, i_l \\ i, j}} |\beta^{ij}_{i_1, \dots, i_l}|^2 \geq 0,$$

and equality is attained if and only if all $\beta^{ij}_{i_1, \dots, i_l}$ are equal to zero and, hence, $u = 0$ in the algebra $\overline{U}$. The theorem is proved.

COROLLARY. *Let* $f(x_1, \dots, x_l)$ *be an arbitrary multilinear polynomial of the algebra* $\mathcal{F}_0$. *Then*

$$\widetilde{\alpha}(f(\overline{x}_1, \dots, \overline{x}_l) \cdot \sigma f(\overline{x}_1, \dots, \overline{x}_l)) \geq 0. \tag{17.1}$$

and equality is attained if and only if $f(x_1, \ldots, x_l) = 0$ in the algebra $\mathcal{F}_0$.

PROOF. As it was noted in Remark 16.3, the expression on the left-hand side of the inequality (17.1) does not depend on the choice of the faithful representation of the Lie algebra $\mathfrak{g}$ and, hence, on the algebra $(\overline{\mathcal{F}}_0, \overline{\mathfrak{L}}_1, \overline{F}_1)$ with three supports. Since all the algebras $\mathcal{F}_0$ in the algebras $(\mathcal{F}_0, \mathfrak{L}_1, F_1)$ (with three supports) constructed for various faithful representations of the Lie algebra $\mathfrak{g}$ are isomorphic, we see that the assertion of the corollary has an invariant meaning and we can prove this for the case when the algebra $(\mathcal{F}_0, \mathfrak{L}_1, F_1)$ with three supports is constructed for some faithful finite-dimensional irreducible representation of the semisimple Lie algebra $\mathfrak{g}$. (If $\mathfrak{g} = \bigoplus_{i=1}^k \mathfrak{g}_i$ is the direct decomposition of the Lie algebra $\mathfrak{g}$ with respect to simple ideals $\mathfrak{g}_i$, then we can take for V the tensor product $\bigotimes_{i=1}^k \mathfrak{g}_i$ of adjoint representations of the Lie algebras $\mathfrak{g}_i$.) In this case we can consider $U = \operatorname{End}_K V$ as a full matrix K-algebra of order $n = \dim_K V$.

By the rank theorem for the full matrix algebra of order n, there exists a multilinear associative polynomial $d(t_1, \ldots, t_{n^2}, z_1, \ldots, z_l)$ that is skew-symmetric with respect to the variables $t_1, \ldots, t_{n^2}$ such that

$$d(a_1, a_2, \ldots, a_{n^2}, b_1, \ldots, b_l) \neq 0$$

for some $a_i, b_j \in U$. Therefore, the ideal of weak Capelli identities of order $m = \dim_K \mathfrak{g} \leq n^2$ is not contained in the ideal of identities of the pair $(U, \mathfrak{g})$, and in the free pair $(F, \mathfrak{L})$ of the variety $\operatorname{var}(U, \mathfrak{g})$ there exists a nonzero multilinear polynomial

$$d_m(x_{l+1}, \ldots, x_{l+m}, x_{l+m+1}, \ldots, x_{l+m+t})$$

that is skew-symmetric with respect to $x_{l+1}, \ldots, x_{l+m}$. Then, by the trace identity (14.1) for the algebra $(K, \mathfrak{g}, U)$ with three supports, for $f \in \mathcal{F}_0$ the element $f d_m$ of the algebra F_1 is equal to a multilinear associative polynomial $g(x_1, \ldots, x_{l+m+t})$. If the element $f \neq 0$ belongs to $\mathcal{F}_0$, then it is obvious that the polynomial g is also nonzero in F. But then $\alpha(\varphi(g)\boldsymbol{\sigma}\varphi(g)) > 0$ by Theorem 17.1 and by Proposition 15.1. On the other hand,

$$\begin{aligned}\alpha(\varphi(g)\boldsymbol{\sigma}\varphi(g)) &= \tilde{\alpha}(\varphi(f)\boldsymbol{\sigma}\varphi(f)\varphi(d_m)\boldsymbol{\sigma}\varphi(d_m)) \\ &= \tilde{\alpha}(\varphi(f)\boldsymbol{\sigma}\varphi(f))\alpha(\varphi(d_m)\boldsymbol{\sigma}\varphi(d_m)),\end{aligned}$$

and, since $\alpha(\varphi(d_m)\boldsymbol{\sigma}\varphi(d_m)) > 0$ by Theorem 17.1 and by Proposition 15.1, we see that the inequality $f(\overline{x}_1, \ldots, \overline{x}_l) \neq 0$ in $\mathcal{F}_0$ implies the relation

$$\alpha(f(\overline{x}_1, \ldots, \overline{x}_l)\boldsymbol{\sigma} f(\overline{x}_1, \ldots, \overline{x}_l)) > 0.$$

As in the proof of Theorem 17.1, the inequality $f(\overline{x}_1, \ldots, \overline{x}_l) \neq 0$ in the algebra $\overline{\mathcal{F}}_0$ is equivalent to the inequality $f(x_1, \ldots, x_l) \neq 0$ in the algebra $\mathcal{F}_0$. The corollary is proved.

§18. Proof of Theorem 12.1

We continue to assume that $\mathfrak{g}$ is a semisimple Lie algebra over an algebraically closed field K, U is a simple associative enveloping K-algebra with nonzero center, V is an irreducible representation of the K-algebra U, and the objects $\mathcal{F}$, $\overline{\mathcal{F}}$, $(\overline{U}, \overline{\mathfrak{g}})$, $(\widetilde{U}, \widetilde{\mathfrak{g}})$, $(\mathcal{F}_0, \mathfrak{L}_1, F_1)$, $(\overline{\mathcal{F}}_0, \overline{\mathfrak{L}}_1, \overline{F}_1)$, $(F, \mathfrak{L})$, and $(\overline{F}, \overline{\mathfrak{L}})$, and the function $\widetilde{\alpha}$ are constructed for the algebras $\mathfrak{g}$ and U and for the representation V in the same way as in §§15 and 16.

In this section we shall prove

LEMMA 18.1. *In the free pair $(F, \mathfrak{L})$ of the variety* $\operatorname{var}(U, \mathfrak{g})$ *there exists a multilinear polynomial $u(x_1, \dots, x_l, x_{l+1}, \dots, x_{2l})$ belonging to the ideal of weak Capelli identities $\mathfrak{K}_m$ $(m = \dim_K \mathfrak{g})$ for which*

$$\alpha(u(\overline{x}_1, \dots, \overline{x}_l, \overline{x}_1, \dots, \overline{x}_l)) \neq 0, \tag{18.1}$$

where $\overline{x}_1, \dots, \overline{x}_l \in \mathfrak{L}_1 \subset \overline{F}_1$.

After this we shall give an algorithm of constructing, for a polynomial u of this kind, a central polynomial of the pair $(U, \mathfrak{g})$ belonging to the set $R_{m.2l}$.

PROOF OF THE LEMMA. If U is a finite-dimensional simple algebra, then, as it was noted in the proof of Corollary of Theorem 17.1, the ideal $\mathfrak{K}_m$ is nonzero and, hence, there exists a multilinear polynomial $d_m(x_1, \dots, x_m, x_{m+1}, \dots, x_l)$ that is skew-symmetric with respect to $x_1, \dots, x_m$ and nonzero in the algebra F. Then

$$\alpha(d_m(\overline{x}_1, \dots, \overline{x}_l)\boldsymbol{\sigma} d_m(\overline{x}_1, \dots, \overline{x}_l)) > 0,$$

by Theorem 17.1, and for a polynomial u we can take

$$d_m(x_1, \dots, x_l)\boldsymbol{\sigma} d_m(x_{l+1}, \dots, x_{l+m}).$$

Let U be an infinite-dimensional simple algebra. Then from Theorem 14.1 it follows that in the ideal $\mathfrak{K}_m$ of the algebra F there exists a nonzero central multilinear polynomial $g(x_1, \dots, x_l)$. Moreover, Remark 14.1 shows that we can choose g in such a way that the polynomial g can be expressed in the algebra $\widetilde{U}$ as a linear combination of products (u_i, v_i), where $u_i, v_i \in \mathfrak{L}$. Then $g(x_1, \dots, x_l) = f(x_1, \dots, x_l)$ for some multilinear polynomial f in the algebra $\mathcal{F}_0$. By Corollary of Theorem 17.1, we have

$$\widetilde{\alpha}(f(\overline{x}_1, \dots, \overline{x}_l)\boldsymbol{\sigma} f(\overline{x}_1, \dots, \overline{x}_l)) > 0. \tag{18.2}$$

Hence, the element

$$u = f(x_1, \dots, x_l)\boldsymbol{\sigma} f(x_{l+1}, \dots, x_{2l}) = g(x_1, \dots, x_l)\boldsymbol{\sigma} f(x_{l+1}, \dots, x_{2l})$$

is not equal to zero in the algebra F_1. By Corollary of Lemma 14.2, the right-hand side of this equality can be rewritten in the form of a multilinear polynomial belonging to $\mathfrak{K}_m$. The inequality (18.2) shows that this associative polynomial is the desired one. The lemma is proved.

REMARK 18.1. It is well known that in the adjoint representation of a semisimple Lie algebra, the Casimir element $\sum_{i=1}^{m}(\operatorname{ad} e_i)^2$ is equal to the unit. Hence, in the algebra U we have the equality

$$[\overline{x}_i, \overline{x}_j, \overline{x}_j] = \left(\overline{x}_j^{(1)}\right)^2 \overline{x}_i. \tag{18.3}$$

Denote by the letter v the multilinear polynomial obtained from the polynomial u of Lemma 18.1 by substitution of commutators $[y_{2i-1}, y_{2i}]$ instead of variables x_i. Then equality (18.3) and Property 2 of the function α show that

$$\begin{aligned}\pm\,\alpha(u(\overline{y}_1, \dots, \overline{y}_l, \overline{y}_1, \dots, \overline{y}_l))\\ = \alpha(u([\overline{y}_1, \overline{y}_2], \dots, [\overline{y}_{2l-1}, \overline{y}_{2l}], [\overline{y}_1, \overline{y}_2], \dots, [\overline{y}_{2l-1}, \overline{y}_{2l}]).\end{aligned}$$

Hence, for the polynomial v all the properties of the polynomial u are satisfied, and, moreover, the polynomial obtained from v by substitution of the unit instead of any of its variables is a zero polynomial.

Denote by $z_1, \dots, z_m, y_1, \dots, y_m$ various free generators of the pair $(F, \mathfrak{L})$ and denote by $\mathfrak{k} = \mathfrak{k}(z_1, y_1, \dots, z_m, y_m)$ the square matrix with the element $(z_i, y_j) \in \mathcal{F}_0$ on the intersection of the ith row and the jth column. Let $c(z_1, \dots, z_m, y_1, \dots, y_m)$ be the determinant of this matrix and let $c_{ij}(z_1, \dots, \widehat{z}_i, \dots, z_m, y_1, \dots, \widehat{y}_j, \dots, y_m)$ be the cofactor of the (i, j)-th element of the matrix $\mathfrak{k}$. Obviously, $c, c_{i,j} \in \mathcal{F}_0$. As it was shown in §14, the elements $z_1, \dots, z_m$ form a basis of the Lie algebra $\mathfrak{g}_1 = \mathcal{F}_1\mathfrak{L}_1$ over the field of quotients $\mathcal{F}_1$ of the algebra $\mathcal{F}_0$. Denote by z_i' $(i = 1, \dots, m)$ the element

$$\sum_{j=1}^{m} c_{ij} y_j, \tag{18.4}$$

which belongs to $\mathcal{F}_0 \cdot \mathfrak{L} \subset \mathfrak{L}_1$.

LEMMA 18.2. *In the algebra $\mathcal{F}_0$ we have the equalities*

$$(z_i, z_j') = \delta_{ij} c(z_1, \dots, z_m, y_1, \dots, y_m),$$

where δ_{ij} is the Kronecker symbol.

PROOF. If $i \neq j$, then (z_i, z_j') is equal to the determinant of the matrix obtained from the matrix $\mathfrak{k}$ by replacing the jth row by the ith row. Then the ith row and the jth row of the new matrix are equal, and its determinant equals zero. If $i = j$ then $(z_i, z_j') = \sum_{j=1}^{m} c_{ij}(z_i, y_j)$, and this sum is exactly the decomposition of the determinant of the matrix $\mathfrak{k}$ with respect to the ith row. The lemma is proved.

Choose a multilinear associative polynomial $u(x_1, \dots, x_{2l})$ such that it belongs to $\mathfrak{K}_m$ and inequality (18.1) holds. In order to construct a central polynomial of the pair $(U, \mathfrak{g})$, we take $2ml$ distinct free generators x_i of the pair $(F, \mathfrak{L})$ and denote them by z_{ij} and y_{ij} $(i = 1, \dots, l;\ j = 1, \dots, m)$. For any fixed index i $(i = 1, \dots, l)$, from the free generators $z_{i1}, \dots, z_{im}, y_{i1}, \dots, y_{im}$ we construct elements $z_{i1}', \dots, z_{im}'$ in the same way as it was done for the elements z_k and y_k $(k = 1, \dots, m)$. Then, by Lemma 18.2, the sets of elements $\{z_{i1}, \dots, z_{im}\}$ and $z_{i1}', \dots, z_{im}'$ are dual bases (up to a nonzero multiple $c_i = c(z_{i1}, \dots, z_{im}, y_{i1}, \dots, y_{im}))$ in the semisimple $\mathcal{F}_1$-algebra $\mathfrak{g}_1 = \mathcal{F}_1\mathfrak{L}_1$. Then, due to the independence of the element u' in Lemma 16.2 of the choice of dual bases (Lemma 16.1 and Remark 16.1), we have

$$u(\overline{x}_1, \dots, \overline{x}_l, \overline{x}_1, \dots, \overline{x}_l) = \\ \left(\overline{x}_1^{(1)}\right)^2 \cdots \left(\overline{x}_l^{(1)}\right)^2 c_1^{-1} \cdots c_l^{-1} \sum_{j_1=1}^{m} \cdots \sum_{j_l=1}^{m} u\big|_{x_1=z_{1j_1}, x_{l+1}=z'_{1j_1}, \dots, x_l=z_{lj_l}, x_{2l}=z'_{lj_l}}.$$

Hence, the element

$$r = \sum_{j_1=1}^{m} \cdots \sum_{j_l=1}^{m} u\Bigg|_{\substack{x_1=z_{1j_1}, x_{l+1}=z'_{1j_1} \\ \cdots \\ x_l=z_{lj_l}, x_{2l}=z'_{lj_l}}} = \alpha(\overline{u})c_1 \cdots c_l \tag{18.5}$$

is nonzero and belongs to the algebra $\mathcal{F}_0 \subset \widetilde{U}$. On the other hand, by the explicit form of the elements z'_{ij} (formula (18.4)), the element r can be represented as a linear combination of polynomials of the form

$$u\big|_{x_1=z_{1j_1}, x_{l+1}=y'_{1j_1}, \dots, x_l=z_{lj_l}, x_{2l}=y'_{lj_l}}$$

of the algebra F with coefficients in $\mathcal{F}_0$. Since the polynomial u belongs to the ideal of weak Capelli identities $\mathfrak{K}_m$, the elements of the form (18.6) also belong to $\mathfrak{K}_m$. From Corollary to Lemma 14.2 we have $\mathcal{F}_0\mathfrak{K}_m \subset \mathfrak{K}_m$; therefore, using identity (14.1) for the algebra $(\mathcal{F}_0, \mathfrak{L}_1, F_1)$ with three supports, we can rewrite the expression $r(\dots, z_{ij}, y_{ij}, \dots)$ in the algebra F_1 in the form of an associative multilinear polynomial b in the algebra F. Since $0 \neq r \in \mathcal{F}_0$, we see that this associative polynomial is central. It remains to note that the expression in the central part of equality (18.5) is skew-symmetric with respect to the sets of variables $\{z_{i1}, \dots, z_{im}\}$ and $\{y_{i1}, \dots, y_{im}\}$, because the determinant $c(z_1, \dots, z_m, y_1, \dots, y_m)$ of the matrix $\mathfrak{k}$ is a skew-symmetric function of its rows and columns. Therefore, the associative polynomial b in the algebra F, which is equal to r, can be chosen in the set $R_{m,2l}$.

REMARK 18.2. If we replace the polynomial u in the proof of Theorem 12.1 by the polynomial v given in Remark 18.1, then identity (18.3) shows that the central polynomial $b \in R_{m,4l}$ constructed has an additional property: $b\big|_{x_i=1}$ is a zero polynomial ($i = 1, \dots, 4lm$). The existence of a central polynomial of this kind will be used in the proof of Theorem 12.1 for a reductive Lie algebra $\mathfrak{g}$.

§19. For an arbitrary simple algebra U, the identities of the pair $(U, \mathfrak{g})$ are determined by its α-function

Let $\mathfrak{g}$, U, K, V,$\mathcal{F}$, $\overline{\mathcal{F}}$, $\widetilde{U}$, $\overline{U}$, $\widetilde{\mathfrak{g}}$, $\overline{\mathfrak{g}}$, $(\mathcal{F}_0, \mathfrak{L}_1, F_1)$, $(\overline{\mathcal{F}}_0, \overline{\mathfrak{L}}_1, \overline{F}_1)$, $(F, \mathfrak{L})$, $(\overline{F}, \overline{\mathfrak{L}})$, and $\widetilde{\alpha}$ have the same meaning as in the previous section.

In order to prove Theorem 16.1 in the general case, it is sufficient to prove that if a multilinear associative polynomial $f_0(t_1, \dots, t_p)$ is not an identity for the pair $(U, \mathfrak{g})$, then there exists a multihomogeneous polynomial h, multilinear with respect to $t_1, \dots, t_p$ and of degree 2 with respect to any other variable on which it depends essentially, such that $\alpha(\varphi(f_0 h)) \neq 0$.

Since the pair $(F, \mathfrak{L})$ is free in the variety $\mathrm{var}(U, \mathfrak{g})$ with free generators $x_1, x_2, \dots$, we see that $f_0(x_1, \dots, x_p) \neq 0$ in the algebra F. Let $V(f_0)$ be an ideal of weak identities generated in F by this polynomial. Then, by Theorem 14.1, $V(f_0)\mathfrak{K}_m$ is nonzero and contains a central multilinear polynomial $g(x_1, \dots, x_l)$, which in view of Remark 14.1 may be considered equal to $f(x_1, \dots, x_l) \in \mathcal{F}_0$. Let us construct from f and g polynomials u, b, and r in the same way as it was done in the proofs of Lemma 18.1 and Theorem 12.1. Obviously, $u \in V(f_0)\mathfrak{K}_m$. Identity (14.1) shows that $b \in (V(f_0)\mathfrak{K}_m) \cap R_{m,2l}$. From formula (18.5) we have

$$\alpha(\varphi(bb)) = \widetilde{\alpha}(\varphi(rr)) = \{\alpha(u(\overline{x}_1, \dots, \overline{x}_{2l}))\}^2 \prod_{i=1}^{l} \alpha(\varphi(c_i c_i)),$$

where c_i is the determinant of the matrix $\mathfrak{k}(z_{i1}, \dots, z_{im}, y_{i1}, \dots, y_{im})$. From the definition of the matrix $\mathfrak{k}$ it is clear that $\sigma\varphi(c_i) = \varphi(c_i)$ and, hence, $\widetilde{\alpha}(\varphi(c_i)\sigma\varphi(c_i)) > 0$ (Corollary of Theorem 17.1).

However, in this case from Proposition 8.1 it follows that $f_0 \notin \mathrm{Ann}_{H_\lambda} \mathfrak{b}$, because $b \in V(f_0)$ and $\alpha(\varphi(bb)) \neq 0$. The theorem is proved.

§20. Some consequences of Theorems 12.1 and 16.1

COROLLARY 1. *Let $\mathfrak{g}_1$ and $\mathfrak{g}_2$ be semisimple Lie algebras over an algebraically closed field K, and let U_1 and U_2 be simple enveloping K-algebras with nonzero centers. Then, if the pairs $(U_1, \mathfrak{g}_1)$ and $(U_2, \mathfrak{g}_2)$ are not isomorphic and if $\dim_K \mathfrak{g}_1 = m_1 \geq \dim_K \mathfrak{g}_2 = m_2$, then there exists a polynomial $p \in R_{m_1,k}$ such that $p = 0$ is an identity for the pair $(U_2, \mathfrak{g}_2)$, but is not an identity for the pair $(U_1, \mathfrak{g}_1)$.*

PROOF. If $m_1 > m_2$, then it is sufficient to take for p any central polynomial of the pair $(U_1, \mathfrak{g}_1)$ belonging to $R_{m_1,k}$. Then $\mathfrak{K}_{m_1}(U_2) = 0$ and $p = 0$ is an identity for the pair $(U_2, \mathfrak{g}_2)$.

If $m = m_1 = m_2$, then the pairs $(U_1, \mathfrak{g}_1)$ and $(U_2, \mathfrak{g}_2)$ have different identities, because they are not isomorphic (see Theorem 5.1). From Theorem 16.1 it follows that then the linear functions α_1 and α_2 constructed for the pairs $(U_1, \mathfrak{g}_1)$ and $(U_2, \mathfrak{g}_2)$ must have different sets of zeros on 2-elements. Moreover, there exists a multihomogeneous 2-element $v(\overline{x}_1, \dots, \overline{x}_l)$ such that $\alpha_2(v) = 0$ and $\alpha_1(v) \neq 0$. Choose a multilinear associative polynomial $u(x_1, \dots, x_l, x_{l+1}, \dots, x_{2l})$ such that $v = u(\overline{x}_1, \dots, \overline{x}_l, \overline{x}_1, \dots, \overline{x}_l)$. By Theorem 12.1, there exists a central polynomial $h \in R_{m,k}$ for the pair $(U_1, \mathfrak{g}_1)$. From the polynomial u, we contruct the element r determined by the first equality in (18.5). Since the polynomial u does not necessarily belong to the ideal of weak Capelli identities $\mathfrak{K}_m$, we see that the element r in $F_1(U_1)$ (or in $F_1(U_2)$) is not necessary expressed as a polynomial in $F(U_1)$ (or in $F(U_2)$). However, if we regard the elements r and h as depending on different variables, then, using identity (14.1), we can construct an associative multilinear polynomial p that is equal to the element hr both in $F_1(U_1)$ and in $F_1(U_2)$. The second equality in (18.5) shows that $p = \alpha_1(v)h \prod_{i=1}^{l} c_i \neq 0$ in $F_1(U_1)$ and that $p = \alpha_2(v)h \prod_{i=1}^{l} c_i = 0$ in $F_1(U_2)$. The corollary is proved.

PROOF OF THEOREM 12.1 FOR A REDUCTIVE LIE ALGEBRA $\mathfrak{h}$. First we note that for any finite-dimensional Lie algebra of dimension m and for any associative polynomial $f(t_1, \ldots, t_l) \in R_{m.k}$ in an enveloping algebra U we have the identity

$$[f(v_1, \ldots, v_{mk}), y] = k \operatorname{tr}(\operatorname{ad} y) f(v_1, \ldots, v_{mk}), \tag{20.1}$$

where $v_1, \ldots, v_{mk}, y \in \mathfrak{h}$. Obviously, the polynomial f can be nonzero only under the substitution of all the elements of the basis of the Lie algebra $\mathfrak{h}$ instead of the skew-symmetric variables. However, in this case the left-hand side of identity (20.1) is equal to

$$\sum_{i=1}^{mk} f(v_1, \ldots, [v_i, y], \ldots, v_{mk})$$

and the discussion used in the proof of identity (14.1) shows that this expression is equal to the right-hand side of relation (20.1). Since in a reductive Lie algebra we have $\operatorname{tr}(\operatorname{ad} y) = 0$, we see that $f(v_1, \ldots, v_{mk})$ belongs to the center of the algebra U, and to prove the theorem it is sufficient to show that $f = 0$ is not an identity for the pair $(U, \mathfrak{h})$ for a suitably chosen $f \in R_{m.k}$. If the reductive Lie algebra $\mathfrak{h}$ is semisimple, then this follows from Theorem 12.1.

Let $\mathfrak{h} = \mathfrak{g}_0 \oplus \mathfrak{g}$, where $\mathfrak{g}_0$ is a one-dimensional Lie algebra and $\mathfrak{g}$ is a semisimple one. Since U is a simple algebra, Lemma 3.5 implies that $\mathfrak{g}_0$ coincides with the field $K \cdot \mathbf{1}$, which is the center of the algebra U. However, in this case the enveloping K-algebra of the Lie algebra $\mathfrak{g}$ is an ideal in U and, hence, coincides with U. Therefore, for the pair $(U, \mathfrak{g})$ we can construct a central multilinear polynomial $b(x_1, \ldots, x_l) \in R_{m-1.k}$ $(l = (m-1)k)$ such that $b\big|_{x_i=1}$ $(i = 1, \ldots, l)$ are zero polynomials (see Remark 18.2). Denote by the letter z_{ij} the variables $x_{(i-1)(m-1)+j}$ $(i = 1, \ldots, k;\ j = 1, \ldots, m-1)$ and suppose that b is skew-symmetric with respect to the sets of variables $\{z_{i1}, \ldots, z_{i(m-1)}\}$ $(i = 1, \ldots, k)$. Denote by the letter h the polynomial belonging to $R_{m.k}$ obtained from the polynomial $bz_{1m} \ldots z_{km}$ by skew-symmetrization with respect to the variables in each set $\{z_{i1}, \ldots, z_{im}\}$ $(i = 1, \ldots, k)$. Let us choose a basis $e_1, \ldots, e_{m-1}$ in the Lie algebra $\mathfrak{g}$ and an element $e_m \in \mathfrak{g}_0$ such that $e_m = 1$. Then, since $b\big|_{x_i=1} = 0$, we see that the expression obtained under the substitution of elements e_j into the polynomial h, instead of the variables z_{ij} $(i = 1, \ldots, k;\ j = 1, \ldots, m)$, is equal to the value of the polynomial b for $z_{ij} = e_j$ $(i = 1, \ldots, k;\ j = 1, \ldots, m-1)$, which is nonzero, because $b = 0$ is not an identity for the pair $(U, \mathfrak{g})$. The theorem is proved.

COROLLARY 2. *Let $\mathfrak{g}$ be a simple noncommutative Lie algebra over an algebraically closed field K of characteristic zero and let U be its simple enveloping algebra with nonzero center. Then in any proper subvariety of pairs of the variety* $\operatorname{var}(U, \mathfrak{g})$ *an identity $h = 0$ is valid, where h is a Lie polynomial such that $h = 0$ is not an identity for the Lie algebra $\mathfrak{g}$.*

PROOF. For an arbitrary polynomial $f \in R_{m.k}$, denote by ${}^n f$ the multilinear polynomial belonging to $R_{m.kn}$ that is equal to the product of n polynomials obtained from f by changing the variables.

Let $\dim_K \mathfrak{g} = m$ and $b \in R_{m.k}$. By Theorem 12.1 applied to the adjoint representation of the Lie algebra $\mathfrak{g}$, there exists $g(x_1, \ldots, x_l) \in R_{m.n}$ $(l = mn)$ such that the Lie polynomial $v = zg(\operatorname{ad} x_1, \ldots, \operatorname{ad} x_l)$ is not an identity for the Lie algebra $\mathfrak{g}$. Let

${}^k g$ and ${}^n b \in R_{m,kn}$ depend on the same skew-symmetric variables. Obviously, the Lie polynomial

$$h = z \cdot {}^k g(\operatorname{ad} x_1, \dots, \operatorname{ad} x_t), \qquad t = knm,$$

and the associative polynomial $p = {}^n bz$ can be simultaneously nonzero only under the substitution of all the elements of a basis of the Lie algebra $\mathfrak{g}$ instead of the skew-symmetric variables. Under this substitution, the polynomials h and p will be equal to $\beta_1 z$ and $\beta_2 z$, respectively, because the polynomial ${}^k g$ is central in the adjoint representation of the Lie algebra $\mathfrak{g}$, and the values of the polynomial ${}^n b$ belong to the center of the algebra U (see formula (20.1)). Moreover, the value of the fraction $\beta = \beta_2/\beta_1$ does not depend on the choice of basis of the Lie algebra $\mathfrak{g}$.

This discussion shows that in the free pair $(F, \mathfrak{L})$ of the variety $\operatorname{var}(U, \mathfrak{g})$ an identity

$$\beta z \cdot {}^k g(\operatorname{ad} x_1, \dots, \operatorname{ad} x_l) = z \cdot {}^n b(x_1, \dots, x_l) \tag{20.2}$$

is valid, where we have $\beta \neq 0$ for any central polynomial $b \in R_{m,k}$ of the pair $(U, \mathfrak{g})$. Therefore, in order to prove Corollary 2 it is sufficient to find an identity $b = 0$ for a proper subvariety $\mathfrak{B}$ of the variety $\operatorname{var}(U, \mathfrak{g})$, where $b \in R_{m,k}$ is a central polynomial for the pair $(U, \mathfrak{g})$ (in this case it is sufficient to take for h the left-hand side of equality (20.2)).

Let $\mathfrak{V}$ be the ideal of identities of the variety of pairs $\mathfrak{B}$. Since $\mathfrak{B}$ is a proper subvariety in $\operatorname{var}(U, \mathfrak{g})$, we see that the ideal $\mathfrak{V}$ is nonzero in F. Then, by Theorem 14.1, we have $\mathfrak{V}\mathfrak{K}_m \neq 0$, and, by Theorem 14.1, the ideal $\mathfrak{V}\mathfrak{K}_m$ contains a multihomogeneous polynomial $v(x_1, \dots, x_l)$, of degree 2 with respect to any variable x_i $(i = 1, \dots, l)$, such that $\alpha(v(\overline{x}_1, \dots, \overline{x}_l)) \neq 0$. Choosing a multilinear polynomial $u(x_1, \dots, x_l, x_{l+1}, \dots, x_{2l}) \in \mathfrak{V}\mathfrak{K}_m$ such that $v = u(x_1, \dots, x_l, x_1, \dots, x_l)$, we can construct (in the same way as in the proof of Theorem 12.1) a central polynomial $b \in R_{m,2l}$ that belongs to the ideal $\mathfrak{V}\mathfrak{K}_m$. The corollary is proved.

§21. A construction of a polynomial mapping that recovers the algebra of commutative polynomials from the Lie algebra of all of its derivations

Let $\mathcal{T}$ be the algebra of commutative polynomials with unit with free generators $x_1, \dots, x_n$ over a field K of characteristic zero and let W_n be the Lie K-algebra of derivations of the algebra $\mathcal{T}$. Any element of the algebra W_n can be represented in the form $\sum_{i=1}^n g_i \partial_i$, where g_i $(i = 1, \dots, n)$ belongs to the algebra $\mathcal{T}$ and ∂_i is the partial derivative with respect to the variable x_i. The Lie algebra W_n can be regarded as a left $\mathcal{T}$-module with respect to the action $g\left(\sum_{i=1}^n g_i \partial_i\right) = \sum_{i=1}^n (g g_i) \partial_i$. Therefore, the natural gradation $\mathcal{T} = \mathcal{T}_0 \oplus \mathcal{T}_1 \oplus \mathcal{T}_2 \oplus \cdots$, where $\mathcal{T}_i$ is the linear space of homogeneous polynomials of degree i, determines uniquely a gradation of the Lie algebra $W_n = L_{-1} \oplus L_0 \oplus L_1 \oplus L_2 \oplus \cdots$, where the linear space L_i consists of derivations of the form $\sum_{j=1}^n g_j \partial_j$ for $g_j \in \mathcal{T}_{j+1}$. Let $w \in L_i$; we set by definition $|w| = i$. It is well known that the algebra W_n is left transitive with respect to this gradation. This means that for any nonzero element $w \in L_j$, $j \geq 0$, we have $[w, L_{-1}] \neq 0$.

A Lie polynomial $f(t_1, \dots, t_l)$ gives an identity $f = 0$ for the Lie algebra W_n if for any $w_1, \dots, w_l \in W_n$ we have $f(w_1, \dots, w_l) = 0$ in the algebra W_n.

The following proposition shows that over a field of characteristic zero there exists an algorithm that distinguishes whether a Lie polynomial $f(t_1, \dots, t_l)$ gives an identity for the algebra W_n or not.

PROPOSITION 21.1. *Let $f(t_1, \dots, t_l)$ be a multilinear Lie polynomial. Then $f = 0$ is an identity for the Lie algebra W_n if and only if for any homogeneous elements $w_1, \dots, w_l$ satisfying the relation $\sum_{i=1}^{l} |w_i| = -1$ we have $f(w_1, \dots, w_l) = 0$ in the algebra W_n.*

PROOF. If $f = 0$ is an identity for W_n, then, in particular, $f(w_1, \dots, w_l) = 0$ in the algebra W_n for any homogeneous elements $w_1, \dots, w_l$ such that $\sum_{i=1}^{l} |w_i| = -1$. This proves the assertion in one direction.

Let $f(t_1, \dots, t_l)$ be not an identity for W_n. By the multilinearity of the polynomial f, there are homogeneous elements $u_i \in W_n$ such that $f(u_1, \dots, u_l) \neq 0$.

Let $\sum_{i=1}^{l} |u_i| = k > -1$. Then, by the left transitiveness of the gradation of the Lie algebra W_n, there exists an element $v \in L_{-1}$ such that

$$0 \neq [f|_{t_1=u_1,\dots,t_l=u_l}, v] = \sum_{i=1}^{l} f|_{t_1=u_1,\dots,t_i=[u_i,v],\dots,t_l=u_l},$$

whence it follows that one of the summands in the right-hand side of this inequality is nonzero. But we have $|u_1| + \dots + |[u_i, v]| + \dots + |u_l| = k - 1$, and inductive arguments imply that there exist homogeneous elements $w_1, \dots, w_l$ of the Lie algebra W_n such that $\sum_{i=1}^{l} |w_i| = -1$ and $f(w_1, \dots, w_l) \neq 0$. The proposition is proved.

REMARK 21.1. Over a field of characteristic zero, any Lie identity is equivalent to a finite set of multilinear identities that are obtained constructively from the original one. By Proposition 21.1, in order to verify whether a given multilinear Lie polynomial f of degree l determines an identity for the Lie algebra W_n, it is sufficient to substitute, instead of its variables, only the elements of some fixed homogeneous basis satisfying $|e_1| + \dots + |e_l| = -1$. Obviously, there is only a finite number of such sets $\{e_1, \dots, e_l\}$. Therefore, it is required to do only a finite number of tests (whether $f(e_1, \dots, e_l)$ is equal to zero or not). This proves the existence of an identity algorithm in the Lie algebra W_n.

We set $m = 2n + n^2$ and denote by the letter A the linear subspace in $\mathrm{End}_K W_n$ generated by the elements of the form $f(\mathrm{ad}\, w_1, \dots, \mathrm{ad}\, w_l)$, where $l = km$, $k = 1, 2, \dots$, f ranges over the set of associative polynomials in $R_{m,k}$, and $w_1, \dots, w_l$ are arbitrary elements of the Lie algebra W_n.

Since for any polynomials $f_1(t_1, \dots, t_{l_1}) \in R_{m,k_1}$ and $f_2(t_{l_1+1}, \dots, t_{l_1+l_2}) \in R_{m,k_2}$ the polynomial $f_1 f_2$ belongs to R_{m,k_1+k_2}, we see that the subspace A is a subalgebra in $\mathrm{End}_K W_n$. The algebra A can be naturally endowed with a gradation $A = \sum_{i=-\infty}^{+\infty} A_i$, where A_i is spanned (as a linear space) by all possible elements of the form $f(\mathrm{ad}\, w_1, \dots, \mathrm{ad}\, w_l)$, where $w_1, \dots, w_l$ are homogeneous elements of the Lie algebra W_n and $\sum_{j=1}^{l} |w_j| = i$ and f is a multilinear polynomial belonging to $R_{m,k}$ $(mk = l)$, $k = 1, 2, \dots$.

PROPOSITION 21.2. *The associative subalgebra $A = \sum_{i=-\infty}^{+\infty} A_i$ in $\mathrm{End}_K W_n$ is graded and has the following properties.*

1. $[A, \mathrm{ad}\, W_n] \subseteq A$, $[A_t, \mathrm{ad}\, L_i] \subseteq A_{t+i}$.
2. $A_t L_i \subset L_{t+i}$.
3. *For any negative integer t we have $A_t = 0$.*

4. $[A, \operatorname{ad} W_n] \neq 0$.

PROOF. The first property of the algebra A follows from the equality

$$[g(\operatorname{ad} w_1, \dots, \operatorname{ad} w_l), \operatorname{ad} v] = \sum_{i=1}^{l} f\big|_{t_1=\operatorname{ad} w_1, \dots, t_i=\operatorname{ad}[w_i, v], \dots, t_l=\operatorname{ad} w_l},$$

where $w_1, \dots, w_l, v$ are homogeneous elements of the Lie algebra W_n and g is an arbitrary multilinear associative polynomial.

The second property of the algebra A is obvious and follows immediately from the definition of the subspace A_i. This proves that the sum of subspaces $\sum_{t=-\infty}^{+\infty} A_t$ is direct.

In order to prove the third and the fourth properties of the algebra A, it is useful to introduce the following agreement. Let $f(t_1, \dots, t_{l_1})$ be a multilinear polynomial in the set $R_{m.k}$ $(l = mk)$; then we shall denote the variable $t_{(i-1)\cdot m+j}$ by z_{ij} $(i = 1, \dots, k;$ $j = 1, \dots, m)$ and suppose that for any $i = 1, \dots, k$ the polynomial f is skew-symmetric with respect to the variables $z_{i1}, \dots, z_{im}$.

Considered as a linear space, A_t is spanned by all elements of the form

$$f(\operatorname{ad} w_1, \dots, \operatorname{ad} w_l),$$

where w_i are homogeneous elements from some fixed homogeneous basis $e_1, e_2, e_3, \dots$ of the Lie algebra W_n, $\sum_{i=1}^{l} |w_i| = t$, and f is a multilinear polynomial belonging to $R_{m.k}$ $(mk = l)$, $k = 1, 2, \dots$. If $t < 0$, then for any substitution in the polynomial f of basis elements instead of variables, in one of the blocks $z_{i1}, \dots, z_{im}$ two equal elements will be substituted. Since the polynomial f is skew-symmetric with respect to these variables, we see that $f(\operatorname{ad} w_1, \dots, \operatorname{ad} w_l) = 0$ for $t < 0$. This proves the third property of the algebra A.

Denote by ∂ the element $\sum_{i=1}^{n} x_i \partial_i$ of the algebra W_n. It is immediately checked that the linear subspace $L = L_{-1} \oplus L_0 \oplus \{x_1\partial, \dots, x_n\partial\}$ is a Lie subalgebra in W_n and is isomorphic to the Lie algebra $\mathrm{sl}(n+1, K)$ of all matrices with zero trace of order $n+1$ over the field K. Since $m = 2n + n^2 = \dim_K L$, we see from Theorem 12.1, applied to the adjoint representation of the semisimple Lie algebra W_n, that there exists an associative multilinear polynomial $f \in R_{m.k}$ such that for some $w_1, \dots, w_l \in L \subset W_n$ we have $f_0 = f(\operatorname{ad} w_1, \dots, \operatorname{ad} w_l) \neq 0$ in the adjoint representation of the Lie algebra W_n. This proves that the algebra A is nonzero.

Suppose that $[A, \operatorname{ad} W_n] = 0$. Then the algebra A belongs to the centroid of the Lie algebra W_n. Since the Lie algebra W_n is centrally prime, we see that $A = K \cdot \mathbf{1}$. In order to lead our assumption to a contradiction, we consider the Lie polynomial

$$\begin{aligned} &g(z, z_{11}, \dots, z_{km}) \\ &\quad \rightleftharpoons f(\operatorname{ad} t_1, \dots, \operatorname{ad} t_l)z - \sum_{i=1}^{m} f\Bigg|_{\substack{z_{1i}=\operatorname{ad} z, z_{pq}=\operatorname{ad} z_{pq} \\ p=1,\dots,k;\, q=1,\dots,m;\, (p,q)\neq(1,i)}} z_{1i}. \end{aligned} \tag{21.1}$$

For any $i > 1$ this polynomial is skew-symmetric with respect to the variables $z_{i1}, \dots, z_{im}$. Moreover, this Lie polynomial is skew-symmetric with respect to the variables $z, z_{11}, \dots, z_{1m}$. However, in this case from Proposition 21.1 it follows that $g = 0$ is an identity for the Lie algebra W_n. Indeed, for any choice of a homogeneous basis $e_1, e_2, \dots$ in the Lie algebra W_n and for any choice of e and e_{ij} $(i = 1, \dots, k;$ $j = 1, \dots, m)$ in this basis, satisfying the relations

$$|e| + \sum_{i=1}^{k} \sum_{j=1}^{m} |e_{ij}| = -1,$$

we have either two equal elements among $e_{i1}, \dots, e_{im}$ for some $i = 2, \dots, k$ or two equal elements among $e, e_{11}, \dots, e_{1m}$. Then, however, by the skew-symmetry of the polynomial g with respect to the corresponding subsets of variables, under the substitution of the element e instead of z and of the elements e_{ij} instead of the variables z_{ij} $(i = 1, \dots, k; j = 1, \dots, m)$, the polynomial g vanishes.

We substitute in the polynomial of the form (21.1) the elements $w_1, \dots, w_l$ instead of the variables $t_1, \dots, t_l$ and an arbitrary element v instead of the variable z. Then this expression vanishes; since $0 \neq f_0 = f(\operatorname{ad} w_1, \dots, \operatorname{ad} w_l)$, and all the elements $f(\operatorname{ad} w_1, \dots, \operatorname{ad} v, \dots, \operatorname{ad} w_m, \dots, \operatorname{ad} w_l)$ belong to A, where we have $A = K$ by assumption, we see that any element v of the algebra W_n is representable in the form of a K-linear combination of elements $w_1, \dots, w_m$ and, hence, $\dim_K W_n \leq m = n^2 + 2n$. This contradicts the fact that the centrally prime Lie K-algebra W_n is infinite-dimensional. Hence, $[A, \operatorname{ad} W_n] \neq 0$, and Proposition 21.2 is proved.

It turns out that all graded associative subalgebras $D = \sum_{i=-\infty}^{+\infty} D_i$ in $\operatorname{End}_K W_n$ sharing Properties 1–3 of the algebra A admit a complete description.

THEOREM 21.1. *Let $D = \sum_{i=-\infty}^{+\infty} D_i$ be a nonzero graded associative subalgebra of the algebra* $\operatorname{End}_K W_n$ *satisfying the following properties*:

1°. $[D, \operatorname{ad} W_n] \subseteq D$, $[D_t, \operatorname{ad} L_i] \subseteq D_{t+i}$.

2°. $D_t L_i \subset L_{t+i}$.

3°. *For any negative integer t we have $D_t = 0$.*

Then the Lie algebra D coincides either with the algebra $\operatorname{End}_{\mathcal{T}} W_n$ *or with the center $\mathcal{T}$ of the algebra* $\operatorname{End}_{\mathcal{T}} W_n$ *or with the field K.*

The proof of this theorem follows from Lemmas 21.1–21.4.

LEMMA 21.1. *Among the graded subalgebras D in* $\operatorname{End}_K W_n$ *satisfying the conditions of Theorem* 21.1 *there exists the largest one $\widetilde{D}$.*

PROOF. For the largest subalgebra, we take the subalgebra $\widetilde{D}$ in $\operatorname{End}_K W_n$ generated by all the graded subalgebras D in $\operatorname{End}_K W_n$ satisfying the conditions of Theorem 21.1. Then we have $\widetilde{D} = \sum_{t=-\infty}^{+\infty} \widetilde{D}_t$, where $\widetilde{D}_t$ is spanned (as a linear space) by elements of the form $d_1 d_2 \cdots d_k$ for homogeneous elements d_i of various subalgebras D such that $t = \sum_{i=1}^{k} |d_i|$. Obviously, Properties 1°–3° are satisfied in $\widetilde{D}$, and the subspaces $\widetilde{D}_t$ $(t = 0, \pm 1, \dots)$ determine a gradation on $\widetilde{D}$. The lemma is proved.

LEMMA 21.2. *Suppose that for a subalgebra D of* $\operatorname{End}_K W_n$ *all the conditions of Theorem* 21.1 *are satisfied. Then the gradation $\sum_{i=0}^{\infty} D_i$ of the algebra D is left transitive,* *i.e.*, $[a, \operatorname{ad} L_{-1}] \neq 0$ *for any nonzero element* $a \in D_t$ *for* $t > 0$. *Moreover, the subalgebra* D_0 *possesses a faithful representation in the space* L_{-1}.

PROOF. Suppose that for some integer $t > 0$ there exists a nonzero element $a \in D_t$ such that $[a, \operatorname{ad} L_{-1}] = 0$. However, in this case, by Property 2°, there exist $j \geq -1$ and an element $w \in L_j$ such that the Lie element aw belongs to L_{j+1} and is nonzero. Since $j + t > -1$, we see that the left transitiveness of the gradation of the Lie algebra W_n implies

$$0 \neq [v, aw] = a[v, w] + [\operatorname{ad} v, a]w = a[v, w]. \tag{21.2}$$

This inequality shows that if $j > -1$, then $0 \neq [v, w] \in L_{j-1}$, and it follows from the inductive arguments that we can assume $j = -1$ and $w \in L_{-1}$. However, in this case we have $[v, w] = 0$ in formula (21.2) and, hence, the relation of the form (21.2) gets the form $0 \neq 0$. The contradiction thus obtained proves that the graded algebra D is left transitive.

Moreover, if $0 \neq a \in D_0$, then Properties 1° and 3°, formula (21.2) and inductive arguments imply the existence of an element $w \in L_{-1}$ such that $aw \neq 0$. The lemma is proved.

COROLLARY 1. *Let D be a graded subalgebra in* $\operatorname{End}_K W_n$ *such that all the conditions of Lemma* 21.2 *are satisfied. Then the algebra D and the subalgebra D_0 have the same identities.*

PROOF. Obviously, any identity for the algebra D is an identity for the algebra D_0.

Since the ground field K is of characteristic zero, we see that, in order to prove Corollary 1, it is sufficient to show that if a multilinear associative polynomial $h(t_1, \dots, t_l)$ is not an identity for the algebra D, then $h = 0$ is not an identity for the algebra D_0. Toward this end, we choose homogeneous elements $a_1, \dots, a_l \in D$ such that $h(a_1, \dots, a_l) \neq 0$ in the algebra D. Let $\sum_{i=1}^{l} |a_i| = t$. By Property 3°, we have $|a_i| \geq 0$ for $i = 1, \dots, l$. If $t = 0$, then any $|a_i| = 0$ and, hence, $h = 0$ is not an identity for the algebra D_0. If $t > 0$, then the left transitivity of the gradation of the algebra D implies the existence of an element $v \in L_{-1}$ such that

$$0 \neq [h(a_1, \dots, a_l), \operatorname{ad} v] = \sum_{i=1}^{l} h\Big|_{t_1 = a_1, \dots, t_i = [a_i, \operatorname{ad} v], \dots, t_l = a_l}.$$

Therefore, one of the summands in the right-hand side of this expression is nonzero. But we have $|a_1| + \cdots + |[a_i, \operatorname{ad} v]| + \cdots + |a_l| = t - 1$, and, by inductive arguments, there are elements $b_1, \dots, b_l \in D_0$ such that $h(b_1, \dots, b_l) \neq 0$ in the algebra D_0. The corollary is proved.

COROLLARY 2. *If for a nonzero graded subalgebra D in* $\operatorname{End}_K W_n$ *all the assumptions of Lemma* 21.2 *are satisfied, then either* $D_0 \simeq \operatorname{End}_K L_{-1}$ *or the subalgebra D_0 is isomorphic to the center of the algebra* $\operatorname{End}_K L_{-1}$.

PROOF. It follows from Property 1° of the algebra D that $[D_0, \operatorname{ad} L_0] \subseteq D_0$. Obviously, as a linear space, the restriction of $\operatorname{ad} L_0$ to L_1 coincides with $\operatorname{End}_K L_{-1}$. By Lemma 21.2, the restriction of D_0 to L_{-1} is faithful. Hence, D_0 is isomorphic to a nonzero subalgebra in $\mathrm{End}_K L_{-1}$ invariant under the adjoint action of $\mathrm{End}_K L_{-1}$. However, the Lie algebra possesses four ideals: $\{0\}$, the center of $\mathrm{End}_K L_{-1}$, matrices with zero trace, and $\mathrm{End}_K L_{-1}$. Among them, only the center of $\mathrm{End}_K L_{-1}$ and the $\mathrm{End}_K L_{-1}$ itself are nonzero associative subalgebras in $\mathrm{End}_K L_{-1}$. The corollary is proved.

The Lie algebra W_n is a free module over the polynomial algebra $\mathcal{T}$. We suppose that the algebra $\mathcal{T}$ is embedded in $\mathrm{End}_K W_n$ in a natural way, and we shall identify $\mathcal{T}$ with the center of $\mathrm{End}_{\mathcal{T}} W_n$.

Obviously, the algebra $\mathcal{T} = \sum_{i=0}^{\infty} \mathcal{T}_i$ is graded and has Properties 1°–3° mentioned in Theorem 21.1 for the algebra D. Therefore, the largest subalgebra $\widetilde{D}$ in $\mathrm{End}_K W_n$ (see Lemma 21.1) contains the subalgebra $\mathcal{T}$.

LEMMA 21.3. *The subalgebra $\mathcal{T}$ belongs to the center of the algebra $\widetilde{D}$.*

PROOF. It suffices to verify that for any homogeneous elements $c \in \mathcal{T}_i = \mathcal{T} \cap \widetilde{D}_i$ and $a \in \widetilde{D}_j$ we have the equality $[a, c] = 0$ in the algebra $\widetilde{D}$. We shall prove it by induction on $t = i + j$. The base of induction is $t = 0$. In this case $c \in K \cdot \mathbf{1}$ and $[c, a] = 0$. Let the equation $[c, a] = 0$ have already been proved for $i + j < l$. Let $i + j = l > 0$. If $[a, c] \neq 0$, then, by the left transitivity of the gradation of the algebra D, there would exist an element $v \in L_{-1}$ such that

$$0 \neq [[a, c], \operatorname{ad} v] = [[a, \operatorname{ad} v], c] + [a, [c, \operatorname{ad} v]].$$

It follows from Property 1° of the algebra $\widetilde{D}$ that $[a, \operatorname{ad} v] \in \widetilde{D}_{j-1}$ and $[c, \operatorname{ad} v] \in \mathcal{T}_{i-1}$ and thus, by the induction assumption, the right-hand side of the expression under consideration is equal to zero. The contradiction $0 \neq 0$ thus obtained shows that the relation $[c, a] = 0$ holds for $i + j = l$, too. The lemma is proved.

COROLLARY. *The largest graded subalgebra $\widetilde{D}$ in* $\mathrm{End}_K W_n$ *(see Lemma* 21.1*) coincides with* $\mathrm{End}_{\mathcal{T}} W_n$.

PROOF. Lemma 21.3 implies the inclusion $\widetilde{D} \subset \mathrm{End}_{\mathcal{T}} W_n$. Obviously, in the basis of free generators $\partial_1, \dots, \partial_n$ of the $\mathcal{T}$-module W_n every element of $\mathrm{End}_{\mathcal{T}} W_n$ can be represented by a matrix of order n over the algebra $\mathcal{T}$. Therefore, the algebra $\mathrm{End}_{\mathcal{T}} W_n$ is graded and has Properties 2° and 3° of the algebra $\widetilde{D}$. Since $[\operatorname{ad} W_n, \mathcal{T}] \subseteq \mathcal{T}$, we see that for any elements $a \in \mathrm{End}_{\mathcal{T}} W_n$, $w \in W_n$, and $c \in \mathcal{T}$

$$[[a, \operatorname{ad} w], c] = [[a, c], \operatorname{ad} w] + [a, [c, \operatorname{ad} w]] = 0.$$

This equality shows that the first property of the algebra $\widetilde{D}$ holds in $\mathrm{End}_{\mathcal{T}} W_n$. Since the algebra $\widetilde{D}$ contains any graded subalgebra of the algebra $\mathrm{End}_K W_n$ having Properties 1°–3°, we see that $\widetilde{D} \supset \mathrm{End}_{\mathcal{T}} W_n$. The corollary is proved.

LEMMA 21.4. *The largest graded subalgebra $\widetilde{D}$ (see Lemma* 21.1*) is generated by the elements of* $[\widetilde{D}_0, \operatorname{ad} W_n]$.

PROOF. Obviously, $[\widetilde{D}_0, \operatorname{ad} L_0]$ generates the subalgebra $\widetilde{D}_0$. It follows from Corollary of Lemma 21.3 that $\widetilde{D} = \mathcal{T} D_0$. But the algebra $\mathcal{T}$ is generated by the subspace $\mathcal{T}_1$. Therefore, it remains to show that $\widetilde{D}_0[\widetilde{D}_0, \operatorname{ad} W_n]\widetilde{D}_0$ contains $\mathcal{T}_1$.

For any $a \in \widetilde{D}_0$ and any

$$w = \sum_{i=1}^{n} f_i \partial_i \in W_n,$$

where $f_i \in \mathcal{T}$, we have the relation

$$[a, \operatorname{ad} w]\partial_i = -\sum_{j=1}^{n} (\partial_i f_j) a \partial_j - (\operatorname{ad} w) a \partial_i = -\sum_{j=1}^{n} \{(\partial_i f_j) a \partial_j - ((a \partial_i) f_j) \partial_j\}.$$

This relation implies that in the basis $\partial_1, \dots, \partial_n$ of the free $\mathcal{T}$-module W_n the elements of $[\widetilde{D}_0, \operatorname{ad} L_1]$ are written as matrices from $\widetilde{D}_1$, and some of these matrices are nonzero. However, then $\widetilde{D}_0[\widetilde{D}_0, \operatorname{ad} L_1]\widetilde{D}_0$ contains a nonzero scalar matrix with elements from $\mathcal{T}_1$. Since the subspace $\mathcal{T}_1$ is an irreducible module with respect to the action of $\operatorname{ad} L_0$ ($[\operatorname{ad} L_0, \mathcal{T}_1] \subset \mathcal{T}_1$) and

$$[[\widetilde{D}_0, \operatorname{ad} L_1], \operatorname{ad} L_0] \subseteq [[\widetilde{D}_0, \operatorname{ad} L_0], \operatorname{ad} L_1] + [\widetilde{D}_0, [\operatorname{ad} L_1, \operatorname{ad} L_0]] \subseteq [\widetilde{D}_0, \operatorname{ad} L_1],$$

we see that the set $\widetilde{D}_0[\widetilde{D}_0, \operatorname{ad} L_1]\widetilde{D}_0$ contains all the scalar matrices with elements from $\mathcal{T}_1$. Therefore, $\widetilde{D}_0 \cdot [\widetilde{D}_0, \operatorname{ad} W_n]\widetilde{D}_0$ contains $\mathcal{T}_1$ and, hence, the algebra $\widetilde{D}$ is generated by the subspace $[\widetilde{D}_0, \operatorname{ad} W_n]$. The lemma is proved.

PROOF of Theorem 21.1. By Lemma 21.1, it suffices to describe the nonzero $\mathbb{Z}$-graded subalgebras $D = \sum_{i=-\infty}^{\infty} D_i$ of the algebra $\widetilde{D} = \sum_{i=-\infty}^{\infty} \widetilde{D}_i$ ($D \cap \widetilde{D}_i = D_i$) satisfying Properties 1°–3°. It follows from Lemma 21.2 and its Corollary 2 that D_0 coincides either with $\widetilde{D}_0$ or with the center of $\widetilde{D}_0$.

If $D_0 = \widetilde{D}_0$, then it follows from Lemma 21.4 and Property 1° of the algebra D that $D = \widetilde{D}$.

If D_0 coincides with the center of the algebra $\widetilde{D}_0$, then, by Corollaries 1 and 2 of Lemma 21.2, the algebra D is commutative and is contained in the center of the algebra $\widetilde{D}$. It follows from Property 1° of the algebra D that this subalgebra is invariant under the action of $\operatorname{ad} W_n$. Since the center of the algebra $\widetilde{D}$ with respect to the action of the Lie algebra $\operatorname{ad} W_n$ is isomorphic to the algebra $\mathcal{T}$ with respect to the action of $\operatorname{Der}_K \mathcal{T} = W_n$, to prove the theorem it remains to find all invariant subalgebras of the algebra $\mathcal{T}$ with respect to $\operatorname{Der}_K \mathcal{T}$. But there are two subalgebras of this kind: the field K and the algebra $\mathcal{T}$ itself. The theorem is proved.

COROLLARY. *Let A be the subalgebra in* $\operatorname{End}_K W_n$ *defined in Proposition* 21.2. *Then the algebra A coincides either with the algebra $\widetilde{D}$ or with the center $\mathcal{T}$ of the algebra $\widetilde{D}$.*

PROOF. Proposition 21.2 shows that the algebra A satisfies all the assumptions of Theorem 21.1. Moreover, Property 4 of the algebra A excludes the case $A = K \cdot \mathbf{1}$. The corollary is proved.

Now we have all the necessary tools to prove Theorem 12.2. Since the algebra A is not equal to the field $K \cdot \mathbf{1}$ (Property 4 of the algebra A), we see that there exists a multilinear polynomial $h(t_1, \dots, t_p) \in R_{m.k}$ ($p = mk$) such that $h(\operatorname{ad} w_1, \dots, \operatorname{ad} w_p) \neq 0$ for some homogeneous elements $w_1, \dots, w_p \in W_n$ satisfying the relation $\sum_{i=1}^{p} |w_i| = 0$ and, moreover, $h(\operatorname{ad} v_1, \dots, \operatorname{ad} v_p) \notin K \cdot \mathbf{1}$ for some $v_1, \dots, v_p \in W_n$.

If the algebra A coincides with the center $\mathcal{T}$ of the algebra $\widetilde{D}$, then the element $h(\operatorname{ad} v_1, \dots, \operatorname{ad} v_p)$ belongs to $\mathcal{T}$ for any $v_1, \dots, v_p \in W_n$, and we can take the polynomial h in the role of f in Theorem 12.2.

If the algebra A coincides with the algebra $\widetilde{D}$, then we can choose the polynomial h in such a way that the values $h(\operatorname{ad} w_1, \dots, \operatorname{ad} w_p)$ range throughout the algebra $\widetilde{D}_0$, when $w_1, \dots, w_p$ range over all the sets of homogeneous elements of the Lie algebra W_n satisfying the relation $\sum_{i=1}^{p} |w_i| = 0$. By Corollary 2 of Lemma 21.2, $\widetilde{D}_0 = \operatorname{End}_K L_{-1}$ and the subalgebra $\widetilde{D}_0$ is isomorphic to the full matrix algebra M_n of order n over the field K. It is well known that for the algebra M_n there exists a multilinear polynomial $f_0(z_1, \dots, z_q)$ such that $f_0(a_1, \dots, a_q)$ is a scalar matrix for any elements $a_1, \dots, a_q \in M_n$ and f_0 determines a nonzero mapping $M_n \otimes \dots \otimes M_n \to M_n$. This follows, in particular, from Theorem 12.1 of the present chapter applied to the natural representation of the reductive Lie algebra M_n.

Denote by h_i the polynomials $h(t_{p(i-1)+1}, \dots, t_{p(i-1)+p})$. We set by definition $l = pq$ and take the polynomial $f_0\big|_{z_1=h_1,\dots,z_q=h_q}$ as f in Theorem 12.2. Obviously, a) the polynomial f belongs to $R_{m,k'}$, where $k' = kq$; b) if the elements $w_1, \dots, w_l \in W_n$ are homogeneous and $\sum_{i=1}^{l} |w_i| = 0$, then we have $f(\operatorname{ad} w_1, \dots, \operatorname{ad} w_l) \in K \cdot \mathbf{1}$, and some of these values are nonzero. Moreover, since $\widetilde{D} \simeq \mathcal{T} \otimes_K M_n$ (Corollary of Lemma 21.3) and the values of the polynomial h_i on $\operatorname{ad} W_n$ belong to the algebra $A \subseteq \widetilde{D}$, we see that $f(\operatorname{ad} w_1, \dots, \operatorname{ad} w_l)$ belongs to the center $\mathcal{T}$ of the algebra $\widetilde{D}$ for any $w_1, \dots, w_l \in W_n$.

Thus, for either alternative ($A = \mathcal{T}$ or $A = \widetilde{D}$) we have constructed a multilinear associative polynomial $f(t_1, \dots, t_l) \in R_{m,k}$ determining a nonzero mapping

$$f \circ \operatorname{ad}\colon W_n \otimes \dots \otimes W_n \to \mathcal{T} \subset \widetilde{D} \subset \operatorname{End}_K W_n.$$

It remains to show that the image of the mapping f coincides with $\mathcal{T}$. It follows from the proof of Proposition 21.2 (Property 4 of the algebra A) that the image of the mapping f is not contained in the field $K \cdot \mathbf{1}$. Since for any elements $w_1, \dots, w_l, w \in W_n$

$$[f(\operatorname{ad} w_1, \dots, \operatorname{ad} w_l), \operatorname{ad} w] = \sum_{i=1}^{l} f\big|_{t_1=\operatorname{ad} w_1,\dots,t_i=\operatorname{ad}[w_i,w],\dots,t_l=\operatorname{ad} w_l},$$

the image of the mapping f is an invariant subspace in $\mathcal{T}$ with respect to the natural action of the Lie algebra $\operatorname{Der}_K \mathcal{T} = W_n$. However, there are only two nonzero invariant subspaces in $\mathcal{T}$: the field K and the whole algebra $\mathcal{T}$. Hence, the image of the mapping f coincides with $\mathcal{T}$. Theorem 21.2 is completely proved.

A straightforward verification shows that for $n = 1$ the mapping

$$f \circ \operatorname{ad}\colon W_1 \otimes W_1 \otimes W_1 \to K[x_1] = \mathcal{T} = \operatorname{End}_{\mathcal{T}} W_n$$

is nonzero for a nonzero skew-symmetric multilinear polynomial f in three variables. Moreover, this mapping is nonzero for any field K whose characteristic differs from two. Therefore, for a Lie algebra of type W_1 (see Definition 44.3), a polynomial $f \in R_{3,1}$ permits to recover the commutative ring of this Lie algebra also for fields of positive characteristic ($\operatorname{char} K \neq 2$). We shall use this fact in §46 in the description of simple Lie algebras satisfying the standard Lie identity of degree 5.

Comments

The main results of the chapter: Theorems 12.1, 12.2, and 16.1 are published in the author's papers [Raz12, Raz14, Raz18]. In the chapter an important problem of the theory of 2-words and $\boldsymbol{\alpha}$-functions, namely, the one on the method of calculation of the $\boldsymbol{\alpha}$-function for irreducible representations of a semisimple Lie algebra, was passed over in silence. Since, by M. Duflo's theorem, any primitive ideal in the universal enveloping algebra $U(\mathfrak{g})$ of the semisimple Lie algebra $\mathfrak{g}$ over an algebraically closed field K of characteristic zero is realized as the annihilator of a quotient of a Verma module (see [Dix, Duf]), we see that the precise setting of this problem is the following.

PROBLEM 3.1. For any representation $\rho_\Lambda \colon \mathfrak{g} \to \operatorname{End}_K V_\Lambda$ of the simple Lie algebra $\mathfrak{g}$ in the Verma module with the highest weight Λ, calculate the value of $\boldsymbol{\alpha}$-function $\alpha_\Lambda \colon B \to K$ for any 2-word $u \in B$ ($\operatorname{char} K = 0$).

Let $\rho \colon \mathfrak{g} \to U(\mathfrak{g})$ be the canonical representation of the semisimple Lie algebra $\mathfrak{g}$ in its universal enveloping algebra and let an $\boldsymbol{\alpha}$-function $\alpha \colon B \to \mathcal{E}$ be constructed with respect to this representation (see §9). By definition, the algebra $\mathcal{E}$ coincides with the subalgebra of the center $\mathfrak{Z}$ of the algebra $U(\mathfrak{g})$ that is generated by $\alpha(B)$. At present, a method of calculation of values of generating elements of the center $\mathfrak{Z}$ in Verma modules for simple Lie algebras $\mathfrak{g}$ of types A, B, C, and D is known (see [Zhe]). Therefore, Problem 3.1 is closely related to the problem of description of the subalgebra $\mathcal{E}$ of the algebra $\mathfrak{Z}$. A coarse information on $\mathcal{E}$ can be extracted from Theorem 16.1: the natural mapping $\operatorname{Spec} \mathfrak{Z} \to \operatorname{Spec} \mathcal{E}$ is onto and one-to-one.

The results of Chapter 6 and the paper [Raz22] show that Theorem 12.2 permits us to obtain a fine information on the properties of Lie algebras of Cartan type. Therefore, the following problem is important in the theory of simple finite-dimensional Lie algebras over fields of positive characteristic.

PROBLEM 3.2. Let $\operatorname{char} K = p > 0$. For what numbers $n \in \mathbb{N}$ there exists an associative multilinear polynomial $f(x_1, \dots, x_l)$ such that all values $f\big|_{\operatorname{ad} W_n(K)}$ generate a nonzero commutative subalgebra in the algebra $\operatorname{End}_K W_n(K)$? (Presently it is known that for $p = 2$ and $n = 1$ there are no polynomials of this kind.)

CHAPTER IV

α-FUNCTIONS RELATED TO FULL MATRIX ALGEBRAS. TRACE IDENTITIES AND CENTRAL POLYNOMIALS OF FULL MATRIX ALGEBRAS M_n AND MATRIX SUPERALGEBRAS $M_{n,k}$

§22. Main results of the chapter

22.1. Main notation.

K is the ground field,

G is an associative Grassmann K-algebra with unit element, defined by a countable set of generators $\widetilde{t}_1, \widetilde{t}_2, \dots$ and defining relations $\widetilde{t}_i\widetilde{t}_j = -\widetilde{t}_j\widetilde{t}_i$, $i, j = 1, 2, \dots$;

G_0 is the K-subalgebra of even elements of the algebra G and its basis consists of all monomials of even degree;

G_1 is the K-subspace of odd elements of the algebra G and its basis consists of all monomials of odd degree;

M_n is the full matrix algebra of all square matrices of order n over the field K;

$\mathrm{tr}\colon M_n \to K$ is the K-linear function that assigns to any matrix $a \in M_n$ the sum of its diagonal elements;

$\mathrm{Sp}\colon M_n \to M_n$ is the K-linear mapping such that $\mathrm{Sp}(a) \rightleftharpoons \mathrm{tr}(a)\cdot \mathbf{1}$, where $a \in M_n$ and $\mathbf{1}$ is the unit matrix;

$M_{n,k}$ is the associative K-subalgebra in $G \otimes_K M_{n+k}$ consisting of all the matrices of the form

$$a = \begin{pmatrix} a_{11} & a_{12} \\ a_{21} & a_{22} \end{pmatrix}, \tag{22.1}$$

where a_{11} and a_{22} are square G_0-matrices of order $n \times n$ and $k \times k$, respectively, a_{21} and a_{12} are rectangular G_1-matrices of type $k \times n$ and $n \times k$, respectively; the algebra $M_{n,k}$ has a natural 2-gradation $M^0_{n,k} \oplus M^1_{n,k}$, where $M^0_{n,k}$ consists of matrices of the form $\begin{pmatrix} a_{11} & 0 \\ 0 & a_{22} \end{pmatrix}$ and $M^1_{n,k}$ of matrices of the form $\begin{pmatrix} 0 & a_{12} \\ a_{21} & 0 \end{pmatrix}$;

$\mathrm{str}\colon M_{n,k} \to G_0$ is the G_0-linear mapping that assigns to any matrix $a \in M_{n,k}$ of the form (22.1) an element of G_0 that is equal to the sum of all diagonal elements of the matrix a_{11} minus the sum of all diagonal elements of the matrix a_{22};

$\mathrm{Sp}\colon M_{n,k} \to M_{n,k}$ is the G_0-linear mapping such that $\mathrm{Sp}(a) \rightleftharpoons \mathrm{str}(a)\cdot\mathbf{1}$, where $a \in M_{n,k}$ and $\mathbf{1}$ is the unit matrix in $M_{n,k}$.

22.2. Main results. In this chapter we consider the problem posed in §11 of Chapter 2 for the case in which the algebraically closed field K is of characteristic zero, S is the set $\{1, 2, 3, \dots\}$, $\rho_n\colon \mathfrak{g}_n \to \mathrm{End}_K V_n$ is the natural irreducible representation of the reductive Lie algebra $\mathfrak{g}_n = \mathrm{End}_K V_n$ in an n-dimensional linear K-space V_n, the nondegenerate symmetric invariant bilinear form $(\,,\,)\colon \mathfrak{g}_n \otimes_K \mathfrak{g}_n \to K$ on $\mathfrak{g}_n$ is given by the formula $(x, y) \rightleftharpoons \operatorname{tr}(xy)$, and the $\boldsymbol{\alpha}$-function $\alpha_n\colon B \to K$ is constructed with respect to the representation ρ_n in the same way as in §9 of Chapter 2. In this case the associative subalgebra U_n generated in $\operatorname{End}_K V_n$ by the Lie algebra $\rho_n(\mathfrak{g}_n) = \mathfrak{g}_n = \operatorname{End}_K V_n$ coincides with $\operatorname{End}_K V_n$ and is isomorphic to the full matrix algebra M_n over the field K.

Let the K-algebra $\mathcal{E}$ and the $\boldsymbol{\alpha}$-function $\alpha\colon B \to \mathcal{E}$ be constructed from the $\boldsymbol{\alpha}$-functions $\alpha_n\colon B \to \mathcal{E}$ as it was recommended in §11. For this concrete $\boldsymbol{\alpha}$-function, in this chapter we shall give answers to all the problems of §11.

Answering these questions, we obtain an important additional and very substantial information about associative algebras with identical relations. Note that the final results expressing this information admit formulations without mention of the $\boldsymbol{\alpha}$-function, although their proofs are completely based on the classification of the lattice of $\eta_{\widetilde{\alpha}}$-closed ideals of the algebra $\mathcal{E}$ for some extension $\widetilde{\alpha}\colon \widetilde{B} \to \mathcal{E}$ of the $\boldsymbol{\alpha}$-function $\alpha\colon B \to \mathcal{E}$ (see Theorems 26.1 and 26.2).

Here are the main results following from this classification:

a) all trace identities of the full matrix algebra M_n over a field of characteristic zero are equivalent to a sole trace identity of degree n—the trace identity expressing the Hamilton–Cayley theorem (see Theorem 27.1);

b) all trace identities of the matrix superalgebra $M_{n,k}$ over a field of characteristic zero are equivalent to trace identities of degree $nk + n + k$ (see Theorem 27.2);

c) for any positive integers n and k, the algebra $M_{n,k}$ over a field of arbitrary characteristic possesses central polynomials, i.e., associative polynomials f such that their values on $M_{n,k}$ belong to the center of the algebra $M_{n,k}$ and the identity $f = 0$ does not hold in $M_{n,k}$ (see Theorem 31.1); in the special case $k = 0$ the algebras M_n and $M_{n,0}$ have the same identities, and we obtain a central polynomial for the full matrix algebra M_n, which gives a solution to a problem of I. Kaplansky (see [Kap3]);

d) in associative algebras over fields of characteristic zero the identity $y^n = 0$ implies the nilpotence identity $x_1x_2\cdots x_{n^2} = 0$ (see Theorem 33.1); this is a substantial amelioration of the bound of nilpotence class given by G. Higman: the identity $y^n = 0$ implies the identity $x_1x_2\cdots x_{2^n} = 0$ (see [Higm]);

e) for any prime $p > 3$, over a field of characteristic p there exist Lie unsolvable associative algebras satisfying the $(p-1)$th Engel identity $[x, \underbrace{y, \dots, y}_{p-1 \text{ times}}] = 0$ and the identity $y^p = 0$ (see Theorem 33.2);

f) for any prime $p > 3$, over a field of characteristic p there exist unsolvable prime varieties $\overline{\mathfrak{B}}_{\gamma-i}$ $(i = 2, 3, \dots, (p-1)/2)$ of associative algebras such that in these algebras the $(p-1)$th Engel identity $[x, \underbrace{y, \dots, y}_{p-1 \text{ times}}] = 0$ holds and $\operatorname{var} M_i \supset \overline{\mathfrak{B}}_{\gamma-i}$, but in $\overline{\mathfrak{B}}_{\gamma-i}$ the standard identity of degree $2(i-1)$ fails (see Theorem 33.3).

We stress that all these results are necessary and appropriate consequences of the intrinsic logic of solution of the problem on the description of the semilattice of η_α-closed ideals for the $\boldsymbol{\alpha}$-function $\alpha\colon B \to \mathcal{E}$, considered in this chapter, constructed by $\boldsymbol{\alpha}$-functions α_n of representations $\rho_n\colon \mathfrak{g}_n \leftrightarrow \operatorname{End}_K V_n$.

§23. Calculation of the $\boldsymbol{\alpha}$-function α_n and the algebra $\mathcal{E}$

Let the ground field K be the complex field.

In this section, following the directions of §9, we shall construct for any irreducible representation $\rho_n\colon \mathfrak{g}_n \to \operatorname{End}_K V_n$ the $\boldsymbol{\alpha}$-function $\alpha_n\colon B \to \mathcal{E}_n \simeq K$ and indicate an algorithm for calculating it. The merit of this algorithm is that it permits us to find the value of the function α_n for a 2-word u directly, not using the representation ρ_n itself. Subsequently, in the category ALPHA, in accordance with the recommendation of §11, we shall construct for $\alpha_n\colon B \to \mathcal{E}_n$ an $\boldsymbol{\alpha}$-function $\alpha\colon B \to \prod_{n=1}^{\infty} \mathcal{E}_n$ (the product of the objects $\alpha_n\colon B \to \mathcal{E}_n$) and calculate the subalgebra $\mathcal{E}$ of the algebra $\prod_{n=1}^{\infty} \mathcal{E}_n$ generated by elements $\alpha(u) \in \prod_{n=1}^{\infty} \mathcal{E}_n$, where u ranges over B. Thus, we shall answer the first two questions posed in §11, because we shall have an explicit description of the algebra and a successful algorithm for calculating the $\boldsymbol{\alpha}$-function $\alpha\colon B \to \mathcal{E}$. This algorithm will force us to extend the signature operations by introducing the formal "trace" function Sp, and finally it will lead us to an extension of the domain of the function α and to a more general treatment of the problem of description of the lattice of η_α-closed ideals of the algebra $\mathcal{E}$.

LEMMA 23.1. *Let $e_1, \dots, e_{n^2}$ be an arbitrary basis of the matrix algebra M_n over an arbitrary field and let $e'_1, \dots, e'_{n^2}$ be the dual basis with respect to the bilinear form determined by the trace $(x, y) \rightleftharpoons \operatorname{tr}(xy)$. Then for any element $a \in M_n$ we have the relation*

$$\sum_{t=1}^{n^2} e_t a e'_t = \operatorname{tr}(a) \cdot \mathbf{1}, \tag{23.1}$$

where $\mathbf{1}$ denotes the unit matrix.

PROOF. Obviously, equality (23.1) is valid in a special case in which we take for a basis $e_1, \dots, e_{n^2}$ the basis of matrix units E_{ij} $(i, j = 1, \dots, n)$. Therefore, it suffices to verify that the element in the left-hand side of (23.1) does not depend on the choice of the basis $e_1, \dots, e_{n^2}$. Let $f_1, \dots, f_{n^2}$ be another basis and $f_i = \sum_{j=1}^{n^2} c_{ji} e_j$; then it is well known that $e'_i = \sum_{j=1}^{n^2} c_{ij} f'_j$. Using these two relations, we obtain

$$\sum_{t=1}^{n^2} f_t a f'_t = \sum_{t=1}^{n^2} \sum_{j=1}^{n^2} c_{jt} e_j a f'_t = \sum_{j=1}^{n^2} e_j a \left(\sum_{t=1}^{n^2} c_{jt} f'_t \right) = \sum_{j=1}^{n^2} e_j a e'_j,$$

and this proves that the element in the left-hand side of equation (23.1) does not depend on the choice of the basis. The lemma is proved.

Let $\boldsymbol{\sigma}$ be an anti-isomorphism of the full matrix algebra M_n over the complex field K defined as the composition of matrix transposition and of complex conjugation. Let $e_1, \dots, e_{n^2}$ be an orthonormal basis of the Lie algebra $M_n = \mathfrak{g}_n$ with respect to the bilinear form $(x, y)_n = \operatorname{tr}(xy)$ and $\boldsymbol{\sigma}(e_i) = e_i$ (we can take the matrix units E_{ii} for the first n vectors and, for example, $\frac{1}{\sqrt{2}}(E_{kl} + E_{lk})$ and $\sqrt{\frac{-1}{2}}(E_{kl} - E_{lk})$ $(k > l)$ for the remaining vectors). Let $\overline{\mathcal{F}}$ be a commutative K-algebra with unit given by generators $\overline{x}_i^{(j)}$ $(i = 1, \dots, n^2;\ j = 1, 2, \dots)$ and defining relations

$$\overline{x}_i^{(j)}\overline{x}_i^{(t)}\overline{x}_i^{(k)} = 0, \qquad \overline{x}_i^{(j)}\overline{x}_i^{(t)} = \delta_{jt}\big(\overline{x}_i^{(1)}\big)^2. \tag{23.2}$$

Denote by $\overline{M}_n$ the matrix algebra $\overline{\mathcal{F}} \otimes_K M_n$ over the ring $\overline{\mathcal{F}}$. We extend the anti-isomorphism $\boldsymbol{\sigma}$ to $\overline{M}_n$ by $\overline{\mathcal{F}}$-linearity. Let us consider in $\overline{M}_n$ the associative K-subalgebra $\overline{F}_n$ generated by the elements $\overline{x}_i = \sum_{j=1}^{n^2} \overline{x}_i^{(j)} e_j$, $i = 1, 2, \dots$. For any matrix $b \in \overline{M}_n$ denote by $\mathrm{Sp}(b)$ the scalar matrix whose diagonal elements (belonging to the algebra $\overline{\mathcal{F}}$) are equal to the trace of the matrix b.

By Lemma 9.1, for any multihomogeneous associative polynomial $u(t_1, \dots, t_l)$ in ${}_\mu H$ we have the following equality in $\overline{M}_n$:

$$\overline{u} \rightleftharpoons u(\overline{x}_1, \dots, \overline{x}_l) = \big(\overline{x}_1^{(1)}\big)^2 \cdots \big(\overline{x}_l^{(1)}\big)^2 u', \tag{23.3}$$

where u' belongs to the center of the algebra M_n, and, by definition, the linear $\boldsymbol{\alpha}$-function α_n on the 2-word $\overline{u}$ is equal to $u' \in \mathcal{E}_n$. Our immediate task is to find relations among the generators $\overline{x}_i$ of the algebra $\overline{F}_n$, which will permit us to find the right-hand side of the equality (23.3) for any $\overline{u}$.

Consider the element $\overline{x}_i a \overline{x}_i$, where a is an arbitrary element of $\overline{M}_n$. By Lemma 23.1 and relations (23.2), we obtain

$$\overline{x}_i a \overline{x}_i = \big(\overline{x}_i^{(1)}\big)^2 \sum_{j=1}^{n^2} e_j a e_j' = \big(\overline{x}_i^{(1)}\big)^2 \,\mathrm{Sp}(a).$$

If $a = \mathbf{1}$ is the unit matrix, then this equality obtains the form

$$\overline{x}_i^2 = n\big(\overline{x}_i^{(1)}\big)^2 \cdot \mathbf{1} \tag{23.4}$$

and, hence,

$$\overline{x}_i a \overline{x}_i = \frac{\overline{x}_i^2}{n}\,\mathrm{Sp}(a), \qquad \overline{x}_i \overline{x}_i = n\frac{\overline{x}_i^2}{n} \tag{23.5}$$

for any $a \in \overline{M}_n$. The equalities (23.5) imply

$$\big[\overline{x}_i^2, \overline{x}_j\big] = 0, \qquad \big[\overline{x}_i a \overline{x}_i, \overline{x}_j\big] = 0, \qquad \overline{x}_i a b \overline{x}_i = \overline{x}_i b a \overline{x}_i. \tag{23.6}$$

A direct calculation shows that

$$\overline{x}_i\overline{x}_j\overline{x}_i\overline{x}_j = \frac{\overline{x}_i^2}{n}\,\mathrm{Sp}(\overline{x}_j)\overline{x}_j = \frac{\overline{x}_i^2}{n}\big(\overline{x}_j^{(1)} + \dots + \overline{x}_j^{(n)}\big)\overline{x}_j = \frac{\overline{x}_i^2}{n}\big(\overline{x}_j^{(1)}\big)^2 = \frac{\overline{x}_i^2}{n}\cdot\frac{\overline{x}_j^2}{n}. \tag{23.7}$$

It follows from the equalities (23.6) and (23.7) that

$$\overline{x}_i\overline{x}_j a \overline{x}_i\overline{x}_j = \overline{x}_i a \overline{x}_j\overline{x}_i\overline{x}_j = \overline{x}_i\overline{x}_j\overline{x}_i\overline{x}_j a = \frac{\overline{x}_i^2}{n}\cdot\frac{\overline{x}_j^2}{n}\cdot a. \tag{23.8}$$

On the other hand, (23.4) implies that

$$\overline{x}_i\overline{x}_j a\overline{x}_i\overline{x}_j = \frac{\overline{x}_i^2}{n}\,\mathrm{Sp}(\overline{x}_j a)\overline{x}_j.$$

Comparing the last two equalities and taking into account that $\frac{\overline{x}_i^2}{n}$ is a nonzero scalar matrix, we see that for any $a \in \overline{M}_n$

$$\mathrm{Sp}(\overline{x}_j a)\overline{x}_j = \frac{\overline{x}_i^2}{n}\cdot a. \tag{23.9}$$

By virtue of what was said above, the following equalities are evident:

$$\mathrm{Sp}(\overline{x}_i^2 a) = \overline{x}_i^2\,\mathrm{Sp}(a), \qquad \mathrm{Sp}(\overline{x}_i^2) = n\overline{x}_i^2. \tag{23.10}$$

Obviously, using relations (23.6), we can express any 2-word $\overline{u}$ as a product of words of two forms: a) $\overline{x}_i\overline{x}_j\overline{x}_i\overline{x}_j$, b) $\overline{x}_t^2$. Comparing relations (23.3) and (23.4), (23.7), we see that

$$\alpha_n(\overline{x}_i\overline{x}_j\overline{x}_i\overline{x}_j) = 1, \qquad \alpha_n(\overline{x}_i^2) = n \tag{23.11}$$

and we can rewrite the equality (23.3) in the form

$$\overline{u} = u(\overline{x}_1,\dots,\overline{x}_l) = \frac{\overline{x}_1^2}{n}\cdots\frac{\overline{x}_l^2}{n}\,\alpha_n(\overline{u}). \tag{23.12}$$

Therefore, from Corollary of Lemma 9.1 and the third-defining property of an $\boldsymbol{\alpha}$-function it follows that for any 2-word $\overline{u} \in B$

$$\alpha_n(\overline{u}) = n^k, \tag{23.13}$$

where the positive integer k depends on the word $\overline{u}$ only. This proves that the $\boldsymbol{\alpha}$-function $\alpha\colon B \to \prod_{i=1}^{\infty}\mathcal{E}_i$, being the product of $\boldsymbol{\alpha}$-functions $\alpha_n\colon B \to \mathcal{E}_n$, on the 2-word $\overline{u}$ takes the value

$$(1^k, 2^k, 3^k, \dots, n^k, \dots) \in \prod_{i=1}^{\infty}\mathcal{E}_i.$$

LEMMA 23.2. *The subalgebra $\mathcal{E}$ generated in $\prod_{n=1}^{\infty}\mathcal{E}_n$ by the values of the function α contains the unit and coincides with the polynomial K-algebra $K[\gamma]$ in one variable $\gamma = (1,\dots,n,\dots) \in \prod_{n=1}^{\infty}\mathcal{E}_n$. Moreover,*

$$\alpha(\overline{x}_i\overline{x}_j\overline{x}_i\overline{x}_j) = 1, \qquad \alpha(\overline{x}_i^2) = \gamma, \tag{23.14}$$

and the linear function $\alpha\colon B \to K[\gamma]$ is uniquely calculated from the relations (23.6).

The proof is obvious.

This lemma gives a description of the K-algebra $\mathcal{E}$ and an algorithm for calculating the $\boldsymbol{\alpha}$-function $\alpha\colon B \to \mathcal{E} = K[\gamma]$.

In the concrete situation $\rho_n\colon \mathfrak{g}_n \to \operatorname{End}_K V_n$ under consideration there are two peculiarities. The first is that $\mathfrak{g}_n = \operatorname{End}_K V_n$, i.e., the Lie algebra coincides with its associative enveloping algebra. Obviously, in this case the ideal of identities $\mathfrak{V}$ of the pair $(U, \mathfrak{g})$, where $U = \mathfrak{g}$, is a T-ideal of the free associative algebra H, i.e., the relation $f(t_1, \dots, t_l) \in \mathfrak{V}$ implies $f(g_1, \dots, g_l) \in \mathfrak{V}$ for any $g_1, \dots, g_l$ in H. Therefore, we are justified in expecting that for the function $\alpha\colon B \to K[\gamma]$ just constructed all θ_α-closed ideals of weak identities will be T-ideals. As we shall see in §§25–27, this is really the case, and this follows from the fact that for the bilinear pairing $\mathfrak{b}\colon {}_\nu H_\lambda \otimes_K {}_\mu H_\lambda \to \mathcal{E} = K[\gamma]$ the third of equalities (23.6) implies the following reinforcement of equality (8.2):

$$\mathfrak{b}\big(f\big|_{x_1 = x_1 v_1, \dots, x_l = x_l v_l}, g\big) = \mathfrak{b}\big(f, g\big|_{x_1 = v_1 x_1, \dots, x_l = v_l x_l}\big),$$

where $x_1, \dots, x_l \in \lambda$, $v_1, \dots, v_l \in \nu \cup \lambda$. Therefore, the study of θ_α-closed ideals of weak identities for our $\boldsymbol{\alpha}$-function α must give an information on varieties of associative algebras.

The second peculiarity is in the following. Although the $\boldsymbol{\alpha}$-function α is calculated by formulas (23.6) and (23.14) not including the Sp sign, nevertheless, the expression $\overline{x}_i a \overline{x}_i$ interprets the function "trace" implicitly and, as we shall see below, it is of interest to make this interpretation explicit. The most powerful arguments in favor of this idea are the results of §§25–28. At this stage, the leading reasons in favor of introducing the trace function Sp are as follows. First, in the calculation of the function $\alpha\colon B \to \mathcal{E}$, the function Sp permits us to use (besides relations (23.6)) the trace relations (23.5), (23.9), and (23.10), which give the same value of the function α; moreover, these formulas permit us to calculate this function also for multihomogeneous trace expressions of order 2 with respect to each variable $\overline{x}_i$. Second, in this calculation of the expression $\mathfrak{b}(f, w)$, where $f \in H_\lambda$, $w \in {}_\mu H_\lambda$, we can choose the element w in such a way that $w = w_1 w_2$ and $\mathfrak{b}(f, w) = \alpha(\overline{h w_2})$, where h is a trace expression of a rather general form.

§24. An algebra of trace polynomials. Main notions

In this section we shall formalize the problem on trace polynomials. Toward this end, we shall construct an algebra of generalized polynomials (an algebra of trace polynomials), which will play for trace identities the same role as the algebra H of noncommutative polynomials plays for identities of associative algebras.

In order to construct this algebra, we need variables $x_1, x_2, \dots$ and a formal function $\operatorname{Sp}(\cdots)$. Let F be the free associative K-algebra with free generators $x_1, x_2, \dots$. By the *free algebra of generalized polynomials* (or *free trace algebra*) with free generators $x_1, x_2, \dots$ we mean an associative algebra $\mathcal{G}$ given by generators $x_1, x_2, \dots$ and the elements $\operatorname{Sp}(a)$, where a ranges over the algebra F, and with defining relations

$$[\operatorname{Sp} a, b] = [\operatorname{Sp} a, \operatorname{Sp} b] = 0, \tag{24.1}$$

$$\operatorname{Sp}(ab) = \operatorname{Sp}(ba), \tag{24.2}$$

$$\operatorname{Sp}(\beta a + \delta b) = \beta \operatorname{Sp}(a) + \delta \operatorname{Sp}(b), \tag{24.3}$$

where a and b range over F and β and δ range over the ground field K. It is clear that the algebra F is contained in $\mathcal{G}$ in a natural manner and the algebra $\mathcal{G}$ is spanned as a linear space by generalized monomials of the form

$$a_0 \operatorname{Sp}(a_1) \cdots \operatorname{Sp}(a_t), \tag{24.4}$$

where $a_0, a_1, \dots, a_t$ are words and $a_1, \dots, a_t$ are nonempty words. Obviously, different monomials of the form (24.4) can be equal to each other in view of relations (24.1) and (24.2). However, this ambiguity can be easily eliminated if we take for a representative of a class of congruent elements of the form (24.4) an element such that the word $a_0 a_1 \cdots a_t$ is maximal from the viewpoint of lexicographic order. Clearly, these monomials form a basis of the trace algebra $\mathcal{G}$.

The elements of the form (24.4) will be called *generalized monomials* and the elements of the algebra $\mathcal{G}$ will be called *trace polynomials* (or *generalized polynomials*). For any generalized monomial we can naturally define its degree with respect to a variable x_i as the number of entries of x_i in the notation of this monomial. Thus, to any generalized monomial in $x_1, \dots, x_l$ we can assign the collection of numbers $(r_1, \dots, r_l)$, where r_i is equal to the degree of this monomial with respect to x_i. This collection of numbers will be called the *type* of the generalized monomial, and a trace polynomial that is a linear combination of generalized monomials of the same type $(r_1, \dots, r_l)$ will be called a *multihomogeneous trace polynomial of type* $(r_1, \dots, r_l)$; for instance, $x_1 x_2 + x_2 \operatorname{Sp}(x_1) + \frac{1}{2} \operatorname{Sp}(x_1 x_2)$ is a multihomogeneous trace polynomial of type $(1, 1)$ and the polynomial $x_1 + x_2$ is not multihomogeneous. A multihomogeneous trace polynomial $f(x_1, \dots, x_l)$ of type $(\underbrace{1, \dots, 1}_{l \text{ times}})$ is called a *multihomogeneous trace polynomial of degree* l.

Definition 24.1. Suppose that in an associative K-algebra U a K-linear function $\operatorname{Sp}\colon U \to U$ satisfying (24.2) and taking its values in the center of the algebra U is given. The formula $f(x_1, \dots, x_l) = 0$, where $f \in \mathcal{G}$, is called a *trace identity* for the algebra U if $f(a_1, \dots, a_l) = 0$ for any $a_1, \dots, a_l \in U$.

Obviously, if $f(x_1, \dots, x_l) = 0$ is a trace identity of the algebra U, then for any $g_1, \dots, g_l \in F$ and $h_1, h_2 \in \mathcal{G}$ the formula $h_1 f(g_1, \dots, g_l) h_2 = 0$ is also a trace identity of the algebra U. This leads to the following definition.

Definition 24.2. A two-sided ideal $\mathfrak{V}$ of the trace algebra $\mathcal{G}$ is said to be a *verbal ideal* (or an *ideal of trace identities*) if for any trace polynomial $f(x_1, \dots, x_l)$ contained in $\mathfrak{V}$ and any polynomials $g_1, \dots, g_l \in F$ the trace polynomial $f(g_1, \dots, g_l)$ is contained in $\mathfrak{V}$. A verbal ideal $\mathfrak{V}$ consisting of those and only those trace polynomials f such that $f = 0$ is a trace identity of the algebra U is called the *ideal of trace identities of the algebra* U.

Definition 24.3. Let P be a set of trace polynomials contained in $\mathcal{G}$. We say that the ideal of trace identities $\mathfrak{V}$ is *generated by* P if $\mathfrak{V}$ is the least ideal of trace identities containing P. A trace identity $f = 0$ is said to be a *consequence* of the trace identities $g = 0$, where g ranges over P, if $f \in \mathfrak{V}$. Two sets of trace polynomials are said to be *equivalent* if the ideals of trace identities generated by them coincide.

Clearly, the functions $\operatorname{Sp}\colon M_n \to M_n$ and $\operatorname{Sp}\colon M_{n,k} \to M_{n,k}$ have their values in the centers of these algebras and satisfy property (24.2). Therefore, we can apply to them

Definitions 24.2 and 24.3 given above, and for algebras M_n and $M_{n,k}$ we can pose the problem of finding all trace identities and the existence problem of a finite base of these trace identities.

Denote by T_l the set of multilinear trace polynomials in variables $x_1, \dots, x_l$ of type $(\underbrace{1, \dots, 1}_{l \text{ times}})$.

LEMMA 24.1. *The dimension of the linear space T_l is equal to $(l+1)!$.*

PROOF. As it was noted above, there is a basis in the space T_l formed by generalized monomials of the form (24.4), where $a_1, \dots, a_t$ are nonempty words, and for any generalized monomial $b_0 \operatorname{Sp}(b_1) \cdots \operatorname{Sp}(b_t)$ congruent to $a_0 \operatorname{Sp}(a_1) \cdots \operatorname{Sp}(a_t)$ with respect to (24.1) and (24.2), the multilinear word $a_0 a_1 \cdots a_t$ is strictly greater than the word $b_0 b_1 \cdots b_t$ with respect to the lexicographic order. Let a_0 be a nonempty word. Then the number of basis generalized monomials of the form $x_i a_0' \operatorname{Sp}(a_1) \cdots \operatorname{Sp}(a_t)$ is equal to $l!$ by inductive argument and, hence, the total number of basis generalized monomials of the form (24.4) in T_l, where a_0 is a nonempty word, is equal to $ll!$. If a_0 is empty, then $a_1 = x_l a_1'$ and the number of basis generalized monomials of this type coincides with the number of basis generalized monomials of the form $a_1' \operatorname{Sp}(a_2) \cdots \operatorname{Sp}(a_t)$, which is equal to $l!$ by inductive argument. Hence, the total number of basis generalized monomials of the form (24.4) equals $(l+1)!$. The lemma is proved.

§25. An auxiliary trace algebra

In this section we extend the domain of the function $\alpha\colon B \to K[\gamma]$ from §23, generalize the problem on η_α-closed ideals of the algebra $\mathcal{E} = K[\gamma]$, and describe some properties of $\theta_{\tilde{\alpha}}$-closed ideals of trace identities.

25.1. The algebra $\mathcal{G}_n$. Let $K, \overline{\mathcal{F}}, \overline{M}_n, \alpha, \alpha_n, \boldsymbol{\sigma}$, and $\overline{x}_i$ have the same meaning as in §23. Denote by $\mathcal{G}_n$ the associative subalgebra in $\overline{M}_n$, generated by all the elements of the form $\overline{x}_i$ $(i = 1, 2, \dots)$ and $\operatorname{Sp}(\overline{x}_{i_1}, \overline{x}_{i_2}, \dots, \overline{x}_{i_k})$, containing the unit.

We note the following properties of the algebra $\mathcal{G}_n$.

1°. Since in the algebra $\overline{M}_n$ for any $a, b \in \overline{M}_n$ we have the equality

$$\operatorname{Sp}(a \operatorname{Sp}(b)) = \operatorname{Sp}(a) \operatorname{Sp}(b), \tag{25.1}$$

the equalities (23.5), (23.6), (23.9), and (23.10) imply the following relations in $\mathcal{G}_n$:

$$\overline{x}_i a \overline{x}_i = \frac{\overline{x}_i^2}{n} \operatorname{Sp}(a), \tag{25.2}$$

$$\operatorname{Sp}(\overline{x}_i a)\overline{x}_i = \frac{\overline{x}_i^2}{n} a, \qquad \operatorname{Sp}(\overline{x}_i a) \operatorname{Sp}(\overline{x}_i b) = \frac{\overline{x}_i^2}{n} \operatorname{Sp}(ab), \tag{25.3}$$

$$\operatorname{Sp}(\overline{x}_i b \overline{x}_i a) = \overline{x}_i b \overline{x}_i \operatorname{Sp}(a), \qquad [\overline{x}_i^2, \overline{x}_j] = 0, \tag{25.4}$$

$$\operatorname{Sp}(\mathbf{1}) = n, \qquad \overline{x}_i^3 = 0, \tag{25.5}$$

where a and b are arbitrary words in generators $\overline{x}_t$, $i, j = 1, 2, \dots$. The last equality in (25.5) follows from the defining relations (23.2) of the algebra $\overline{\mathcal{F}}$. From equalities (25.2)–(25.5) it follows that for any multihomogeneous trace polynomial $f(x_{i_1}, \dots, x_{i_k}, x_{j_1}, \dots, x_{j_l})$ of degree 2 with respect to each of the variables $x_{i_1}, \dots, x_{i_k}$ and of degree 1 with respect to each of the variables $x_{j_1}, \dots, x_{j_l}$ we have the relation

$$\overline{f} \rightleftharpoons f(\overline{x}_{i_1}, \dots, \overline{x}_{i_k}, \overline{x}_{j_1}, \dots, \overline{x}_{j_l}) = \left\{\prod_{t=1}^{k} (\overline{x}_{i_t}^2/n)\right\} g(\overline{x}_{j_1}, \dots, \overline{x}_{j_l}), \tag{25.6}$$

where $g(x_{j_1}, \dots, x_{j_l})$ is a multilinear trace polynomial.

2°. The special case $l = 0$ of (25.6) is of particular interest. In this case

$$\overline{f} = \widetilde{\alpha}_n(f)(\overline{x}_{i_1}^2/n) \cdots (\overline{x}_{i_k}^2/n), \tag{25.7}$$

where $\widetilde{\alpha}(f) \in K$ is a number uniquely determined by the trace polynomial f. Equality (23.12) implies that for a multihomogeneous trace polynomial f of type $(2, \dots, 2)$ we have $\alpha_n : B \to \mathcal{E}_n \simeq K$. Therefore, formula (25.7) determines an extension of the $\boldsymbol{\alpha}$-function $\alpha_n : B \to \mathcal{E}_n \simeq K$ to a linear function $\widetilde{\alpha}_n : \widetilde{B} \to K$, where $\widetilde{B}$ is the linear K-space generated by all the multihomogeneous trace polynomials f of type $(2, \dots, 2)$. Obviously, we have

$$\operatorname{tr}(f(\overline{x}_{i_1}, \dots, \overline{x}_{i_k})) = n\widetilde{\alpha}_n(f)(\overline{x}_{i_1}^{(1)})^2 \cdots (\overline{x}_{i_k}^{(1)})^2. \tag{25.8}$$

Let $u(x_1, \dots, x_l)$ be a multihomogeneous trace polynomial. Denote by $\overline{u}$ the matrix $u(\overline{x}_1, \dots, \overline{x}_l) \in \mathcal{G}_n \subset \overline{M}_n$. Then the element standing in it at the intersection of ith row and jth column has the form

$$a_{ij} = \sum_{r_1, \dots, r_l} \beta_{r_1, \dots, r_l}^{ij} \overline{x}_1^{(r_1)} \cdots \overline{x}_l^{(r_l)},$$

where $\beta_{r_1, \dots, r_l}^{ij} \in K$. Taking into account the following equality that is a consequence of (23.2):

$$\overline{x}_1^{(j_1)} \cdots \overline{x}_l^{(j_l)} \overline{x}_1^{(t_1)} \cdots \overline{x}_l^{(t_l)} = \delta_{j_1, \dots, j_l}^{t_1, \dots, t_l} \left(\overline{x}_1^{(1)}\right)^2 \cdots \left(\overline{x}_l^{(1)}\right)^2,$$

where $\delta_{j_1, \dots, j_l}^{t_1, \dots, t_l}$ is the Kronecker symbol, we see that a direct calculation of the trace of the matrix $f = \overline{u}\sigma(\overline{u})$ gives

$$\operatorname{tr}(f) = \left(\overline{x}_1^{(1)}\right)^2 \cdots \left(\overline{x}_l^{(1)}\right)^2 \sum_{i, j, r_1, \dots, r_l} \left|\beta_{r_1, \dots, r_l}^{ij}\right|^2, \tag{25.9}$$

and $\operatorname{tr}(f) = 0$ if and only if $\overline{u} = 0$. On the other hand, it is easy to see that the algebra $\mathcal{G}_n$ is invariant under the anti-isomorphism $\boldsymbol{\sigma}$, by the choice of the basis $e_1, \dots, e_{n^2}$ of the algebra M_n. Indeed,

$$\begin{gathered} \boldsymbol{\sigma}(\overline{x}_i) = \sum_{j=1}^{n^2} \overline{x}_i^{(j)} \boldsymbol{\sigma}(e_j) = \overline{x}_i, \qquad \boldsymbol{\sigma}(\overline{x}_{i_1} \overline{x}_{i_2} \cdots \overline{x}_{i_t}) = \overline{x}_{i_t} \cdots \overline{x}_{i_2} \overline{x}_{i_1}, \\ \boldsymbol{\sigma}(\beta a_0 \operatorname{Sp}(a_1) \cdots \operatorname{Sp}(a_t)) = \overline{\beta} \boldsymbol{\sigma}(a_0) \operatorname{Sp}(\boldsymbol{\sigma}(a_1)) \cdots \operatorname{Sp}(\boldsymbol{\sigma}(a_t)), \end{gathered} \tag{25.10}$$

where $a_0, a_1, \ldots, a_t$ are words in $\overline{x}_i$. Hence, $\boldsymbol{\sigma}(\overline{u}) \in \mathcal{G}_n$, and the trace experssion $f(\overline{x}_1, \ldots, \overline{x}_l)$ is multihomogeneous of type $(2, \ldots, 2)$ and formula (25.7) is valid for it. Due to (25.8), the assertion expressed by the equality (25.9) can be reformulated in the following way: for any multilinear trace polynomial u we have

$$n\widetilde{\alpha}(\overline{u}\boldsymbol{\sigma}(\overline{u})) \geq 0, \tag{25.11}$$

and equality is attained if and only if $\overline{u} = 0$. But by the defining relations (23.2) we have

$$u(\overline{x}_1, \ldots, \overline{x}_l) = \sum_{r_1, \ldots, r_l} \overline{x}_1^{(r_1)} \cdots \overline{x}_l^{(r_l)} u(e_{r_1}, \ldots, e_{r_l}),$$

and $\overline{u} = 0$ if and only if all $u(e_{r_1}, \ldots, e_{r_l})$ are zero. Since $\overline{u}$ is a multilinear trace polynomial, we see that these equalities mean that the trace identity $u = 0$ is valid in M_n.

Therefore, (25.11) gives a criterion for verifying whether $u = 0$ is a trace identity in M_n or not. Note that in application of this criterion we do not need the algebras M_n and $\overline{M}_n$ *per se*, because we can define the anti-isomorphism $\boldsymbol{\sigma}$ formally by (25.10), and we need only relations (25.2)–(25.5) to calculate the function $\widetilde{\alpha}_n$.

Now we pass to the construction of an auxiliary algebra, which will permit us to extend the $\boldsymbol{\alpha}$-function $\alpha \colon B \to K[\gamma]$ to a function $\widetilde{\alpha} \colon \widetilde{B} \to K[\gamma]$, preserving the main properties of an $\boldsymbol{\alpha}$-function, and to forget the algebra M_n completely, operating only with trace polynomials.

25.2. The algebra $\widetilde{\mathcal{G}}(\gamma)$. Let K be an arbitrary field, let $\overline{K} = K(\gamma)$ be a rational function field over K, and let $F(\gamma)$ and $\mathcal{G}(\gamma)$ be the free associative algebra and the algebra of trace polynomials over the field $\overline{K}$ with free generators $x_1, x_2, \ldots$. Adjoin a unit element to the algebra $\mathcal{G}(\gamma)$ and denote by $\widetilde{\mathcal{G}}(\gamma)$ the quotient algebra of $\mathcal{G}(\gamma) + \overline{K}$ by the defining relations

$$x_i a x_i = \frac{x_i^2}{\gamma} \operatorname{Sp}(a), \tag{25.12}$$

$$\operatorname{Sp}(x_i a) x_i = \frac{x_i^2}{\gamma} a, \qquad \operatorname{Sp}(x_i a) \operatorname{Sp}(x_i b) = \frac{x_i^2}{\gamma} \operatorname{Sp}(ab), \tag{25.13}$$

$$[x_i^2, x_j] = 0, \qquad \operatorname{Sp}(x_i b x_i a) = x_i b x_i \operatorname{Sp}(a), \tag{25.14}$$

$$w = 0, \qquad \operatorname{Sp}(1) = \gamma, \tag{25.15}$$

where $i = 1, 2, \ldots$, a and b are arbitrary elements of $F(\gamma)$, and w is an arbitrary generalized monomial of degree greater than two with respect to some of its variables.

For any set λ we shall denote by $|\lambda|$ the cardinality of this set.

Let μ and λ be arbitrary finite or empty subsets of the set of variables $\{x_1, x_2, \ldots\}$ such that $\mu \cap \lambda = \varnothing$. Denote by ${}_\mu\mathcal{G}_\lambda(\gamma)$ the subspace of multihomogeneous trace polynomials in $\mathcal{G}(\gamma)$ of degree 2 with respect to all the variables from μ, of degree 1 with respect to all the variables from λ, and of degree 0 with respect to the remaining variables. We set by definition $\mathcal{G}_\lambda = {}_\varnothing\mathcal{G}_\lambda$, ${}_\mu\mathcal{G} = {}_\mu\mathcal{G}_\varnothing$, ${}_\mu\mathcal{G}_\lambda = \mathcal{G} \cap {}_\mu\mathcal{G}_\lambda(\gamma)$, and ${}_\varnothing\mathcal{G}_\varnothing(\gamma) = K(\gamma)$. In the special case $\mu = \varnothing$ and $\lambda = \{x_1, \ldots, x_l\}$ the space $\mathcal{G}_\lambda$ was denoted in §24 by T_l, and we shall use this notation along with $\mathcal{G}_\lambda$.

The structure of the algebra $\widetilde{\mathcal{G}}(\gamma)$ is described by the following lemma.

LEMMA 25.1. a) *As a linear space, the algebra $\widetilde{\mathcal{G}}(\gamma)$ is spanned by generalized polynomials of degree less than three with respect to any variable x_i.*

b) *As linear spaces, the sets of multilinear polynomials of the algebras $\mathcal{G}(\gamma)$ and $\widetilde{\mathcal{G}}(\gamma)$ are isomorphic.*

c) *For any trace polynomial $f \in {}_\mu\mathcal{G}_\lambda(\gamma)$ there exists a multilinear trace polynomial $g \in \mathcal{G}_\lambda(\gamma)$ such that in $\widetilde{\mathcal{G}}(\gamma)$ we have the equality*

$$f = \left(\prod_{x_i \in \mu} \frac{x_i^2}{\gamma}\right) g, \tag{25.16}$$

where the product is extended over all x_i in the set μ.

d) *Representation* (25.16) *is unique in the following sense: if there exists another representation $f = \{\prod_{x_i \in \mu} x_i^2/\gamma\} g_1$ of this kind, then $g = g_1$ in $\mathcal{G}(\gamma)$.*

e) *If $f \in {}_\mu\mathcal{G}_\lambda$, then the trace polynomial $g \in \mathfrak{G}_\lambda(\gamma)$ in equality* (25.16) *can be expressed as a linear combination of generalized monomials with coefficients in the polynomial algebra $K[\gamma]$.*

PROOF. The assertion a) is valid by the first relation in (25.15); the statement b) is valid, because the defining relations (25.12)–(25.14) are of degree ≥ 2 with respect to some variable. Obviously, the formulas (25.12)–(25.15) permit us to obtain the representation (25.16) and they imply that if the coefficients of f do not depend on γ then the coefficients of the trace polynomial g are polynomials in γ. This proves assertions c) and e).

Clearly, it suffices to prove the uniqueness of the representation (25.16) when K is the complex field.

Let $f = (\prod_{x_i \in \mu} \frac{x_i^2}{\gamma}) g_1$ be another representation and $g \neq g_1$ in $\mathcal{G}(\gamma)$. Since the coefficients of the generalized polynomial $g - g_1$ are rational functions of the variable γ, we see that there exist infinitely many positive integers n such that for $\gamma = n$ the expression makes sense and is nonzero. Denote by $\overline{f}$, $\overline{g}$, and $\overline{g}_1$ the elements of the algebra $\mathcal{G}_n$ obtained from f, g, and g_1, respectively, by the replacement $\gamma \mapsto n$ and $x_i \mapsto \overline{x}_i$ $(i = 1, 2, \dots)$. Then, because relations (25.2)–(25.5) for calculation of the function $\widetilde{\alpha}_n$ are obtained by the same replacement in the defining relations (25.12)–(25.15), we see that

$$\begin{aligned}\widetilde{\alpha}_n((\overline{g} - \overline{g}_1)\boldsymbol{\sigma}(\overline{g} - \overline{g}_1)) &= \widetilde{\alpha}_n\left(\left(\prod_{x_i \in \mu} (x_i^2/n)\right)(g - g_1)\boldsymbol{\sigma}(g - g_1)\right) \\ &= \widetilde{\alpha}_n((\overline{f} - \overline{f}_1)\boldsymbol{\sigma}(\overline{g} - \overline{g}_1)) = 0.\end{aligned}$$

But the criterion (25.11) shows that $(g - g_1)|_{\gamma = n} = 0$ is a nontrivial trace identity in the algebra M_n even if n is greater than the degree t of the generalized polynomial $g - g_1$. In order to lead this assertion to a contradiction, it remains to show that if $h(x_1, \dots, x_n) = 0$ is a multilinear trace identity in the algebra M_{l+1}, then the trace polynomial h is identically zero. In its turn, it suffices to find, for any basis generalized monomial in the space T_l, a collection of matrix units in M_{l+1} such that only the given monomial is nonzero on this collection. Let this monomial have the form

$$w = a_0 \operatorname{Sp}(a_1) \cdots \operatorname{Sp}(a_t),$$

where $a_0a_1\cdots a_t = x_1x_2\cdots x_l$, and the length of the word a_i is equal to r_i. We put the first r_0 variables equal to $E_{12}, E_{23}, \dots, E_{r_0,r_0+1}$ and put the variables corresponding to the word a_i $(i=1,\dots,t)$ successively equal to $E_{s+1,s+2}, E_{s+2,s+3}, \dots, E_{s+r_i-1,s+r_i}$, $E_{s+r_i,s+r_i+1}$, where $s=\sum_{j=0}^{i-1} r_j + 1$. It can be verified directly that on the given collection of the matrix units the monomial w does not vanish and it is the only one with this property, up to congruences (24.1) and (24.2). This implies that the monomial w occurs in the polynomial h with zero coefficient. Since w is not a bit better than any other basis generalized monomial occuring in h, we see that $h\equiv 0$. The lemma is proved.

We notice now an equality that is valid in $\widetilde{\mathcal{G}}(\gamma)$, which will be often useful in the sequel. Let $a,b,c,h,h_1,h_2\in\widetilde{\mathcal{G}}(\gamma)$; then formulas (25.12)–(25.15) imply

$$\begin{aligned} bx_iahx_i &= b\frac{x_i^2}{\gamma}\operatorname{Sp}(ah) = \frac{x_i^2}{\gamma}\operatorname{Sp}(ha)b = x_ihax_ib, \\ \operatorname{Sp}(h_1bx_iah_2)x_i &= \frac{x_i^2}{\gamma}ah_2h_1b = \operatorname{Sp}(h_1x_ih_2)ax_ib, \\ \operatorname{Sp}(h_1bx_iah_2)\operatorname{Sp}(x_ic) &= \operatorname{Sp}(h_1x_ih_2)\operatorname{Sp}(bx_iac). \end{aligned} \tag{25.17}$$

It follows from these equalities that for any multilinear trace polynomials f and g in $x_1,\dots,x_n$ we have the equality $f|_{x_i=bx_ia}\cdot g = f\cdot g|_{x_i=ax_ib}$. Moreover,

$$\begin{aligned} &f(x_1a_1,\dots,x_la_l)\cdot g(x_1,\dots,x_l) \\ &\quad = f(x_1,\dots,x_l)\cdot g(a_1x_1,\dots,a_lx_l), \end{aligned} \tag{25.18}$$

$$f\big|_{x_1=a_1x_1,\dots,x_l=a_lx_l}\cdot g = f\cdot g\big|_{x_1=x_1a_1,\dots,x_l=x_la_l}. \tag{25.19}$$

25.3. Extension of the domain of the α-function $\alpha\colon B\to K[\gamma]$. Denote by $\pi(f)$ the multilinear trace polynomial g uniquely determined by equality (25.16). Obviously, the mapping $\pi\colon {}_\mu\mathcal{G}_\lambda(\gamma)\to\mathcal{G}_\lambda(\gamma)$ is linear. For the special case in which f is a multihomogeneous trace polynomial of type $(2,\dots,2)$, we shall denote $\pi(f)$ by $\widetilde{\alpha}(f)$. In this case $\widetilde{\alpha}(f)$ belongs to $K(\gamma)$.

The linear subspace of $\mathcal{G}$ equal to $\sum_\mu {}_\mu\mathcal{G}_\varnothing$ is called the *space of generalized 2-elements* and will be denoted by $\widetilde{B}$. We identify the space B of 2-elements with a subspace of $\widetilde{B}$ by corresponding to each 2-word in variables $\overline{x}_i$ the same word in variables x_i. By assertions d) and e) of Lemma 25.1, $\widetilde{\alpha}$ is a K-linear function on ${}_\mu\mathcal{G}$ with values in the polynomial algebra $K[\gamma]$. We extend it to the whole space $\widetilde{B}$ by K-linearity. It is clear that relations (23.6), which permit computing the **α**-function $\alpha\colon B\to K[\gamma]$ (see Lemma 23.2), follow from the defining relation (25.12). Hence, the function $\widetilde{\alpha}\colon\widetilde{B}\to K[\gamma]$ must take the same values on 2-words as the function α, and therefore, $\widetilde{\alpha}$ is an extension of the **α**-function $\alpha\colon B\to K[\gamma]$. Obviously, for the function $\widetilde{\alpha}$ on the generalized 2-elements we have all three characteristic equalities that hold for any **α**-function on 2-elements (see §7). In fact, Property 1 of **α**-function can be strengthened in our case. Indeed, it follows from equalities (25.17) of the algebra $\widetilde{\mathcal{G}}(\gamma)$ that for any generalized 2-element of the form $u(\dots,bx_ia,\dots,x_i,\dots)$ in $K[\gamma]$ we have the equality

$$\widetilde{\alpha}(u(\dots,bx_ia,\dots,x_i,\dots)) = \widetilde{\alpha}(u(\dots,x_i,\dots,ax_ib,\dots)), \tag{25.20}$$

where a and b are arbitrary (possibly empty) words.

25.4. A bilinear pairing $\widetilde{\mathfrak{b}}\colon {}_\nu\mathcal{G}_\lambda \otimes {}_\mu\mathcal{G}_\lambda \to K[\gamma]$. Denote by $\overline{x}_i$, equally with x_i, the images of the generators x_i of the algebra $\mathcal{G}(\gamma) + K(\gamma)$ under the canonical homomorphism $\mathcal{G}(\gamma) + K(\gamma) \to \widetilde{\mathcal{G}}(\gamma)$. Define, via the function $\widetilde{\alpha}$, a $K(\gamma)$-bilinear pairing $\widetilde{\mathfrak{b}}\colon {}_\nu\mathcal{G}_\lambda \otimes {}_\mu\mathcal{G}_\lambda \to K[\gamma]$ by setting

$$\begin{aligned}(25.21)\quad \widetilde{\mathfrak{b}}(h(t_1,\dots,t_l,z_1,\dots,z_k),g(t_1,\dots,t_l,y_1,\dots,y_m)) \\ \rightleftharpoons \widetilde{\alpha}\big(h(\overline{x}_1,\dots,\overline{x}_l,\overline{x}_{i_1},\dots,\overline{x}_{i_k})\cdot g(\overline{x}_1,\dots,\overline{x}_l,\overline{x}_{j_1},\dots,\overline{x}_{j_m})\big),\end{aligned}$$

where $h \in {}_\nu\mathcal{G}_\lambda$, $g \in {}_\mu\mathcal{G}_\lambda$, $t_i \in \lambda$, $z_i \in \nu$, $y_j \in \mu$ and there are no equal symbols among the $\overline{x}_1,\dots,\overline{x}_l,\overline{x}_{i_1},\dots,\overline{x}_{i_k},\overline{x}_{j_1},\dots,\overline{x}_{j_m}$. Property 2 of the function $\widetilde{\alpha}$ guarantees that our pairing is well defined. Formulas (25.17) show that if $h = f|_{x_1=b_1a_1,\dots,x_l=b_lx_la_l}$, then in $K[\gamma]$ we have the equality

$$(25.22)\qquad \widetilde{\mathfrak{b}}\big(f|_{x_1=b_1x_1a_1,\dots,x_l=b_lx_la_l}, g\big) = \widetilde{\mathfrak{b}}\big(f, g|_{x_1=a_1x_1b_1,\dots,x_l=a_lx_lb_l}\big).$$

LEMMA 25.2. *The diagram*

$$\begin{array}{ccc} {}_\nu\mathcal{G}_\lambda(\gamma)\otimes_\mu\mathcal{G}_\lambda(\gamma) & \xrightarrow{\ \widetilde{\mathfrak{b}}\ } & K(\gamma) \\ \big\downarrow{\scriptstyle \pi\otimes\pi} & & \big\| \\ \mathcal{G}_\lambda(\gamma)\otimes\mathcal{G}_\lambda(\gamma) & \xrightarrow[\ \widetilde{\mathfrak{b}}\ ]{} & K(\gamma) \end{array}$$

is commutative.

THE PROOF follows from assertion e) of Lemma 25.1.

By assertion a) of Lemma 25.1, the restriction of the form $\widetilde{\mathfrak{b}}$ to ${}_\nu\mathcal{G}_\lambda \otimes {}_\mu\mathcal{G}_\lambda$ takes its values in the polynomial ring $K[\gamma]$. Denote by $E(\gamma)$ some polynomial in $K[\gamma]$ and by $\{E(\gamma)\}$ the ideal generated by $E(\gamma)$ in $K(\gamma)$. Denote by $\widetilde{\mathfrak{b}}_E$ the bilinear pairing on ${}_\nu\mathcal{G}_\lambda \otimes {}_\mu\mathcal{G}_\lambda$ defined by the formula (25.21) and taking its values in the algebra $K[\gamma]/\{E(\gamma)\}$.

LEMMA 25.3. *Let $E(\gamma)$ be a fixed polynomial, and let*

$$\operatorname{Ann}_{\mathcal{G}_\lambda}\widetilde{\mathfrak{b}}_E = \{f \in \mathcal{G}_\lambda \mid \widetilde{\mathfrak{b}}_E(f,\mathcal{G}_\lambda) = 0 \text{ in } K[\gamma]/\{E(\gamma)\}\}.$$

Then for any finite subset μ in $K[\gamma]/\{E(\gamma)\}$ we have $\widetilde{\mathfrak{b}}_E(\operatorname{Ann}_{\mathcal{G}_\lambda}\widetilde{\mathfrak{b}}_E, {}_\mu\mathcal{G}_\lambda) = 0$.

PROOF. Assertion a) of Lemma 25.1 implies $\pi({}_\mu\mathcal{G}_\lambda) \subseteq K[\gamma]\mathcal{G}_\lambda$. Therefore, the assertion of the lemma follows from Lemma 25.2.

LEMMA 25.4. *Let V be an ideal of trace identities in the algebra $\mathcal{G}$ generated by $\operatorname{Ann}_{T_l}\widetilde{\mathfrak{b}}_E$ for all $l = 1,2,\dots$. Then $V \cap T_k = \operatorname{Ann}_{T_k}\widetilde{\mathfrak{b}}_E$ for $k = 1,2,\dots$.*

PROOF. Actually, we are to prove that if $f \in \operatorname{Ann}_{T_l}\widetilde{\mathfrak{b}}_E$, $g \in T_k$ $(k \geq l)$ and g belongs to the ideal of trace identities $V(f)$ generated by the trace polynomial $f(x_1,\dots,x_l)$, then $g \in \operatorname{Ann}_{T_k}\widetilde{\mathfrak{b}}_E$. Definitions 24.2 and 24.3 and the equality $bf(a_1,\dots,a_l) = f(a_1,\dots,a_l)b + \sum_{i=1}^{l} f(a_1,\dots,[b,a_i],\dots,a_l)$, where $a_1,\dots,a_l,b$ are words, imply that V is generated in $\mathcal{G}$ as a right ideal by trace polynomials of the form $f(a_1, \dots, a_l)$, where $a_1, \dots, a_l$ range over all nonempty words in the letters $x_1, x_2, \dots$. Hence, $g \in V(f) \cap T_k$ is representable in the form of a linear combination of trace polynomials $f(a_1, \dots, a_l)h \in T_k$. Therefore, by the second property of the function $\widetilde{\alpha}$, in order to prove that $g \in \operatorname{Ann}_{T_k} \widetilde{\mathfrak{b}}_E$, it suffices to consider the case $g = f(x_1 a_1, \dots, x_l a_l)h$. Applying successively (25.21), (25.22), and Lemma 25.3, we see that for any v in the set T_k we have

$$\widetilde{\mathfrak{b}}_E(g, v) = \widetilde{\mathfrak{b}}_E\left(f, hv|_{x_1 = a_1 x_1, \dots, x_l = a_l x_l}\right) = 0.$$

The lemma is proved.

25.5. The closure operations $\theta_{\widetilde{\alpha}}$ and $\eta_{\widetilde{\alpha}}$ and their relationships with θ_α and η_α. We identify the free generators $t_1, t_2, \dots$ of the free associative algebra H with the free generators $x_1, x_2, \dots$ of the algebra F and we shall consider that ${}_\mu F_\lambda \rightleftharpoons F \cap {}_\mu \mathcal{G}_\lambda = {}_\mu H_\lambda$. Then it follows from Definition 25.21 that $\widetilde{\mathfrak{b}}$ is an extension of the bilinear form $\mathfrak{b}\colon {}_\nu H_\lambda \otimes {}_\mu H_\lambda \to K[\gamma]$ determined by the $\boldsymbol{\alpha}$-function $\alpha\colon B \to K[\gamma]$ (see formula (8.1)). Since each ideal in the algebra $K[\gamma]$ is principal and it is determined by a suitable polynomial $E(\gamma)$, Lemma 25.4 shows that to any ideal $I = \{E(\gamma)\}$ of the algebra $K[\gamma]$ we can uniquely assign the greatest ideal of trace identities $V_{E(\gamma)}$ in the trace algebra $\mathcal{G}$ such that $V_{E(\gamma)} \cap \mathcal{G}_\lambda = \operatorname{Ann}_{\mathcal{G}_\lambda} \widetilde{\mathfrak{b}}_E$. Therefore, by analogy with the mappings Φ_α and Ψ_α, we can introduce mappings $\Phi_{\widetilde{\alpha}}$ and $\Psi_{\widetilde{\alpha}}$.

Denote by $\Psi_{\widetilde{\alpha}}$ the mapping that assigns to each ideal $I = \{E(\gamma)\}$ of the algebra $K[\gamma]$ the ideal of trace identities $V_{E(\gamma)}$ in the algebra $\mathcal{G}$.

Denote by $\Phi_{\widetilde{\alpha}}$ the mapping that assigns to each ideal of trace identities V the least ideal I of the algebra $K[\gamma]$ containing all the sets $\widetilde{\mathfrak{b}}(V \cap \mathcal{G}_\lambda, {}_\mu\mathcal{G}_\lambda)$, where μ and λ are arbitrary finite subsets of $\{x_1, x_2, \dots\}$. Lemma 25.2 implies that $\Phi_{\widetilde{\alpha}}(V)$ is generated by elements of the sets $\widetilde{\mathfrak{b}}(V \cap \mathcal{G}_\lambda, \mathcal{G}_\lambda)$, where λ ranges over all finite subsets of $\{x_1, x_2, \dots\}$.

We put $\eta_{\widetilde{\alpha}} \rightleftharpoons \Phi_{\widetilde{\alpha}} \circ \Psi_{\widetilde{\alpha}}$ and $\theta_{\widetilde{\alpha}} \rightleftharpoons \Psi_{\widetilde{\alpha}} \circ \Phi_{\widetilde{\alpha}}$.

PROPOSITION 25.1. *Let the groundfield K be infinite. For any ideals of weak identities V_1, V_2, and V in the trace algebra $\mathcal{G}$ and any ideals I_1, I_2, and I of the algebra $K[\gamma]$ we have the following relations*:

$$\Phi_{\widetilde{\alpha}}(V_1 + V_2) = \Phi_{\widetilde{\alpha}}(V_1) + \Phi_{\widetilde{\alpha}}(V_2), \tag{25.23}$$

$$\Psi_{\widetilde{\alpha}}(I_1 \cap I_2) = \Psi_{\widetilde{\alpha}}(I_1) \cap \Psi_{\widetilde{\alpha}}(I_2), \tag{25.24}$$

$$\theta_{\widetilde{\alpha}}(V) \supseteq V, \qquad \eta_{\widetilde{\alpha}}(I) \subseteq I; \tag{25.25}$$

$$\Phi_{\widetilde{\alpha}}\Psi_{\widetilde{\alpha}}\Phi_{\widetilde{\alpha}} = \Phi_{\widetilde{\alpha}}, \qquad \Psi_{\widetilde{\alpha}}\Phi_{\widetilde{\alpha}}\Psi_{\widetilde{\alpha}} = \Psi_{\widetilde{\alpha}}; \tag{25.26}$$

$$\theta_{\widetilde{\alpha}}^2 = \theta_{\widetilde{\alpha}}, \qquad \eta_{\widetilde{\alpha}}^2 = \eta_{\widetilde{\alpha}}. \tag{25.27}$$

The proof is similar to the proof of Proposition 10.1.

This proposition shows that $\theta_{\widetilde{\alpha}}$ defines a closure operation on the set of all ideals of trace identities, and V is a $\theta_{\widetilde{\alpha}}$-closed ideal if and only if $V = \Psi_{\widetilde{\alpha}}(I)$; moreover, $\eta_{\widetilde{\alpha}}$ defines a closure operation on the set of all ideals of the algebra $K[\gamma]$, and an ideal I is $\eta_{\widetilde{\alpha}}$-closed if and only if $I = \Phi_{\widetilde{\alpha}}(V)$. In the next section we shall give a complete description of the lattice $\Xi_{\widetilde{\alpha}}$ (of the lattice $\Theta_{\widetilde{\alpha}}$) of $\theta_{\widetilde{\alpha}}$-closed (of $\eta_{\widetilde{\alpha}}$-closed) ideals, respectively, and give answers to the first five problems of §11 for $\Xi_{\widetilde{\alpha}}$, $\Theta_{\widetilde{\alpha}}$, $\theta_{\widetilde{\alpha}}$, and $\eta_{\widetilde{\alpha}}$.

The next assertion expresses the relationship between θ_α-closed and $\theta_{\widetilde{\alpha}}$-closed ideals in F and $\mathcal{G}$, respectively, and between η_α-closed and $\eta_{\widetilde{\alpha}}$-closed ideals in $K[\gamma]$.

PROPOSITION 25.2. *For any ideal I in $K[\gamma]$ and for any T-ideal (i.e., any ideal of identities of some associative algebra) $\mathfrak{V}$ in F the following relations are valid:*

$$\Psi_\alpha(I) = \Psi_{\widetilde{\alpha}}(I) \cap F, \tag{25.28}$$

$$\Phi_\alpha(\mathfrak{V}) = \Phi_{\widetilde{\alpha}}(\mathfrak{V}\mathcal{G} + \mathfrak{V}), \tag{25.29}$$

i.e., any η_α-closed ideal in $K[\gamma]$ is $\eta_{\widetilde{\alpha}}$-closed, and any θ_α-closed ideal of weak identities is a T-ideal and is the intersection of a $\theta_{\widetilde{\alpha}}$-closed ideal of weak identities in $\mathcal{G}$ with the algebra F.

The proof is obvious.

25.6. Properties of the bilinear form $\widetilde{\mathfrak{b}}_E : T_l \otimes T_l \to K[\gamma]$.

LEMMA 25.5. *Let the ground field K be of characteristic zero. Then we have $\operatorname{Ann}_{T_l} \widetilde{\mathfrak{b}}_{\gamma-n} = V_{n,0} \cap T_l$, where $V_{n,0}$ is the ideal of trace identities of the full matrix algebra M_n.*

PROOF. Let first K be the complex field. Denote by the letter ψ the homomorphism of the algebra $\mathcal{G}$ to the algebra $\mathcal{G}_n$ such that $\psi f(x_1, \dots, x_l) = f(\overline{x}_1, \dots, \overline{x}_l)$ and denote by the letter σ the anti-isomorphism of the algebra $\mathcal{G}$ determined by formulas (25.10), where we are to replace $\overline{x}_i$ by x_i. Since for $\gamma = n$ formulas (25.12)–(25.15) for computation of the function $\widetilde{\alpha}$ become formulas for computation of $\widetilde{\alpha}_n$, we see that

$$\widetilde{\mathfrak{b}}_{\gamma-n}(u, v) = \widetilde{\alpha}(\psi(u)\psi(v)) \tag{25.30}$$

for any $u, v \in T_l$. Therefore, by the criterion (25.11), we have $\widetilde{\mathfrak{b}}_{\gamma-n}(u, \sigma(u)) \geq 0$, and equality is attained if and only if the multilinear trace identity $u = 0$ holds in the algebra M_n.

If K is an arbitrary field of characteristic zero, then in the writing of a concrete trace identity of the algebra M_n or in that of an element of $\operatorname{Ann}_{T_l} \widetilde{\mathfrak{b}}_{\gamma-n}$ only finitely many elements of the field K participate. Since the subfield generated by these elements can be embedded in the complex field, we see that the lemma is valid for any field K of characteristic zero. The lemma is proved.

LEMMA 25.6. *The bilinear form $\widetilde{\mathfrak{b}}_E : T_l \otimes T_l \to K[\gamma]$ is symmetric.*

PROOF. It is clear that it suffices to prove the assertion of the lemma for the case $E(\gamma) = 0$. In this case we have $\widetilde{\mathfrak{b}}_0(u, v)|_{\gamma=n} = \widetilde{\mathfrak{b}}_{\gamma-n}(u, v)$. Formulas (25.30) and (25.8) show that $\widetilde{\mathfrak{b}}_0(u, v)|_{\gamma=n} = \widetilde{\mathfrak{b}}_0(v, u)|_{\gamma=n}$ for $n = 1, 2, \dots$. Since for fixed $u, v \in T_l$ the expressions $\widetilde{\mathfrak{b}}_0(u, v)$ and $\widetilde{\mathfrak{b}}_0(v, u)$ are polynomials coinciding on an infinite set, we see that $\widetilde{\mathfrak{b}}_0(u, v) = \widetilde{\mathfrak{b}}_0(v, u)$. The lemma is proved.

Formulas (25.13) and (25.21) imply

$$\widetilde{\mathfrak{b}}(a_0 \operatorname{Sp}(a_1) \cdots \operatorname{Sp}(a_t), \prod_{j=1}^{l} \operatorname{Sp}(x_j)) = \gamma^t, \tag{25.31}$$

where $a_0, \dots, a_t$ are words and $a_0 \operatorname{Sp}(a_1) \cdots \operatorname{Sp}(a_t) \in T_l$.

25.7. Endowing of T_l with the structure of a group algebra. Now we clarify what ideals of trace identities $V_{E(\gamma)}$ are determined by the bilinear form $\widetilde{\mathfrak{b}}_E$.

LEMMA 25.7. *Let an operation $\circ$ be given on the set T_l by the formula*

$$(25.32)\qquad f\circ g \rightleftharpoons \pi\left(f\cdot g\Big|_{\substack{x_1=y_1x_1\\ \cdots\\ x_l=y_lx_l}}\right)\Bigg|_{\substack{y_1=x_1\\ \cdots\\ y_l=x_l}} = \pi\left(f\Big|_{\substack{x_1=y_1x_1\\ \cdots\\ x_l=y_lx_l}}\cdot g\right)\Bigg|_{\substack{y_1=x_1\\ \cdots\\ y_l=x_l}},$$

where $f,g\in T_l$ and for convenience we denote by $y_1,\dots,y_l$ variables x_i such that $x_1,\dots,x_l,y_1,\dots,y_l$ are distinct. Then this operation is well defined and with respect to this operation the set $G(l)$ of generalized monomials in T_l is a group and the linear space T_l is the group algebra of the group $G(l)$. The elements $D_i = x_i\operatorname{Sp}(x_1)\cdots\widehat{\operatorname{Sp}(x_i)}\cdots\operatorname{Sp}(x_l)$ $(i=1,\dots,l)$ are generators of this group, and the unit of $G(l)$ is the generalized monomial $e=\prod_{j=1}^{l}\operatorname{Sp}(x_j)$.

PROOF. We note first that the second equality in (25.32) is valid by (25.18) and by assertion e) of Lemma 25.1. This proves that $\circ$ is well defined by (25.32). We show that the operation $\circ$ is associative in $T_l(\gamma)$, i.e., that for any $f,g,h\in T_l(\gamma)$ we have

$$(25.33)\qquad (f\circ g)\circ h = f\circ(g\circ h).$$

By definition (25.32) and by equality (25.18), the left-hand side of the last relation is equal to

$$\pi\left(\left(fg\Big|_{\substack{x_1=y_1x_1\\ \cdots\\ x_l=y_lx_l}}\right)h\Big|_{\substack{x_1=z_1y_1\\ \cdots\\ x_l=z_ly_l}}\right)\Bigg|_{\substack{z_1=x_1\\ \cdots\\ z_l=x_l}} = \pi\left(\left(fg\Big|_{\substack{x_1=y_1\\ \cdots\\ x_l=y_l}}\right)h\Big|_{\substack{x_1=z_1x_1y_1\\ \cdots\\ x_l=z_lx_ly_l}}\right)\Bigg|_{\substack{z_1=x_1\\ \cdots\\ z_l=x_l}}$$

$$= \pi\left(f\left(\left(g\Big|_{\substack{x_1=y_1\\ \cdots\\ x_l=y_l}}\right)h\Big|_{\substack{x_1=x_1y_1\\ \cdots\\ x_l=x_ly_l}}\right)\Bigg|_{\substack{x_1=z_1x_1\\ \cdots\\ x_l=z_lx_l}}\right)\Bigg|_{\substack{z_1=x_1\\ \cdots\\ z_l=x_l}}.$$

By definition (25.22), the last expression is equal to the right-hand side of (25.33). This proves (25.33).

Applying the equalities (25.13) successively, we obtain

$$\prod_{j=1}^{l}\operatorname{Sp}(x_j)\circ f = \pi\left(\prod_{j=1}^{l}\operatorname{Sp}(x_jy_j)\cdot f\right)\Bigg|_{\substack{y_1=x_1\\ \cdots\\ y_l=x_l}} = \left(f\Big|_{\substack{x_1=y_1\\ \cdots\\ x_l=y_l}}\right)\Bigg|_{\substack{y_1=x_1\\ \cdots\\ y_l=x_l}} = f$$

and, therefore, $e=\prod_{j=1}^{l}\operatorname{Sp}(x_j)$ is a left unit; in the same way we check that e is a right unit of the algebra $T_l(\gamma)$. Similarly, we have

$$(25.34)\qquad f\circ D_i = \pi(fy_ix_i)|_{y_i=x_i}.$$

If $f = a_0\prod_{j=1}^{t}\operatorname{Sp}(x_j)$ is a generalized polynomial, then, considering the two cases $a_0=a_0'x_ia_0''$ and $a_k=x_ia_k'$ with the use of formulas (25.12) and (25.13), we obtain, respectively,

$$(25.35)\qquad \left\{a_0\prod_{j=1}^{t}\operatorname{Sp}(a_j)\right\}\circ D_i = \begin{cases} a_0'\operatorname{Sp}(x_ia_0'')\prod_{j=1}^{t}\operatorname{Sp}(a_j), & \text{for } a_0=a_0'x_ia_0'',\\ a_0x_ia_k'\prod_{\substack{j=1\\ j\neq k}}^{t}\operatorname{Sp}(a_j), & \text{for } a_k=x_ia_k',\end{cases}$$

and this implies that $G(l)$ is invariant under the multiplication by D_i. From (25.35) it follows immediately that

$$D_i \circ D_i = e,$$

$$\left\{g(x_1,\dots,x_k)\prod_{j=k+1}^{l}\mathrm{Sp}(x_j)\right\}\circ D_{k+1} = g(x_1,\dots,x_k)x_{k+1}\prod_{j=k+2}^{l}\mathrm{Sp}(x_j),$$

$$\left\{g(x_1,\dots,x_k)\prod_{j=k+1}^{l}\mathrm{Sp}(x_j)\right\}\circ D_{k+1}\circ D_{k+2}\circ\dots\circ D_{k+t}\circ D_{k+1} \tag{25.36}$$

$$= g(x_1,\dots,x_k)\,\mathrm{Sp}(x_{k+1}\cdots x_{k+t})\prod_{j=k+t+1}^{l}\mathrm{Sp}(x_j),$$

where g is a multilinear generalized polynomial in $x_1,\dots,x_{k+t}$. The latter two equalities show that e can be transformed into any generalized polynomial by successive multiplication by elements D_i. Hence, any generalized polynomial in $G(l)$ is representable as a product of second order elements D_i with respect to the operation $\circ$. This shows that $G(l)$ is a group and the elements D_i are generators of it. Now it is obvious that T_l is the group algebra of this group and the operation $\circ$ is well defined on T_l. The lemma is proved.

The next lemma expresses the relationship between the operation $\circ$ and the symmetric bilinear form $\widetilde{\mathfrak{b}}_E\colon T_l\otimes T_l\to K[\gamma]$.

LEMMA 25.8. *For any elements $f, g, h\in T_l(\gamma)$ we have the equality*

$$\widetilde{\mathfrak{b}}_E(f\circ h, g) = \widetilde{\mathfrak{b}}_E(f, h\circ g). \tag{25.37}$$

PROOF. By Lemma 25.7, it suffices to consider the case $h = D_i$ $(i = 1,\dots,l)$. Using formula (25.34), we obtain

$$\begin{aligned}\widetilde{\mathfrak{b}}_E(f\circ D_i, g) &= \widetilde{\alpha}(\pi(fy_ix_i)|_{y_i=x_i}g) = \widetilde{\alpha}(\pi(fy_ix_i)g|_{x_i=y_i})\\ &= \widetilde{\alpha}(f\pi(y_ix_ig)|_{x_i=y_i}) = \widetilde{\alpha}(f\pi(x_iy_ig)|_{y_i=x_i}) = \widetilde{\mathfrak{b}}_E(f, D_i\circ g).\end{aligned}$$

The lemma is proved.

Lemmas 25.7 and 25.8 imply the following property of $\theta_{\widetilde{\alpha}}$-closed ideals of trace identities.

COROLLARY. $\mathrm{Ann}_{T_k}\widetilde{\mathfrak{b}}_E$ *is a two-sided ideal of the group algebra T_k $(k = 1, 2,\dots)$.*

Let S_{l+1} be the group of all permutations acting from the right on the set $\{x_0, x_1,\dots, x_l\}$. Each permutation can be uniquely decomposed into the product of independent cycles. Denote by $(x_{i_1}x_{i_2}\cdots x_{i_k})$ the cycle mapping $x_{i_1}\mapsto x_{i_2}\mapsto\cdots\mapsto x_{i_k}\mapsto x_{i_1}$. Moreover, if a word a is equal to $x_{i_1}x_{i_2}\cdots x_{i_k}$, then denote the cycle $(x_{i_1}x_{i_2}\cdots x_{i_k})$ by (a).

LEMMA 25.9. *The group $G(l)$ is isomorphic to the group S_{l+1} of all permutations acting from the right on a set with $l+1$ elements. The isomorphism is realized by the mapping*

$$\varphi\{a_0 \operatorname{Sp}(a_1)\cdots \operatorname{Sp}(a_t)\} = (x_0 a_0)(a_1)\cdots(a_t).$$

PROOF. The mapping φ is well defined, because independent cycles commute with each other, and $(x_{i_1}x_{i_2}x_{i_3}\cdots x_{i_p})$ and $(x_{i_2}x_{i_3}\cdots x_{i_p}x_{i_1})$ are notations of the same cycle. By Lemma 24.1, the mapping φ is one-to-one. In order to prove the lemma, it suffices to show that for any $g \in G(l)$

$$\varphi(g)\varphi(D_i) = \varphi(g \circ D_i). \tag{25.38}$$

However,

$$\{(x_0a_0)(a_1)\cdots(a_t)\}(x_0x_i) = \begin{cases} (x_0a_0')(x_ia_0'')(a_1)\cdots(a_t), & \text{for } a_0 = a_0'x_ia_0'', \\ (x_0a_0x_ia_k')(a_1)\cdots(\widehat{a}_k)\cdots(a_t), & \text{for } a_k = x_ia_k'. \end{cases}$$

Comparing this with (25.35), we obtain (25.38). The lemma is proved.

This lemma implies that the inverse element to $g \in G(l)$ with respect to the operation $\circ$ is $\boldsymbol{\sigma}(g)$. Lemmas 25.9 and 25.8 and formula (25.31) show that

$$\widetilde{\mathfrak{b}}_E(s,\delta) = \gamma^m, \tag{25.39}$$

where $m+1$ is the number of independent cycles taking part in the decomposition of the permutation $s \circ \delta$.

From now on, we shall make no difference between the group algebras $K[S_{l+1}]$ and T_l of the groups S_{l+1} and $G(l)$, respectively.

Now we collect together all substantial results obtained in this section.

PROPOSITION 25.3. *The linear space T_l of multilinear trace polynomials in the algebra $\mathcal{G}$ can be endowed with the operation $\circ$ introduced by formula (25.32), and with respect to this operation the set $G(l)$ of all generalized monomials in T_l turns into a group isomorphic to the group of permutations S_{l+1}, and T_l turns into the group algebra of this group. Moreover, for any polynomial $E(\gamma)$ formula (25.21) determines a symmetric bilinear form $\widetilde{\mathfrak{b}}_E$ on T_l, which is associative with respect to the operation $\circ$ and takes its values in the algebra $K[\gamma]/\{E(\gamma)\}$. To any such bilinear form there corresponds a $\theta_{\widetilde{\alpha}}$-closed ideal of trace identities $V_{E(\gamma)} \subset \mathcal{G}$ such that for any $l = 1, 2, 3, \ldots$*

$$V_{E(\gamma)} \cap T_l = \operatorname{Ann}_{T_l} \widetilde{\mathfrak{b}}_E$$

and this set is an ideal of the group algebra T_l. If the ground field K is of characteristic zero and

$$E(\gamma) = \gamma - n,$$

then the $\theta_{\widetilde{\alpha}}$-closed ideal of trace identities $V_{\gamma-n}$ coincides with the ideal of trace identities $V_{n.0}$ of the algebra M_n.

§26. Classification of $\eta_{\tilde{\alpha}}$-closed ideals and ideals of trace identities V such that $V \cap T_l$ is a two-sided ideal in T_l over fields of characteristic zero

In this section we give a complete description of the lattice $\Xi_{\tilde{\alpha}}$ (respectively, the lattice $\Theta_{\tilde{\alpha}}$) of $\eta_{\tilde{\alpha}}$-closed (respectively, of $\theta_{\tilde{\alpha}}$-closed) ideals of the algebra $K[y]$ (respectively, ideals of weak identities in the algebra $\mathcal{G}$) and give answers to the third, fourth, and fifth problems of §11 for $\Phi_{\tilde{\alpha}}$, $\Psi_{\tilde{\alpha}}$, $\theta_{\tilde{\alpha}}$, and $\eta_{\tilde{\alpha}}$.

Proposition 25.3 shows that to this end, we must apply the classical Young theory describing irreducible representations and simple (minimal) two-sided ideals of the group algebra $K[S_{l+1}]$ of the symmetric group over fields of characteristic zero. We use the exposition of this theory given in §28, Chapter 4, of the book of Curtis and Reiner [CR].

We cite without proof the results of the afore-mentioned section that will be used directly.

By a *Young table* $\mathbf{D}$ of type $(n_1, \dots, n_k)$ we mean a table in which the number of cells in the ith row is equal to n_i, where $n_1 \geq n_2 \geq \cdots \geq n_k$. A Young table $\mathbf{D}$ filled arbitrarily by numbers from 1 to $l+1 = n_1 + n_2 + \cdots + n_k$ is called a *Young diagram*. Let $R(\mathbf{D})$ be the subgroup in S_{l+1} consisting of all elements p that permute numbers in each row of the diagram $\mathbf{D}$, and let $C(\mathbf{D})$ be the subgroup consisting of all elements q that permute numbers in each column of the same diagram $\mathbf{D}$.

PROPOSITION 26.1. *The set of Young tables* $\mathbf{D}$ *of type* $(n_1, \dots, n_k)$, *where* $n_1 + n_2 + \cdots + n_k = l + 1$, *is in a one-to-one correspondence with the set of simple two-sided ideals of the group algebra* $K[S_{l+1}]$. *The ideal corresponding to the Young table* $\mathbf{D}$ *is generated, as a two-sided ideal, by the element* $e(\mathbf{D}) = \mathcal{P}(\mathbf{D}) \circ \mathcal{Q}(\mathbf{D})$, *where* $\mathcal{P} = \sum_{p \in R(\mathbf{D})} p$ *and* $\mathcal{Q} = \sum_{q \in C(\mathbf{D})} \operatorname{sign}(q) q$ (*the sums are taken over all elements of the subgroups* $R(\mathbf{D})$ *and* $C(\mathbf{D})$, *respectively, and* $\operatorname{sign}(q) = \pm 1$ *depending on the parity of the permutation* q) *for some diagram obtained from* $\mathbf{D}$. *A permutation* $s \in S_{l+1}$ *is representable in the form* $s = p \circ q$, *where* $p \in R(\mathbf{D})$ *and* $q \in C(\mathbf{D})$, *if and only if the elements of any row of the diagram* $\mathbf{D}$ *belong to distinct columns of the diagram* $\mathbf{D}s$, *and this representation is unique. For any* $x \in K[S_{l+1}]$ *we have* $\mathcal{P} \circ x \circ \mathcal{Q} = \nu \cdot \mathcal{P} \circ \mathcal{Q}$, *where* ν *is some number depending on* x. *In particular*, $e(\mathbf{D}) \circ (\mathbf{D}) = \frac{(l+1)!}{m_{\mathbf{D}}} e(\mathbf{D})$, *where* $m_{\mathbf{D}}$ *is the dimension of the irreducible representation corresponding to the table* $\mathbf{D}$.

Now we have all necessary tools to describe all ideals of trace identities V in $\mathcal{G}$ such that $V \cap T_l$ is a two-sided ideal of the group algebra T_l for any positive integer l. In particular, this will give us a description of all $\theta_{\tilde{\alpha}}$-closed ideals of weak identities in the algebra $\mathcal{G}$.

It is completely obvious that the set $V \cap T_l$ is a direct sum of simple ideals, each of which is determined by some Young table $\mathbf{D}$, but the following is not clear *a priori*.

a) Let V be a verbal ideal generated by the set of multilinear trace polynomials that is a two-sided ideal in T_l. Is it true that $V \cap T_{l+k}$ is a two-sided ideal in T_{l+k} $(k = 1, 2, 3, \dots)$?

b) What are all the Young tables $\mathbf{D}_1$ and $\mathbf{D}_2$ such that $V_{\mathbf{D}_1} \supseteq V_{\mathbf{D}_2}$, where $V_{\mathbf{D}_i}$ is the ideal of trace identities generated by the set of multilinear trace polynomials that is a simple two-sided ideal in T_{l_i} corresponding to the Young table $\mathbf{D}_i$?

c) Do the ideals of trace identities of the form under consideration satisfy the ascending chain condition?

The following lemma answers the first question.

LEMMA 26.1. *Let an ideal of trace identities V be generated by the set of trace polynomials of degree l that forms a two-sided ideal in T_l with respect to the operation $\circ$. Let $f \in T_l$ generate the two-sided ideal $V \cap T_l$. Then for any $k \geq 0$ the space $V \cap T_{l+k}$ is a two-sided ideal in T_{l+k} and the element $f(x_1, \ldots, x_l) \prod_{i=1}^{k} \mathrm{Sp}(x_{l+i})$ generates this two-sided ideal.*

PROOF. First we establish that $V \cap T_{l+k}$ is a two-sided ideal. To this end, it suffices to check that $V \cap T_{l+k}$ is stable under multiplication by the generators $\mathbf{D}_i$. In the proof of Lemma 25.4 it was shown that the linear space $V \cap T_{l+k}$ is spanned by the elements $g_1 = g(a_1, \ldots, a_l)a_{l+1} \prod_{j=1}^{r} \mathrm{Sp}(b_j)$, where $g \in V \cap T_l$ and a_i and b_j are words. Making use of (25.34), (25.18), (25.19), and (25.35), we obtain

$$g_1 \circ \mathbf{D}_i = \pi(g(a_1, \ldots, a_l)a_{l+1} \prod_{j=1}^{r} \mathrm{Sp}(b_j)y_i x_i)|_{y_i = x_i}$$

$$= \begin{cases} g(a_1, \ldots, a_l)a_{l+1}x_i b_s' \prod_{j=1, j\neq s}^{r} \mathrm{Sp}(b_j), & \text{for } b_s = x_i b_s', \\ g(a_1, \ldots, a_l)a_{l+1}' \mathrm{Sp}(x_i a_{l+1}'') \prod_{j=1}^{r} \mathrm{Sp}(b_j), & \text{for } a_{l+1} = a_{l+1}' x_i a_{l+1}'', \\ (g(a_1, \ldots, a_l) \circ \mathbf{D}_s)\Big|_{\substack{x_j = a_j (j \neq s) \\ x_s = a_{l+1} x_i a_s''}} \cdot a_s' \prod_{j=1}^{r} \mathrm{Sp}(b_j), & \text{for } s \leq l \text{ and } a_s = a_s' x_i a_s'', \end{cases}$$

whence it follows that $V \cap T_{l+k}$ is a right ideal. Similarly, we can check that $V \cap T_{l+k}$ is a left ideal.

Let us prove the second assertion of the lemma. Taking into account relation (25.36) and what was said above, it suffices to show that the element g_1 defined by the formula $g_1 = g(a_1, \ldots, a_l) \prod_{j=1}^{k-r+1} \mathrm{Sp}(x_{i_j})$, where $g \in V \cap T_l$, is contained in the two-sided ideal generated by the elements

$$f(x_1, \ldots, x_{l+k}) = f(x_1, \ldots, x_l) \prod_{j=1}^{k} \mathrm{Sp}(x_{l+j}).$$

It is well known that if $\delta = (y_{11} \cdots y_{1r})(y_{21} \cdots y_{2s}) \cdots (y_{m1} \cdots y_{mt})$ is the decomposition of $\delta \in S_n$ into independent cycles, then for any permutation $\delta_1 \in S_n$ we have

$$\delta_1^{-1} \circ \delta \circ \delta_1 = (\delta_1(y_{11}) \cdots \delta_1(y_{1r})) \cdots (\delta_1(y_{m1}) \cdots \delta_1(y_{mt})).$$

Therefore, we can map the element g_1, by an appropriate conjugation, to the element

$$g_2 = g(a_1' x_1, \ldots, a_l' x_l) \prod_{j=k+l}^{l+r} \mathrm{Sp}(x_j).$$

Obviously, the element $g_3 = g(x_1, \ldots, x_l) \cdot \prod_{j=1}^{k} \mathrm{Sp}(x_{l+j})$ belongs to the two-sided ideal generated by f_1. Let

$$s = x_1 a_1' \cdots x_l a_l' \prod_{j=k+l}^{l+r} \mathrm{Sp}(x_j).$$

Then, using equalities (25.13), (25.18), and (25.19), we obtain

$$g_3 \circ s = \pi\left(g(x_1y_1, \dots, x_ly_l) \prod_{j=1}^{k} \mathrm{Sp}(x_{l+j}y_{l+j})s \right)\Bigg|_{\substack{y_1=x_1\\ \dots\\ y_l=x_l}}$$

$$= \pi\left(g(x_1, \dots, x_l)y_1x_1a_1'y_2x_2a_2' \cdots y_lx_la_l' \prod_{j=l+k}^{l+r} \mathrm{Sp}(x_j) \right)\Bigg|_{\substack{y_1=x_1\\ \dots\\ y_l=x_l}}$$

$$= \pi\left(g(a_1'x_1, \dots, a_1'x_l) \prod_{j=l+k}^{l+r} \mathrm{Sp}(x_j)y_1x_1y_2x_2 \cdots y_lx_l \right)\Bigg|_{\substack{y_1=x_1\\ \dots\\ y_l=x_l}} = g_2 \circ t,$$

where $t = x_1x_2 \cdots x_l \,\mathrm{Sp}(x_{l+1}) \cdots \mathrm{Sp}(x_{l+k})$. Thus, $g_2 \circ t$ belongs to the two-sided ideal generated by f_1 and, hence, the elements $g_2 = (g_2 \circ t) \circ t^{-1}$ and g_1 belong to the same ideal. The lemma is proved.

DEFINITION 26.1. Let $\mathbf{D}$ be an arbitrary Young table consisting of $l+1$ cells. Denote by $V_{\mathbf{D}}$ the ideal of trace identities generated by the set of multilinear trace polynomials of degree l that forms a simple two-sided ideal in the group algebra T_l corresponding to the Young table $\mathbf{D}$ (i.e., $V_{\mathbf{D}} \cap T_l$ is the simple ideal in T_l with the symmetry type $\mathbf{D}$ and $V_{\mathbf{D}}$ is the least ideal of trace identities in $\mathcal{G}$ containing $V_{\mathbf{D}} \cap T_l$). The ideal $V_{\mathbf{D}}$ is called the *ideal of trace identities with the symmetry type* $\mathbf{D}$.

REMARK 26.1. By Lemma 26.1, $V_{\mathbf{D}} \cap T_k$ is a two-sided ideal in the group algebra T_k for any Young table $\mathbf{D}$ and for any $k = 1, 2, \dots$. Moreover, any ideal of trace identities V in $\mathcal{G}$ such that $V_{\mathbf{D}} \cap T_k$ forms a two-sided ideal in T_k for any $k = 1, 2, \dots$ can be represented in the form

$$V = 0 + \sum_{V_{\mathbf{D}} \subseteq V} V_{\mathbf{D}}, \tag{26.1}$$

where the sum is taken over all ideals $V_{\mathbf{D}}$ contained in V. By Corollary 25.8 and Proposition 25.3, any $\theta_{\tilde{\alpha}}$-closed ideal of trace identities V in $\mathcal{G}$ has the form $V_{E(\gamma)} = \Psi_{\tilde{\alpha}}(\{E(\gamma)\})$, and $V_{E(\gamma)} \cap T_k$ is an ideal in T_k for $k = 1, 2, \dots$. Therefore, the decomposition (26.1) is valid for any $\theta_{\tilde{\alpha}}$-closed ideal of trace identities $V = V_{E(\gamma)}$.

LEMMA 26.2. *Let a Young table $\mathbf{D}_1$ of type $(n_1, \dots, n_k)$ be contained as a subtable in a Young table $\mathbf{D}_2$ of type $(n_1', \dots, n_k')$, i.e., $n_i \le n_i'$ for each i, and let*

$$l_1 + 1 = \sum_{i=1}^{k} n_i, \qquad l_2 + 1 = \sum_{i=1}^{k} n_i'.$$

Denote by $V_{\mathbf{D}_i}$ $(i = 1, 2)$ the ideal of trace identities of the algebra $\mathcal{G}$ generated by the set of multilinear trace polynomials of degree l_i that form a two-sided ideal in T_{l_i} corresponding to the Young table $\mathbf{D}_i$. Then $V_{\mathbf{D}_1} \supseteq V_{\mathbf{D}_2}$.

PROOF. We fill the table $\mathbf{D}_2$ by the numbers from 1 to $l_2 + 1$ so that the numbers from 1 to $l_1 + 1$ belong to the table $\mathbf{D}_1$. Let $e(\mathbf{D}_i) = \mathcal{P}_i \circ \mathcal{Q}_i$ $(i = 1, 2)$ have the same meaning as in Proposition 26.1; then $e(\mathbf{D}_i)$ generates the two-sided ideal $V_{\mathbf{D}_i} \cap T_{l_i}$. We can regard the group S_{l_1+1} of permutations acting on $\{x_0, x_1, \dots, x_{l_1}\}$ as naturally embedded in the group S_{l_2+1} acting on $\{x_0, x_1, \dots, x_{l_2}\}$. Toward this end, we must identify S_{l_1+1} with the subgroup of S_{l_2+1} that leaves the elements $x_{l_1+1}, \dots, x_{l_2}$ fixed. Then $R(\mathbf{D}_2) \supseteq R(\mathbf{D}_1)$ and $C(\mathbf{D}_2) \supseteq C(\mathbf{D}_1)$. In the same way, we can regard the group ring T_{l_1} as embedded in T_{l_2}. In the sense of this embedding, the element $e(\mathbf{D}_1)$ generates $V_{\mathbf{D}_1} \cap T_{l_2}$ as a two-sided ideal by Lemma 26.1. On the other hand,

$$\mathcal{P}_2 \circ e(\mathbf{D}_1) \circ \mathcal{Q}_2 = \mathcal{P}_2 \circ \mathcal{P}_1 \circ \mathcal{Q}_1 \circ \mathcal{Q}_2 = |\mathcal{P}_1| \cdot |\mathcal{Q}_1| \, e(\mathbf{D}_2),$$

where $|\mathcal{P}_1|$ and $|\mathcal{Q}_1|$ are the orders of the groups $R(\mathbf{D}_1)$ and $C(\mathbf{D}_1)$, respectively. Therefore, the generator $e(\mathbf{D}_2)$ of the ideal $V_{\mathbf{D}_2} \cap T_{l_2}$ belongs to the ideal $V_{\mathbf{D}_1} \cap T_{l_2}$. Now from the definition of $V_{\mathbf{D}_2}$ and $V_{\mathbf{D}_1}$ and from Lemma 26.1 it follows that $V_{\mathbf{D}_1} \supseteq V_{\mathbf{D}_2}$. The lemma is proved.

Now we turn to the bilinear form $\widetilde{\mathfrak{b}}_E$ defined on $T_l = K[S_{l+1}]$ and having its values in the ring $K_E = K[\gamma]/\{E(\gamma)\}$.

Denote by τ_t the sum of all permutations in S_{l+1}, each of which is decomposed into exactly t independent cycles. Let $d \rightleftharpoons \gamma \sum_{s \in S_{l+1}} \widetilde{\mathfrak{b}}_E(s, 1)s$, where the sum is taken over all permutations $s \in S_{l+1}$. By formula (25.31), $d = \sum_{t=1}^{l+1} \gamma^t \tau_t$ and d is an element of the center of the algebra $T_l(\gamma)$. If $g = \sum_{\delta \in S_{l+1}} \beta_\delta \delta$, then we denote by $\boldsymbol{\sigma}(g)$ the element $\sum_{\delta \in S_{l+1}} \beta_\delta \cdot \delta^{-1}$. Then, taking into account that the bilinear form $\widetilde{\mathfrak{b}}_E$ is associative, we have

$$\begin{aligned}(\gamma^{-1} d) \circ g &= \sum_s \sum_\delta \beta_\delta \widetilde{\mathfrak{b}}_E(s, 1) s\delta = \sum_\delta \sum_{r = s\delta} \beta_\delta \widetilde{\mathfrak{b}}_E(r, \delta^{-1}) r \\ &= \sum_r \widetilde{\mathfrak{b}}_E\Big(r, \sum_\delta \beta_\delta \delta^{-1}\Big) r = \sum_{s \in S_{l+1}} \widetilde{\mathfrak{b}}_E(s, \boldsymbol{\sigma}(g)) s,\end{aligned}$$

whence it follows that $g \in \operatorname{Ann}_{T_l} \widetilde{\mathfrak{b}}_E$ if and only if $(\gamma^{-1} d) \circ \boldsymbol{\sigma}(g) = 0$ in $K_E[S_{l+1}]$. However, since $\boldsymbol{\sigma}(d) = d$, we see that $0 = \boldsymbol{\sigma}(\gamma^{-1} d) \circ \boldsymbol{\sigma}(g) = g \circ (\gamma^{-1} d) = (\gamma^{-1} d) \circ g$, and we have proved

LEMMA 26.3. *A trace polynomial g is in $\operatorname{Ann}_{T_l} \widetilde{\mathfrak{b}}_E$ if and only if $(\gamma^{-1} d) \circ g = 0$ in $K_E[S_{l+1}]$.*

Let $\mathbf{1}_\mathbf{D}$ be the unit element of the simple ideal corresponding to a Young table $\mathbf{D}$. Then, since d is an element of the center of the algebra $T_l(\gamma)$, we see that $d \circ \mathbf{1}_\mathbf{D} = \mathbf{D}(\gamma) \mathbf{1}_\mathbf{D}$; moreover, for any element a of this ideal we have

$$d \circ a = d \circ (\mathbf{1}_\mathbf{D} \circ a) = \mathbf{D}(\gamma) a, \tag{26.2}$$

where $\mathbf{D}(\gamma)$ is the polynomial depending only on the Young table. If $\chi_\mathbf{D}$ is the character of the irreducible representation of the group S_{l+1} corresponding to $\mathbf{D}$ and $m_\mathbf{D}$ is the dimension of this representation, then

$$\mathbf{D}(\gamma) = \sum_{t=1}^{l+1} \gamma^t \frac{\chi_\mathbf{D}(\tau_t)}{m_\mathbf{D}}. \tag{26.3}$$

It follows from the definition of the element d that $\gamma^{-1} d \in K[\gamma][S_{l+1}]$, and formula (26.3) implies that γ divides all the polynomials $\mathbf{D}(\gamma)$.

LEMMA 26.4. *Let $E(\gamma)$ be some polynomial and let $V_{E(\gamma)} = \Psi_{\tilde{\alpha}}(\{E(\gamma)\})$ be the $\theta_{\tilde{\alpha}}$-closed ideal of trace identities corresponding to this polynomial. Let $V_{\mathbf{D}}$ be the ideal of trace identities generated by the set of multilinear trace polynomials of degree l that forms a two-sided ideal in T_l corresponding to the Young table $\mathbf{D}$. Then $V_{E(\gamma)} \supseteq V_{\mathbf{D}}$ if and only if the polynomial $\gamma E(\gamma)$ divides $\mathbf{D}(\gamma)$.*

PROOF. Let a be an arbitrary element of the set $V_{\mathbf{D}} \cap T_l$ that generates the trace ideal $V_{\mathbf{D}}$. Then, by Lemma 26.3, $a \in \operatorname{Ann}_{T_l} \widetilde{\mathfrak{b}}_E = V_{E(\gamma)} \cap T_l$ if and only if $(\gamma^{-1}d) \circ g = 0$ in the ring $K_E[S_{l+1}]$. From formula (26.2) it follows that this is equivalent to the condition that $\gamma E(\gamma)$ divides $\mathbf{D}(\gamma)$. The lemma is proved.

COROLLARY. *Let $V_{\mathbf{D}_1}$ and $V_{\mathbf{D}_2}$ have the same meaning as in Lemma 26.2, and let $\mathbf{D}_1(\gamma)$ and $\mathbf{D}_2(\gamma)$ be the polynomials defined by formula (26.3). Then $\mathbf{D}_1(\gamma)$ divides $\mathbf{D}_2(\gamma)$.*

PROOF. Put $E(\gamma) \rightleftharpoons \gamma^{-1}\mathbf{D}_1(\gamma)$. Then we have $V_{\mathbf{D}_1} \supseteq V_{\mathbf{D}_2}$ by Lemma 26.2 and $V_{E(\gamma)} \supseteq V_{\mathbf{D}_1}$ by Lemma 26.4. Since $V_{E(\gamma)} \supseteq V_{\mathbf{D}_2}$, we see that $\gamma E(\gamma) = \mathbf{D}_1(\gamma)$ divides $\mathbf{D}_2(\gamma)$ by Lemma 26.4.

Remark 26.1, Lemma 26.4, and its Corollary show that in order to find an explicit description of $\eta_{\tilde{\alpha}}$-closed ideals in $K[\gamma]$ it is necessary to find an explicit formula for the polynomials $\mathbf{D}(\gamma)$, or, to be more precise, a formula of decomposition of $\mathbf{D}(\gamma)$ into prime factors.

LEMMA 26.5. *Let $\mathbf{D}$ be the rectangular Young table whose length of rows is equal to l and the length of columns is equal to k. Then the sum of the roots of the polynomial $\mathbf{D}(\gamma)$ is equal to*

$$l\binom{k}{2} - k\binom{l}{2} \qquad \left(\binom{1}{2} \rightleftharpoons 0\right).$$

PROOF. The leading coefficient of the polynomial $\mathbf{D}(\gamma)$ in formula (26.3) is equal to $\chi_{\mathbf{D}}(\mathbf{1})/m_{\mathbf{D}} = 1$. Hence, the sum of the roots of the polynomial $\mathbf{D}(\gamma)$, taken with opposite sign, is equal to $v = \chi_{\mathbf{D}}(\tau_{lk-1})/m_{\mathbf{D}}$. Obviously, τ_{lk-1} is the sum of all transpositions $t_{ij} \in S_{lk}$, and we can compute v by the formula

$$ve(\mathbf{D}) = \tau_{lk-1} \circ e(\mathbf{D}) = \mathcal{P}(\mathbf{D}) \circ \tau_{lk-1} \circ \mathcal{Q}(\mathbf{D}),$$

where $e(\mathbf{D})$, $\mathcal{P}$, and $\mathcal{Q}$ have the same meaning as in Proposition 26.1 for some Young diagram corresponding to the table $\mathbf{D}$.

First we compute $\mathcal{P} \circ t_{ij} \circ \mathcal{Q}$, where t_{ij} is the transposition of indices i and j. From Proposition 26.1 it follows that this element is equal to $n\mathcal{P} \circ \mathcal{Q}$. It is clear that the number n is equal to the coefficient of $\mathbf{1}$ in the expression $\mathcal{P} \circ t_{ij} \circ \mathcal{Q}$, and, hence, $n = \sum \operatorname{sign}(s'')$, where the sum is taken over all pairs (s', s''), where $s' \in R(\mathbf{D})$ and $s'' \in C(\mathbf{D})$, such that $s' \circ t_{ij} \circ s'' = \mathbf{1}$ (or $t_{ij} = (s')^{-1} \circ (s'')^{-1}$). Then, by Proposition 26.1, the elements belonging to the same row of the diagram $\mathbf{D}$ belong to distinct columns of the diagram $\mathbf{D} \circ t_{ij}$. However, this is possible if and only if the indices i and j belong either to the same row or to the same column of the diagram $\mathbf{D}$. In the first case $\mathcal{P} \circ t_{ij} = \mathcal{P}$ and in the second $t_{ij} \circ \mathcal{Q} = -\mathcal{Q}$. By the uniqueness of the representation $t_{ij} = (s')^{-1} \circ (s'')^{-1}$ we have

$$\mathcal{P} \circ t_{ij} \circ \mathcal{Q} = \begin{cases} \mathcal{P} \circ \mathcal{Q}, & \text{for } t_{ij} \in R(\mathbf{D}), \\ -\mathcal{P} \circ \mathcal{Q}, & \text{for } t_{ij} \in C(\mathbf{D}), \\ 0, & \text{for } t_{ij} \notin R(\mathbf{D}) \cup C(\mathbf{D}). \end{cases}$$

This implies $\nu = \nu_1 - \nu_2$, where ν_1 is the number of transpositions in $R(\mathbf{D})$ and ν_2 is the number of transpositions in $C(\mathbf{D})$. Obviously, $\nu_1 = k\binom{l}{2}$ and $\nu_2 = l\binom{k}{2}$, whence follows the formula we need. The lemma is proved.

We fill the Young table of type $(n_1, \dots, n_k)$ by integers, putting in the jth cell of the ith row $(1 \le j \le n_i)$ the number $i - j$, for instance,

0	−1	−2	−3	−4	−5
1	0	−1	−2	−3	−4
2	1	0	−1		
3	2				

Consider the polynomial

$$\prod_{(i,j)\in\mathbf{D}} (\gamma - (i - j)), \tag{26.4}$$

where the product is extended over all cells (i, j) of the Young table $\mathbf{D}$.

LEMMA 26.6. *The polynomial* $\mathbf{D}(\gamma)$ *given by formula* (26.3) *coincides with the polynomial* (26.4).

PROOF. Note first that the coefficient of the leading term of the polynomial $\mathbf{D}(\gamma)$ is equal to $\chi_{\mathbf{D}}(\mathbf{1})/m_{\mathbf{D}} = 1$. We shall prove the assertion of the lemma by induction on the number of cells in the Young table $\mathbf{D}$. The base of induction is the table with one cell. In this case formulas (26.3) and (26.4) coincide. Suppose that for the Young tables with the number of cells $< n$ the lemma is proved. Let $\mathbf{D}$ be a table of type $\{n_1, \dots, n_k\}$. If $\mathbf{D}$ is not a rectangular table, then we can obtain two distinct Young tables $\mathbf{D}'$ and $\mathbf{D}''$ by deleting from $\mathbf{D}$ exposed rim cells (k, n_k) and (r, n_r), respectively, where $r < k$. Denote by $\mathbf{D}_0$ the Young table obtained by deleting from the table $\mathbf{D}$ both these cells simultaneously. By the induction assumption, we can suppose that $\mathbf{D}'(\gamma)$, $\mathbf{D}''(\gamma)$, and $\mathbf{D}_0(\gamma)$ are expressed by formula (26.4). From Corollary of Lemma 26.4 it follows that

$$\mathbf{D}'(\gamma) = \mathbf{D}_0(\gamma)(\gamma - (r - n_r)), \qquad \mathbf{D}''(\gamma) = \mathbf{D}_0(\gamma)(\gamma - (k - n_k))$$

and $\mathbf{D}'(\gamma) \mid \mathbf{D}(\gamma)$ and $\mathbf{D}''(\gamma) \mid \mathbf{D}(\gamma)$. Since $r < k$ and $n_k < n_r$, we see that $n_r - r > n_k - k$ and these numbers are distinct, hence, $\mathbf{D}(\gamma)$ is equal to the product $\mathbf{D}_0(\gamma)(\gamma - (r - n_r))(\gamma - (k - n_k))$, and the assertion is proved for the case of a nonrectangular table $\mathbf{D}$.

If $\mathbf{D}$ is a rectangular table with the length of rows equal to l and the height of columns equal to k, then we can obtain only the equality $\mathbf{D}(\gamma) = \mathbf{D}'(\gamma)(\gamma - \xi)$, where

$\mathbf{D}'$ is the Young table obtained from $\mathbf{D}$ by deleting the only exposed rim cell (k, l), and ξ is a number. By the induction assumption, $\mathbf{D}'(\gamma)$ is expressed by formula (26.4) for the table $\mathbf{D}'(\gamma)$. Therefore, the sum of the roots of the polynomial $\mathbf{D}'(\gamma)$ is equal to $\sum' = l\binom{k}{2} - k\binom{l}{2} - (k-l)$. Then $\sum' + \xi$ is the sum of the roots of the polynomial $\mathbf{D}(\gamma)$, which is equal to $l\binom{k}{2} - k\binom{l}{2}$ by Lemma 26.5. Direct verification shows that for the rectangular Young table $\mathbf{D}$ the sum of the roots of the polynomial (26.4) is also equal to $l\binom{k}{2} - k\binom{l}{2}$, whence $\xi = k - l$ follows. The lemma is proved.

COROLLARY 1. *Let* $\mathbf{D}$ *be an arbitrary Young table,* $E(\gamma) = \gamma^{-1}\mathbf{D}(\gamma)$, *and* $V_{E(\gamma)} = \Psi_{\tilde{\alpha}}(\{E(\gamma)\})$. *Then* $V_{E(\gamma)} = V_{\mathbf{D}}$. *Moreover, for any Young table* $\mathbf{D}'$ *the inclusion* $V_{\mathbf{D}} \supseteq V_{\mathbf{D}'}$ *takes place if and only if* $\mathbf{D}$ *is contained as a subtable in* $\mathbf{D}'$.

PROOF. By Lemma 26.4, $V_{E(\gamma)} \supseteq V_{\mathbf{D}}$, and $V_{E(\gamma)} \supseteq V_{\mathbf{D}'}$ if and only if $\gamma E(\gamma) = \mathbf{D}(\gamma)$ divides $\mathbf{D}'(\gamma)$. Formula (26.4) implies that $\mathbf{D}(\gamma)$ divides $\mathbf{D}'(\gamma)$ if and only if $\mathbf{D}$ is contained in $\mathbf{D}'$ as a subtable. However, by Lemma 26.2, if $\mathbf{D}$ is a subtable of $\mathbf{D}'$, then $V_{\mathbf{D}} \supseteq V_{\mathbf{D}'}$. Hence, $V_{E(\gamma)} \supseteq V_{\mathbf{D}'}$ if and only if $V_{\mathbf{D}} \supseteq V_{\mathbf{D}'}$. Both the assertions of this corollary follow now from Remark 26.1 and from the decomposition (26.1) of the ideal $V_{E(\gamma)}$. The corollary is proved.

COROLLARY 2. *Let* $E(\gamma)$ *be an arbitrary polynomial and let* $V_{E(\gamma)} = \Psi_{\tilde{\alpha}}(\{E(\gamma)\})$. *Then we have only two possibilities*:

a) *there is a noninteger root of the polynomial* $\gamma E(\gamma)$; *in this case* $V_{E(\gamma)} = 0$;

b) *all the roots of the polynomial* $\gamma E(\gamma)$ *are integers*; *in this case* $V_{E(\gamma)} = V_{\mathbf{D}_E}$, *where* $\mathbf{D}_E$ *is the unique minimal Young table among the Young tables* $\mathbf{D}$ *such that* $\gamma E(\gamma)$ *divides* $\mathbf{D}(\gamma)$.

PROOF. From Remark 26.1 and Lemma 26.4 it follows that $V = 0 + \sum_{\mathbf{D}\in Q_E} V_{\mathbf{D}}$, where the sum is expanded over all Young tables from the set

$$Q_E = \{\,\mathbf{D} \mid \mathbf{D}(\gamma) \text{ is divisible by } \gamma E(\gamma)\,\}.$$

It is clear that in the case a) the set Q_E is empty and $V_{E(\gamma)} = 0$. Formula (26.4) implies that in the case b) there is a unique table $\mathbf{D}_E$ in Q_E such that $\mathbf{D}_E(\gamma)$ divides any polynomial $\mathbf{D}(\gamma)$ for $\mathbf{D} \in Q_E$, and this table is the smallest one in the set Q_E. Therefore, $V_{\mathbf{D}_E} \supseteq V_{\mathbf{D}}$, where $\mathbf{D}$ is an arbitrary Young table in Q_E, and in the sum under consideration $\sum_{\mathbf{D}\in Q_E}$ we can keep only the term $V_{\mathbf{D}_E}$. The corollary is proved.

The following theorem summarizes the results of Lemmas 26.1–26.6 and of Proposition 25.3.

THEOREM 26.1. *The linear space* T_m *of multilinear trace polynomials in the algebra* $\mathcal{G}$ *can be endowed with the operation* $\circ$ *introduced by formula* (25.32), *and with respect to this operation the set* $G(m)$ *of all generalized monomials in* T_m *turns into a group isomorphic to the group of permutations* S_{m+1}, *and* T_m *turns into the group algebra of this group. Moreover, for any polynomial* $E(\gamma)$ *formula* (25.21) *determines a symmetric bilinear form* $\widetilde{\mathfrak{b}}_E$ *on* T_m, *which is associative with respect to the operation* $\circ$ *and takes its values in the algebra* $K[\gamma]/\{E(\gamma)\}$. *For this form we have*

$$\widetilde{\mathfrak{b}}_E(s, \mathbf{1}) = \gamma^t,$$

where $s = a_0 \operatorname{Sp}(a_1)\cdots\operatorname{Sp}(a_t)$, $\mathbf{1} = \operatorname{Sp}(x_1)\cdots\operatorname{Sp}(x_m)$, *and* $t+1$ *is the number of independent cycles taking part in the decomposition of the permutation* s. *To any such bilinear form there corresponds a* $\theta_{\widetilde{\alpha}}$*-closed ideal of trace identities* $V_{E(\gamma)} = \Psi_{\widetilde{\alpha}}(\{E(\gamma)\})$ *of the algebra* $\mathcal{G}$ *such that for any* $m = 1, 2, \dots$ *we have* $V_{E(\gamma)} \cap T_m = \operatorname{Ann}_{T_m} \widetilde{\mathfrak{b}}_E$, *and this set is an ideal of the group algebra* $T_m = K[S_{m+1}]$. *Moreover, for the element* $d = \sum_{t=1}^{m+1} \gamma^t \tau_t$ *of the group algebra* $K[\gamma][S_{m+1}]$, *where* τ_t *is the sum of all permutations in* S_{m+1}, *each of which is decomposable into exactly* t *independent cycles, the equality* $(\gamma^{-1} d) \circ g = 0$, $g \in T_m$, *is valid in* $K_E[S_{m+1}]$ *if and only if* $g \in \operatorname{Ann}_{T_m} \widetilde{\mathfrak{b}}_E$.

To every Young table $\mathbf{D}$ *of type* $(n_1, \dots, n_k)$ *with an arbitrary number of cells* $l+1 = n_1 + \cdots + n_k$ *we associate a polynomial* $\mathbf{D}(\gamma)$ *given by formula* (26.4) *and an ideal of trace identities* $V_{\mathbf{D}}$ *with symmetry type* $\mathbf{D}$ (*see Definition* 26.1). *Then, if the characteristic of the ground field* K *is zero, we have the following additional properties*:

a) $V_{\mathbf{D}} = V_{E(\gamma)}$, *where* $E(\gamma) = \gamma^{-1}\mathbf{D}(\gamma)$.

b) *The mapping* $\mathbf{D} \to V_{\mathbf{D}}$ *is one-to-one, and* $V_{\mathbf{D}_1} \supseteq V_{\mathbf{D}_2}$ *if and only if the Young table* $\mathbf{D}_1$ *is contained in the Young table* $\mathbf{D}_2$ *as a subtable.*

c) *In the group algebra* $T_l(\gamma) = K(\gamma)[S_{l+1}]$ *we have the equality*

$$\sum_{t=1}^{l+1} \gamma^t \tau_t \rightleftharpoons d = \sum_{\mathbf{D}} \mathbf{D}(\gamma) \mathbf{1}_{\mathbf{D}}, \tag{26.5}$$

where the sum is taken over all Young tables $\mathbf{D}$ *containing* $l+1$ *cells, and* $\mathbf{1}_{\mathbf{D}}$ *is the central idempotent of the simple ideal corresponding to the Young table* $\mathbf{D}$.

d) $V_{E(\gamma)} = V_{\mathbf{D}_E}$, *where* $\mathbf{D}_E$ *is the unique minimal Young table in the set of Young tables* $Q_E \rightleftharpoons \{\mathbf{D} \mid \gamma E(\gamma) \text{ divides } \mathbf{D}(\gamma)\}$.

e) *For any ideal of trace identities* V *such that the space* $V \cap T_m$ *forms a two-sided ideal in the group algebra* T_m *for any* $m = 1, 2, \dots$, *the subset* P *of minimal Young tables in the set* $P(V) = \{\mathbf{D} \mid V_{\mathbf{D}} \subseteq V\}$ *is finite and*

$$V = \sum_{\mathbf{D} \in P} V_{\mathbf{D}}.$$

The last assertion of the theorem follows from Remark 26.1 and from the next lemma.

LEMMA 26.7. *In any infinite subset of the set of all Young tables there exist two distinct Young tables that are comparable with respect to the inclusion relation.*

PROOF. Let us consider the topological abelian group $\mathfrak{z}$ whose elements are all functions from the set of all integers $\mathbb{Z}$ to $\mathbb{Z}$ and whose zero neighborhoods are of the form $\mathfrak{z}_n = \{ f \in \mathfrak{z} \mid f(m) = 0,\ m = 0, \pm 1, \dots, \pm n \}$. Define a partial order on $\mathfrak{z}$ by setting $f_1 \leq f_2$ if and only if $f_1(j) \leq f_2(j)$ for any $j \in \mathbb{Z}$. To any Young table $\mathbf{D}$ we assign a function $g_{\mathbf{D}} \in \mathfrak{z}$ such that $g_{\mathbf{D}}(m)$ is equal to the multiplicity of the root m of the polynomial $\mathbf{D}(\gamma)$. It follows from the definition of the Young table and from the explicit form (26.4) of the polynomial $\mathbf{D}(\gamma)$ that

$$\begin{aligned} &g_{\mathbf{D}}(m) \geq g_{\mathbf{D}}(m+1) \text{ for } m \geq 0, \\ &g_{\mathbf{D}}(m-1) \leq g_{\mathbf{D}}(m) \text{ for } m \leq 0; \end{aligned} \tag{26.6}$$

moreover, we have $g_{\mathbf{D}}(m) = 0$ $(|m| > N)$ for sufficiently large $N = N(\mathbf{D})$. Obviously, the mapping $\mathbf{D} \to g_{\mathbf{D}}$ is monotone, and thus it suffices to show that there are two comparable functions in the infinite sequence $g_{\mathbf{D}_1}, g_{\mathbf{D}_2}, g_{\mathbf{D}_3}, \dots$.

Let $r = \max(n, l)$, where n is the length of the first row and l is the length of the first column of the table $\mathbf{D}_1$. If $g_{\mathbf{D}_k}(0) > r$ for some k, then $\mathbf{D}_1$ is contained in $\mathbf{D}_k$, and everything is proved.

Let $g_{\mathbf{D}_k}(0) \leq r$ for $k = 1, 2, \dots$. By (26.6), all the functions $g_{\mathbf{D}_k}$ are bounded. Hence, there exists a limit point $g \in \mathfrak{z}$ for the function sequence $\{g_{\mathbf{D}_i}\}$. Since convergence in the topology under consideration implies pointwise convergence of functions, we have the following conditions for the function g: $g(0) \leq r$, $g(m) \geq g(m+1) \geq 0$ for $m \geq 0$, $0 \leq g(m-1) \leq g(m)$ for $m \leq 0$. Hence, there exist numbers M, m_+ and m_- such that $g(j) = m_+$ for $j > M$ and $g(j) = m_-$ for $j < -M$. Since g is a limit point for the sequence $\{g_{\mathbf{D}_i}\}$, there exists a function $g_{\mathbf{D}_k}$ such that $g - g_{\mathbf{D}_k} \in \mathfrak{z}_M$. Since the support of the function $g_{\mathbf{D}_k}$ is finite, there exists some number $M' > M$ such that $g_{\mathbf{D}_k}(m) = 0$ for $|m| > M'$. We take $\mathbf{D}_l \in \{\mathbf{D}_i\}$ so that $g_{\mathbf{D}_l} - g \in \mathfrak{z}_{M'}$ for $l > k$. Obviously, $g_{\mathbf{D}_k}(m) \leq g_{\mathbf{D}_l}(m)$ for all $m = 0, \pm 1, \pm 2, \dots$. Hence, the Young table $\mathbf{D}_k$ is contained in the Young table $\mathbf{D}_l$. The lemma is proved.

The following theorem is, in essence, a reformulation of our Theorem 26.1 in terms of $\eta_{\widetilde{\alpha}}$-mappings, $\Psi_{\widetilde{\alpha}}$-mappings, $\theta_{\widetilde{\alpha}}$-mappings, and $\Phi_{\widetilde{\alpha}}$-mappings, and answers the third, the fourth, and the fifth questions on the function $\widetilde{\alpha}\colon \widetilde{B} \to K[\gamma]$.

THEOREM 26.2. *The lattice of all $\eta_{\widetilde{\alpha}}$-closed ideals of the algebra $K[\gamma]$ (of $\theta_{\widetilde{\alpha}}$-closed ideals of trace identities of the algebra $\mathcal{G}$) is anti-isomorphic to the lattice of all Young tables with the inclusion relation, completed by the maximal element $\mathbf{D}_\infty$. Moreover,*

a) *all nontrivial $\eta_{\widetilde{\alpha}}$-closed ideals of the algebra $K[\gamma]$ have the form $\{\gamma^{-1}\mathbf{D}(\gamma)\}$, where the polynomials $\mathbf{D}(\gamma)$ are constructed from the Young table $\mathbf{D}$ by formula* (26.4);

b) *all nontrivial $\theta_{\widetilde{\alpha}}$-closed ideals of trace identities of the algebra $\mathcal{G}$ have the form $V_{\mathbf{D}}$* (*see Definition* 26.1);

c) *for any ideal $\{E(\gamma)\}$ of the algebra $K[\gamma]$ we have*

$$\Psi_{\widetilde{\alpha}}\{E(\gamma)\} = \begin{cases} V_{\mathbf{D}_E} & \text{for } Q_E \neq \varnothing, \\ 0 & \text{for } Q_E = \varnothing; \end{cases}$$

$$\eta_{\widetilde{\alpha}}\{E(\gamma)\} = \begin{cases} \gamma^{-1}\mathbf{D}_E(\gamma) & \text{for } Q_E \neq \varnothing; \\ 0 & \text{for } Q_E = \varnothing. \end{cases}$$

where $Q_E = \{\mathbf{D} \mid \gamma E(\gamma) \text{ divides } \mathbf{D}(\gamma)\}$, and $\mathbf{D}_E$ is the smallest Young table in Q_E.

We note two obvious consequences of Theorem 26.1.

COROLLARY 1. *Let $I_{\mathbf{D}}$ be a simple two-sided ideal of the group algebra $T_l = K[S_{l+1}]$ of the symmetric group S_{l+1} corresponding to the Young table $\mathbf{D}$. Then $a \in I_{\mathbf{D}}$ if and only if for any $s \in S_{l+1}$ we have the equality $\gamma\widetilde{\mathfrak{b}}(a, s) = \beta_s \mathbf{D}(\gamma)$, where β_s belongs to the ground field K of characteristic zero.*

COROLLARY 2. *Let $\mathbf{D}_1$ and $\mathbf{D}_2$ be incomparable tables in the lattice of all Young tables relative to the partial order by inclusion, and let $\mathbf{D} = \mathbf{D}_1 \cup \mathbf{D}_2$. Then $V_{\mathbf{D}_1} \cdot V_{\mathbf{D}_2} \subseteq V_{\mathbf{D}}$.*

PROOF. By formula (25.24), the intersection of two $\theta_{\widetilde{\alpha}}$-closed ideals of trace identities is a $\theta_{\widetilde{\alpha}}$-closed ideal. Therefore, $V_{\mathbf{D}_1} \cap V_{\mathbf{D}_2} = V_{\mathbf{D}_1 \cup \mathbf{D}_2} \supseteq V_{\mathbf{D}_1} \cdot V_{\mathbf{D}_2}$. The corollary is proved.

§27. Description of trace identities in the full matrix algebras M_n and in the matrix superalgebras $M_{n,k}$

In this section we state the answer to the sixth question of §11 concerning the function $\widetilde{\alpha}\colon \widetilde{B} \to K[\gamma]$ and, over fields of characteristic zero, for any Young table $\mathbf{D}$ we show explicitly classical associative trace algebras whose ideal of trace identities coincides with the ideal $V_{\mathbf{D}}$.

27.1 The Hamilton–Cayley trace identity and trace identities of the algebra M_n. If $E(\gamma) = \gamma - n$, then the ideal of trace identities $V_{\gamma-n}$ coincides with the ideal of trace identities $V_{n,0}$ of the full matrix algebra M_n (see Proposition 25.3). It follows from assertion d) of Theorem 26.1 and from formula (26.4) that $V_{\gamma-n} = V_{\mathbf{D}^{(n)}}$, where $D^{(n)}$ is a one-column Young table with $n+1$ cells. Definition 26.1 of the ideal $V_{\mathbf{D}}$ implies that $V_{\mathbf{D}^{(n)}}$ as ideal of trace identities is generated by the trace polynomials in $V_{\mathbf{D}^{(n)}} \cap T_n$. However, by Proposition 26.1, $V_{\mathbf{D}^{(n)}} \cap T_n$ is a one-dimensional ideal of the group algebra $T_n = K[S_{n+1}]$, and the element

$$e(\mathbf{D}^{(n)}) = \sum_{s\in S_n} \operatorname{sign}(s)s = (-1)^n \sum_{G(l)} (-1)^t a_0 \operatorname{Sp}(a_1)\cdots \operatorname{Sp}(a_t)$$

is a generator of this ideal. This proves that $V_{n,0} = V_{\mathbf{D}^{(n)}}$, and that all trace identities for the algebra M_n follow from a single multilinear trace identity $e(\mathbf{D}^{(n)}) = 0$ of degree n.

It turns out that this identity is closely related to the Hamilton–Cayley theorem, which states that the characteristic polynomial of an arbitrary matrix y vanishes under the substitution of y in it, i.e., $\{\operatorname{Det}(t\cdot 1 - y)\}|_{t=y} = 0$. However,

$$\operatorname{Det}(t\cdot 1 - y) = \sum_{i=0}^{n} (-1)^i \sigma_i(\beta_1, \dots, \beta_n) t^{n-i},$$

where $\beta_1, \dots, \beta_n$ are the eigenvalues of the matrix y and σ_i are the elementary symmetric functions. It is well known that, over a field of characteristic zero, any $\sigma_i(\beta_1, \dots, \beta_n)$ can be expressed in terms of the polynomials $\beta_1^k + \cdots + \beta_n^k = \operatorname{Sp}(y^k)$ $(k = 1, 2, \dots)$. Hence,

$$\sigma_i(\beta_1, \dots, \beta_n) = g_i(\operatorname{Sp}(y), \dots, \operatorname{Sp}(y^n)) \tag{27.1}$$

and the Hamilton–Cayley theorem gives the trace identity

$$f_n(y) \rightleftharpoons \sum_{i=0}^{n} (-1)^i y^{n-i} g_i(\operatorname{Sp}(y), \dots, \operatorname{Sp}(y^n)) = 0.$$

We shall call this identity the *Hamilton–Cayley trace identity*. The recurrent formulas for computing the trace polynomial $f_n(y)$ are as follows:

$$f_1(y) = y - \operatorname{Sp}(y), \qquad f_n(y) = y f_{n-1}(y) - \tfrac{1}{n} \operatorname{Sp}(f_{n-1}(y)y).$$

We do not need these formulas. For our purposes, it suffices to know that f_n has the form

$$f_n(y) = y^n + \sum_{r_0+\cdots+r_t=n} \beta_{r_0,\dots,r_t}\, y^{r_0} \operatorname{Sp}(y^{r_1}) \cdots \operatorname{Sp}(y^{r_t}) \tag{27.2}$$

for some rational coefficients depending on n only, and this follows by comparing the degrees of both sides of equality (27.1).

THEOREM 27.1. *All trace identities of the full matrix algebra M_n over a field of characteristic zero are consequences of the Hamilton–Cayley trace identity: $f_n = 0$.*

The proof follows immediately from the fact that the degree of f_n equals n and the total linearization of f_n is a nontrivial multilinear polynomial in $x_1, \dots, x_n$, and, by the one-dimensionality of $V_{\mathbf{D}^{(n)}} \cap T_n$, this linearization must coincide with $e(\mathbf{D}^{(n)})$, up to a factor.

REMARK 27.1. Denote by φ_1 the automorphism of the generalized polynomial algebra $\mathcal{G}$ such that

$$\varphi_1(a_0 \operatorname{Sp}(a_1) \cdots \operatorname{Sp}(a_t)) = (-1)^t a_0 \operatorname{Sp}(a_1) \cdots \operatorname{Sp}(a_t), \tag{27.3}$$

and by φ_2 the automorphism of the group algebra $T_l = K[S_{l+1}]$ for which $\varphi_2(s) = \operatorname{sign}(s)s$. It is easily shown that φ_1 transforms ideals of trace identities into ideals of trace identities, and the restriction of φ_1 to any T_l coincides with $(-1)^l\varphi_2$. Since φ_2 transforms any simple ideal in T_l corresponding to the Young table $\mathbf{D}$ into the ideal corresponding to the table $\mathbf{D}^*$ obtained by transposition from $\mathbf{D}$, we see that $\varphi_1(\mathbf{D}) = \mathbf{D}^*$. However, in this case it follows from formula (27.3) that the ideal of trace identities $V_{\mathbf{D}^{(n)*}}$, where $\mathbf{D}^{(n)*}$ consists of a single row with $n+1$ cells, is the ideal of trace identities $V_{0,n}$ for the full matrix algebra M_n, on which the value of the "trace" function $\operatorname{Sp}(a)$ is equal to a scalar matrix whose elements on the main diagonal are $-\sum_{i=1}^n a_{ii}$. Moreover, the ideal of trace identities $V_{\mathbf{D}^{(n)*}}$ is generated by a single element $\varphi_1(f_n(y))$ having the form

$$\varphi_1(f_n(y)) = y^n + \sum_{r_0+\cdots+r_t=n} |\beta_{r_0,\dots,r_t}|\, y^{r_0} \operatorname{Sp}(y^{r_1}) \cdots \operatorname{Sp}(y^{r_t}). \tag{27.4}$$

27.2. Model algebras for ideals of trace identities $V_{\mathbf{D}}$. The set of all Young tables forms a lattice relative to the natural inclusion relation such that the equality $\mathbf{D} = \bigcup_{i=1}^l \mathbf{D}_i$ is equivalent to the equality $V_{\mathbf{D}} = \bigcap_{i=1}^l V_{\mathbf{D}_i}$ (see Theorem 26.2 and Proposition 25.1). The rectangular Young tables are the indecomposable elements of this lattice with respect to the join operation. Let $\mathbf{D}$ be any Young table and let $\mathbf{D} = \mathbf{D}_1 \cup \cdots \cup \mathbf{D}_l$ be a decomposition of it into rectangular Young tables. Theorem 26.1 implies that then $V_{\mathbf{D}} = V_{\mathbf{D}_1} \cap \cdots \cap V_{\mathbf{D}_l}$. Hence, if for some associative algebra U_i with function Sp_i its ideal of trace identities coincides with $V_{\mathbf{D}_i}$, then for the direct sum $\sum_{i=1}^l U_i$ with the function Sp defined as $\operatorname{Sp}(a_1 + \cdots + a_l) = \operatorname{Sp}_1(a_1) + \cdots + \operatorname{Sp}_l(a_l)$, where $a_i \in U_i$, the ideal of trace identities coincides with $\bigcap_{i=1}^l V_{\mathbf{D}_i} = V_{\mathbf{D}}$. Therefore, for an explicit construction of a nontrivial trace algebra model for which the ideal of trace identities coincides with $V_{\mathbf{D}}$, where $\mathbf{D}$ is an arbitrary Young table, it suffices to show a model of this kind for any rectangular Young table. This is given by

THEOREM 27.2. *Let the ground field K be of characteristic zero and let $V_{n,k}$ be the ideal of trace identities for the algebra $M_{n,k}$ that is contained in $\mathcal{G}$. Then for any $l = 1, 2, \dots$ the space $V_{n,k} \cap T_l$ is a two-sided ideal in the group algebra T_l with respect to the operation $\circ$, and $V_{n,k} = V_{\mathbf{D}}$, where $\mathbf{D}$ is the rectangular Young table with $n+1$ rows and $k+1$ columns. In particular, all trace identities of the algebra $M_{n,k}$ are consequences of the subset of multilinear trace identities of the algebra $M_{n,k}$ of degree $n \cdot k + n + k$.*

The proof of this theorem will be given in §§28–30.

§28. Three lemmas

28.1. A ramification lemma. As usual, the (i, j)-cell of the Young table $\mathbf{D}$ is called an exposed rim cell if the cells $(i + 1, j)$ and $(i, j + 1)$ do not belong to $\mathbf{D}$. For any exposed rim cell (i, j) of the table $\mathbf{D}$ we denote by $\mathbf{D}_{ij}$ the Young table obtained from $\mathbf{D}$ by omitting this cell. The set of all Young tables $\mathbf{D}_{ij}$ obtained from $\mathbf{D}$ by omitting an exposed rim cell is denoted by $\overline{\mathbf{D}}$.

In the algebra $\mathcal{G}(\gamma)$ we consider the subspaces $T_l(\gamma)$ and $T_{l-1}(\gamma)$ as bimodules with respect to the right and the left action of group algebras $T_l = K[S_{l+1}]$ and $T_{l-1} = K[S_l]$, respectively. Let us define a linear mapping $\varkappa_l \colon T_l \to T_{l-1}(\gamma)$ by setting $\varkappa_l(f) \rightleftharpoons \pi(f \operatorname{Sp}(x_i))$ for any multilinear trace polynomial f belonging to T_l. By Lemma 25.1 and relations (25.13) and (25.15), we may regard $\varkappa_i(f)$ as obtained from f under the composition of two mappings: replacing the variable x_i in the trace polynomial f by 1 and then replacing all occurrences of the element $\operatorname{Sp}(1)$ in the expression just obtained by the variable γ.

We shall be interested in the structure of the T_{l-1}-bimodule $\varkappa_l(I_{\mathbf{D}})$, where $I_{\mathbf{D}}$ is the simple ideal of the group algebra $T_l = K[S_{l+1}]$ corresponding to the Young table $\mathbf{D}$.

LEMMA 28.1. *Let $I_{\mathbf{D}}$ be a simple ideal of the group algebra $T_l = K[S_{l+1}]$, and let the $I_{\mathbf{D}_{ij}}$ be all the simple ideals of the group algebra $T_{l-1} = K[S_l]$ corresponding to the Young tables $\mathbf{D}_{ij}$ obtained from $\mathbf{D}$ by omitting an exposed rim cell. Then*

$$\varkappa_l(I_{\mathbf{D}}) = \sum_{\mathbf{D}_{ij} \in \overline{\mathbf{D}}} (\gamma - (i - j)) I_{\mathbf{D}_{ij}}. \tag{28.1}$$

THE PROOF of the lemma will be effected in three steps. First we show that $\varkappa_l(I_{\mathbf{D}})$ is a T_{l-1}-bimodule, second we show that the left-hand side of equality (28.1) is contained in the the right-hand side of the same equality, and third we show that for any table $\mathbf{D}_{ij} \in \overline{\mathbf{D}}$ we have the inclusion $\varkappa_l(I_{\mathbf{D}}) \supseteq (\gamma - (i - j)) I_{\mathbf{D}_{ij}}$.

Step I. Since $I_{\mathbf{D}}$ is a two-sided ideal, we see in particular that for any $g \in T_{l-1}$ and $f \in I_{\mathbf{D}}$ the trace polynomial $h = f \circ (g \operatorname{Sp}(x_l))$ belongs to $I_{\mathbf{D}}$. From equalities (25.32), (25.13), and (25.15) we have

$$\begin{aligned} \pi(h \operatorname{Sp}(x_l)) &= \pi\left(\pi\left(fg \Big|_{\substack{x_1 = y_1 x_1 \\ \cdots \\ x_{l-1} = y_{l-1} x_{l-1}}} \right) \Big|_{\substack{y_1 = x_1 \\ \cdots \\ y_{l-1} = x_{l-1}}} \operatorname{Sp}(x_l) \right) \\ &= \pi\left(\pi(f \operatorname{Sp}(x_l)) g \Big|_{\substack{x_1 = y_1 x_1 \\ \cdots \\ x_{l-1} = y_{l-1} x_{l-1}}} \right) \Big|_{\substack{y_1 = x_1 \\ \cdots \\ y_{l-1} = x_{l-1}}} = \pi(f \operatorname{Sp}(x_l)) \circ g. \end{aligned}$$

This equality implies that the linear subspace $\varkappa_l(I_{\mathbf{D}})$ is preserved under right multiplication by the elements of the group algebra T_{l-1}. Similarly we prove that this subspace is preserved under left multiplication by the elements of T_{l-1}.

Step II. For any Young table $\mathbf{D}'$ and any simple ideal $I_{\mathbf{D}'}$ of the group algebra $T_{l-1} = K[S_l]$ we denote by $\mathbf{1}_{\mathbf{D}'}$ the central idempotent of this ideal. Then we have the decomposition $\mathbf{1} = \sum_{\mathbf{D}'} \mathbf{1}_{\mathbf{D}'}$ for the unit element $\mathbf{1}$ of the algebra T_{l-1}, where the sum is taken over all Young tables $\mathbf{D}'$ with l cells. Therefore, any element $h \in T_{l-1}(\gamma)$ can be written in the form

$$h = \sum_{\mathbf{D}'} h(\mathbf{D}'), \tag{28.2}$$

where $h(\mathbf{D}')$ denotes the element $\mathbf{1}_{\mathbf{D}'} \circ h$.

Let h be an arbitrary element of $\varkappa_l(I_{\mathbf{D}})$. Then $h = \pi(f\ \mathrm{Sp}(x_l))$, where $f \in I_{\mathbf{D}} \subset T_l$. By assertion c) of Lemma 25.1 and formulas (25.13) and (25.15) it follows that $h = \gamma h_1 + h_2$, where $h_1, h_2 \in T_{l-1}$, and therefore, in the decomposition (28.2) we have

$$h(\mathbf{D}') = \gamma h_1(\mathbf{D}') + h_2(\mathbf{D}'), \tag{28.3}$$

where $h_1(\mathbf{D}'), h_2(\mathbf{D}') \in I_{\mathbf{D}'} \subset T_{l-1}$. However, in this case $h(\mathbf{D}') = \pi(g\ \mathrm{Sp}(x_l))$ for some $g \in I_{\mathbf{D}}$, and for any permutation $s \in S_l$ we have

$$\widetilde{\mathfrak{b}}(h(\mathbf{D}'), s) = \widetilde{\alpha}(\pi(g\ \mathrm{Sp}(x_l))s) = \widetilde{\alpha}(gs\ \mathrm{Sp}(x_l)) = \widetilde{\mathfrak{b}}(g, s\ \mathrm{Sp}(x_l)). \tag{28.4}$$

Since $g \in I_{\mathbf{D}}$, by Corollary 1 of Theorem 26.1 the right-hand side of this equality is $\beta_s \gamma^{-1}\mathbf{D}(\gamma)$, where $\beta_s \in K$. Since the form $\widetilde{\mathfrak{b}}$ is nondegenerate, we see that for $h(\mathbf{D}') \neq 0$ there exists a permutation $s \in T_{l-1}$ such that $\widetilde{\mathfrak{b}}(h(\mathbf{D}'), s) \neq 0$. Therefore, from Corollary 1 of Theorem 26.1 and equality (28.4), applied to this permutation s, we obtain

$$(\beta_1\gamma + \beta_2)\gamma^{-1}\mathbf{D}'(\gamma) = \widetilde{\mathfrak{b}}(h(\mathbf{D}'), s) = \beta_s \mathbf{D}(\gamma)\gamma^{-1},$$

where $\beta_s, \beta_1, \beta_2 \in K$ and $\beta_1\gamma + \beta_2 \neq 0$. From the explicit formula (26.4) for the polynomials $\mathbf{D}'(\gamma)$ and $\mathbf{D}(\gamma)$ it follows that such an equality is possible only when the table $\mathbf{D}'$ belongs to the set $\overline{\mathbf{D}}$. Therefore, all nonzero elements $h(\mathbf{D}')$ in the decomposition (28.2) of the element $h \in \varkappa_l(I_{\mathbf{D}})$ have a Young table $\mathbf{D}'$ belonging to the set $\overline{\mathbf{D}}$. Suppose that $\mathbf{D}' = \mathbf{D}_{ij}$, where (i, j) is an exposed rim cell of the table $\mathbf{D}$. Then equality (28.3) can be rewritten in the form

$$h(\mathbf{D}_{ij}) = (\gamma - (i - j))h_1(\mathbf{D}_{ij}) + h_3(\mathbf{D}_{ij}),$$

where $h_3(\mathbf{D}_{ij}) = h_2(\mathbf{D}_{ij}) + (i - j)h_1(\mathbf{D}_{ij})$. Then for any $s \in S_l$ relation (28.4) gives

$$\widetilde{\mathfrak{b}}(h_3(\mathbf{D}_{ij}), s) = \widetilde{\mathfrak{b}}(g, s\ \mathrm{Sp}(x_l)) - (\gamma - (i - j))\widetilde{\mathfrak{b}}(h_1(\mathbf{D}_{ij}), s).$$

Corollary 1 of Theorem 26.1 shows that the right-hand side of this equality is divisible by the polynomial $\mathbf{D}(\gamma)\gamma^{-1}$ of degree $l - 1$, while the left-hand side is a polynomial of degree $< l - 1$. Therefore, $\widetilde{\mathfrak{b}}(h_3(\mathbf{D}_{ij}), s) = 0$ for any permutation $s \in S_l$. Since the bilinear form $\widetilde{\mathfrak{b}}$ is nondegenerate, we have $h_3(\mathbf{D}_{ij}) = 0$. Thus, if $h \in \varkappa_l(I_{\mathbf{D}})$, then

$$h = \sum_{\mathbf{D}_{ij} \in \overline{\mathbf{D}}} (\gamma - (i - j))h_1(\mathbf{D}_{ij}).$$

This proves the inclusion

$$\varkappa_l(I_{\mathbf{D}}) \subseteq \sum_{\mathbf{D}_{ij} \in \overline{\mathbf{D}}} (\gamma - (i - j))I_{\mathbf{D}_{ij}}.$$

Step III. Let us fix an arbitrary exposed rim cell (i, j) in the Young table $\mathbf{D}$ and arrange the elements $x_0, x_1, \dots, x_l$ in $\mathbf{D}$ in such a way that x_l is placed in this cell. For the Young diagram $\mathbf{D}$ thus obtained we construct the Young symmetrizer $e(\mathbf{D}) = \mathcal{P}(\mathbf{D}) \circ \mathcal{Q}(\mathbf{D})$, where $\mathcal{P}(\mathbf{D}) = \sum_{p\in R} p$ and $\mathcal{Q}(\mathbf{D}) = \sum_{q\in C} \operatorname{sign}(q)q$ ($\operatorname{sign}(q)$ is the sign of the permutation q). Proposition 26.1 implies that if a permutation is representable in the form $p \circ q$, where $p \in R(\mathbf{D})$ and $q \in C(\mathbf{D})$, then this decomposition is unique. Therefore,

$$e(\mathbf{D}) = \sum \pm a_0 \operatorname{Sp}(a_1) \cdots \operatorname{Sp}(a_t), \tag{28.5}$$

where the sum is extended over all multilinear generalized monomials in T_l which can be represented in the form $p \circ q$ ($p \in R$, $q \in C$). As we showed earlier,

$$\varkappa_l(e(\mathbf{D})) = \gamma h_1 + h_2, \tag{28.6}$$

where $h_1, h_2 \in T_{l-1}$. From formulas (25.13) and (25.15) it follows that all the generalized monomials which occur in the expression for h_1 are obtained from those and only those generalized monomials on the right-hand side of equality (28.5) which have the form $s = b_0 \operatorname{Sp}(b_1) \operatorname{Sp}(b_2) \cdots \operatorname{Sp}(b_{t-1}) \operatorname{Sp}(x_l)$. The isomorphism φ of Lemma 25.9, by means of which we achieved the identification of the algebra T_l with respect to the operation $\circ$ with the group algebra $K[S_{l+1}]$ of the symmetric group S_{l+1} acting from the right on the set $\{ x_0, x_1, \dots, x_l \}$, shows that then the permutation s acts identically on the element x_l. However, in this case the equality $s = p \circ q$, where $p \in R(\mathbf{D})$ and $q \in C(\mathbf{D})$, is possible only in the case when p and q both leave x_l fixed. This proves that the element h_1 in formula (28.6) coincides with the Young symmetrizer $e(\mathbf{D}_{ij})$, where in the Young diagram $\mathbf{D}_{ij}$ the elements $x_0, x_1, \dots, x_{l-1}$ are placed in the same cells as in the diagram $\mathbf{D}$. Since $e(\mathbf{D}_{ij}) \in I_{\mathbf{D}_{ij}}$, we see that, repeating the argument in Step II, we obtain

$$\varkappa_l(e(\mathbf{D})) = (\gamma - (i - j))\, e(\mathbf{D}_{ij}).$$

This equality implies that the intersection $\varkappa_l(I_{\mathbf{D}}) \cap (\gamma - (i - j))I_{\mathbf{D}_{ij}}$ is nonzero for any Young table $\mathbf{D}_{ij}$ from the set $\overline{\mathbf{D}}$. This together with Steps I and II proves the lemma.

28.2. Full matrix algebras over a field in the variety var $M_{n.k}$. Let $\mathfrak{B}$ be an arbitrary nontrivial variety of associative K-algebras and let M_t be the full matrix algebra over a field, belonging to $\mathfrak{B}$, of maximal possible order t. Let F be the relatively free associative algebra of the variety $\mathfrak{B}$ with countably many generators. From the general theory of PI-algebras (see the paper [Am]) it follows that the Jacobson radical $\operatorname{Rad}(F)$ of the algebra F coincides with the Baer radical of this algebra and the quotient algebra $f/\operatorname{Rad}(F)$ is relatively free in the variety var M_t generated by the full matrix algebra M_t over an appropriate field. The number t is sometimes called the complexity of the variety $\mathfrak{B}$.

We denote by $F_{n.k}$ the relatively free associative algebra with countably many generators of the variety generated by the algebra $M_{n.k}$.

LEMMA 28.2. *The algebra* $F_{n.k}/\operatorname{Rad} F_{n.k} \rightleftharpoons E$ *is relatively free in the variety* var M_t, *where* t *is the maximum of the numbers* n *and* k.

PROOF. Since the full matrix algebras M_n and M_k are subalgebras of $M_{n.k}$ and the free algebra of the variety var M_t has zero Jacobson radical, it follows that var $M_t \subseteq$

var E. In order to prove the inverse inclusion, we show that E can be embedded in a full direct product of an infinite number of algebras $M^0_{n,k}$.

PROPOSITION 28.1. *Let $l = 2nk + 1$. Then for any $x \in M^1_{n,k}$ and $y_0, y_1, \dots, y_l \in M_{n,k}$ we have in $M_{n,k}$ the equality*

$$y_0 x y_1 x y_2 \cdots y_{l-1} x y_l = 0. \tag{28.7}$$

FOR THE PROOF of this assertion it suffices to establish it for the case of ground field of characteristic zero. Then identity relation (28.7) is equivalent to the multilinear identity relation

$$\sum_{\sigma \in S_l} y_0 x_{\sigma(1)} y_1 x_{\sigma(2)} y_2 \cdots y_{l-1} x_{\sigma(l)} y_l = 0, \tag{28.8}$$

where the sum is taken over all permutations σ acting on the set $\{1, 2, \dots, l\}$ and $x_1, \dots, x_l \in M^1_{n,k}$. Since relation (28.8) is multilinear with respect to the variables x_i, it suffices to check it for the x_i of the form $\widetilde{t}_j E_{r,s}$, where the $\widetilde{t}_j$ are generators of the Grassmann algebra G and $E_{r,s}$ are the matrix units ($r \leq n \;\;\&\;\; s > n$ or $r > n \;\;\&\;\; s \leq n$). Since $l = 2nk + 1$, we can find two elements x_i and x_j for which $x_i = \widetilde{t}_{i_1} E_{r,s}$ and $x_j = \widetilde{t}_{j_1} E_{r,s}$. However, for any $a \in M_{n,k}$

$$x_i a x_j = \widetilde{t}_{i_1} E_{r,s} a \widetilde{t}_{j_1} E_{r,s} = -\widetilde{t}_{j_1} E_{r,s} a \widetilde{t}_{i_1} E_{r,s} = -x_j a x_i, \tag{28.9}$$

and, by the symmetry of the expression on the left-hand side of equality (28.8) with respect to x_i and x_j, it equals to itself with the opposite sign. This establishes (28.8). The proposition is proved.

By the Birkhoff theorem on free algebras (see §49) the algebra $F_{n,k}$ can be embedded in a full direct product M of an infinite number of copies of the algebra $M_{n,k}$. Since the algebra $M_{n,k}$ is 2-graded, the algebra M inherits this 2-grading and $M = M^0 + M^1$. By Proposition 28.1 the identity relation (28.7) is valid in the algebra M for any elements $x \in M^1$ and $y_0, y_1, \dots, y_l \in M$. But this means that the two-sided ideal I_x generated in M by the element $x \in M^1$ is nilpotent. Hence, the ideal I in M generated by M^1 is the sum of nilpotent ideals. Therefore, $F_{n,k} \cap I \subset \operatorname{Rad} F_{n,k}$ and $F_{n,k}/(F_{n,k} \cap I)$ is contained in $M/I \simeq M^0/(M^0 \cap I)$. Since M^0 is a full direct product of an infinite number of copies of $M^0_{n,k}$ and var $M^0_{n,k}$ = var M_t, it follows that $E \in$ var M^0 = var M_t. Hence, var $M_t \supset$ var E. The lemma is proved.

28.3. Matrix superalgebras in the variety var $M_{n,k}$**.** The next lemma gives necessary and sufficient conditions for any polynomial identity holding in $M_{n,k}$ to hold in $M_{p,q}$.

LEMMA 28.3. *Let the ground field K be of characteristic zero. Then for the inclusion* var $M_{n,k} \supseteq$ var $M_{p,q}$ *to be valid, where $n \geq k$ and $p \geq q$, it is necessary and sufficient that $n \geq p \;\;\&\;\; k \geq q$.*

PROOF. The sufficiency of the condition is obvious.

Necessity. Let var $M_{n,k} \supseteq$ var $M_{p,q}$ and $n \geq k$ and $p \geq q$. This means that any polynomial identity taking place in $M_{n,k}$ takes place in $M_{p,q}$. Then from Lemma 28.2 it follows that $n \geq p$.

Suppose that $q > k$. Then $n > k$, and from the obvious relation $M_{k+1,k+1} \in \operatorname{var} M_{p,q}$ it follows that it suffices to indicate an associative polynomial f for which the identity $f = 0$ holds in $M_{n,k}$ $(n > k)$ but fails to hold in $M_{k+1,k+1}$, to have a contradiction to our assumption $(q > k)$.

Let $c(y_1, \dots, y_l)$ be a central polynomial for the full matrix algebra M_{k+1} over the field K (its existence follows from Theorem 12.1 applied to the natural representation of the reductive Lie algebra M_{k+1}). We put

$$f \rightleftharpoons \sum_{\delta \in S_m} \operatorname{sign}(\delta) c x_{\delta(1)} c x_{\delta(2)} c x_{\delta(3)} c \cdots c x_{\delta(m)} c,$$

where $m > (2n+1)(l \cdot 2nk + 1)$.

We shall show that f is the desired polynomial.

Step I. The identity $f = 0$ does not hold in the algebra $M_{k+1,k+1}$ $(k \geq 0)$.

Indeed, by setting $b_i = \begin{pmatrix} b_i' & 0 \\ 0 & b_i' \end{pmatrix} \in M^0_{k+1,k+1}$, where $b_i' \in M_{k+1}$, we can force the value $c(b_1, \dots, b_l)$ to be equal to the unit matrix. Therefore, if $f = 0$ is an identity in $M_{k+1,k+1}$, then in this algebra we have the standard identity of degree m. However, for $k \geq 0$ the algebra $M_{k+1,k+1}$ contains a subalgebra that is isomorphic to the Grassmann algebra G, and in the latter algebra, over a field of characteristic zero, no standard identity holds.

Step II. The identity $f = 0$ holds in the algebra $M_{n,k}$.

Let $b_j = b_j^0 + b_j^1$, where $b_j^0 \in M^0_{n,k}$ and $b_j^1 \in M^{1}_{n,k}$, $j = 1, \dots, l$. Then we have

$$c(b_1, \dots, b_l) = e_0 + e_1,$$

where e_1 belongs to the ideal I generated in $M_{n,k}$ by the elements $b_1^1, \dots, b_l^1$, and $e_0 = c(b_1^0, \dots, b_l^0)$ has the form $\begin{pmatrix} e_{11} & 0 \\ 0 & e_{22} \end{pmatrix}$, where e_{11} and e_{22} are square G_0-matrices of orders $n \times n$ and $k \times k$, respectively. Since any central polynomial for M_{k+1} vanishes on M_k, we see that e_{22} is the zero matrix. Therefore, for any $a = \begin{pmatrix} a_{11} & a_{12} \\ a_{21} & a_{22} \end{pmatrix} \in M_{n,k}$ the equality $e_0 a e_0 = \begin{pmatrix} e_{11}a_{11}e_{11} & 0 \\ 0 & 0 \end{pmatrix}$ holds. Since in the algebra M_n the standard identity of degree $2n$ is valid, it follows that for any elements $a_1, \dots, a_{2n} \in M_{n,k}$ in the algebra $M_{n,k}$ we have the equality

$$\sum_{\sigma \in S_{2n}} \operatorname{sign}(\sigma) e_0 a_{\sigma(1)} e_0 a_{\sigma(2)} e_0 \cdots e_0 a_{\sigma(2n)} e_0 = 0. \tag{28.10}$$

By Proposition 28.1, the ideal I generated by $b_1^1, \dots, b_l^1$ is nilpotent and $I^t = 0$ for $t > l \cdot 2nk$. Hence, for any $a_1, \dots, a_t \in M_{n,k}$ for $t \geq l \cdot 2nk + 1$ we have

$$e_1 a_1 e_1 a_2 e_1 a_3 e_1 \cdots e_1 a_t e_1 = 0. \tag{28.11}$$

Equalities (28.10) and (28.11) show that whenever the variables y_j and x_i are replaced by elements b_j and a_i, respectively $(i = 1, \dots, m)$, the polynomial f vanishes on $M_{n,k}$ for $m > (2n+1)(2nkl+1)$.

Thus, the identity $f = 0$ holds in $M_{n,k}$ and fails to hold in $M_{p,q}$, and the assumption $q > k$ leads to a contradiction to $\operatorname{var} M_{n,k} \geq \operatorname{var} M_{p,q}$. The lemma is proved.

§29. c-dual sets in the algebra $M_{n,k}$

In the following definition, by an indexing of the elements of a set P by the elements of another set I we mean a surjective map $I \to P$, associating with each element $\nu \in I$ an element $e_\nu \in P$; that is, a single element of P may have several indices.

DEFINITION 29.1. Let M be a module over a commutative and associative K-algebra Φ and let a Φ-bilinear (symmetric) form $(\,,\,)$ be given on M taking values in Φ. Let $c \in \Phi$. Subsets P and P' of M are said to be *c-dual* with respect to the form $(\,,\,)$ if there exists an indexing of the elements of P and P' by the elements of some set I such that for any element a belonging to M we have the equality

$$\sum_{\nu\in I}(e_\nu, a)e'_\nu = c\cdot a = \sum_{\nu\in I}(a, e'_\nu)e_\nu, \tag{29.1}$$

where $e_\nu \in P$, $e'_\nu \in P'$, and in both sums only finitely many elements of the form (e_ν, a) and (a, e'_ν) are nonzero. If $P = P'$ and for any $\nu \in I$ we have $e_\nu = e'_\nu$, then P is said to be *c-normalized*. If $c \neq 0$, we shall call c dual sets *nontrivially c-dual*.

REMARK 29.1. It is clear that if P_i and P'_i are c_i-dual ($i = 1, 2$), then the sets $P_1 \cup P_2$ and $P'_1 \cup P'_2$ are $(c_1 + c_2)$-dual. Moreover, if P and P' are c-dual, then for any element $e \in \Phi$ the sets P and eP' are ec-dual. Therefore, the set of all normalizations $c \in \Phi$ for which there exist c-dual sets forms an ideal of the algebra Φ.

By setting $\Phi = G_0$, $M = M_{n,k}$, and $(x, y) = \operatorname{str}(xy)$, we are in a position to apply Definition 29.1 to the algebra $M_{n,k}$.

LEMMA 29.1. *For the indexed sets $P = \{e_1, \dots, e_t\}$ and $P' = \{e'_1, \dots, e'_t\}$ to be c-dual in $M_{n,k}$ it is necessary and sufficient that for any $a \in M_{n,k}$ the relation*

$$\sum_{i=1}^{t} e_i a e'_i = c\operatorname{str}(a)\cdot \mathbf{1}, \tag{29.2}$$

which is equivalent to the first equality of (29.1), *be valid.*

PROOF. Let $E_{r,s}$ $(r, s = 1, 2, \dots, n+k)$ be the matrix units in $M_{n,k}$. Then

$$e_i = \sum_{m,l=1}^{n+k} c_i^{ml} E_{ml}, \qquad e'_i = \sum_{m,l=1}^{n+k} \tilde{c}_i^{ml} E_{ml}.$$

In equalities (29.2) and (29.1), let us put the element a equal to $t_{pq}E_{pq}$, where $ct_{pq} \neq 0$ and $t_{pq} \in G_0$ if $p, q \le n$ or $p, q > n$, and $t_{pq} \in G_1$ otherwise. Then equality (29.2) is equivalent to the following relations among c, c_i^{ml}, and $\tilde{c}_i^{ml}$:

$$\sum_{i=1}^{t} c_i^{mp}\tilde{c}_i^{ql} = \begin{cases} 0, & \text{for } (m, p) \neq (l, q), \\ c, & \text{for } (m, p) = (l, q) \ \ \& \ \ p, q \le n, \\ -c, & \text{for } (m, p) = (l, q) \ \ \& \ \ p, q > n. \end{cases} \tag{29.3}$$

And the first of relations (29.1) is equivalent to

$$\sum_{i=1}^{t} c_i^{qp} \tilde{c}_i^{ml} = \begin{cases} 0, & \text{for } (m,l) \neq (p,q), \\ c, & \text{for } (m,l) = (p,q) \ \& \ p,q \leq n, \\ -c, & \text{for } (m,l) = (p,q) \ \& \ q \leq n \ \& \ p > n, \\ -c, & \text{for } (m,l) = (p,q) \ \& \ p,q > n, \\ c, & \text{for } (m,l) = (p,q) \ \& \ q > n \ \& \ p \leq n. \end{cases}$$

By replacing $q \to m$ and $m \to q$ in the last equalities, we find that, up to indexing, they coincide with (29.3). This establishes that the first equality of (29.1) is equivalent to (29.2). Now it follows easily from (29.3) that $\sum_{i=1}^{t} e_i' a e_i = c \cdot \operatorname{str}(a) \cdot \mathbf{1}$. Therefore, the first and the second relations of (29.1) are equivalent to each other in the algebra $M_{n,k}$. The lemma is proved.

LEMMA 29.2. *For any monomial $c \in G_1 \cdot G_1 \subset G_0$ of even degree in the algebra $M_{n,k}$ there exist c-dual sets of cardinality $t = (n+k)^2$.*

PROOF. By the condition of the lemma, $c = q_1 q_2$, where q_1 and q_2 belong to the odd component G_1 of the Grassmann algebra G. Let $c = c_1 c_2$, where $c_1, c_2 \in G_0$ (e.g., $c_1 = 1$ and $c_2 = c$). We put

$$\widetilde{E}_{ij} \rightleftharpoons \begin{cases} c_1 E_{ij}, & \text{for } i,j \leq n \vee i,j > n, \\ q_1 E_{ij}, & \text{for } (i > n \ \& \ j \leq n) \vee (i \leq n \ \& \ j > n), \end{cases} \tag{29.4}$$

$$\widetilde{E}'_{ij} \rightleftharpoons \begin{cases} c_2 E_{ji}, & \text{for } i,j \leq n, \\ -c_2 E_{ji}, & \text{for } i,j > n, \\ q_2 E_{ji}, & \text{for } i > n \ \& \ j \leq n, \\ -q_2 E_{ji}, & \text{for } i \leq n \ \& \ j > n. \end{cases} \tag{29.5}$$

It is immediately checked that $\sum_{i,j=1}^{n+k} \widetilde{E}_{ij} a \widetilde{E}'_{ij} = c \operatorname{str}(a) \cdot \mathbf{1}$. Therefore, by Lemma 29.1 the sets $P = \{\widetilde{E}_{ij}\}$ and $P' = \{\widetilde{E}'_{ij}\}$ are c-dual. The lemma is proved.

From this lemma and Remark 29.1 it follows that c-dual sets exist for any $c \in G_0$ with zero constant term.

LEMMA 29.3. *If $c = a_1 a_2 a_3 a_4 \in G_0$, where $a_i \in G_1$ $(i = 1,2,3,4)$, then there exists a c-normalized set in $M_{n,k}$ of cardinality $(n+k)^2$.*

PROOF. Let $c_1 = a_1 a_2$, $c_2 = a_3 a_4$, $q_1 = a_1$, and $q_2 = a_2 a_3 a_4$. For any positive integers i and j $(1 \leq j \leq i \leq n+k)$, we put

$$e_{ij}^{+} \rightleftharpoons \begin{cases} \sqrt{1/2}(c_1 E_{ij} + c_2 E_{ji}), & \text{for } i,j \leq n, \\ \sqrt{-1/2}(c_1 E_{ij} + c_2 E_{ji}), & \text{for } i,j > n, \\ \sqrt{1/2}(q_1 E_{ij} + q_2 E_{ji}), & \text{for } i > n \ \& \ j \leq n; \end{cases}$$

$$e_{ij}^{-} \rightleftharpoons \begin{cases} \sqrt{-1/2}(c_1 E_{ij} - c_2 E_{ji}), & \text{for } i,j \leq n, \\ \sqrt{1/2}(c_1 E_{ij} - c_2 E_{ji}), & \text{for } i,j > n, \\ \sqrt{-1/2}(q_1 E_{ij} - q_2 E_{ji}), & \text{for } i > n \ \& \ j \leq n, \\ 0, & \text{for } i = j. \end{cases}$$

From these equalities and equalities (29.4) and (29.5) it follows that

$$e_{ij}^{+}ae_{ij}^{+} + e_{ij}^{-}ae_{ij}^{-} = \widetilde{E}_{ij}a\widetilde{E}'_{ij} + \widetilde{E}_{ji}a\widetilde{E}'_{ji}.$$

Therefore, in $M_{n,k}$ we have the equality

$$\sum_{i\geq j=1}^{n+k} (e_{ij}^{+}ae_{ij}^{+} + e_{ij}^{-}ae_{ij}^{-}) = c\ \mathrm{str}(a)\cdot \mathbf{1},$$

which, by Lemma 29.1, shows that the set $P = \{e_{ij}^{+}, e_{ij}^{-} \mid 1 \leq j \leq i \leq n+k\}$ is c-normalized. The lemma is proved.

§30. Trace identities of the superalgebra $M_{n,k}$

30.1 The algebra $\mathcal{G}_{n,k}$. Let us choose in the even component of the Grassmann algebra a countable set of monomials $c_1, c_2, \ldots$. We shall suppose that each monomial c_i is of degree four and that any product of a finite number of distinct monomials c_i is nonzero. For each integer $i \geq 0$ we construct a c_i-normalized set $e_1^{(i)}, \ldots, e_t^{(i)}$ (this exists by Lemma 29.3).

The construction given below of the algebra $\mathcal{G}_{n,k}$ for $M_{n,k}$ is similar to the construction of the algebra $\mathcal{G}_n$ for the full matrix algebra M_n (see §25).

Let $\overline{\mathcal{F}}$ be a commutative K-algebra with unit element, given by generators $\overline{x}_i^{(j)}$ $(j = 1, \ldots, t;\ i = 1, 2, \ldots)$ and defining relations

$$\overline{x}_i^{(j)}\overline{x}_i^{(m)} = \delta_{jm}(\overline{x}_i^{(1)})^2, \qquad (\overline{x}_i^{(j)})^3 = 0, \tag{30.1}$$

where δ_{jm} in the Kronecker symbol. We denote by $\overline{M}_{n,k}$ the algebra $\overline{\mathcal{F}} \otimes M_{n,k}$ and by $\mathcal{G}_{n,k}$ the associative subalgebra of $\overline{M}_{n,k}$ containing the unit element and generated by the elements $\overline{x}_i = \sum_{j=1}^{t} \overline{x}_i^{(j)} e_i^{(j)}$ $(i = 1, 2, \ldots)$ and all elements of the form $\mathrm{Sp}\,(\overline{x}_{i_1}\cdots\overline{x}_{i_k}) \rightleftharpoons \mathrm{str}\,(\overline{x}_{i_1}\cdots\overline{x}_{i_k})\cdot\mathbf{1}$ (here the supertrace str is extended from $M_{n,k}$ to $\overline{M}_{n,k}$ by $\overline{\mathcal{F}}$-linearity).

LEMMA 30.1. *For any elements $a, b \in \mathcal{G}_{n,k}$ in the algebra $\overline{M}_{n,k}$ we have*

$$\overline{x}_i a \overline{x}_i = \delta_i\,\mathrm{Sp}(a), \tag{30.2}$$

$$\mathrm{Sp}(\overline{x}_i a)\overline{x}_i = \delta_i a, \qquad \mathrm{Sp}(\overline{x}_i a)\,\mathrm{Sp}(\overline{x}_i b) = \delta_i\,\mathrm{Sp}(ab), \tag{30.3}$$

$$[\overline{x}_i^2, \overline{x}_j] = 0, \qquad \mathrm{Sp}(\overline{x}_i b \overline{x}_i a) = \overline{x}_i b \overline{x}_i\,\mathrm{Sp}(a), \tag{30.4}$$

$$w^3 = \mathbf{1}, \qquad \mathrm{Sp}(\mathbf{1}) = (n-k)\cdot\mathbf{1}, \tag{30.5}$$

where δ_i denotes the element of $\overline{\mathcal{F}} \otimes_K G_0$ equal to $\left(\overline{x}_i^{(1)}\right)^2 c_i$ and w is an arbitrary word in $\overline{x}_j$ containing some letter $\overline{x}_l$ three times.

PROOF. Relation (30.2) follows from the definition of a c_i-normalized set, Lemma 29.1, and the defining relations (30.1). The first assertion of (30.3) is the definition of c_i-normalized set in another notation (see equalities (30.1)). The second relation of (30.3) follows from the first by identity (25.1), which is valid in $M_{n,k}$. Equalities (30.5) are clear. The first equality of (30.4) follows from (30.2), and the second follows from (30.2), (25.1), and $\overline{\mathcal{F}}$-linearity of the function str on $\overline{M}_{n,k}$. The lemma is proved.

LEMMA 30.2. *Let $h(x_1, \dots, x_l)$ be a multilinear trace polynomial. Then the trace identity $h = 0$ is valid in the algebra $M_{n.k}$ if and only if the equality*

$$h(\overline{x}_1, \dots, \overline{x}_l) = 0$$

is valid in the algebra $\mathcal{G}_{n,k}$.

PROOF. Clearly the algebras $M_{n,k}$ and $\overline{M}_{n,k}$ satisfy the same trace identities. Therefore, if $h = 0$ is a trace identity in $M_{n,k}$, then (30.6) holds in $\overline{M}_{n,k}$.

Suppose, conversely, that (30.6) holds. In order to prove that the trace identity $h = 0$ is valid in the algebra $M_{n,k}$ it is sufficient to show that $h(a_1, \dots, a_l) = 0$, where $a_1, \dots, a_l$ are any elements in $M_{n,k}$ whose notation does not involve the generators of the Grassmann algebra occurring in the expressions of $c_1, c_2, \dots, c_l$. Multiplying equation (30.6) successively by $\operatorname{Sp}(\overline{x}_i a_i)$ $(i = 1, 2, \dots, l)$ and applying relation (30.3), we obtain

$$\left(\prod_{i=1}^{l} \delta_i\right) \cdot h(a_1, \dots, a_l) = 0,$$

whence the desired equality $h(a_1, \dots, a_l) = 0$ now follows, because $\delta_1, \dots, \delta_l$ and $a_1, \dots, a_l$ depend on different generators $\widetilde{t}_j$ of the Grassmann algebra. The lemma is proved.

PROPOSITION 30.1. *Let f be a trace polynomial of the algebra $\mathcal{G}$ contained in ${}_\mu\mathcal{G}_\lambda$ and let $f = 0$ be a trace identity of the algebra $M_{n.k}$. Then the multilinear trace polynomial $g = \pi(f)$ in the algebra $M_{n.k}$ satisfies the identity $g|_{\gamma=n-k} = 0$.*

PROOF. For the sake of definiteness, let $\mu = \{x_{l+1}, \dots, x_p\}$ and $\lambda = \{x_1, \dots, x_l\}$. By Lemma 25.1, the coefficients of the polynomial $g \in T_\lambda(\gamma)$ are contained in the polynomial algebra $K[\gamma]$, and f and g are related by (25.16). From the definition of the algebra $\overline{\mathcal{G}}(\gamma)$ it follows that this equality is a consequence of relations (25.12)–(25.15). Since their exact analogs—relations (30.2)–(30.5)—are valid in $\mathcal{G}_{n.k}$, we see that the trace identity $f = 0$, which is valid in the algebra $M_{n.k}$, implies

$$\left(\prod_{i=l+1}^{p} \delta_i\right) g\big|_{\gamma=n-k}(\overline{x}_1, \dots, \overline{x}_l) = 0.$$

By the condition imposed on $c_1, c_2, \dots$ this implies that in $\mathcal{G}_{n.k}$ the relation $g|_{\gamma=n-k}(\overline{x}_1, \dots, \overline{x}_l) = 0$ is valid. The assertion now follows from Lemma 30.2.

COROLLARY 1. *Let $V_{n.k}$ be an ideal of trace identities of the algebra $M_{n.k}$. Then for any $l = 1, 2, \dots$ the subspace $V_{n.k} \cap T_l$ forms a two-sided ideal of the group algebra $T_l = K[S_{l+1}]$ with respect to the operation $\circ$.*

PROOF. The isomorphism φ of Lemma 25.9 shows that the operation $\circ$ given by formula (25.32) does not depend on γ. Let $f \in T_l \cap V_{n.k}$. Then for any $g \in T_l$ the relation

$$f(x_1, \dots, x_l) g(y_1 x_1, \dots, y_l x_l) = 0$$

is a trace identity of the algebra $M_{n.k}$. From Proposition 30.1 it follows that then $\pi(f g|_{x_1 = y_1 x_1, \dots, x_l = y_l x_l}) = 0$ is a trace identity of the algebra $M_{n.k}$. By the change of variables $y_1 \to x_1, y_2 \to x_2, \dots, y_l \to x_l$ in this identity, we see by (25.32) that $f \circ g = 0$ is a trace identity in $M_{n.k}$. Similarly it is shown that $g \circ f = 0$ is also a trace identity in $M_{n.k}$. The corollary is proved.

COROLLARY 2. *For any two ideals of trace identities $V_1, V_2 \subset \mathcal{G}$, if $V_1V_2 \subseteq V_{n,k}$, then either $V_1 \subseteq V_{n,k}$ or $V_2 \subseteq V_{n,k}$.*

PROOF. Assume that neither of the ideals V_1 and V_2 is contained in $V_{n,k}$. This means that there exist trace polynomials $f \in V_1$ and $g \in V_2$ such that $f = 0$ and $g = 0$ are not trace identities in the algebra $M_{n,k}$. By the condition of the lemma,

$$h \rightleftharpoons x_i f(x_1, \dots, x_l)x_{l+1}x_i x_j g(x_{l+2}, \dots, x_m)x_{m+1}x_j = 0$$

is an identity in $M_{n,k}$. If f and g are multilinear, then by formula (25.12) and Lemma 25.1, $\pi(h) = \operatorname{Sp}(fx_{l+1})\operatorname{Sp}(gx_{m+1})$, and, by Proposition 30.1, $\pi(h) = 0$ is a trace identity in the algebra $M_{n,k}$. However, the bilinear form $(x, y) = \operatorname{str}(xy)$ is nondegenerate. Therefore, there exist elements

$$a_1, \dots, a_l, a, a_{l+2}, \dots, a_m, b \in M_{n,k}$$

such that the elements $\operatorname{str}(f(a_1, \dots, a_l)a)$ and $\operatorname{str}(f(a_{l+2}, \dots, a_m)b)$ are nonzero and depend on different generators of the Grassmann algebra. But then their product is nonzero, which contradicts the fact that $f, g \notin V_{n,k}$. This proves the assertion of the corollary in the case where the characteristic of the ground field is zero.

If the characteristic of the field K is different from zero, then there exist ideals of trace identities such that all their multilinear polynomials belong to $V_{n,k}$, and formally we cannot use Proposition 30.1. Nevertheless, in this case the identity

$$\operatorname{Sp}(f(x_1, \dots, x_l)x_{l+1})\operatorname{Sp}(g(x_{l+2}, \dots, x_m)x_{m+1}) = 0$$

still holds in $M_{n,k}$. In order to check this, it is sufficient to apply formula (29.2) to the expression

$$\left(\sum_{i=1}^{t} e_i f(a_1, \dots, a_l)ae_i'\right)\left(\sum_{j=1}^{t} v_j g(a_{l+1}, \dots, a_{m-1})bv_j'\right),$$

where $\{e_i\}$ and $\{e_i'\}$ are c_1-dual, $\{v_i\}$ and $\{v_i'\}$ are c_2-dual, and c_1, c_2 and a, b, a_i $(i = 1, \dots, m)$ depend on different generators of the Grassmann algebra. The corollary is proved.

REMARK 30.1. Lemmas 30.1 and 30.2, Proposition 30.1, and its corollaries are valid over fields of arbitrary characteristic. As the end of the proof of Corollary 2 shows, the algebra $\mathcal{G}_{n,k}$ is used in the proof of these facts not so much in a substantial way but merely as an artifice to allow us to avoid the summation sign in expressions of the form

$$\sum_{i=1}^{t} f(e_i, e_i', x_3, \dots, x_l)$$

in "contracting" of a trace polynomial f with respect to two variables x_1 and x_2 by using c-dual sets $\{e_i\}$ and $\{e_i'\}$. As Lemma 29.2 shows, c-dual sets in $M_{n,k}$ exist for any $c \in G_1G_1$ over fields of any characteristic. Therefore, the proofs given above can be interpreted in this language of "contractions". From our viewpoint, the use of $\mathcal{G}_{n,k}$ and relations (30.2)–(30.5) gives the statements of results and their proofs greater clarity.

30.2. The proof of Theorem 27.2. This proof proceeds in three steps, assuming the ground field to be of characteristic zero.

By Corollary 1 of Proposition 30.1, for the ideal of trace identities $V_{n,k}$ of the algebra $M_{n,k}$ the subspace $V_{n,k} \cap T_l$ is an ideal of the algebra T_l with respect to the operation $\circ$ for any $l = 1, 2, \dots$. Therefore, by assertion e) of Theorem 26.1,

$$V_{n,k} = \sum_{\mathbf{D} \in P} V_{\mathbf{D}}, \tag{30.7}$$

where the sum is taken over a finite set P of Young tables $\mathbf{D}$ that are pairwise incomparable with respect to inclusion. The set P is determined uniquely by the ideal $V_{n,k}$ and consists of all minimal Young tables $\mathbf{D}$ for which the inclusion $V_{\mathbf{D}} \subseteq V_{n,k}$ takes place. This proves the uniqueness of decomposition (30.7).

As the first step in the proof of Theorem 27.2, we shall show that all Young tables $\mathbf{D}$ that occur in decomposition (30.7) are rectangular; in the second step we shall show that the number p of rows and the number q of columns in any such table satisfy the relation

$$p - q = n - k. \tag{30.8}$$

Since any rectangular tables satisfying this relation are comparable in the lattice of all Young tables, it follows from assertion b) of Theorem 26.1 that decomposition (30.7) obtains the form $V_{n,k} = V_{\mathbf{D}}$, where $\mathbf{D}$ is the rectangular Young table with p rows and q columns. As the third step we shall show, using Lemma 28.2, that the maximum of the numbers p and q coincides with the maximum of the numbers $n+1$ and $k+1$. Together with equality (30.8), this will give the assertion of Theorem 27.2.

Step 1. Suppose that some Young table $\mathbf{D}$ occurring in decomposition (30.7) is not rectangular. Then in the lattice of all Young tables we have a decomposition $\mathbf{D} = \mathbf{D}_1 \cup \mathbf{D}_2$, where $\mathbf{D}_1$ and $\mathbf{D}_2$ are incompareble Young tables. From Corollary 2 of Theorem 26.1 we have $V_{\mathbf{D}_1} V_{\mathbf{D}_2} \subseteq V_{\mathbf{D}} \subseteq V_{n,k}$. Then by Corollary 2 of Proposition 30.1 we have one of the following inclusions: either $V_{\mathbf{D}_1} \subseteq V_{n,k}$ or $V_{\mathbf{D}_2} \subseteq V_{n,k}$. Since $\mathbf{D}_1 \subset \mathbf{D}$ and $\mathbf{D}_2 \subset \mathbf{D}$, by Theorem 26.1 we have either $V_{n,k} \supseteq V_{\mathbf{D}_1} \supset V_{\mathbf{D}}$ or $V_{n,k} \supseteq V_{\mathbf{D}_2} \supset V_{\mathbf{D}}$, and in each case we obtain a contradiction to the uniqueness of decomposition (30.7). This shows all Young tables in (30.7) to be rectangular.

Step 2. Let $\mathbf{D}$ be any rectangular Young table that occurs in decomposition (30.7), with p rows and q columns. It has a unique exposed rim cell (p, q). From the uniqueness of decomposition (30.7) it follows that $V_{\mathbf{D}_{p,q}}$ is not contained in $V_{n,k}$, where the Young table $\mathbf{D}_{p,q}$ is obtained from the table $\mathbf{D}$ by removing the exposed rim cell (p, q).

Let $l = pq - 1$. By Definition 26.1, the ideal of trace identities $V_{\mathbf{D}}$ is generated by the set $I_{\mathbf{D}} = V_{\mathbf{D}} \cap T_l$ and $I_{\mathbf{D}}$ is a simple ideal in T_l corresponding to the Young table $\mathbf{D}$. By Lemma 28.1, for any multilinear polynomial $f \in I_{\mathbf{D}}$ we have $\varkappa_l(f) = (\gamma - (p-q))g$, where $g \in I_{\mathbf{D}_{p,q}} \subset T_{l-1}$. Since $f = 0$ is a trace identity of the algebra $M_{n,k}$, by Corollary 1 of Proposition 30.1

$$\pi(f\, \mathrm{Sp}(x_l))|_{\gamma=n-k} = (\varkappa_l(f))|_{\gamma=n-k} = ((n-k) - (p-q))\, g = 0$$

is a trace identity in $M_{n,k}$. Since $\varkappa_l(I_{\mathbf{D}}) = (\gamma - (p-q)) I_{\mathbf{D}_{p,q}}$, we see that $I_{\mathbf{D}_{p,q}} \subset V_{n,k}$ for $p - q \neq n - k$. By Definition 26.1, the subset $I_{\mathbf{D}_{p,q}}$ of T_{l-1} generates the ideal

$V_{\mathbf{D}_{p,q}}$ of trace identities; therefore, for $p - q \neq n - k$ this ideal would be contained in $V_{n,k}$. However, we noted above that $V_{\mathbf{D}_{p,q}} \not\subseteq V_{n,k}$. This proves equality (30.8) for any Young table in decomposition (30.7). Since the rectangular tables satisfying (30.8) are comparable with respect to inclusion, decomposition (30.8) must have the form $V_{n,k} = V_{\mathbf{D}}$ by Theorem 26.1.

Step 3. Denote by m the maximum of the numbers $n+1$ and $k+1$. From formula (30.8) it follows that Theorem 27.2 is equivalent to the assertion $m = r$, where r the maximum of the numbers p and q.

We shall prove the equality $m = r$ by induction on the number $e = \min\{n, k\}$.

The base of induction is $e = 0$. In this case either $n = 0$ or $k = 0$; in the second case the equality $m = r$ follows from the fact that the ideal of trace identities of the full matrix algebra M_n coincides with the ideal $V_{\mathbf{D}}$, where the Young table $\mathbf{D}$ consists of one column with $n+1$ cells. In the first case the equality $m = r$ follows from Remark 27.1.

Suppose we have shown that $m = r$ for $e < e_0$, i.e., $V_{n,k} = V_{\mathbf{D}}$ for $e < e_0$, where the Young table $\mathbf{D}$ is rectangular and contains $n+1$ rows and $k+1$ columns. Let us show that the same is true for $e = e_0$. Denote by $\mathbf{D}_r$ the Young table with r cells posed in a single row for $q - p > 0$, or in a single column for $p - q \geq 0$. Obviously, $\mathbf{D}_r \subseteq \mathbf{D}$, where the rectangular table $\mathbf{D}$ contains p rows and q columns ($r = \max\{p, q\}$). Therefore, by Theorem 26.4, we have $V_{\mathbf{D}_r} \supseteq V_{\mathbf{D}}$. But $V_{\mathbf{D}_r}$ is the ideal of trace identities of the full matrix algebra M_{r-1} (taking as a trace, as usual, the function tr for $r = p \geq q$ and the function $-$ tr for $p < q = r$). It follows that the relatively free algebra $F_{n,k}$ of the variety generated by the algebra $M_{n,k}$ has an epimorphism onto the algebra M_{r-1}. By Lemma 28.2, this means that $r - 1 \leq m - 1$, i.e., $r \leq m$.

Suppose that $r < m$. Then we have by relation (30.8) that $p \leq n$ and $q \leq k$, and $\min(p, q) < e_0$. Therefore, by induction assumption, $V_{\mathbf{D}} = V_{n,k}$ coincides with the ideal of trace identities of the algebra $M_{n-i,k-i}$ for some $i > 0$, i.e., the trace identities of $M_{n-i,k-i}$ and $M_{n,k}$ coincide. But then the relatively free associative algebras $F_{n-i,k-i}$ and $F_{n,k}$ of varieties var $M_{n-i,k-i}$ and var $M_{n,k}$, respectively, are isomorphic. This contradicts Lemma 28.2. Hence, $r = m$ is valid also in the case $e = e_0$. Theorem 27.2 is proved.

§31. Central polynomials in the algebras M_n and $M_{n,k}$

In this section we shall prove the following

THEOREM 31.1. *Over a field K of arbitrary characteristic, for any nonnegative integers n and k there exist multilinear central polynomials for the algebra $M_{n,k}$.*

We recall that an associative polynomial f is called a *central polynomial* for the algebra $M_{n,k}$ if all its values on the elements of $M_{n,k}$ belong to the center of the algebra $M_{n,k}$ and the identity $f = 0$ does not hold in $M_{n,k}$.

A special case of this theorem for $k = 0$ gives the existence of a central polynomial for the full matrix algebra M_n over an arbitrary field K.

We shall give two proofs of Theorem 31.1. The first one is based on the notion of a weak identity and in fact repeats the proof of existence of a central polynomial for the full matrix algebra M_t given in the paper [Raz3], while the second one uses the notion of c-dual sets and is similar to the construction of a central polynomial for the full matrix algebra M_t expounded in the paper [Raz7]. Both proofs lead to an explicit construction of a central polynomial.

31.1. Weak identities in the algebras $M_{n,k}$ and an existence criterion of central polynomials for the algebras $M_{n,k}$. Let F, $\mathcal{G}$, $\mathcal{G}(\gamma)$, $V_{n,k}$, $\mathcal{G}_{n,k}$, φ, $\circ$, T_l, and ${}_\mu\mathcal{G}_\lambda$ have the same meaning as in §§25–30.

Having in mind the isomorphism φ (see Lemma 25.9), we continue to identify the space of multilinear trace polynomials T_l depending on $x_1, \dots, x_l$ with the group algebra $K[S_{l+1}]$ of the symmetric group acting from the right on the set of elements $\{x_1, \dots, x_l\}$. It is clear that under this identification the space $F \cap T_l$ coincides with the subspace of $K[S_{l+1}]$ spanned by all permutations that are complete cycles.

We define on $F \cap T_l$ $(l = 1, 2, \dots)$ a linear operator A_i (or A_{x_i}), where $i \leq l$, by the formula $A_i(ax_ib) = bx_ia$ for all $ax_ib \in F \cap T_l$. Denote by τ_{0i} the permutation in S_{l+1} that is the transposition of the elements x_0 and x_i. In can be directly verified that for any polynomial f in $F \cap T_l$ we have the equality $A_i f = \tau_{0i} \circ f \circ \tau_{0i}$. This formula and Corollary 1 of Proposition 30.1 imply

LEMMA 31.1. *Let the ground field K be of arbitrary characteristic, and let $f \in F \cap T_l$. Then the identity $f = 0$ holds in the algebra $M_{n,k}$ if and only if the identity $A_i f = 0$ holds in this algebra.*

Denote by $\mathrm{sl}(n, k, K)$ the K-linear subspace in $M_{n,k}$ consisting of the elements a such that $\mathrm{str}(a) = 0$.

DEFINITION 31.1. Let $g(x_1, \dots, x_t) \in F$. The formula $g = 0$ is called a *weak identity* in the algebra $M_{n,k}$ if the identity $g = 0$ holds in the pair $(M_{n,k}, \mathrm{sl}(n, k, K))$. A weak identity $g = 0$ is said to be an *essentially weak identity* if the formula $g = 0$ is not a polynomial identity of the algebra $M_{n,k}$.

PROPOSITION 31.1. *The existence of multilinear central polynomials for the algebra $M_{n,k}$ is equivalent to the existence of essentially weak identities for the algebra $M_{n,k}$.*

The proof of this proposition will be given in a completely constructive manner. We shall show how to construct an essentially weak identity from a central polynomial and how to construct a central polynomial from an essentially weak identity.

I. NECESSITY. It is clear that $A_i(y_1ax_iby_2) = by_2x_iy_1a$, whence for any $f \in F \cap T_l$ it follows that

$$A_i(y_1 f y_2) = (A_i f)|_{x_i = y_2x_iy_1}, \qquad A_i([f, y]) = (A_i f)|_{x_i = [y, x_i]}. \tag{31.1}$$

Let $f \in F \cap T_l$ be a central polynomial for $M_{n,k}$. This is equivalent to $f = 0$ not being an identity of the algebra $M_{n,k}$ and $[f, x_{l+1}] = 0$ being an identity of the algebra $M_{n,k}$. Applying to this identity the operator A_i $(i = 1, \dots, l)$ and using (31.1) and Lemma 31.1, we see that $(A_i f)|_{x_i = [x_{l+1}, x_i]} = 0$ is an identity of the algebra $M_{n,k}$. Since the commutator subalgebra $[M_{n,k}, M_{n,k}]$ of the Lie algebra $M_{n,k}$ almost coincides with $\mathrm{sl}(n, k, K)$, this implies that $A_i f = 0$ is a weak identity of the algebra $M_{n,k}$. Since the identity $f = 0$ does not hold in $M_{n,k}$, the identity $A_i f = 0$ also does not hold in $M_{n,k}$ by Lemma 31.1. Thus, $A_i f = 0$ is an essentially weak identity for any multilinear central polynomial $f \in F \cap T_l$.

In fact, the polynomial $g = A_i f$ vanishes under the substitution of any element of $\mathrm{sl}(n, k, K)$ instead of variable x_i and of any elements of the algebra $M_{n,k}$ instead of other variables. As we shall see later, if for a certain multilinear polynomial g this condition is satisfied and $g = 0$ is not an identity of $M_{n,k}$, then we can completely reverse the above argument for the polynomial f and find that $f = A_i g$ is a central polynomial.

II. SUFFICIENCY. Let $h \in F \cap T_l$ and let $h = 0$ be an essentially weak multilinear identity of the algebra $M_{n,k}$. Then it is possible to choose a positive integer m such that: a) $1 < m \leq l$; b) for any $v_1, \dots, v_m \in \mathrm{sl}(n, k, K)$ and $a_{m+1}, \dots, a_t \in M_{n,k}$ the element $h(v_1, \dots, v_m, a_{m+1}, \dots, a_t)$ is equal to zero in $M_{n,k}$; c) there exist elements $w_1, \dots, w_{m-1} \in \mathrm{sl}(n, k, K)$ and $a_m, a_{m+1}, \dots, a_t \in M_{n,k}$ such that $h(w_1, \dots, w_m, a_{m+1}, \dots, a_t) \neq 0$ in $M_{n,k}$. However, in this case the multilinear polynomial

$$g = h\big|_{x_1=[x_1,x_{t+1}],x_2=[x_2,x_{t+2}],\dots,x_{m-1}=[x_{m-1},x_{t+m-1}]}$$

belongs to $F \cap T_q$, $q = t+m-1$, and has the property that $g = 0$ is not an identity of the algebra $M_{n,k}$ (this follows from the property c) of the number m and from the fact that the subspaces $[M_{n,k}, M_{n,k}]$ and $\mathrm{sl}(n, k, K)$ almost coincide) and $g|_{x_m=[x_{t+m},x_m]} = 0$ is an identity of the algebra $M_{n,k}$ (the property b) of the number m). Then from Lemma 31.1 and formula (31.1) it follows that the identity $A_m g = 0$ does not hold in $M_{n,k}$ and $[A_m g, x_{t+m}] = 0$ is an identity of the algebra $M_{n,k}$. This proves that $A_m g$ is a multilinear central polynomial for $M_{n,k}$. The proposition is proved.

We decompose the construction of central polynomials for the algebras $M_{n,k}$ into two cases: $\mathrm{char}\, K > 0$ and $\mathrm{char}\, K = 0$.

31.2. Existence of multilinear essentially weak identities for the algebras $M_{n,k}$ over fields of positive characteristic.

REMARK 31.1. Let a multilinear polynomial identity have the form

$$f \rightleftharpoons \sum_{h_1,\dots,h_{q+1}} h_1 y_1 h_2 y_2 \dots h_q y_q h_{q+1} = 0, \tag{31.2}$$

where $f \in F \cap T_l$, and $y_1, \dots, y_q$ are mutually distinct fixed variables from the set $\{ x_1, \dots, x_l \}$ that do not occur in the notation of the polynomials $h_1, \dots, h_{q+1}$. Denote by $\tau_{0,i}$ a permutation in S_{l+1} that is the transposition of x_0 and y_i. Using formulas (25.12), (25.13), (25.32), and the isomorphism φ, we can immediately verify that

$$\tau_{0,q} \circ \tau_{0,q-1} \circ \dots \circ \tau_{0,1} \circ f = \sum_{h_1,\dots,h_{q+1}} \mathrm{Sp}(h_1 y_1) \, \mathrm{Sp}(h_2 y_2) \cdots \mathrm{Sp}(h_q y_q) h_{q+1}.$$

Therefore, by Corollary 1 of Proposition 30.1, a multilinear polynomial identity of the form (31.2) holds in the algebra $M_{n,k}$ if and only if in $M_{n,k}$ we have a trace identity of the form

$$\sum_{h_1,\dots,h_{q+1}} \mathrm{Sp}(h_1 y_1) \, \mathrm{Sp}(h_2 y_2) \cdots \mathrm{Sp}(h_q y_q) h_{q+1} = 0. \tag{31.3}$$

Denote by d_m a multilinear associative polynomial of the form

$$\sum_{\sigma\in S_m} x_{\sigma(1)}y_1x_{\sigma(2)}y_2\cdots x_{\sigma(m)}y_m,$$

where $y_1,\dots,y_m$ are the variables $x_{m+1},\dots,x_{2m}$, respectively, and the sum is extended over all permutations σ of the set $\{1,2,\dots,m\}$.

LEMMA 31.2. *Let the characteristic of the ground field K equals $p>0$. Then for $m=(p-1)(n^2+k^2-1)+2nk+1$ the formula $d_m=0$ is an essentially weak identity of the algebra $M_{n,k}$. Moreover, for $q=(p-1)(n^2+k^2)+2nk$ the identity $d_q=0$ does not hold in the algebra $M_{n,k}$, and $d_{q+1}=0$ is an identity in $M_{n,k}$.*

PROOF. The algebra $M_{n,k}$ has a natural embedding in the tensor product $G\otimes_K M_{n+k}$ of the Grassmann algebra G and the full matrix algebra M_{n+k}. Denote by E_{ij} the matrix units of the algebra M_{n+k} (i.e., E_{ij} is the matrix whose element on the intersection of the ith row and of the jth column is the unit of the field K, and all other elements of it are equal to zero). Let E_1 be the set of all matrix units E_{ij} such that $i,j\le n$ or $i,j>n$, let E_2 be the set of all matrix units E_{ij} such that $i\le n$ & $j>n$ or $i>n$ & $j\le n$, and let E_1' be the set consisting of all elements of the form E_{ij}, where $i\ne j$ & $(i,j\le n$ or $i,j>n)$, and also of the elements $E_{i-1,i-1}-E_{ii}$, where $2\le i\le n$ or $n+1<i\le n+k$, and of the element $E_{nn}+E_{n+1,n+1}$. It is clear that $\mathrm{sl}(n,k,K)$ and $M_{n,k}$ as G_0-modules are generated by the elements of the sets $E'=\bigcup_{i=1}^{\infty}\tilde{t}_iE_2\cup E_1'$ and $E=\bigcup_{i=1}^{\infty}\tilde{t}_iE_2\cup E_1$, respectively, where $\tilde{t}_i$ are the generators of the Grassmann algebra.

Let $m=(p-1)(n^2+k^2-1)+2nk+1$. By the multilinearity of the polynomial d_m, in order to verify that $d_m=0$ is a weak identity in $M_{n,k}$ it suffices to show that the polynomial d_m vanishes if we substitute any elements of the set E' for the variables of it. The number m is chosen so that for any such substitution either p of the different variables in the set $\{x_1,\dots,x_m\}$ take the same value in E_1' or two distinct variables x_i and x_j take values of the form $\tilde{t}_{i_1}E_{rs}$ and $\tilde{t}_{j_1}E_{rs}$, respectively. In the first case we obtain an expression that vanishes over a field of characteristic p, because each term of the same type will appear $p!$ times in the expression. In the second case, using equality (28.9) as in the proof of Lemma 28.2, we conclude that the expression obtained after the substitution is equal to itself with the opposite sign (and over a field of characteristic 2 is equal to zero). This proves that $d_m=0$ is a weak identity.

Replacing in this argument m, E', and E_1' by $q+1$, E, and E_1, respectively, we see that $d_{q+1}=0$ is an identity of the algebra $M_{n,k}$.

Since $q>m$, in order to prove that $d_m=0$ is an essentially weak identity it suffices to show that the identity $d_q=0$ does not hold in the algebra $M_{n,k}$. Toward this end, by Remark 31.1, where we are to take the polynomial f equal to d_q, it suffices to verify that the following trace identity does not hold in $M_{n,k}$:

$$\sum_{\sigma\in S_q}\mathrm{Sp}(x_{\sigma(1)}y_1)\cdots\mathrm{Sp}(x_{\sigma(q)}y_q)=0, \tag{31.4}$$

where the sum is extended over all permutations σ of the set $\{1,\dots,q\}$. Let us choose in some way for any matrix unit $E_{ij}\in E_2$ two generators ξ_{ij} and η_{ij} of the Grassmann algebra. For any integer $t = 0, 1, 2, \ldots, p-2$ we substitute in the left-hand side of the equality different matrix units from the set E_1 instead of variables x_i (similarly y_i), where $(n^2+k^2)t \le i \le (n^2+k^2)(t+1)$, and substitute ξ_{ij} and η_{ij} instead of the remaining $2nk$ variables x_ν (similarly y_ν) so that different variables x_ν correspond to different matrix units in E_2. It is clear that then the left-hand side of relation (31.4) is equal to

$$\pm(p-1)!^{n^2+k^2} \cdot \prod_{E_{ij} \in E_2} \xi_{ij}\eta_{ij},$$

and we can choose ξ_{ij} and η_{ij} so that their product be nonzero. The lemma is proved.

Proposition 31.1 and Lemma 31.2 prove Theorem 31.1 for fields of positive characteristic.

31.3. The construction of polynomial c-dual sets and central polynomials for the algebra $M_{n,k}$ over fields of positive characteristic.

LEMMA 31.3. *Let p, q, n, and k have the same meaning as in Lemma* 31.2. *Then in the algebra $M_{n,k}$ we have the trace identity*

$$\sum_{\sigma \in S_{q+1}} \operatorname{Sp}(x_{\sigma(1)}y_1x_{\sigma(2)}y_2 \cdots x_{\sigma(q)}y_q)x_{\sigma(q+1)} = 0. \tag{31.5}$$

PROOF. By Lemma 31.2, the identity $d_{q+1} = 0$ is satisfied in $M_{n,k}$. Then the relation $y_0d_{q+1}|_{y_{q+1}=1} = 0$ is an identity in the algebra $M_{n,k}$. We put the polynomial f equal to $y_0d_{q+1}|_{y_{q+1}=1}$ in Remark 31.1. Applying the transposition τ_{0q} of the variables x_0 and y_q to the identity (31.2), we see that $\tau_{0q} \circ f = 0$ is a trace identity in the algebra $M_{n,k}$. As in Remark 31.1, by using formulas (25.12), (25.13), (25.32), and the isomorphism φ, we see that $\tau_{oq} \circ f$ coincides with the left-hand side of equality (31.5). The lemma is proved.

Replacing in formula (31.5) the variable x_{q+1} by an arbitrary element a and using the obvious equality $\operatorname{Sp}(bx_idy_0) = \operatorname{Sp}(dy_0bx_i)$ and the definition of the operator A_i, we rewrite this equality in the following way:

$$-\operatorname{Sp}(d_qy_0)a = \sum_{i=1}^{q} \operatorname{Sp}\left((A_id_q)|_{x_i=y_0}a\right)x_i. \tag{31.6}$$

Denoting by c the polynomial $-\operatorname{Sp}(d_qy_0)$ and comparing this identity with (29.1), we see that the sets $\{ (A_id_q)|_{x_i=y_0} \mid i = 1, \ldots, q \}$ and $\{ x_1, \ldots, x_q \}$ with the natural indexing are c-dual. To be more precise, if we replace all variables in identity (31.6) by arbitrary elements of the algebra $M_{n,k}$ so that $x_i \to e_i' \in M_{n,k}$ and denote by c_0 and e_i the values of polynomials $-\operatorname{Sp}(d_qy_0)$ and $(A_id_q)|_{x_i=y_0}$, then the sets $\{e_i \mid i = 1, \ldots, q\}$ and $\{e_i' \mid i = 1, \ldots, q\}$ will be c_0-dual in $M_{n,k}$. However, in this case Lemma 29.1 implies that we have the following trace identity in $M_{n,k}$:

$$\sum_{i=1}^{q} x_ia(A_id_q)|_{x_i=y_0} = -\operatorname{Sp}(a)\cdot\operatorname{Sp}(d_qy_0).$$

By Lemma 31.2, $d_q = 0$ is not an identity in $M_{n,k}$ and, because the bilinear form $(x, y) = \operatorname{str}(xy)$ is nondegenerate on $M_{n,k}$, we see that the right-hand side of this equality is not identically zero on $M_{n,k}$. This proves that the multilinear polynomial on the left-hand side of the equality is central for the algebra $M_{n,k}$.

31.4. Existence of multilinear essentially weak identities for the algebras $M_{n,k}$ over fields of characteristic zero. Let A be an arbitrary (associative) K-algebra with unit, let V_0 and V_1 be two finite-dimensional subspaces of A, and let $\{e_1, \dots, e_r\}$ and $\{v_1, \dots, v_s\}$ be bases of these subspaces. We put $\widetilde{A} = G \otimes_K A$, $\widetilde{V}_0 = G_0 \otimes_K V_0$, $\widetilde{V}_1 = G_1 \otimes_K V_1$, and $\widetilde{V} = \widetilde{V}_0 + \widetilde{V}_1$. For any fixed integer $l \in \{0, 1, 2, \dots\}$ consider the set of associative polynomials of the form

$$f(x_1, \dots, x_l, y_1, \dots, y_l) = \sum_{\sigma \in S_l} \beta_\sigma x_{\sigma(1)} y_1 x_{\sigma(2)} y_2 \cdots x_{\sigma(l)} y_l, \tag{31.7}$$

where the sum is taken over all permutations σ of the set $\{1, 2, \dots, l\}$, and $\beta_\sigma \in K$. The set of polynomials of this kind forms a left S_l-module with respect to the natural action of permutations on the set $\{x_1, \dots, x_l\}$. Let $\mathbf{D}$ be a Young table with l cells. We place the variables $x_1, \dots, x_l$ in this table in some way and then we construct, from the Young diagram $\mathbf{D}$ thus obtained, the groups $R(\mathbf{D})$ and $C(\mathbf{D})$, the elements $\mathcal{P}(\mathbf{D}), \mathcal{Q}(\mathbf{D}) \in K[S_l]$, and the Young symmetrizer $e(\mathbf{D}) = \mathcal{P}(\mathbf{D}) \circ \mathcal{Q}(\mathbf{D})$, as it was done in §28. Denote by $f_{\mathbf{D}}(x_1, \dots, x_l, y_1, \dots, y_l)$ the multilinear polynomial $e(\mathbf{D}) x_1 y_1 \cdots x_l y_l$. Denote by the symbol $\widetilde{\mathbf{D}}$ the Young table $\mathbf{D}$ of type $(n_1, \dots, n_k)$ filled by the elements x_j so that all cells of the same row are occupied by equal variables. For definiteness, we assume that the variable x_j is placed in the jth row of the table $\widetilde{\mathbf{D}}$. Denote by $f_{\widetilde{\mathbf{D}}}(x_1, \dots, x_k, y_1, \dots, y_l)$ $(l = n_1 + \dots + n_k)$ the polynomial obtained from $f_{\mathbf{D}}$ by substituting the element x_i $(i = 1, \dots, k)$ instead of the variables x_j standing in the ith row of the diagram $\mathbf{D}$ and by division of the expression thus obtained by the number $\prod_{i=1}^k n_i!$. Since the polynomial $f_{\mathbf{D}}$ is symmetric with respect to any group of variables x_j from the same row of the diagram $\mathbf{D}$, we see that in the incancellable notation of the polynomial $f_{\widetilde{\mathbf{D}}}$ any monomial participates with coefficients $0, \pm 1$, and the total linearization of the polynomial $f_{\widetilde{\mathbf{D}}}$ with respect to the variables $x_1, \dots, x_k$ coincides with $f_{\mathbf{D}}$, up to renotation of variables . It is clear that if $\mathbf{D}$ is a rectangular Young table, then for any $j = 1, \dots, k$ and any $\beta_1, \dots, \beta_k \in K$ we have

$$\beta_j^{n_1} f_{\widetilde{\mathbf{D}}} = f_{\widetilde{\mathbf{D}}}|_{x_j = \beta_1 x_1 + \beta_2 x_2 + \dots + \beta_j x_j + \dots + \beta_k x_k}. \tag{31.8}$$

PROPOSITION 31.2. *Let $\mathbf{D}$ be a rectangular Young table containing $r+1$ rows and $s+1$ columns. Then for any elements $w_1, \dots, w_{r+1} \in \widetilde{V}$ and $a_1, \dots, a_l \in \widetilde{A}$ $(l = (r+1)(s+1))$ the expression*

$$f_{\widetilde{\mathbf{D}}}(w_1, \dots, w_{r+1}, a_1, \dots, a_l) \tag{31.9}$$

vanishes in the algebra $\widetilde{A}$.

PROOF. It is clear that it suffices to prove this assertion for the case when the ground field is infinite. Any element $w_i \in \widetilde{V}$ is representable in the form

$$w_i^0 + w_i^1, \quad \text{where} \quad w_i^0 = \sum_{j=1}^r \xi_i^{(j)} e_j \in \widetilde{V}_0 \quad (\xi_i^{(j)} \in G_0), \quad w_i^1 \in \widetilde{V}_1.$$

Since the ground field is infinite, it suffices to verify only that the expression (31.9) vanishes for elements $\xi_i^{(j)} \in K$ $(i = 1, \dots, r+1;\ j = 1, \dots, r)$. Therefore, we can suppose that all $w_i^0 \in \mathbf{1} \otimes V_0$. Since $\dim_K V_0 = r$, there exists a nontrivial linear combination $\sum_{i=1}^{r+1} \beta_i w_i^0$ that is equal to zero. Suppose that $\beta_j \neq 0$. Then $w = \sum_{i=1}^{r+1} \beta_i w_i \in \widetilde{V}_1$, and from equality (31.8) it follows that

$$\beta_j^{s+1} f_{\widetilde{\mathbf{D}}}(w_1, \dots, w_{r+1}, a_1, \dots, a_l) = f_{\widetilde{\mathbf{D}}}\big|_{\substack{x_1 = w_1, \dots, x_j = w, \dots, x_{r+1} = w_{r+1} \\ y_1 = a_1, \dots, y_l = a_l}}.$$

However, the polynomial $f_{\widetilde{\mathbf{D}}}$ is multihomogeneous and of degree $s+1$ with respect to the variable x_j. Therefore, to complete the proof it remains to show that for any $b_0, b_1, \dots, b_{s+1} \in \widetilde{A}$ and $w \in \widetilde{V}$ we have in the algebra $\widetilde{A}$ the identity

$$b_0 w b_1 w b_2 w \cdots b_s w b_{s+1} = 0. \tag{31.10}$$

Let us represent the element w in the form $\sum_{j=1}^{s} \eta_j v_j$, where $v_1, \dots, v_s$ is a basis of the space V_1 and $\eta_j \in G_1$. Then the element on the left-hand side of equality (31.10) will be equal to

$$\sum b_0 \eta_{j_1} v_{j_1} b_1 \eta_{j_2} v_{j_2} b_2 \cdots b_s \eta_{j_{s+1}} v_{j_{s+1}} b_{s+1},$$

where the sum is extended over all collections of indices $\{ j_1, j_2, \dots, j_{s+1} \}$ from the set $\{ 1, 2, 3, \dots, s \}$. However, in any such collection there must be at least two equal indices. Since $\eta_j^2 = 0$ $(j = 1, \dots, s)$, this gives equality (31.10). The proposition is proved.

COROLLARY 1. *Let $r = n^2 + k^2 - 1$, $s = 2nk$, and let $\mathbf{D}$ be a rectangular Young diagram with $r+1$ rows and $s+1$ columns. Then the weak identities $f_{\mathbf{D}} = 0$ and $f_{\widetilde{\mathbf{D}}} = 0$ are valid in the algebra $M_{n,k}$.*

COROLLARY 2. *Let $r = n^2 + k^2$, $s = 2nk$, and let the rectangular Young table $\mathbf{D}$ contains $r + 1$ rows and $s + 1$ columns. Then $f_{\mathbf{D}} = 0$ and $f_{\widetilde{\mathbf{D}}} = 0$ are identities of the algebra $M_{n,k}$.*

PROOF. Let E_1, E_1', and E_2 have the same meaning as in the proof of Lemma 31.2. Putting in Proposition 31.2 the algebra A equal to $M_{n,k}$ and identifying V_0 and V_1 with the K-subspaces spanned by E_1' and E_2, respectively, we obtain the assertion of Corollary 1. Replacing E_1' by E_1 in the preceding sentence, we obtain the proof of Corollary 2.

LEMMA 31.4. *Suppose that for $r, s, \mathbf{D}$ all assumptions of Proposition 31.2 are satisfied and the ground field is of characteristic zero. Then $f_{\mathbf{D}} = 0$ is an essentially weak identity of the algebra $M_{n,k}$.*

PROOF. In Remark 31.1 we put the polynomial f equal to $f_{\mathbf{D}}$ and the number q equal to $l = (r+1)(s+1)$. Then, in order to prove that $f_{\mathbf{D}} = 0$ is not an identity of $M_{n,k}$, it suffices to show that the following trace identity does not hold in $M_{n,k}$:

$$e(\mathbf{D}) \operatorname{Sp}(x_1 y_1) \operatorname{Sp}(x_2 y_2) \cdots \operatorname{Sp}(x_l y_l) = 0, \tag{31.11}$$

where $l = (n^2 + k^2)(s+1)$. We assume that the variable $x_{(j-1)(r+1)+i}$ stands in the ith row of the jth column of the diagram $\mathbf{D}$.

Since the expression on the left-hand side of formula (31.11) is symmetric with respect to any group of variables belonging to the same row of the diagram $\mathbf{D}$, we see that after the replacement of all variables of the ith row by x_i $(i = 1, \dots, n^2+k^2)$ and the substitution $y_{(j-1)(r+1)+i} \to y_i$ $(i = 1, 2, \dots, s+1)$ in this expression, the left-hand side of (31.11) becomes equal to

$$(s+1)!^{n^2+k^2}\{\operatorname{Det} e\}^{s+1},$$

where e is a square matrix of order n^2+k^2 having $\operatorname{Sp}(x_i, y_j)$ in the (i, j)th position. Since the restriction of the bilinear form $(x, y) = \operatorname{str}(xy)$ to a K-subalgebra in $M_{n,k}$ with basis of matrix units E_{ij}, where $i, j \le n$ or $i, j > n$, is nondegenerate and e is the matrix of this bilinear form, we see that $\operatorname{Det} e \ne 0$ and the trace identity (31.11) does not hold in $M_{n,k}$. The lemma is proved.

This lemma together with Proposition 31.1 proves Theorem 31.1 for the case of fields of characteristic zero.

LEMMA 31.5. *Let $r = n^2+k^2$, $s = 2nk$, let $\mathbf{D}$ be a rectangular table containing $r+1$ rows and $s+1$ columns, and let $\mathbf{D}_0$ be the Young table obtained from $\mathbf{D}$ by omitting the exposed rim cell $(r+1, s+1)$. Then the identity $f_{\tilde{\mathbf{D}}_0}$ does not hold in the algebra $M_{n,k}$ over a field of arbitrary characteristic.*

PROOF. We put in Remark 31.1 the polynomial f equal to $f_{\tilde{\mathbf{D}}_0}$ and the number q equal to $l = (r+1)(s+1)$. Then, in order to prove that $f_{\tilde{\mathbf{D}}_0}$ is not an identity for $M_{n,k}$ it is sufficient to show that the trace identity

$$\prod_{\nu=1}^{s+1} \operatorname{Det} c_\nu = 0 \tag{31.12}$$

does not hold in $M_{n,k}$, where the matrices $c_1, \dots, c_s$ are of order $r+1$, the matrix c_{r+1} is of order r, and on the intersection of the ith row and jth column there stands the element $\operatorname{Sp}(x_i, y_{\nu,j})$ (here $y_{\nu,j}$ denotes the variable $y_{(\nu-1)(r+1)+j}$). In the algebra $M^0_{n,k}$ we choose elements $w^0_1, \dots, w^0_r, u_1, \dots, u_r$ such that the determinant of the matrix of order r composed of the elements $\operatorname{str}(w^0_i u_j)$ is equal to the unit of the algebra. Let $w, w_1, \dots, w_s$ be arbitrary elements of $M_{n,k}$. We substitute in the right-hand side of equality (31.12) the elements $w^0_1, \dots, w^0_r$ instead of $x_1, \dots, x_r$, respectively, the elements $u_1, \dots, u_r$ instead of $y_{\nu,1}, \dots, y_{\nu,r}$, the element w instead of x_{r+1}, and the element w_ν instead of $y_{\nu,r+1}$ $(\nu = 1, \dots, s)$. Then the expression thus obtained is equal to the element $\operatorname{str}(ww_1)\cdots\operatorname{str}(ww_s)$, which is nonzero for an appropriate choice of $w, w_1, \dots, w_s$. The lemma is proved.

31.5. Construction of polynomial c-dual sets and central polynomials over fields of characteristic zero. In this subsection we consider in more detail the idea that permitted, in subsection 31.3, the explicit construction of a central polynomial over fields of positive characteristic using the notion of c-dual sets and independent of Proposition 31.1.

LEMMA 31.6. *Let a multihomogeneous associative polynomial g have the form*

$$g(x_1, \dots, x_\nu, y) \rightleftharpoons \sum_{i=1}^{\nu} h_i y x_i, \tag{31.13}$$

and let g be multilinear with respect to two variables x_t and y. Let the identity $g = 0$ be valid in $M_{n,k}$. Then the trace identities

$$(31.14) \qquad - \operatorname{Sp}(h_t y)a = \sum_{i=1, i\neq t}^{v} \operatorname{Sp}\left((A_t h_i)|_{x_t = y} a\right) x_i,$$

$$(31.15) \qquad \sum_{i=1, i\neq t}^{v} (A_t h_i)|_{x_t=y} a x_i = - \operatorname{Sp}(h_t y) \operatorname{Sp}(a)$$

are valid in $M_{n,k}$.

Moreover, the polynomial on the left-hand side of the trace identity (31.15) is central for the algebra $M_{n,k}$ if and only if the polynomial identity $h_t = 0$ does not hold in $M_{n,k}$.

PROOF. We make use of Remark 31.1 by setting in it $q = 1$ and $y_1 = y$ and by identifying g with the polynomial f. Then from the fact that $g = 0$ is an identity in $M_{n,k}$ it follows that $\tau_{0,1} \circ g = \sum_{i=1}^{v} \operatorname{Sp}(h_i y) x_i = 0$ is a trace identity for the algebra $M_{n,k}$, multilinear with respect to the variables y and x_t. This means that the variable x_t occurs exactly once in all monomials of the polynomials h_i ($i = 1, \dots, \widehat{t}, \dots, v$). Replacing x_t by a in the equality thus obtained and using the relations $\operatorname{Sp}(c x_t d y) = \operatorname{Sp}(d y c x_t)$ and $A_t(c x_t d) = d x_t c$, we prove the trace identity (31.14). A comparison of the trace identity (31.14) with equality (29.1) leads to a natural interpretation of this identity. Let us denote $-\operatorname{Sp}(h_t y)$ by c, x_i by e_i', and $(A_t h_i)|_{x_t=y}$ by e_i. Then for any substitution of elements of the algebra $M_{n,k}$ instead of the variables $x_1, \dots, \widehat{x_t}, \dots, x_v$ the sets of values $\{ \widetilde{e}_i' \mid i = 1, \dots, \widehat{t}, \dots, v \}$ and $\{ \widetilde{e}_i \mid i = 1, \dots, \widehat{t}, \dots, v \}$ of the polynomials e_i' and e_i are $\widetilde{c}$-dual sets, where $\widetilde{c}$ is the value of the polynomial c. Therefore, identity (31.15) follows from (31.14) by Lemma 29.1, and formally we can regard the sets of polynomials $P = \{ (A_t h_i)|_{x_t=y} \mid i = 1, \dots, \widehat{t}, \dots, v \}$ and $P' = \{ x_i \mid i = 1, \dots, \widehat{t}, \dots, v \}$ as $-\operatorname{Sp}(h_t y)$-dual sets for the algebra $M_{n,k}$.

The last assertion of the lemma follows from (31.15) and from the fact that the polynomial h_t does not depend on the variable y and the bilinear form $(x, y) = \operatorname{str}(xy)$ is nondegenerate on $M_{n,k}$. The lemma is proved.

Lemma 31.6 and its proof permit us to construct polynomial c-dual sets formally from any identity $g = 0$, where g has the form (31.13), and no restrictions are imposed on the characteristic of the ground field. For these c-dual sets to be nontrivial, it is necessary and sufficient that $h_t = 0$ not be an identity in $M_{n,k}$. The following argument shows how we can construct central polynomials and nontrivial c-dual sets, using Lemma 31.6, from any identity for the algebra $M_{n,k}$ of the form

$$(31.16) \qquad \sum_{\sigma \in S_m} \beta_\sigma x_{\sigma(1)} y_1 x_{\sigma(2)} y_2 \cdots x_{\sigma(m-1)} y_{m-1} x_{\sigma(m)} = 0,$$

where the sum is taken over all permutations of the set $\{1, 2, \dots, m\}$ and the β_s are elements of the field K, not all equal to zero.

Denote the left-hand side of identity (31.16) by g and y_{m-1} by y. Choose a variable $x_t \in \{ x_1, \dots, x_m \}$ such that for some permutation $\sigma \in S_m$ we have $\beta_\sigma \neq 0$ and $\sigma(m) = t$. Then g is representable in the form $\sum_{i=1}^{m} h_i y x_i$, g is multilinear with respect to y and x_t, and we can apply Lemma 31.6. If the identity $h_t = 0$ does not hold in $M_{n,k}$, then formulas (31.14) and (31.15) give nontrivial $-\operatorname{Sp}(h_t y)$-dual sets and a central polynomial for the algebra $M_{n,k}$. In the contrary case, the identity $h(x_1, \ldots, \widehat{x}_t, \ldots, x_m) = 0$ has the same form as identity (31.16), but depends on lesser number of variables. Then we can put $g = h_t$ and $y = y_{m-2}$ and apply Lemma 31.6 again. In a certain number of steps, this descending process will give nontrivial c-dual polynomial sets and a central polynomial for the algebra $M_{n,k}$ over a field of arbitrary characteristic.

To complete the picture, it remains to show that over a field of characteristic zero, for the total linearization of the identity $f_{\widetilde{\mathbf{D}}}\big|_{y_l=1}$, where $l = (r+1)(l+1)$ and $f_{\widetilde{\mathbf{D}}}$ is the polynomial defined in Corollary 2 of Proposition 31.2, the process of construction of a central polynomial and $-\operatorname{Sp}(h_t y)$-dual sets will be interrupted at the first step. Indeed, in this case

$$f_{\widetilde{\mathbf{D}}}(x_1, \ldots, x_{n^2+k^2+1}, y_1, \ldots, y_{l-1}, 1) = \sum_{i=1}^{n^2+k^2+1} h_i y_{l-1} x_i, \tag{31.17}$$

where the polynomial $h_i(x_1, \ldots, x_{n^2+k^2+1}, y_1, \ldots, y_{l-1})$ is multihomogeneous and of degree $2nk$ with respect to the variable x_i and degree $2nk+1$ with respect to the variables $x_1, \ldots, \widehat{x}_i, \ldots, x_{n^2+k^2+1}$, and $h_i y_{l-1}$ coincides, up to sign, with the polynomial $f_{\widetilde{\mathbf{D}}_i}$, where $\widetilde{\mathbf{D}}_i$ denotes the diagram corresponding to the Young table $\mathbf{D}_0$ obtained from $\mathbf{D}$ by omitting the exposed rim cell $(n^2+k^2+1, 2nk+1)$ and filled by the variables $x_1, \ldots, x_{n^2+k^2+1}$ in such a way that the variable x_i stands in the last row. Therefore, $h_{n^2+k^2+1} y_{l-1} = f_{\widetilde{\mathbf{D}}_0}$. By Lemma 31.5, the identity $f_{\widetilde{\mathbf{D}}_0} = 0$ does not hold in the algebra $M_{n,k}$. But then $h_{n^2+k^2+1} = 0$ is not an identity in $M_{n,k}$. Partially linearizing the identity $f_{\widetilde{\mathbf{D}}}|_{y_l=1} = 0$ with respect to the variable $x_{n^2+k^2+1}$, using a new variable x_t and applying Lemma 31.6 to this linearization, which is multilinear with respect to y and x_t, where $y \rightleftarrows y_{l-1}$, we obtain obviously nontrivial $-\operatorname{Sp}\left(f_{\widetilde{\mathbf{D}}_0}\right)$-dual polynomial sets of cardinality n^2+k^2+1 and a multihomogeneous central polynomial.

§32. A description of the lattice of η_α-closed ideals in $K[\gamma]$

In this section it is assumed that the ground field K is of characteristic zero.

We continue to assume that the free associative K-algebra F with free generators $x_1, x_2, \ldots$ is embedded in the free trace algebra $\mathcal{G}$. Proposition 25.2 and Theorems 26.1 and 26.2 imply that any nonzero θ_α-closed ideal in F has the form $F \cap V_{\mathbf{D}}$, and all nonzero η_α-closed ideals in $K[\gamma]$ are among the ideals $\{\gamma^{-1}\mathbf{D}(\gamma)\}$, where $\mathbf{D}$ ranges over all Young tables; moreover, for any Young table $\mathbf{D}$ the ideal $F \cap V_{\mathbf{D}}$ is θ_α-closed. In this section we shall clarify for which Young tables $\mathbf{D}_1$ and $\mathbf{D}_2$ the equality $F \cap V_{\mathbf{D}_1} = F \cap V_{\mathbf{D}_2}$ takes place and show that $\Phi_\alpha(F \cap V_{\mathbf{D}}) = \{\gamma^{-1}\mathbf{D}(\gamma)\}$ for a symmetric Young table $\mathbf{D}$. Thus, the lattice of all η_α-closed ideals in $K[\gamma]$ will be completely described.

Denote by $\mathbf{D}^*$ the table obtained by transposition from the Young table $\mathbf{D}$ and by $\mathbf{D}_1 \cup \mathbf{D}_2$ the least Young table containing the Young tables $\mathbf{D}_1$ and $\mathbf{D}_2$.

THEOREM 32.1. *Let the ground field K be of characteristic zero. Then the lattice of all η_α-closed ideals of the algebra $K[\gamma]$ (of θ_α-closed ideals of trace identities of the algebra*

F) is anti-isomorphic to the lattice of all symmetric Young tables with the inclusion relation, completed by a maximal element. Moreover:

a) *for any Young table* $\mathbf{D}$

$$F \cap V_{\mathbf{D}} = F \cap V_{\mathbf{D}^*} = F \cap V_{\mathbf{D}\cup\mathbf{D}^*}; \tag{32.1}$$

b) *for any symmetric Young table* $\mathbf{D}$ *we have* $\Phi_\alpha(F \cap V_{\mathbf{D}}) = \{\gamma^{-1}\mathbf{D}(\gamma)\}$;

c) *if* $\mathbf{D}_1$ *and* $\mathbf{D}_2$ *are symmetric Young tables, then* $F \cap V_{\mathbf{D}_1} \supseteq F \cap V_{\mathbf{D}_2}$ *if and only if* $\mathbf{D}_1 \subseteq \mathbf{D}_2$;

d) *the zero ideal and the ideals* $\{\gamma^{-1}\mathbf{D}(\gamma)\}$, *where* $\mathbf{D}$ *ranges over all symmetric Young tables* $\mathbf{D}$, *exhaust all* η_α*-closed ideals in* $K[\gamma]$.

PROOF. Let φ_1 be the automorphism of the trace algebra $\mathcal{G}$ defined in Remark 27.1 by formula (27.3). It is clear that $\varphi_1^2 = \varphi_1$ and φ_1 acts as an identity on the subalgebra F. As it was noted in Remark 27.1, $\varphi_1(V_{\mathbf{D}}) = V_{\mathbf{D}^*}$, where $\mathbf{D}$ is any Young table. Therefore, $\varphi_1(V_{\mathbf{D}}) = \varphi_1(F) \cap \varphi_1(V_{\mathbf{D}}) = F \cap V_{\mathbf{D}^*} = F \cap V_{\mathbf{D}}$ and, because $V_{\mathbf{D}} \cap V_{\mathbf{D}^*} = V_{\mathbf{D}\cup\mathbf{D}^*}$, this proves the equalities of assertion a).

Thus, all nonzero θ_α-closed ideals in F have the form $F \cap V_{\mathbf{D}}$, where $\mathbf{D}$ is a symmetric Young table, and all nonzero η_α-closed ideals in $K[\gamma]$ are included among the ideals $\{\gamma^{-1}\mathbf{D}(\gamma)\}$, where $\mathbf{D}$ ranges over all symmetric Young tables. Therefore, for any symmetric Young table $\mathbf{D}$ the following equalities are valid:

$$\Phi_\alpha(F \cap V_{\mathbf{D}}) = \{\gamma^{-1}\mathbf{D}'(\gamma)\}, \tag{32.2}$$

$$\eta_\alpha(\{\gamma^{-1}\mathbf{D}(\gamma)\}) = \Phi_\alpha(F \cap V_{\mathbf{D}}) = \{\gamma^{-1}\mathbf{D}'(\gamma)\}, \tag{32.3}$$

$$\begin{aligned} F \cap V_{\mathbf{D}} = \theta_\alpha(F \cap V_{\mathbf{D}}) &= \Psi_\alpha\Phi_\alpha(F \cap V_{\mathbf{D}}) \\ &= F \cap \Psi_\alpha(\{\gamma^{-1}\mathbf{D}'(\gamma)\}) = F \cap V_{\mathbf{D}'}, \end{aligned} \tag{32.4}$$

where $\mathbf{D}'$ is some symmetric Young table. Equality (32.3) and the second relation of (10.3) imply that $\mathbf{D}(\gamma)$ divides $\mathbf{D}'(\gamma)$ and, hence, $\mathbf{D}$ is contained in $\mathbf{D}'$ as a subtable.

Therefore, in order to prove assertion b) it suffices to deduce the coincidence of the tables $\mathbf{D}$ and $\mathbf{D}'$ from equality (32.4) in the case $\mathbf{D} \subseteq \mathbf{D}'$.

Toward this end, we shall use the model considerations of §27. Theorem 27.2 and the results of §27 imply that the ideal $V_{\mathbf{D}}$ is the ideal of trace identities of the direct sum $\sum M_{n,k}$, where the sum is taken over all pairs of numbers n, k for which $(n+1, k+1)$ is an exposed rim cell of the Young table $\mathbf{D}$. Therefore, the ideal $F \cap V_{\mathbf{D}}$ is the ideal of identities for this direct sum. Since $\mathbf{D}$ is a symmetric Young table and the algebras $M_{n,k}$ and $M_{k,n}$ have the same polynomial identities, we see that $F \cap V_{\mathbf{D}}$ is the ideal of identities for the algebra $M(\mathbf{D}) = \sum_{n \ge k} M_{n,k}$, where the direct sum is extended over all pairs of numbers n, k for which $n \ge k$ and $(n+1, k+1)$ is an exposed rim cell of the table $\mathbf{D}$. Similarly, $F \cap V_{\mathbf{D}'}$ is the ideal if identities for the algebra $M(\mathbf{D}') = \sum_{n' \ge k'} M_{n',k'}$, where $(n'+1, k'+1)$ are exposed rim cells of $\mathbf{D}'$.

The fact that $\mathbf{D}$ is contained in $\mathbf{D}'$ means that all exposed rim cells of $\mathbf{D}$ belong to $\mathbf{D}'$ and, hence, for any direct summand $M_{n,k}$ of $M(\mathbf{D})$ there exists a direct summand $M_{n',k'}$ of $M(\mathbf{D}')$ such that $n' \ge n$ and $k' \ge k$. Equality (32.4) means that in the algebras $M(\mathbf{D})$ and $M(\mathbf{D}')$ the same polynomial identities take place. We shall show that in this case $M(\mathbf{D})$ and $M(\mathbf{D}')$ must have the same direct summands (this is equivalent to the coincidence of the tables $\mathbf{D}$ and $\mathbf{D}'$).

Assume the contrary and suppose that in $M(\mathbf{D}')$ there exists a summand $M_{n',k'}$ not belonging to $M(\mathbf{D})$. Then from the definition of $M(\mathbf{D})$ and from the fact that $\mathbf{D} \subseteq \mathbf{D}'$ it follows that for any direct summand $M_{n,k}$ of $M(\mathbf{D})$ we have either $k < k'$ or $n < n'$. In this case Lemma 28.3 implies that the ideal of identities $I_{n,k} = F \cap V_{n,k}$ of the algebra $M_{n,k}$ is not contained in the ideal of identities $I_{n',k'}$ of the algebra $M_{n',k'}$. However, then the ideal $\prod_{M_{n,k} \in M(\mathbf{D})} I_{n,k} \rightleftharpoons I$ is contained in the ideal of identities of the algebra $M(\mathbf{D}')$. Thus, I belongs to $F \cap V_{\mathbf{D}}$, but it does not belong to $F \cap V_{\mathbf{D}'}$, which contradicts equality (32.4). The contradiction thus obtained proves the equality $\mathbf{D} = \mathbf{D}'$ and assertion b) of the theorem.

Assertion b) implies that for symmetric Young tables $\mathbf{D}_1$ and $\mathbf{D}_2$ the inclusion of θ_α-closed ideals $F \cap V_{\mathbf{D}_1} \supseteq F \cap V_{\mathbf{D}_2}$ is possible if and only if $\mathbf{D}_2$ contains $\mathbf{D}_1$ as a subtable. This proves assertion c). The remaining assertions of the theorem are now clear.

§33. Consequences of the classification of $\theta_{\tilde{\alpha}}$-closed ideals of trace identities related to varieties of associative nil-algebras

It is clear that in any variety of associative algebras consisting of nil-algebras we have the identity $y^n = 0$ for some n.

33.1. Varieties of nil-algebras over fields of characteristic zero.

THEOREM 33.1. *In any associative algebra over a field of characteristic zero in which the identity $y^n = 0$ is valid, the identity $x_1x_2\cdots x_{n^2} = 0$ is valid.*

PROOF. It suffices to show that in the free associative algebra F of countable rank the ideal generated by the elements of the form g^n, where G ranges over F, contains the element $x_1x_2\cdots x_{n^2}$. The algebra F is contained in the trace algebra $\mathcal{G}$. Consider the ideal of trace identities $V = V_{\mathbf{D}^{(n)}} + V_{\mathbf{D}^{(n)*}}$, where the Young table $\mathbf{D}^{(n)}$ consists of one column with $n+1$ cells. Let $\mathbf{D}$ be an arbitrary Young table containing n^2+1 cells. Obviously, $\mathbf{D}$ contains either $\mathbf{D}^{(n)}$ or $\mathbf{D}^{(n)*}$ as a subtable. Hence, by assertion b) of Theorem 26.1, $V_{\mathbf{D}}$ is contained either in $V_{\mathbf{D}^{(n)}}$ or in $V_{\mathbf{D}^{(n)*}}$. Therefore, $V_{\mathbf{D}} \subseteq V$. Moreover, V contains the ideal $V_{n^2} = \sum V_{\mathbf{D}}$, where $\mathbf{D}$ ranges over all Young tables containing n^2+1 cells. From Definition 26.1 of the ideals $V_{\mathbf{D}}$ it follows that V_{n^2} contains all multilinear trace polynomials of degree n^2. Therefore, the element $x_1x_2\cdots x_{n^2}$ is contained in $V_{\mathbf{D}^{(n)}} + V_{\mathbf{D}^{(n)*}}$. Theorem 28.1 and Remark 27.1 show that the ideal of trace identities V is generated by the trace polynomials $f_n(y)$ and $\varphi_1(f_n(y))$ (see formulas (27.2) and (27.4)). Therefore, for some $a_i, b_i, c_i, d_i \in \mathcal{G}$ and $g_i, h_i \in F$ we have the following equality in $\mathcal{G}$:

$$x_1x_2\cdots x_{n^2} = \sum_i a_i f_n(g_i) b_i + \sum_i c_i \varphi_1(f_n(h_i)) d_i. \tag{33.1}$$

Since no nonzero element of the algebra F can be expressed in $\mathcal{G}$ as a linear combination of generalized monomials with the symbol Sp in their notation, the explicit form of the trace polynomials f_n and $\varphi_1(f_n)$ and a comparison of the terms of equality (33.1) not containing Sp show that in F we have the equality

$$x_1x_2\cdots x_{n^2} = \sum_i a_i' g_i^n b_i + \sum_i c_i' h_i^n d_i',$$

where a_i', b_i', c_i', d_i' are those generalized monomials on the right-hand side of the equality (33.1) in whose notation the symbol Sp does not occur. The theorem is proved.

33.2. Examples of nonsolvable $(p-1)$th Engel varieties of associative algebras over fields of characteristic $p \geq 5$. If the characteristic of the ground field K is positive, Lemma 25.5 does not take place. However, $\theta_{\widetilde{\alpha}}$-closed ideals of trace identities $V_{\gamma-i}$ give interesting examples of varieties of associative algebras and of Lie algebras.

THEOREM 33.2. *Let the ground field K be of characteristic $p > 3$ and $E(\gamma) = \prod_{i=1}^{p-1}(\gamma - i)$. Then the ideal of trace identities $V_{E(\gamma)} = \Psi_{\widetilde{\alpha}}(\{E(\gamma)\})$ contains the polynomial $[x, \underbrace{y, \dots, y}_{p-1 \text{ times}}]$ and does not contain the polynomials $[x_1, x_2, \dots, x_k]$ ($k = 1, 2, \dots$), i.e., the $(p-1)$th Engel identity does not imply the identity of Lie nilpotency. Moreover, in the $(p-1)$th Engel algebra $\mathcal{G}/V_{E(\gamma)}$ the identity $y^p = 0$ is valid, and this algebra is nonsolvable.*

PROOF. Since $[x_1, \dots, x_k]$ is a multilinear polynomial, we see that, by equality $\operatorname{Ann}_{T_k} \widetilde{\mathfrak{b}}_E = V_{E(\gamma)} \cap T_k$, it suffices to show that $[x_1, x_2, \dots, x_k] \notin \operatorname{Ann}_{T_k} \widetilde{\mathfrak{b}}_E$. From the definition of $\widetilde{\mathfrak{b}}_E$ and from (25.12) and (24.2) it follows that

$$\begin{aligned}\widetilde{\mathfrak{b}}_E([x_1, \dots, x_k], x_k \cdots x_1) &= \widetilde{\alpha}([\overline{x}_1, \dots, \overline{x}_k]\overline{x}_k \cdots \overline{x}_1) \\ &= \widetilde{\alpha}\left([\overline{x}_1, \dots, \overline{x}_k]\gamma\left(\frac{\overline{x}_k^2}{\gamma}\right)\overline{x}_{k-1} \cdots \overline{x}_1\right) \\ &= \gamma\widetilde{\alpha}([\overline{x}_1, \dots, \overline{x}_k]\overline{x}_{k-1} \cdots \overline{x}_1) \\ &= \gamma^{k-2}\widetilde{\alpha}([\overline{x}_1, \overline{x}_2]\overline{x}_2\overline{x}_1) = (\gamma^2 - 1)\gamma^{k-2},\end{aligned} \tag{33.2}$$

and, since $E(\gamma)$ does not divide this polynomial for $p > 3$, we see that $[x_1, \dots, x_k] \notin V_{E(\gamma)}$.

The $(p-1)$th Engel identity $x(\operatorname{ad} y)^{p-1} = 0$ over a field of characteristic $p > 0$ is equivalent to the multilinear identity

$$\sum_{\sigma \in S_p} x_{\sigma(1)} x_{\sigma(2)} \cdots x_{\sigma(p)} = 0, \tag{33.3}$$

where the sum is taken over all permutations σ of the set $\{1, 2, \dots, p\}$. By Lemma 25.9, which identifies T_p with respect to the operation $\circ$ and $K[S_{p+1}]$ with the help of the isomorphism φ, the polynomial on the left-hand side of equality (33.3) is equal to the sum τ_1 of all permutations from S_{p+1} that are cycles of length $p+1$. From Theorem 26.1 it follows that in order to prove the inclusion $\tau_1 \in V_{E(\gamma)} \cap T_p = \operatorname{Ann}_{T_p} \widetilde{\mathfrak{b}}_E$, it suffices to verify that $\gamma^{-1} d \circ \tau_1 = 0$ in the group ring $K_E[S_{p+1}]$, where $K_E = K[\gamma]/\{E(\gamma)\}$.

First we find the value of the element $\gamma^{-1} d \circ \tau_1$ in the case when the characteristic of the ground field is equal to zero. In this case a comparison of terms with the first degree of the variable γ in formula (26.4) together with formula (26.3) for the polynomial $\mathbf{D}(\gamma)$ allows us to deduce that

$$\tau_1 = \sum \beta_{\mathbf{D}} \cdot \mathbf{1}_{\mathbf{D}}, \tag{33.4}$$

where the sum is taken over all Young tables of the form

$$\mathbf{D} = \begin{array}{|c|c|c|c|c|} \hline 0 & -1 & -2 & \cdots & -k \\ \hline 1 \\ \cline{1-1} \cdots \\ \cline{1-1} p-k \\ \cline{1-1} \end{array}$$

and

(33.5) $$\beta_{\mathbf{D}} = (-1)^{p-k}(p-k)!\,k!$$

LEMMA 33.1. *For any Young table* $\mathbf{D}$ *of the form* (33.5) *and elements* $\beta_{\mathbf{D}}$ *of the form* (33.5) *over a field of characteristic zero we have*

$$\beta_{\mathbf{D}} \cdot \mathbf{1}_{\mathbf{D}} = (-1)^{p-k}\frac{1}{p+1}\sum_{\mathbf{D}'} e(\mathbf{D}'),$$

where the sum is taken over all standard diagrams $\mathbf{D}'$ *obtained from the table* $\mathbf{D}$, *and* $e(\mathbf{D}')$ *are the Young symmetrizers of these diagrams.*

PROOF. A Young diagram $\mathbf{D}'$ is called *standard* if the numbers $0, 1, \dots, p$ are posed in $\mathbf{D}$ in such a way that they increase in any row and in any column, if we look on the table $\mathbf{D}$ from left to right and from top to bottom. It is well known (see [Jam]) that the number of various standard diagrams constructed from the table $\mathbf{D}$ coincides with the dimension $m_{\mathbf{D}}$ of the irreducible representation of the symmetric group S_{p+1} with the symmetry type $\mathbf{D}$. From Proposition 26.1 it follows that $m_{\mathbf{D}}^2$ coincides with the dimension of the simple ideal $I_{\mathbf{D}}$, and the elements $\frac{m_{\mathbf{D}}}{(p+1)!}e(\mathbf{D}')$ are the idempotents of $I_{\mathbf{D}}$. Since for any two distinct standard diagrams $\mathbf{D}'$ and $\mathbf{D}''$ corresponding to the table $\mathbf{D}$ of the form (33.5) there exists an element $k \in \{1, 2, \dots, p\}$ such that the transposition t_{0k} belongs to $R(\mathbf{D}'') \cap C(\mathbf{D}')$, we see that

$$e(\mathbf{D}') \circ e(\mathbf{D}'') = \mathcal{P}(\mathbf{D}') \circ \mathcal{Q}(\mathbf{D}') \circ t_{ik} \circ t_{ik} \circ \mathcal{P}(\mathbf{D}'') \circ \mathcal{Q}(\mathbf{D}'') = -e(\mathbf{D}') \circ e(\mathbf{D}'').$$

Hence, $e(\mathbf{D}') \circ e(\mathbf{D}'') = 0$, and the set $\left\{\frac{m_{\mathbf{D}}}{(p+1)!}e(\mathbf{D}')\right\}$, where $\mathbf{D}'$ ranges over all standard diagrams of the Young table $\mathbf{D}$, is a full system of mutually orthogonal idempotents of the ideal $I_{\mathbf{D}}$. Therefore,

$$\mathbf{1}_{\mathbf{D}} = \frac{m_{\mathbf{D}}}{(p+1)!}\sum_{\mathbf{D}'} e(\mathbf{D}').$$

It is clear that the number of standard diagrams $m_{\mathbf{D}}$ for the tables $\mathbf{D}$ of the form (33.5) is equal to $\binom{p}{k} = \frac{p!}{k!(p-k)!}$ and $\beta_{\mathbf{D}}\frac{m_{\mathbf{D}}}{(p+1)!} = \frac{(-1)^{p-k}}{(p+1)}$. This proves the lemma.

This lemma and equalities (33.5) and (26.5) imply that

(33.6) $$\gamma^{-1}d \circ \tau_1 = \sum_{\mathbf{D}} \gamma^{-1}\mathbf{D}(\gamma) \cdot (-1)^{p-k} \cdot \frac{1}{p+1}\sum_{\mathbf{D}'} e(\mathbf{D}'),$$

where the first sum is taken over all Young tables $\mathbf{D}$ of the form (33.5) and the second sum is extended over all standard diagrams of these tables. It is clear that equality (33.6) is valid also for fields of positive characteristic. It remains to note that all polynomials $\gamma^{-1}\mathbf{D}(\gamma)$, where $\mathbf{D}$ has the form (33.5), are divisible by $E(\gamma) = \prod_{i=1}^{p-1}(\gamma - i)$ over a field of characteristic $p > 0$. Therefore, $\gamma^{-1}d \circ \tau_1 = 0$ in $K_E[S_{p+1}]$. This proves that the $(p-1)$th Engel Lie polynomial belongs to $V_{E(\gamma)}$, and identity (33.3) holds in the algebra $\mathcal{G}/V_{E(\gamma)}$.

Since all multilinear identities that are consequences of the identity $y^n = 0$ follow from the identity (33.3), the definition of the ideal $V_{E(\gamma)} = \Psi_{\widetilde{\alpha}}(\{E(\gamma)\})$ implies $y^p \in V_{E(\gamma)}$. The nonsolvability of the algebra $\mathcal{G}/V_{E(\gamma)}$ follows from the fact that it is not nilpotent and from the Higgins theorem (see §6) on the nilpotency of solvable $(p-1)$th Engel Lie algebras. The theorem is proved.

As a consequence of this theorem for fields K of characteristic $p > 3$, we obtain for $n = 2, 3, \dots, \frac{p-1}{2}$ new (different) examples of prime (in the sense of Kemer, see [Kem1, Kem2]) varieties of associative algebras $\mathfrak{B}_{\gamma^2 - n^2}$ whose ideal of identities is $\Psi_{\widetilde{\alpha}}(\{\gamma^2 - n^2\})$.

DEFINITION 33.1. A variety of associative algebras is said to be *prime* if in the free algebra of this variety of countable rank the equality $I_1 I_2 = 0$, where I_1 and I_2 are T-ideals, imply that either $I_1 = 0$ or $I_2 = 0$.

THEOREM 33.3. *Let the ground field K be of characteristic $p > 3$. Suppose that $n \in \{1, 2, \dots, p-1\}$ and $V_{\gamma - n} = \Psi_{\widetilde{\alpha}}(\{\gamma - n\})$. Then:*

a) $V_{\gamma - n} \cap F = V_{\gamma + n} \cap F = \Psi_{\widetilde{\alpha}}(\{\gamma^2 - n^2\})$, *i.e., the algebras $\mathcal{G}/V_{\gamma - i}$ and $\mathcal{G}/V_{\gamma + i}$ have the same identities and generate the same variety of associative algebras $\mathfrak{B}_{\gamma^2 - n^2}$;*

b) *for any two trace ideals V_1 and V_2 of the trace algebra $\mathcal{G}$ the inclusion $V_1 V_2 \subseteq V_{\gamma - n}$ implies that either $V_1 \subseteq V_{\gamma - n}$ or $V_1 \subseteq V_{\gamma - n}$; in particular, the variety $\mathfrak{B}_{\gamma^2 - n^2} =$* var$\{\mathcal{G}/V_{\gamma - n}\}$ *is prime in the sense of Definition* 33.1;

c) $\mathfrak{B}_{\gamma^2 - n^2} \subset \operatorname{var} M_n$, *i.e., in the algebra $\mathcal{G}/V_{\gamma - n}$ all identities of the full matrix algebra M_n over some infinite field of characteristic p are satisfied;*

d) *the variety $\mathfrak{B}_{\gamma^2 - n^2}$ is $(p-1)$th Engel and for $n = 2, 3, \dots, \frac{(p-1)}{2}$ it is nonsolvable;*

e) *for $n = 2, 3, \dots, \frac{(p-1)}{2}$ in the algebra $\mathcal{G}/V_{\gamma - n}$ the standard identity of degree $2(n-1)$ is not satisfied.*

PROOF. Let φ_1 and φ_2 have the same meaning as in Remark 27.1. Then $\varphi_1^2 = 1$, $\varphi_1 \colon F \to F$ is the identity mapping, and from formula (25.39) for the computation of $\widetilde{\mathfrak{b}}_E$ it follows that if $\widetilde{\mathfrak{b}}(s, \delta) = \gamma^t$, then $\widetilde{\mathfrak{b}}(\varphi_1(s), \varphi_1(\delta)) = (-\gamma)^t$. Therefore, $\varphi_1(V_{\gamma - n}) = V_{\gamma + n}$ and $\varphi_1(V_{\gamma + n}) = V_{\gamma - n}$. These equalities together with formula (25.28) of Proposition 25.2 prove assertion a).

Since by the definition of the ideal $V_{\gamma - n}$ this ideal is the greatest among the ideals of trace identities V for which $V \cap T_k = V_{\gamma - n} \cap T_k$ $(k = 1, 2, \dots)$, we see that in order to prove assertion b) it suffices to show that if $f \in T_l$ and $g \in T_m$ and $f, g \notin V_{\gamma - n}$, then $h \rightleftharpoons f(x_1, \dots, x_l) g(x_{l+1}, \dots, x_{l+m}) \notin V_{\gamma - n}$. However, for these f and g there exist $f' \in T_l$ and $g' \in T_m$ for which

$$\widetilde{\mathfrak{b}}_{\gamma - n}(f, f') = \beta_1 \in K[\gamma]/\{\gamma - n\} \simeq K \ni \beta_2 = \widetilde{\mathfrak{b}}_{\gamma - n}(g, g')$$

and $\beta_1 \beta_2 \neq 0$. However, in this case for $h' \rightleftharpoons g'(x_{l+1}, \dots, x_{l+m}) f'(x_1, \dots, x_l)$ the value $\widetilde{\mathfrak{b}}_{\gamma - n}(h, h') = \beta_1 \beta_2$ is nonzero. This proves assertion b).

Assertion c) follows from the fact that $\widetilde{\mathfrak{b}}_{\gamma-n}(u,v) = \widetilde{\alpha}_n(uv)$ and the computation of the $\boldsymbol{\alpha}$-function α_n can be realized in the algebra $\mathcal{G}_n$ (see §25), which is contained in $\overline{\mathcal{F}} \otimes_K M_n$ and in which, therefore, all identities of the full matrix algebra M_n over an infinite field of characteristic p are valid.

Assertion d) follows from formula (33.2) and Theorem 33.2.

Denote by St_{2l} the standard polynomial of degree $2l$. In order to prove assertion d), we must show that $\mathrm{St}_{2(n-1)} \notin \mathrm{Ann}_{T_{2(n-1)}} \widehat{\mathcal{G}}_{\gamma-n}$. A direct calculation, using the assertion of Lemma 25.1 and the defining relations (25.12)–(25.15) of the algebra $\widetilde{\mathcal{G}}(\gamma)$, gives the equality

$$\pi(\mathrm{St}_{2l}(x_1,\dots,x_{2l})x_{2l}\cdots x_1) = (\gamma^2 - l^2)\pi(\mathrm{St}_{2(l-1)}(x_1,\dots,x_{2(l-1)})x_{2(l-1)}\cdots x_1),$$

and we obtain from this by induction

$$\widetilde{\mathfrak{b}}(\mathrm{St}_{2l}, x_{2l}\cdots x_1) = (\gamma^2 - l^2)\cdots(\gamma^2 - 2^2)(\gamma^2 - 1^2),$$

$$\widetilde{\mathfrak{b}}_{\gamma-n}(\mathrm{St}_{2(n-1)}, x_{2(n-1)}\cdots x_1) = \prod_{i=1}^{(n-1)} (n^2 - i^2).$$

For $n = 1, 2, \dots, (p-1)/2$ this number is not divisible by p. The theorem is proved.

Comments

The main results of this chapter are exposed from a unified viewpoint in the author's papers [Raz7, Raz17, Raz23]. Theorem 27.1 was first proved in my paper [Raz7] in 1974 and was later recovered by Procesi [Pro1]. The first proofs of existence of a central polynomial for the algebra M_n were independently obtained by Formanek [Fo] and the author [Raz1]. Theorems 26.1, 26.2, 33.1, and 33.2 are originated from the paper [Raz23]. The proof of Theorem 31.1 shows that the existence of central polynomials in the algebras $M_{n,k}$ is related to the fact that T-ideals of the varieties var $M_{n,k}$ are invariant (in its multilinear part) with respect to operators A_i (see §31). The ideals of that kind are called the stable ones by V. N. Latyshev. Recently Okhitin [Okh] proved that for any PI-algebra with the stable T-ideal there exist central polynomials. The proof of Theorems 26.1, 26.2, and 27.2 shows that the theory of ideals of trace identities $V_{n,k}$ of the trace algebras $M_{n,k}$ is closely related to the representation theory of the symmetric group S_l, and these theories complete each other. This relationship permits to obtain Theorem 27.2 for fields of characteristic zero as a consequence of the Young theory of minimal ideals of the group algebra $K[S_{l+1}]$. It is not excluded that this relationship can move forward the representation theory of the symmetric group over fields of positive characteristic (see Corollary 1 of Proposition 30.1 and Theorem 27.2).

PROBLEM 4.1. Let $V_{n,k}$ be the ideal of trace identities of the algebra $M_{n,k}$ over a field K of positive characteristic. Describe for $l = nk + n + k$ the two-sided ideal $V_{n,k} \cap T_l$ of the group algebra $T_l = K[S_{l+1}]$ of the symmetric group S_{l+1} in group-theoretic terms.

The terms "prime variety of associative algebras" and "prime T=ideal" are introduced in the theory of PI-algebras by A. R. Kemer, and he also proved that all varieties of this kind of associative algebras over fields of characteristic zero have the form $\operatorname{var} M_{n,k}$ and $\operatorname{var} G \otimes M_n$ $(n, k = 1, 2, \dots)$ (see [Kem2, Kem3]). Theorem 33.3 shows that this result cannot be extended for fields of positive characteristic. The nature of prime varieties of associative algebras mentioned in Theorem 33.3 remains unclear. A solution of the following problem should lead to understanding of this nature.

PROBLEM 4.2. Describe all prime varieties of associative algebras over fields of positive characteristic.

CHAPTER V

THE α-FUNCTION RELATED TO REPRESENTATIONS OF THE SIMPLE THREE-DIMENSIONAL LIE ALGEBRA $\mathfrak{g}$ AND ITS APPLICATIONS TO VARIETIES OF GROUPS AND ASSOCIATIVE ALGEBRAS

In this chapter $\mathfrak{g}$ denotes the simple three-dimensional Lie algebra over a field K, $U(\mathfrak{g})$ denotes its universal enveloping algebra, and $U_{\mathrm{id}}(\mathfrak{g})$ its universal enveloping algebra with unit.

§34. Preliminaries

34.1. A summary of the results of the chapter. We consider the problem posed in §11 of Chapter 2 for the special case in which the ground field K is algebraically closed and of characteristic zero, the bilinear form $(\,,\,)\colon \mathfrak{g}\otimes_K\mathfrak{g}\to K$ equals a half of the Killing form, the set of all irreducible representations $\rho_s\colon \mathfrak{g}\to \mathrm{End}_K V_s$ of the Lie algebra $\mathfrak{g}$ in linear spaces V_s is taken for S, the $\boldsymbol{\alpha}$-function $\alpha_s\colon B\to K$ is constructed from the representation ρ_s in the same way as in §9, and the algebra $\mathcal{E}$ and the $\boldsymbol{\alpha}$-function $\alpha\colon B\to\mathcal{E}$ are determined by the set S as it was recommended in §11 of Chapter 2. We shall show that in this case the algebra $\mathcal{E}$ is isomorphic to the algebra of polynomials in one variable γ without the constant term, any ideal in this algebra is $\eta_{\underline{\alpha}}$-closed, and the extension of the $\boldsymbol{\alpha}$-function α to the space of generalized 2-elements $\widetilde{B}$ (see §16) is a natural minimal extension, which permits us to compute the values of the $\boldsymbol{\alpha}$-function α by simple recurrent formulas. As a result, we shall obtain:

1) the finite basis property for the identities of the Lie algebra $\mathfrak{g}$;

2) the Specht property for the variety of pairs $\mathrm{var}(U(\mathfrak{g}),\mathfrak{g})$;

3) finite bases of identities for arbitrary irreducible representations of the Lie algebra $\mathfrak{g}$ over fields of characteristic zero;

4) nonsolvability of varieties of locally finite groups of exponent 4 and of prime exponent p for $p>3$;

4) a finite basis of identities for the full matrix algebra of second order M_2 over a field of characteristic zero.

34.2. Irreducible representations and primitive ideals of the algebra $U(\mathfrak{g})$. In each simple three-dimensional Lie algebra $\mathfrak{g}$ over an algebraically closed field K we can choose a basis e_1, e_2, e_3 such that

$$(34.1)\qquad [e_1,e_2]=\sqrt{-1}e_3,\qquad [e_2,e_3]=\sqrt{-1}e_1,\qquad [e_3,e_1]=\sqrt{-1}e_2.$$

Define a bilinear form $(\,,\,)\colon \mathfrak{g}\otimes_K\mathfrak{g}\to K$ by the formula

$$(34.2)\qquad (e_i,e_j)=\delta_{ij}\qquad (\delta_{ij}\text{ is the Kronecker symbol}).$$

It can be verified directly that

$$(34.3) \qquad 2(v_1, v_2) = \operatorname{tr}(\operatorname{ad} v_1 \operatorname{ad} v_2) = \mathfrak{k}(v_1, v_2) \qquad (v_1, v_2 \in \mathfrak{g})$$

and for any element $v \in \mathfrak{g}$ the characteristic polynomial for the linear transformation $\operatorname{ad} v \colon \mathfrak{g} \to \mathfrak{g}$ has the form

$$(34.4) \qquad \operatorname{Det}(t \cdot \mathbf{1} - \operatorname{ad} v) = t^3 - (v, v)t.$$

The element $g \in \mathfrak{g}$ is called *regular* if $(g, g) \neq 0$. From (34.4) it follows that for any regular element $g \in \mathfrak{g}$ the element $h = h(g) = g/\sqrt{(g, g)}$ in the adjoint representation possesses three eigenvectors corresponding to the eigenvalues $0, \pm 1$. Normalizing these vectors, we see that in the Lie algebra $\mathfrak{g}$ for any regular element g we can construct a basis h, e_+, e_- for which

$$(34.5) \qquad [h, e_+] = e_+, \qquad [h, e_-] = -e_-, \qquad [e_+, e_-] = h = g/\sqrt{(g, g)}.$$

With respect to the bilinear form $(\, , \,)$, this basis is dual to the basis h, e_-, e_+. Therefore, by Lemma 16.2 and Proposition 16.1, the element

$$(34.6) \qquad \gamma \rightleftharpoons h^2 + e_+e_- + e_-e_+ = h^2 + h + 2e_-e_+$$

belongs to the center of the algebra $U(\mathfrak{g})$ and does not depend on the choice of dual bases in $\mathfrak{g}$. This element γ is called the *Casimir element*.

Suppose that a linear space V is endowed with the structure of a $\mathfrak{g}$-module. An element $v \in V$ is called a *highest vector* of the $\mathfrak{g}$-module V if $v \neq 0$ & $e_+v = 0$ and v is an eigenvector of the operator h. The eigenvalue $\Lambda(h)$ of the operator h in the equality $hv = \Lambda(h)v$ is said to be the *highest weight*. The next assertion gives sufficiently many examples of irreducible $\mathfrak{g}$-modules.

PROPOSITION 34.1. *Let $\Lambda(h)$ be an arbitrary element of the field K, let V_Λ be a linear space with a countable basis $v_0, v_1, v_2, \dots$, and let the structure of a $\mathfrak{g}$-module on V_Λ be given by the formulas*

$$(34.7) \qquad \begin{cases} e_-v_i = v_{i+1}, & hv_i = (\Lambda(h) - i)v_i, \\ e_+v_0 = 0, & e_+v_{i+1} = (i+1)(\Lambda(h) - i/2)v_i \end{cases} \qquad (i = 0, 1, 2, \dots).$$

Then v_0 is a highest vector of the $\mathfrak{g}$-module V_Λ, $\Lambda(h)$ is the highest weight of it, and the Casimir operator $\gamma\,|_{V_\Lambda}$ is equal to $\Lambda(h)(\Lambda(h)+1)$. Moreover, if $\operatorname{char}(K) = 0$, then the $\mathfrak{g}$-module V_Λ is irreducible if and only if $\Lambda(h) \notin \{k/2 \mid k = 0, 1, 2, \dots\}$, and for $\Lambda(h) = k/2$ $(k \in \mathbb{N})$ the subspace $V' = \sum_{j=k+1}^{+\infty} Kv_j$ is an irreducible $\mathfrak{g}$-submodule in V, isomorphic to $V_{-\Lambda-1}$, and the quotient module V_Λ/V' is an irreducible finite-dimensional $\mathfrak{g}$-module of dimension $k+1$.

PROOF. The fact that formulas (34.7) endow V with a structure of $\mathfrak{g}$-module and that $\gamma\,|_{V_\Lambda} = \Lambda(\Lambda+1)$ can be verified directly (see relations (34.5) and (34.6)).

Applying the operators e_+^j to any nonzero vector $v \in V$, we can obtain, up to a scalar multiple, a highest vector v_0 for $\Lambda \neq k/2$ $(k \in \mathbb{N})$, and for $\Lambda = k/2$ we can obtain in such a way either the vector v_0 or v_{k+1}. In the latter case $e_+v_{k+1} = 0$, $hv_{k+1} = (-\Lambda(h) - 1)v_{k+1}$, and v_{k+1} is a highest vector with the highest weight equal to $-\Lambda - 1$. The assertion of the proposition is now obvious.

COROLLARY. *Let the ground field K be algebraically closed and of characteristic zero, and let V be an irreducible $\mathfrak{g}$-module of dimension $k+1$. Then $V \simeq V_\Lambda/V'$, where $\Lambda = k/2$, and the characteristic polynomial for an arbitrary operator $g \in \mathfrak{g}$ has the form*

$$\text{(34.8)} \qquad \operatorname{Det}(t \cdot \mathbf{1} - g) = \begin{cases} \prod_{q=1}^{l} \left(t^2 - \left(q - \frac{1}{2}\right)^2 (g, g)\right), & \textit{for } k = 2l - 1, \\ t \prod_{q=1}^{l} (t^2 - q^2 (g, g)), & \textit{for } k = 2l. \end{cases}$$

PROOF. Let $g \in \mathfrak{g}$ be an arbitrary regular element. From it, we construct the elements h, e_+, e_- related by formulas (34.5). Let w be an arbitrary eigenvector of the operator h and $hw = \beta w$ ($\beta \in K$). Then $he_+w = e_+hw + [h, e_+]w = (\beta + 1)e_+w$ and e_+w is also an eigenvector with the eigenvalue $\beta + 1$. Moreover,

$$\text{(34.9)} \qquad he_+^j w = (\beta + j)e_+^j w, \qquad he_-^j w = (\beta - j)e_-^j w \qquad (j = 0, 1, 2, \dots).$$

Since the space V is finite-dimensional, this means that for some $v_0 = e_+^j w \neq 0$ we have the equalities

$$e_+v_0 = 0, \qquad hv_0 = \Lambda v_0 \qquad (\Lambda \in K)$$

and v_0 is a highest vector in V. Then (34.9) and (34.5) imply that for the vectors $v_i = e_-^i v_0$ ($i = 1, 2, \dots$) we have the equalities

$$e_-v_i = v_{i+1}, \qquad hv_i = (\Lambda - i)v_i,$$
$$e_+v_{i+1} = (i+1)\left(\Lambda - \tfrac{i}{2}\right)v_i \qquad (i = 0, 1, \dots).$$

Comparing these relations with (34.7) we see that the $\mathfrak{g}$-module V_Λ possesses a nonzero homomorphism into an irreducible finite-dimensional $\mathfrak{g}$-module V. From Proposition 34.1 it follows that this is possible only for $\Lambda = k/2$. Hence, $V \simeq V_\Lambda/V'$, the operator $h = g/\sqrt{(g,g)}$ is semisimple, and $k/2 - j$ ($j = 0, 1, \dots, k$) are all its eigenvalues. This proves formula (34.8) for the case of any regular element $\in \mathfrak{g}$. However, regular elements form an open everywhere dense set in $\mathfrak{g}$ with respect to the Zariski topology. Hence, (34.8) holds for any $g \in \mathfrak{g}$ as well. The corollary is proved.

By Theorem 54.1, the Casimir element γ acts in an arbitrary irreducible $\mathfrak{g}$-module V as a scalar operator $\delta \cdot \mathbf{1}$, where δ is an element of the algebraically closed field K that depends on V. Proposition 34.1 and its Corollary show that, conversely, for any $\delta \in K$ there exists an irreducible representation ρ of the Lie algebra $\mathfrak{g}$ for which $\rho(\gamma) = \delta \cdot \mathbf{1}$. Denote by I_δ the ideal of the universal enveloping algebra $U_{\mathrm{id}}(\mathfrak{g})$ (with unit) generated by $\gamma - \delta \cdot \mathbf{1}$. Then it follows from what was just said above that for any $\delta \in K$ in the algebra $U_{\mathrm{id}}(\mathfrak{g})$ there exists a primitive ideal I containing I_δ for which the quotient algebra $U_{\mathrm{id}}(\mathfrak{g})/I$ is primitive.

PROPOSITION 34.2. *Let I be an arbitrary primitive ideal of the algebra $U_{\mathrm{id}}(\mathfrak{g})$, and let the ground field K be algebraically closed and of characteristic zero. Then the ideal I coincides with the annihilator of one of the $\mathfrak{g}$-modules V_Λ and V_Λ/V'. Moreover, if $I = \operatorname{Ann} V_\Lambda$, then $I = I_{\Lambda(\Lambda+1)}$, and the quotient algebra $U_{\mathrm{id}}(\mathfrak{g})/I$ is simple if and only if $\Lambda \neq k/2$ ($k = 0, 1, \dots$).*

This statement can be verified by a simple direct calculation, and we leave it to the reader.

Now we consider the case in which $\operatorname{char} K = p$ and $p > 2$. In this case from formula (34.4) and from the Hamilton–Cayley theorem it follows that $(\operatorname{ad} g)^p = (g, g)^{(p-1)/2} \operatorname{ad} g$ for any $g \in \mathfrak{g}$. Since $(h, h) = 1$, $(e_+, e_+) = 0$, and $(e_-, e_-) = 0$, formula (34.4) implies that in the universal enveloping algebra $U(\mathfrak{g})$ we have the equalities

$$[e_+^p, g_0] = (\operatorname{ad} e_+)^p g_0 = 0, \qquad [e_-^p, g_0] = (\operatorname{ad} e_-)^p g_0 = 0,$$
$$[h^p - h, g_0] = ((\operatorname{ad} h)^p - \operatorname{ad} h) g_0 = 0 \qquad (g_0 \in \mathfrak{g}).$$

Hence, the center of the algebra $U(\mathfrak{g})$ contains, besides the Casimir element, also the elements $e_+^p, e_-^p, h^p - h$.

We shall describe all irreducible $\mathfrak{g}$-modules V over an algebraically closed field K. By Theorem 54.1, the elements of the center act as scalars on V. Therefore,

$$e_+^p = \beta_+^p \cdot \mathbf{1}, \qquad e_-^p = \beta_-^p \cdot \mathbf{1}, \qquad h^p - h = (\beta_0^p - \beta_0) \cdot \mathbf{1} \tag{34.10}$$

for some $\beta_+, \beta_-, \beta_0 \in K$. These equalities and Theorem 52.1 imply that a basis of the associative subalgebra U generated in $\operatorname{End}_K V$ by operators of $\mathfrak{g}$ consists of the monomials

$$e_-^{m_-} e_+^{m_+} h^{m_0} \qquad (0 \le m_0, m_+, m_- \le p - 1).$$

Therefore, $\dim_K U < \infty$ and the module V must be finite-dimensional. Then by equalities (34.10) we conclude that the operator e_+ (operator e_-) possesses one eigenvalue β_+ (β_-), and all eigenvalues of the operator h have the form $\beta_0 + j$, where $j = 0, 1, \dots, p-1$. Let $\gamma|_V = \delta \cdot \mathbf{1}$ as above.

If $\beta_+ = 0$ & $\beta_- = 0$, then the argument used in the proof of Corollary of Proposition 34.1 permits us to find the highest eigenvector $v_0 \in V$ and show that

$$\dim_K V \le p, \qquad V \simeq V_\Lambda / V', \qquad \Lambda = (\dim_K V - 1)/2,$$
$$h^p - h = 0, \qquad \delta = \Lambda(\Lambda + 1).$$

Consider the case $\beta_+ \ne 0$. Let v_0 be an eigenvector of the operator h corresponding to the eigenvalue β_0. Then for the vectors $v_j = e_+^j v_0 / \beta_+^j$ $(j = 0, 1, \dots, p-1)$ we have the equalities

$$h v_j = (\beta_0 + j) v_j, \qquad e_+ v_j = \beta_+ v_{j+1} \qquad (j = 0, 1, \dots, p-1, v_p \rightleftharpoons v_0). \tag{34.11}$$

Applying the operator e_- to the second one of these equalities and using (34.6), we obtain

$$e_- v_{j+1} = ((\delta - (\beta_0 + j)(\beta_0 + j + 1)/(2\beta_+)) v_j \qquad (j = 0, 1, \dots, p-1). \tag{34.12}$$

These equalities show that the vectors $v_0, \dots, v_{p-1}$ are linearly independent and the subspace spanned by them is stable under the operators from $\mathfrak{g}$. Since the $\mathfrak{g}$-module V is irreducible, this means that $\dim_K V = p$.

A direct verification shows that formulas (34.11) and (34.12) allow us to define a p-dimensional representation of $\mathfrak{g}$ for any β_0, β_+, δ with $\beta_+ \ne 0$. The case $\beta_- \ne 0$ is treated similarly, and we obtain the following assertion.

PROPOSITION 34.3. *If the ground field K is algebraically closed and* $\operatorname{char} K = p > 2$, *then the dimension of any irreducible representation of the Lie algebra $\mathfrak{g}$ does not exceed the number p, and for any homomorphism $\varkappa$ of the center $\mathfrak{Z}$ of the algebra $U(\mathfrak{g})$ into the field K there exist*:

a) *only one irreducible representation for* $\varkappa(e_+^p) \neq 0 \vee \varkappa(e_-^p) \neq 0$,

b) *exactly two irreducible representations for* $\varkappa(e_+^p) = 0 \;\;\&\;\; \varkappa(e_-^p) = 0$, *for which the operator $a \in \mathfrak{Z}$ acts as* $\varkappa(a) \cdot \mathbf{1}$.

§35. Computation of the α-function $\alpha\colon B \to \mathcal{E}$ and of the algebra $\mathcal{E}$

The success of our activities with the function α from Chapter 4 was predetermined by relations (23.5)–(23.10), which forced us to extend the signature by introducing the function Sp. This new signature turned out to be convenient, because of our ability, due to Proposition 25.1, to define a mapping $\pi\colon {}_\mu\mathcal{G}_\lambda \to \mathcal{G}_\lambda$. On one hand, this mapping permitted us to compute the function α recursively, and on the other hand, since the diagram in Lemma 25.2 is commutative, it allowed us to identify the quotient space of $F \cap {}_\mu\mathcal{G}_\lambda$ modulo the kernel of the pairing $\widetilde{\mathfrak{b}}$ with a subspace of $\mathcal{G}_\lambda$, which, in view of the finite-dimensionality of $\mathcal{G}_\lambda$, assured the effectiveness of the check of the membership of an element to $\operatorname{Ann}_{T_\lambda} \widetilde{\mathfrak{b}}_E$. It turns out that a largely similar construction is possible also for the **α**-function constructed from representations of the simple three-dimensional Lie algebra $\mathfrak{g}$. We shall deviate from the natural line of discussion explaining the reasons for extending the signature in this case and we shall formally show the way of computation of the function $\widetilde{\alpha}\colon B \to \mathcal{E}$, where B is the space of generalized 2-elements defined in §16, in the case of an arbitrary representation of the three-dimensional Lie algebra $\mathfrak{g}$.

35.1. Some identities of the algebra $\{K, \mathfrak{g}, U(\mathfrak{g})\}$ with three supports and of the pair $(U(\mathfrak{g}), \mathfrak{g})$. Let the field K be algebraically closed, let the bilinear form on $\mathfrak{g}$ be given by the formula $(x, y) = \frac{1}{2}\mathfrak{k}(x, y) = 2\operatorname{tr}(\rho_2(xy))$, where $\mathfrak{k}$ is the Killing form and ρ_2 is a two-dimensional representation of the Lie algebra $\mathfrak{g}$, and let e_1, e_2, e_3 be a basis of the algebra $\mathfrak{g}$. It can be checked directly that for any x, y, z from the simple three-dimensional Lie algebra $\mathfrak{g}$ we have

$$[x, y, z] = (y, z)x - (x, z)y. \tag{35.1}$$

Let $\rho\colon \mathfrak{g} \to \operatorname{End}_K V$ be an arbitrary faithful representation of the Lie algebra $\mathfrak{g}$ and U an associative subalgebra in $\operatorname{End}_K V$ generated by $\rho(\mathfrak{g})$. In the same way as in §15, let us consider the algebra with three supports $\{K, \mathfrak{g}, U\}$, where the operation $k\colon \mathfrak{g} \otimes_K \mathfrak{g} \to K$ must be replaced by $(\ ,\ )\colon \mathfrak{g} \otimes_K \mathfrak{g} \to K$, and, using the commutative algebra $\mathcal{F}$ and the algebras $\widetilde{U} = \mathcal{F} \otimes U$ and $\widetilde{\mathfrak{g}} = \mathcal{F} \otimes_K \mathfrak{g}$, let us construct the free pair $(F, \mathfrak{L})$ with the set of free generators $X = \{x_i = \sum_{j=1}^3 x_i^{(j)} \mid i = 1, 2, 3, \dots\}$ and the free algebra with three supports $\{\mathcal{F}_0, \mathfrak{L}_1, F_1\}$ with free generators $\{\varnothing, X, \varnothing\}$ of the varieties $\operatorname{var}(U, \mathfrak{g})$ and $\operatorname{var}\{K, \mathfrak{g}, U\}$, respectively.

LEMMA 35.1. *The identities*

$$([x, y], z)v = (v, z)[x, y] - (y, v)[x, z] + (x, v)[y, z], \tag{35.2}$$

$$(y, y)[x_i, x_j] = [x_i, y, y, x_j] + [y, x_j, x_i, y] \tag{35.3}$$

are consequences of identity (35.1) *and are therefore satisfied in the algebra with three supports* $\{K, \mathfrak{g}, U\}$ *(and in* $\{\mathcal{F}_0, \mathfrak{L}_1, F_1\}$*) for any* $x, y, z, v \in \mathfrak{g}$.

PROOF. Applying identity (35.1) in two ways to the commutator of weight four:

$$[[x, y], v, z] = (v, z)[x, y] - ([x, y], z)v = (y, v)[x, z] - (x, v)[y, z],$$

we obtain (35.2). Applying (35.1) to the commutators $[x_i, y, y]$ and $[y, x_j, x_i]$, we see that the right-hand side of equality (35.3) is equal to

$$(y, y)[x_i, x_j] - (x_i, y)[y, x_j] + (x_j, x_i)[y, y] - (y, x_i)[x_j, y] = (y, y)[x_i, x_j].$$

The lemma is proved.

LEMMA 35.2. *Let* y *be an arbitrary element of the Lie algebra* $\mathfrak{L}$. *Suppose that* $g(x_1, \dots, x_l)$ *is an arbitrary multilinear polynomial of the associative algebra* F *that is skew-symmetric with respect to the variables* x_i *and* x_j. *Then in the algebra with three supports* $\{K, \mathfrak{g}, U\}$ *(and in* $\{\mathcal{F}_0, \mathfrak{L}_1, F_1\}$*) we have the identity*

$$(y, y)g = \tfrac{1}{2}(g|_{x_i=[x_i, y, y]} + g|_{x_j=[x_j, y, y]} + g|_{x_i=[x_i, x_j, y], x_j=y}). \tag{35.4}$$

PROOF. Applying (35.1) to the commutators $[x_i, y, y]$, $[x_j, y, y]$, and $[x_i, x_j, y]$, we find that the right-hand side of identity (35.4) is equal to

$$\tfrac{1}{2}((y, y)g - (x_i, y)g|_{x_i=y, x_j=x_j} + (y, y)g$$
$$- (x_j, y)g|_{x_i=x_i, x_j=y} + (x_j, y)g|_{x_i=x_i, x_j=y} - (x_i, y)\cdot g|_{x_i=x_j, x_j=y}).$$

In this expression the fourth and the fifth terms have different signs, and the sum of the second and the sixth terms is equal to zero, because g is skew-symmetric with respect to x_i and x_j. The lemma is proved.

We put by definition

$$[u, v] * (x, y) \rightleftharpoons \tfrac{1}{2}([u, x, y, v] + [u, y, x, v] + [x, v, u, y] + [y, v, u, x]), \tag{35.5}$$

$$\langle v_1, v_2, v_3 \rangle \rightleftharpoons \sum_{\sigma \in S_3} \operatorname{sign}(\sigma) v_{\sigma(1)} v_{\sigma(2)} v_{\sigma(3)}, \tag{35.6}$$

$$v * \langle v_1, v_2, v_3 \rangle \rightleftharpoons \sum_{\sigma \in S_3} \operatorname{sign}(\sigma)[v, v_{\sigma(1)}, v_{\sigma(2)}, v_{\sigma(3)}], \tag{35.7}$$

where x, y, u, v, v_i are some formal variables.

REMARK 35.1. The right-hand side of (35.5) coincides, up to a change of variables, with a half of the total linearization of the Lie polynomial on the right-hand side of (35.3).

REMARK 35.2. In the Lie algebra $\mathfrak{g}$, let us choose a basis h, e_+, e_- satisfying relations (34.5). It can be checked directly that in $U(\mathfrak{g})$

$$\langle h, e_+, e_- \rangle = h^2 + e_+e_- + e_-e_+ = h^2 + h + 2e_-e_+. \tag{35.8}$$

Comparing this equality with (34.6), we see that this element coincides with the Casimir element γ. From the results of §34 it follows that for any irreducible representation ρ of the Lie algebra $\mathfrak{g}$ over an algebraically closed field K this element defines a scalar operator $\delta \cdot \mathbf{1}$, where $\delta \in K$; in particular, if an irreducible $\mathfrak{g}$-module V has a highest vector v ($hv = \nu v$, $e_+ v = 0$), then $\delta = \nu(\nu + 1)$, and if, moreover, $\dim_K V = n < \infty$, then $\nu = (n-1)/2$ and $\delta = (n^2 - 1)/4$. For the adjoint representation we have $V = \mathfrak{g}$, $n = 3$, and $\delta = 2$. Therefore, for any $g \in \mathfrak{g}$ we have $g * \langle h, e_+, e_- \rangle = 2g$ and $([h, e_+], e_-) = 1$, and in any pair $(U, \mathfrak{g})$ obtained from an irreducible representation ρ the following identities hold:

$$\langle x_1, x_2, x_3 \rangle = \delta([x_1, x_2], x_3), \tag{35.9}$$

$$\delta[y * \langle x_1, x_2, x_3 \rangle] = 2\delta([x_1, x_2], x_3)y = 2y \cdot \langle x_1, x_2, x_3 \rangle, \tag{35.10}$$

where δ is the value of the Casimir element γ of this representation.

LEMMA 35.3. *In the pair* $(U, \mathfrak{g})$ (*and in* $(F, \mathfrak{L})$) *the following identities hold*:

$$[x, y, z][u, v] = x([u, v] * (y, z)) - y([u, v] * (x, z)), \tag{35.11}$$

$$[z * \langle x_1, x_2, x_3 \rangle]y = z[y * \langle x_1, x_2, x_3 \rangle], \tag{35.12}$$

$$\langle x_1, x_2, x_3 \rangle y = y \langle x_1, x_2, x_3 \rangle. \tag{35.13}$$

PROOF. Since the algebra F lies in the algebra F_1, we can use identities (35.1) and (35.3) and Remark 35.1 to prove identity (35.11). Identity (35.13) is verified by the direct substitution of the elements of the orthonormal basis e_1, e_2, e_3 instead of x_1, x_2, x_3. Then, up to a multiple, $\langle e_1, e_2, e_3 \rangle$ is the Casimir element $e_1^2 + e_2^2 + e_3^2$ belonging to the center of the algebra U. Identity (35.12) is obtained from the Schur lemma and identity (35.13) applied to the adjoint representation of the algebra $\mathfrak{g}$. The lemma is proved.

LEMMA 35.4. *In the pair* $(U, \mathfrak{g})$ (*and in* $(F, \mathfrak{L})$) *all Capelli identities of order four are valid.*

PROOF. Since the Lie algebra $\mathfrak{g}$ is three-dimensional, we see that an arbitrary polynomial that is skew-symmetric and multilinear with respect to certain four its variables vanishes under the substitution of elements of the Lie algebra $\mathfrak{g}$ instead of the variables. The lemma is proved.

LEMMA 35.5. *In the Lie algebra* $\mathfrak{g}$ (*and in* $\mathfrak{L}$) *the following identities hold*:

$$[z * \langle x_1, x_2, x_3 \rangle, y] = [z * \langle x_1, x_2, x_3 \rangle]|_{z=[z,y]}, \tag{35.14}$$

$$[z, y, y, x, y] = [z, y, x, y, y], \tag{35.15}$$

$$[z * \langle x_1, x_2, x_3 \rangle]|_{z=[z,y]} = 2\{[z, y] * (x_1, x_3)\}|_{x_1=[x_1,x_2]}. \tag{35.16}$$

PROOF. Since the Lie algebra $\mathfrak{g}$ is embedded in U, we can use the identities of the pair $(U, \mathfrak{g})$ and of the algebra with three supports $\{K, \mathfrak{g}, U\}$. Identity (35.14) is a consequence of (35.10). Identity (35.16) is obtained by successive application of the first equality of (35.10), the identity (35.3), and Remark 35.1 to the left-hand side of the first equality of (35.10). Applying identity (35.1) to the subcommutators $[z, y, x]$ and $[[z, y], y, x]$ on both sides of (33.15), we see that both sides are equal to $(x, y)[z, y, y]$. The lemma is proved.

LEMMA 35.6. *Any element of the algebra $\mathcal{F}_0$ can be represented as a linear combination of elements of two types:*
the first type: $(x_{i_1}, x_{j_1})\cdots(x_{i_k}, x_{j_k})$,
the second type: $(x_{i_1}, x_{j_1})\cdots(x_{i_k}, x_{j_k})([x_i, x_j], x_m)$.

PROOF. Identity (35.1) and the invariance of the form $(\ ,\ )$ imply that any element of $\mathcal{F}_0$ can be represented as a polynomial in (v, x_i), where v is a Lie commutator of weight 1 or 2. If v_1 and v_2 are commutators of weight 2, then $(v_1, x_i)(v_2, x_j) = (v_1, (v_2, x_j)x_i)$; applying (35.2) to the element $(v_2, x_j)x_i$ and using the invariance of the form $(\ ,\ )$ and (35.1), we see that $(v_1, (v_2, x_j)x_i)$ can be expressed via elements of the first type. This shows that any element of $\mathcal{F}_0$ is expressed as a linear combination of products of parentheses, and in each product the square brackets occur at most once. The lemma is proved.

35.2. The auxiliary algebra with three supports $\{\overline{\mathcal{F}}_0, \overline{\mathfrak{L}}, \overline{F}_1\}$. Using the quotient algebra $\overline{\mathcal{F}}$ of the algebra $\mathcal{F}$ with respect to the defining relations (16.2) and the orthonormal basis e_1, e_2, e_3 of the algebra $\mathfrak{g}$, we construct from the representation ρ (in the same way as it was done in §16) the algebras $\overline{\mathfrak{g}}$ and $\overline{U}$, the auxiliary pair $(\overline{F}, \overline{\mathfrak{L}})$ with the set of generators $\overline{X} = \{\overline{x}_i = \sum_{j=1}^{3}\overline{x}_i^{(j)}e_j \mid i = 1, 2, \dots\}$, and the algebra with three supports $\{\overline{\mathcal{F}}_0, \overline{\mathfrak{L}}_1, \overline{F}_1\}$ with the set of generators $\{\varnothing, \overline{X}, \varnothing\}$, where $(\ ,\ ) = \frac{1}{2}\mathfrak{k}$.

LEMMA 35.7. *Let ρ be an irreducible representation of the Lie algebra $\mathfrak{g}$ and let $\delta \in K$ be the value of the Casimir element* (35.8) *in this representation. Let $\overline{A}$ be an associative subalgebra in $\overline{U}$ equal to $\overline{\mathcal{F}}_0 + \overline{F}_1$. Then for any $\overline{x}_i \in \overline{X}$, $w, v \in \overline{\mathfrak{g}}$ (or $w, v \in \overline{\mathfrak{L}}_1$), and $a \in \overline{A}$ in the algebras $\overline{A}$ and U the following relations hold*:

$$\overline{x}_i^2 = \delta\big(\overline{x}_i^{(1)}\big)^2 \qquad (\overline{x}_i, \overline{x}_i) = 3\big(\overline{x}_i^{(1)}\big)^2 = 3\frac{\overline{x}_i^2}{\delta}, \tag{35.17}$$

$$\begin{cases} (\overline{x}_i, v)\overline{x}_i = \big(\overline{x}_i^{(1)}\big)^2 v = \frac{\overline{x}_i^2}{\delta}v, \\ (\overline{x}_i, v)(\overline{x}_i, w) = \frac{\overline{x}_i^2}{\delta}(w, v), \end{cases} \tag{35.18}$$

$$[\overline{x}_i, \overline{x}_j, \overline{x}_k]a\overline{x}_i = (\overline{x}_j, \overline{x}_k)\overline{x}_i a\overline{x}_i - \overline{x}_j a\overline{x}_k\frac{\overline{x}_i^2}{\delta}, \tag{35.19}$$

$$[v, \overline{x}_i, \overline{x}_i] = (\overline{x}_i, \overline{x}_i)v - (\overline{x}_i, v)\overline{x}_i = 2\frac{\overline{x}_i^2}{\delta}v, \tag{35.20}$$

$$\overline{x}_i v\overline{x}_i = (\delta - 1)(\overline{x}_i^2/\delta)v. \tag{35.21}$$

PROOF. Equalities (35.17) are clear. Relations (35.18) express the fact that $v = \sum_{i=1}^{3}(v, e_i)e_i$, where e_1, e_2, e_3 is an orthonormal basis. Equalities (35.19) and (35.20) follow from (35.18) with the help of (35.1), and (35.20) implies (35.21). The lemma is proved.

For any two disjoint finite subsets μ and λ of the set $\overline{X}$ denote by ${}_{\mu}\overline{A}_{\lambda}$ the subspace in $\overline{A}$ consisting of multihomogeneous elements of degree two with respect to each variable $\overline{x}_i \in \mu$ and of degree one with respect to each variable $\overline{x}_j \in \lambda$.

LEMMA 35.8. *For each $f \in {}_\mu\overline{A}_\lambda$ there exists $g \in \overline{A}_\lambda$ such that*

$$f = \prod_{\overline{x}_i \in \mu} \frac{\overline{x}_i^2}{\delta} g \tag{35.22}$$

and this representation is unique in the algebra $\overline{A}$. Moreover, this equality is a consequence of relations (35.17)–(35.21) *and identity* (35.1), *and the scalar coefficients in g are polynomial functions in δ, depending only on the notation of the element f.*

PROOF. Since the algebra $\overline{\mathcal{F}}_0$ is a homomorphic image of the algebra $\mathcal{F}_0$, we see that, by Lemma 35.6, it suffices to consider the case in which $f = f_0 a$, where a is a word and f_0 is an element of the first or the second type in $\overline{x}_i \in \overline{X}$. Let $\overline{x}_i \in \mu$. Two cases are possible: a) $\overline{x}_i$ occurs in f_0, b) both occurrences of the letter $\overline{x}_i$ are contained in a. In the first case, using (35.18), we have $f = f' \frac{\overline{x}_i^2}{\delta}$, and (35.22) follows from inductive arguments. In the second case (35.19) implies

$$\overline{x}_i \overline{x}_j \overline{x}_k b \overline{x}_i = [\overline{x}_i, \overline{x}_j, \overline{x}_k] b \overline{x}_i + \cdots = (\overline{x}_j, \overline{x}_k) \overline{x}_i b \overline{x}_i + \overline{x}_j b \overline{x}_k \frac{\overline{x}_i^2}{\delta} + \cdots,$$

where the ellipses denote the terms for which the distance between $\overline{x}_i$'s is lesser that in the original word. In the first summand the distance between the letters $\overline{x}_i$ has diminished. This process of reconciliation the letters $\overline{x}_i$ to each other will give a possibility of applying (35.21) and again will lead to the equality $f = \frac{\overline{x}_i^2}{\delta} f'$. Therefore, (35.22) can be obtained by an inductive agrument, with the help of (35.17)–(35.21) and (35.1) only. All other assertions of the lemma are now obvious.

This lemma shows that for any $\rho = \rho_s$ the polynomial g on the right-hand side of (35.22) depends only on the value of the Casimir element $\delta = \rho_s(\gamma)$. Denote g by $\pi_\delta(f)$. From Lemma 35.8 it follows that the mapping $\pi_\delta : {}_\mu\overline{A}_\lambda \to \overline{A}_\lambda$ can be computed with the help of relations (35.17)–(35.21). By the definition of the $\boldsymbol{\alpha}$-function $\alpha_s : B \to K$, constructed from an irreducible representation ρ_s, and by the definition of the extension $\widetilde{\alpha}_s : \widetilde{B} \to K$ of this function (see §16) it is clear that the mapping $\pi_\delta : {}_\mu\overline{A} \to K$ coincides with $\widetilde{\alpha}_s$ and permits the computation of $\widetilde{\alpha}_s$ via the relations mentioned above. In these computations the role of the algebra $\overline{A}_s$ is mainly similar to the role of the algebra $\mathcal{G}_n$ in §25.

35.3. Auxiliary associative algebras $A(\gamma)$ and $\overline{A}(\gamma)$. From the realization of the computation indicated of the function $\widetilde{\alpha}$ it follows that in these computations we can use, side by side with the relations of Lemma 35.7, any identities satisfied in the algebra with three supports $\{K, \mathfrak{g}, U_s\}$ and all the more in $\{K, \mathfrak{g}, U(\mathfrak{g})\}$. Therefore, the same goes also for the function $\alpha: B \to \mathcal{E}$ constructed in accordance with the recommendations of §11 from $\boldsymbol{\alpha}$-functions α_s corresponding to the irreducible representations ρ_s. It is clear that $\mathcal{E}$ is isomorphic to the ideal of the polynomial algebra $K[\gamma]$ generated by γ.

We shall present a method of computing the extension $\widetilde{\alpha}: \widetilde{B} \to \mathcal{E}$ of the function α, constructed in an obvious way from the extensions $\widetilde{\alpha}_s : \widetilde{B} \to K$. Toward this end, we take for the representation ρ the natural representation of $\mathfrak{g}$ in its universal enveloping algebra $U(\mathfrak{g})$ and construct for the pair $(U(\mathfrak{g}), \mathfrak{g})$, using the commutative algebras $\overline{\mathcal{F}}$ and $\mathcal{F}$, the objects $\widetilde{\mathfrak{g}}$, $\widetilde{U}(\mathfrak{g})$, $\overline{\mathfrak{g}}$, $\overline{U}(\mathfrak{g})$, $(F, \mathfrak{L})$, $(\overline{F}, \overline{\mathfrak{L}})$, $(\mathcal{F}_0, \mathfrak{L}_1, F_1)$, and $(\overline{\mathcal{F}}_0, \overline{\mathfrak{L}}_1, \overline{F}_1)$. By construction, $F_1 \subset \widetilde{U}(\mathfrak{g})$, $\overline{F}_1 \subset \overline{U}(\mathfrak{g})$. Denote by $A(\gamma)$ the associative subalgebra in $\widetilde{U}(\mathfrak{g})$ generated by the Casimir element γ, $\gamma\mathcal{F}_0$, and F_1, and denote by $\overline{A}(\gamma)$ the image of $A(\gamma)$ under the homomorphism $\psi\colon \widetilde{U}(\mathfrak{g}) \to \overline{U}(\mathfrak{g})$.

It is completely clear that if we replace δ by γ, then for the elements of the algebras $\overline{A}(\gamma)$ and $\overline{\mathcal{F}}_0$ relations (35.17)–(35.21) are satisfied in the algebra $\overline{U}(\mathfrak{g})$, and the assertions of Lemmas 35.7 and 35.8 remain valid for the algebra $\overline{A}(\gamma)$. This permits us to define well a mapping $\pi\colon {}_\mu A_\lambda(\gamma) \to A_\lambda(\gamma)$ by setting $\pi(f) \rightleftharpoons g$, where $\psi(f)$ and $\psi(g)$ are interrelated by formula (35.22). It is clear that the mapping $\pi\colon F_1 \cap {}_\mu A(\gamma) \to \mathcal{E}$ coincides with $\widetilde{\alpha}\colon {}_\mu \widetilde{B} \to K$, and therefore the function $\widetilde{\alpha}$ can be computed via relations of the form (35.17)–(35.21) with permissible usage of any identities of $\{K, \mathfrak{g}, U(\mathfrak{g})\}$. In these computations, the algebra $\overline{A}(\gamma)$ is an analog of the algebra $\mathcal{G}(\gamma)$ of Chapter 4.

35.4. The bilinear pairing $\widetilde{\mathfrak{b}}\colon {}_\mu A_\lambda \otimes {}_\nu A_\lambda \to \mathcal{E}$. The function $\widetilde{\alpha}$ just constructed permits us to define a bilinear pairing $\widetilde{\mathfrak{b}}$ by formula (25.21). This pairing gives on $T_\lambda \rightleftharpoons F_1 \cap A_\lambda(\gamma)$ a bilinear form, which will be denoted by $\widetilde{\mathfrak{b}}_0$. (In the special case $\lambda = \{x_1, \dots, x_l\}$ we shall denote T_λ by T_l.) In the same way as in §25, for the function $\widetilde{\alpha}$ just constructed we can define the mappings $\Phi_{\widetilde{\alpha}}$ and $\Psi_{\widetilde{\alpha}}$ and the closures $\theta_{\widetilde{\alpha}}$ and $\eta_{\widetilde{\alpha}}$, with the help of $\widetilde{\mathfrak{b}}$, and we can show that the following equalities hold for them:

$$\Psi_{\widetilde{\alpha}}(I) \cap I = \Psi_\alpha(I), \qquad \Phi_{\widetilde{\alpha}}(\mathfrak{V} + \mathcal{F}_0\mathfrak{V}) = \Phi_\alpha(\mathfrak{V}), \tag{35.23}$$

where I is an arbitrary ideal in $\mathcal{E}$ and $\mathfrak{V}$ is an arbitrary ideal of weak identities in F ($(F, \mathfrak{L})$ is a free pair of $\mathrm{var}(U(\mathfrak{g}), \mathfrak{g})$).

PROPOSITION 35.1. *The diagram*

$$\begin{array}{ccc} {}_\mu A_\lambda[\gamma] \otimes {}_{\mu'} A_\lambda[\gamma] & \xrightarrow{\widetilde{\mathfrak{b}}} & \mathcal{E} \\ {\scriptstyle \pi\otimes\pi}\downarrow & & \parallel \\ A_\lambda[\gamma] \otimes A_\lambda[\gamma] & \xrightarrow{\widetilde{\mathfrak{b}}_0} & \mathcal{E} \end{array}$$

is commutative.

The proof is obvious.

Denote by $\{E(\gamma)\}$ the ideal in $\mathcal{E}$ generated by the polynomial $E(\gamma) \in \mathcal{E}$. Since the algebra $\mathcal{E}$ is singly generated (by γ), any ideal I in $\mathcal{E}$ is principal, i.e., $I = \{E(\gamma)\}$ for some polynomial $E(\gamma)$ divisible by γ. Using the natural homomorphism $\mathcal{E} \to K_E \simeq \mathcal{E}/\{E(\gamma)\}$, we can assume that $\mathfrak{b}$ takes its values in K_E. We shall denote this pairing by $\widetilde{\mathfrak{b}}_E$. Let $\mathrm{Ann}_{T_l} \widetilde{\mathfrak{b}}_E \rightleftharpoons \{ f \in T_l \mid \widetilde{\mathfrak{b}}_E(f, g) = 0 \text{ for any } g \in T_l \}$. Since the diagram (35.24) is commutative, we have

COROLLARY. $\Psi_{\widetilde{\alpha}}(\{E(\gamma)\}) \cap T_l = \mathrm{Ann}_{T_l} \widetilde{\mathfrak{b}}_E$.

This shows that in order to justify that the element $f \in T_l$ belongs to a $\theta_{\widetilde{\alpha}}$-closed ideal $\Psi_{\widetilde{\alpha}}(\{E(\gamma)\})$ it is sufficient to make a finite number of checks.

35.5. η_α-closed ideals of the algebra $\mathcal{E}$. From the results of §37 it will follow that almost all ideals of weak identities in F are ideals in F_1 and therefore the mappings $\Phi_{\tilde{\alpha}}$ and $\Psi_{\tilde{\alpha}}$ differ slightly from Φ_α and Ψ_α. Moreover, the lattice Ξ_α of η_α-closed ideals in $\mathcal{E}$ permits a simple description.

PROPOSITION 35.2. *Any ideal in the algebra $\mathcal{E}$ is η_α-closed.*

PROOF. Choose different variables $y, z_{1i}, z_{2i}, z_{3i} \in X$ $(i = 1, 2, \dots)$. For any number $\beta_i \in K$ we put

$$v_i \rightleftharpoons \langle z_{1i}, z_{2i}, z_{3i}\rangle y - \frac{\beta_i}{2} y * \langle z_{1i}, z_{2i}, z_{3i}\rangle \in F.$$

For any collection of numbers $\beta_1, \beta_2, \dots, \beta_n \in K$ define in the algebra F recursively, beginning from $h_0 = h_0' = y$, the multilinear polynomials $h_1, \dots, h_n \in F$ and $h_1', \dots, h_n' \in \mathfrak{L}$ by setting

$$h_i = h_{i-1}\big|_{y=v_i}, \qquad h_i' = h_{i-1}'\big|_{y=[y*\langle z_{1i}, z_{2i}, z_{3i}\rangle]/2}.$$

Let $E(\gamma) = \gamma \prod_{i=1}^n (\gamma - \beta_i)$. We shall show that for the ideal of weak identities $\mathfrak{V}$ generated in F by the element h_n the image $\Phi_\alpha(\mathfrak{B})$ is equal to $\{E(\gamma)\}$. From formula (35.10), which holds in $A(\gamma)$ for $\delta = \gamma$, we have in the algebra $\widetilde{U}$ the equalities

$$h_n = E(\gamma) \prod_{i=1}^n ([z_{1i}, z_{2i}], z_{3i}) y = E(\gamma) h_n', \tag{35.25}$$

which imply that $h_n \in \operatorname{Ann}_{t_l} \widetilde{\mathfrak{b}}_E$ $(l = 3n + 1)$. This means that $\Phi_\alpha(\mathfrak{V}) \subseteq \{E(\gamma)\}$ (see Corollary of Proposition 35.1). From (35.18), from the invariance of the form $(\,,\,)$ and from (35.20) and (35.17) it follows that

$$\pi(y([x_1, x_2], x_3)^2) = \pi(y([x_1, x_2], [x_1, x_2])) = -\pi(y([x_2, x_1, x_1], x_2)) = -6y.$$

Therefore, from (35.25) and an inductive argument it follows that $\widetilde{\mathfrak{b}}(h_n, h_n') = (-6)^n E(\gamma)$ and $\Phi_\alpha(\mathfrak{V})$ contains $E(\gamma)$ over fields whose characteristic differs from 2 and 3. (In fact, $E(\gamma) \in \Phi_\alpha(\mathfrak{V})$ for char $K = 3$ as well.) The proposition is proved.

From the argument just finished it follows that we can take the pair $(U(\mathfrak{g})/I, \mathfrak{g})$, where I is the largest ideal in $U(\mathfrak{g})$ for which $I \cap \mathcal{E} = \{E(\gamma)\}$, as a model of a θ_α-closed ideal of weak identities $\Psi_\alpha(\{E(\gamma)\})$.

§36. A basis of identities for the Lie algebra $\mathfrak{g}$

In this section the ground field K is of characteristic zero and the absolutely free Lie algebra $\mathfrak{M}$ has the free generators $x_1, x_2, \dots$.

THEOREM 36.1. *The set of identities of the simple three-dimensional Lie algebra $\mathfrak{g}$ possesses a finite basis. All identities of the Lie algebra $\mathfrak{g}$ are equivalent to identities* (35.14)–(35.16).

The proof of this theorem is based on two lemmas. In these lemmas for any polynomial $f(x_1, \dots, x_l)$ and any permutation $\sigma \in S_l$ of the set $\{1, 2, \dots, l\}$ we denote by the polynomial $f(x_{\sigma(1)}, \dots, x_{\sigma(l)})$ by $\sigma(f)$. Let τ_{ij} $(i \neq j)$ denote the transposition of the indices $i, j \in \{1, 2, \dots, l\}$.

LEMMA 36.1. *In the variety of Lie algebras determined by identities* (35.14)–(35.16) *the following identity holds for any multilinear Lie polynomial* $f(x_1, \dots, x_l)$, $l > 4$:

$$f - \tau_{ij}(f) = \beta\left\{(1 - \tau_{ij}) \sum_{\sigma} [x_i, x_{i_{\sigma(1)}}, \dots, x_{i_{\sigma(l-2)}}, x_j]\right\} + g\big|_{x_i = [x_i, x_j]}, \tag{36.1}$$

where β *is some number, the sum is extended over all permutations* σ *of the set* $\{1, \dots, \widehat{i}, \dots, \widehat{j}, \dots, l\}$, *and* $g(x_1, \dots, \widehat{x_j}, \dots, x_l)$ *is a multilinear Lie polynomial. Moreover,* $\beta = 0$ *if the degree of* f *is odd.*

LEMMA 36.2. *For any multilinear Lie identity* $f(x_1, \dots, x_l) = 0$ $(l > 4)$ *satisfied in the Lie algebra* $\mathfrak{g}$ *there exists a multilinear Lie polynomial of the form* $g(x_1, \dots, \widehat{x_j}, \dots, x_l)$ *such that the identity*

$$f - \tau_{ij}(f) = g\big|_{x_i = [x_i, x_j]} \tag{36.2}$$

is a consequence of identities (35.14)–(35.16).

In §6 we considered in detail heuristic arguments involving the mapping π and the $\boldsymbol{\alpha}$-function, suggesting the statement of Lemma 36.1 and a plan for proving Theorem 36.1. We omit them now.

First we show how we can derive Lemma 36.2 and Theorem 36.1 from Lemma 36.1.

THE PROOF OF LEMMA 36.2. If the degree of f is odd, then the assertion of the lemma follows directly from Lemma 36.1. If the degree of f is even, then, by Lemma 36.1, there exist β and g such that identity (36.1) is a consequence of identities (35.14)–(35.16). We put in it $x_i = e_+$ and $x_j = e_-$ and set all other variables equal to h (see Remark 35.2). Since identities (35.14)–(35.16) are satisfied in $\mathfrak{g}$, identity (36.1) must become an equality in $\mathfrak{g}$ under this substitution. However, $f = 0$ is an identity in $\mathfrak{g}$ by assumption of the lemma, hence, the left-hand side of (36.1) vanishes. Since $[e_+, e_-] = h$, $g\big|_{x_i = [x_i, x_j]}$ vanishes. Therefore, we have $0 = \beta\,(l-2)! \cdot 2h$ in $\mathfrak{g}$, whence we conclude that $\beta = 0$. The lemma is proved.

THE PROOF OF THE THEOREM will be given by induction on the degree l of a multilinear identity of the algebra $\mathfrak{g}$. The base of induction is $l = 4$. Since over a field of characteristic zero any nontrivial Lie identity of degree 4 implies the 3rd Engel identity or the metabelian identity, which are not satisfied in $\mathfrak{g}$, we see that all identities of this kind are trivial.

Suppose that for multilinear identities of the algebra $\mathfrak{g}$ of degree less than l it is proved that they are consequences of (35.14)–(35.16). Let a multilinear identity $f(x_1, \dots, x_l) = 0$ hold in $\mathfrak{g}$; then, by Lemma 36.2, for a certain Lie polynomial $g(x_1, \dots, \widehat{x_j}, \dots, x_l)$ identity (36.2) will be a consequence of identities (35.14)–(35.16) and, hence, it must hold in $\mathfrak{g}$. Since $f = 0$ is an identity for $\mathfrak{g}$, we see that $g\big|_{x_i = [x_i, x_j]} = 0$ is an identity for $\mathfrak{g}$. However, the commutator subalgebra of $\mathfrak{g}$ coincides with $\mathfrak{g}$. Therefore, $g = 0$ is an identity of $\mathfrak{g}$ of degree $l - 1$, and it follows from (35.14)–(35.16) by the induction assumption. Hence, the identity $f - \tau_{ij}(f) = 0$ follows from (35.14)–(35.16); moreover, for any permutation σ of the set $\{1, \dots, l\}$ the identity $f = \sigma(f)$ is also a consequence of these identities. But then the same is true for the identity $l!\, f = \sum_{\sigma \in S_l} \sigma(f) = 0$. The theorem is proved.

It is clear from the argument just given that if the assertion of Lemma 36.1 is proved for all $l \leq l_0$, then all identities of degree $\leq l_0$ for the algebra $\mathfrak{g}$ are consequences of identities (35.14)–(35.16).

THE PROOF OF LEMMA 36.1 will be given by induction on the degree l of a polynomial f. First we note that the identity

$$[z, x](\operatorname{ad} y)^{2k+1} = [z(\operatorname{ad} y)^{2k+1}, x] + [z, x(\operatorname{ad} y)^{2k+1}] \tag{36.3}$$

is a consequence of identity (35.15), and in any Lie algebra we have the equality

$$k!\,[a, v_1, \dots, v_k] = \sum_{\sigma \in S_k} [a, v_{\sigma(1)}, \dots, v_{\sigma(k)}] + \sum_{w_i} [a, w_1, \dots, w_{k-1}], \tag{36.4}$$

where the first sum is extended over all permutations of the set $\{1, \dots, k\}$ and w_i are commutators in $v_1, \dots, v_k$. We shall assume that f is a commutator, $i = l$, $j = l-1$, and for convenience we write $x_l = z$, $x_{l-1} = x$, and $\tau_{l,l-1} = \tau$.

The base of induction is $l = 5$. For $f = [[z, x], \dots]$, $[z, x_1, x, \dots]$ the assertion of the lemma is obvious. Let $f = [z, x_1, x_2, x, x_3]$; then we have

$$f - \tau(f) = [x_1 * \langle z, x, x_2 \rangle, x_3] - \{[x_1, z, x_2, x_3] + [x_1, x_2, z, x_3]\}|_{z=[z,x]},$$

where the second summand has the required form, and the first one can be transformed to the required form by using (35.14)–(35.16). Equality (36.4) implies that it remains to consider the following two cases.

a) $f = \sum_{\sigma \in S_3} [z, x_{\sigma(1)}, x_{\sigma(2)}, x_{\sigma(3)}, x]$. In this case identity (36.1) for $\beta = 0$ is obtained by the total linearization of (36.3) for $k = 1$.

b) $f = [[x_i, x_j], z, x_t, x]$. Then $f - \tau(f) = -[x_i, x_j] * \langle z, x, x_t \rangle + \{[[x_i, x_j], z, x_t] + [x_i, x_j, x_t, z]\}|_{z=[z,x]}$. The latter term has the required form, and the first one can be transformed to the required form by using (35.16). The base of induction is proved.

Assume that the lemma is proved for the cases in which the degree of f is less than l. Let the degree of f be equal to l.

First we consider the case when l is even. If $f = [z, x_1, \dots, x_t, x, x_{t+1}, \dots, x_{l-2}]$, $t < l-2$, then from the induction assumption it follows that for the commutator $f' = [z, x_1, \dots, x_t, x, x_{t+1}, \dots, x_{l-3}]$ of odd weight there exists a Lie polynomial g' satisfying identity (36.1) for the polynomial f' with $\beta = 0$. Then $g = [g', x_{l-2}]$ satisfies (36.1) for $\beta = 0$. Equality (36.4) implies that it remains to consider the following two cases.

a) $f = \sum_{\sigma \in S_{l-2}} [z, x_{\sigma(1)}, \dots, x_{\sigma(l-2)}, x]$. In this case (36.1) is valid for $\beta = 1$ and $g = 0$.

b) $f' = [z, x_1, \dots, [x_i, x_{i+1}], \dots, x_{l-2}, x]$. Then for the commutator of odd degree $f' = [z, x_1, \dots, x_{i-1}, x_i, x_{i+2}, \dots, x_{l-2}, x]$ identity (36.1) is satisfied with $\beta = 0$ for some multilinear Lie polynomial $g'(z, x_1, \dots, \widehat{x}_{i+1}, \dots, x_{l-2})$. This implies that the polynomial f satisfies identity (36.1) with $\beta = 0$ and $g = g'|_{x_i = [x_i, x_{i+1}]}$.

REMARK 36.1. The assertion of the lemma has already been proved for $l = 6$. As was mentioned above, this implies that all identities of degree 6 that hold in $\mathfrak{g}$ follow from (35.14)–(35.16). Therefore, in the proof of the remaining part of the lemma we can use the following two identities for the algebra $\mathfrak{g}$, which are implied by the identities (35.1) and (35.3) of the algebra with three supports $\{K, \mathfrak{g}, U(\mathfrak{g})\}$:

$$\begin{aligned}[z,y,y,y,x,v] &= [[z,y,x],v](y,y)\\ &= [z,y,x,y,y,v]+[y,v,[z,y,x],y],\end{aligned} \tag{36.5}$$

$$\begin{aligned}[z,v,x,y,y,y] &= [[z,v,x](y,y),y]\\ &= [z,y,y,v,x,y]+[y,v,z,y,x,y],\end{aligned} \tag{36.6}$$

Consider the case in which l is odd. In this case in (36.1) β must be equal to zero. If $f=[z,\dots,x,\dots,x_{l-3},x_{l-2}]$, then identity (36.1) with $\beta=0$ follows easily by the induction assumption.

If $f=[z,x_1,\dots,x_{l-3},x,x_{l-2}]$, then, by the induction assumption,

$$f-\tau(f)=\beta'\Big\{(1-\tau)\sum_{\sigma\in S_{l-3}}[z,x_{\sigma(1)},\dots,x_{\sigma(l-3)},x,x_{l-2}]\Big\}+[g'|_{z=[z,x]},x_{l-2}]$$

The second term has the required form, and, since l is odd and $l>6$, we see that, applying the linearization of the following Lie identity, which is implied by (36.5):

$$[z(\operatorname{ad}y)^{2k},x,x_{l-2}]=[z(\operatorname{ad}y)^{2(k-1)},x,y,y,x_{l-2}]+[y,x_{l-2},[z(\operatorname{ad}y)^{2(k-1)},x],y],$$

and using the induction assumption, we see that the first term can also be transformed to the required form.

Equality (36.4) implies that it remains to consider the following two cases.

a) $f=\sum_{\sigma\in S_{l-2}}[z,x_{\sigma(1)},\dots,x_{\sigma(2k+1)},x]$, $l-2=2\cdot k+1$. In this case identity (36.1) for $\beta=0$ is obtained by the total linearization of identity (36.3), which follows from (35.15).

b) $f'=[z,x_1,\dots,,x_{l-3},x]\,|_{x_i=[x_i,x_{l-2}]}$. Then by the induction assumption

$$f-\tau(f)=\beta'\Big\{(1-\tau)\sum_{\sigma\in S_{l-3}}[z,x_{\sigma(1)},\dots,x_{\sigma(l-3)},x]\Big\}\Big|_{x_i=[x_i,x_{l-2}]}+g'|_{z=[z,x];x_i=[x_i,x_{l-2}]}.$$

The latter summand has the required form. The first term can be rewritten in the form $\beta'\sum[[x_i,x_{l-2}],\dots]$. Denote this element by a. It is clear that a is skew-symmetric with respect to z and x and symmetric with respect to $x_1,\dots,\widehat{x_i},\dots,x_{l-3}$. Applying (36.6), for $t\geq 3$ we have

$$[[u,v,z](\operatorname{ad}y)^t,x]=[[u,y,y,v,z](\operatorname{ad}y)^{t-2},x]+[[y,v,u,y,z](\operatorname{ad}y)^{t-2},x].$$

Continuing the application of (36.6) to each term on the right-hand side of the equality, if this is possible, we finally obtain the following Lie identity, which follows from (35.14)–(35.16):

$$[[u,v,z](\operatorname{ad}y)^t,x]=\sum_i\pm[w_i',z,y,y,x]+\sum_i\pm[u_i',v_i',z,y,x], \tag{36.7}$$

where w_i', u_i', v_i' are commutators in u, v, y. The total linearization of (36.7) with respect to the variable y shows that a is representable in the form

$$(1-\tau)\Big\{\sum_i \pm[w_i, z, x_t, x_r, x] + \sum_i \pm[u_i, v_i, z, x_t, x] + [b, [z, x]]\Big\},$$

where w_i, u_i, v_i are commutators and b is a Lie polynomial. However, for commutators of weight 5 identity (36.1) holds with $\beta = 0$. Therefore, (36.1) holds also for the element a with $\beta = 0$, and, hence, also for f. The lemms is proved.

REMARK 36.2. From the papers [Klya, Dre3] it follows that the space of multilinear Lie polynomials of degree 5 in $x_1, \dots, x_5$ in the free Lie algebra $\mathfrak{M}$ is decomposed with respect to the action of the symmetric group S_5 into the direct sum of 5 irreducible S_5-submodules corresponding to the Young tables of types $(4, 1)$, $(3, 2)$, $(3, 1, 1)$, $(2, 2, 1)$, and $(2, 1, 1, 1)$, and to multilinear Lie identities of degree 5 of the algebra $\mathfrak{g}$ there correspond tables of types $(2, 1, 1, 1)$ and $(3, 1, 1)$. It is clear that (35.15) corresponds to the symmetry type $(3, 1, 1)$, (35.14) implies (35.15) for $z = y$ and $x_1 = y$, and the identity with the symmetry type $(2, 1, 1, 1)$ also follows from (35.14). Therefore, all identities of degree 5 for the algebra $\mathfrak{g}$ follow from (35.14).

§37. Finite basis property for the identities of subvarieties of pairs in var$(U(\mathfrak{g}), \mathfrak{g})$ over fields of characteristic zero

In this section the ground field K is of characteristic zero and $\mathfrak{V}_{\mathfrak{g}}$ is the ideal of weak identities of the pair $(U(\mathfrak{g}), \mathfrak{g})$.

37.1. A minimality condition for subvarieties of var$(U(\mathfrak{g}), \mathfrak{g})$.

THEOREM 37.1. *Any ascending chain of ideals of weak identities containing the ideal $\mathfrak{V}_{\mathfrak{g}}$ stabilizes at a finite step.*

PROOF. Suppose that the free pair $(F, \mathfrak{L})$ of the variety var$(U(\mathfrak{g}), \mathfrak{g})$ and the free algebra with three supports $\{\mathcal{F}_0, \mathfrak{L}_1, F_1\}$ of the variety var$(K, \mathfrak{g}, U(\mathfrak{g}))$ are constucted in the same way as in §35, and

$$\mathfrak{V}_1 \subseteq \mathfrak{V}_2 \subseteq \mathfrak{V}_3 \subseteq \cdots \tag{37.1}$$

is an ascending chain of ideals of weak identities in the algebra F. Let $\mathfrak{V}$ be the ideal of weak identities in F determined by the identity of commutativity. It is clear that to prove that the chain of ideals (37.1) is stabilized it suffices to prove the stabilization of the chain of ideals $\mathfrak{V}_i \cap \mathfrak{V}$, since the quotient algebra $F/\mathfrak{V}$ is commutative and any ideal of weak identities in $F/\mathfrak{V}$ is generated over a field of characteristic zero by a single polynomial x^m for some $m \in \mathbb{N}$. Therefore, we suppose in the sequel that $\mathfrak{V}_i \subseteq \mathfrak{V}$ $(i = 1, 2, \dots)$.

LEMMA 37.1. *For any ideal of weak identities $\mathfrak{V}'$ contained in $\mathfrak{V}$ the linear subspace $\mathfrak{V}' + \mathcal{F}_0\mathfrak{V}'$ of the algebra F_1 coincides with $\mathfrak{V}'$, i.e., any ideal of weak identities contained in $\mathfrak{V}$ is an $\mathcal{F}_0$-module.*

PROOF. Since the characteristic of the ground field K is equal to zero, it suffices to show that for any multilinear polynomial $g(x_1, \dots, x_l) \in \mathfrak{V}' \subset \mathfrak{V}$ and for the element $f = (x_{l+1}, x_{l+2})$ we have $fg \in \mathfrak{V}'$. By Lemma 35.2, for any transposition τ_{ij} $(i, j \in \{1, 2, \ldots, l\})$ and any permutation $\sigma \in S_l$ we have the relation

$$f\{(1 - \tau_{ij})\sigma(g)\} = r_{\sigma,i,j}, \tag{37.2}$$

where $r_{\sigma,i,j} \in \mathfrak{V}'$. Since the two-sided ideal in $K[S_l]$ generated by the elements $1 - \tau_{ij}$ $(i, j \in \{1, \ldots, l\})$ is spanned as a linear space by elements $\sigma - \delta$, where $\sigma, \delta \in S_l$, we see that (37.2) implies that for any permutation $\sigma \in S_l$ we have $f\{g - \sigma(g)\} = r_\sigma$, where $r_\sigma \in \mathfrak{V}'$. Taking the sum of these equalities over all $\sigma \in S_l$, we obtain

$$f\left\{l!\,g - \sum_{\sigma \in S_l} \sigma(g)\right\} = \sum_{\sigma \in S_l} r_\sigma \in \mathfrak{V}'.$$

However, for any multilinear polynomial $g \in \mathfrak{V}$ we have the equality $\sum_{\sigma \in S_l} \sigma(g) = 0$. This proves the lemma.

PROPOSITION 37.1. *If the ground field is of characteristic zero, then, modulo the Capelli identities of order* $m + 1$ (*see* §12), *any set of identities is equivalent to some set of identities in* m *variables.*

PROOF. Denote by $\mathfrak{K}_{m+1}$ the ideal of weak Capelli identities of order $m + 1$ in the absolutely free (associative) algebra H. Since the ground field is of characteristic zero, any set of identities is equivalent to some set of multilinear identities. Therefore, to prove Proposition 37.1 it suffices to indicate, for any multilinear polynomial $f \in H$, a set of polynomials in m variables that generate $\mod \mathfrak{K}_{m+1}$ the same ideal of weak identities as the polynomial f. Let us endow the subspace of all multilinear polynomials H_λ $(\lambda = \{x_1, \ldots, x_l\})$ with the structure of S_l-module with respect to the natural action of the symmetric group S_l on lower indices $1, 2, \ldots, l$. It is clear that in this case for any multilinear polynomial $g \in H_\lambda$ and for any element $a = \sum_{\sigma \in S_l} \beta_\sigma \sigma$ of the group algebra $K[S_l]$ the identity

$$a(g) \rightleftharpoons \sum_{\sigma \in S_l} \beta_\sigma g(x_{\sigma(1)}, \ldots, x_{\sigma(l)}) = 0$$

is a consequence of the identity $g = 0$. Let $\mathbf{D}$ be a Young table of type $(n_1, \ldots, n_t)$ and $l = n_1 + \cdots + n_t$, and let $\mathbf{D}'$ be a Young diagram obtained by filling the table $\mathbf{D}$ by numbers $1, 2, \ldots, l$. Denote by $R(\mathbf{D}')$ the subgroup of S_l whose elements map the numbers of any row of the diagram $\mathbf{D}'$ into the numbers of the same row and denote by $C(\mathbf{D}')$ the subgroup of S_l whose elements produce the same with the numbers standing in the columns. We put $\mathcal{P}(\mathbf{D}') = \sum_{\sigma \in R(\mathbf{D}')} \sigma$ and $\mathcal{Q}(\mathbf{D}') = \sum_{\sigma \in C(\mathbf{D}')} \operatorname{sign}(\sigma)\sigma$, where the sums are extended over all elements of the groups $R(\mathbf{D}')$ and $C(\mathbf{D}')$. From the structure theory of the group algebra $K[S_l]$ (see Proposition 26.1) it follows that for some rationals $\beta(\mathbf{D}')$ depending only on the Young table $\mathbf{D}$ we have the equality

$$\mathbf{1} = \sum_{\mathbf{D}'} \beta(\mathbf{D}')\mathcal{P}(\mathbf{D}')\mathcal{Q}(\mathbf{D}'),$$

where the sum is taken over all Young diagrams $\mathbf{D}'$ and Young tables $\mathbf{D}$ with l cells. This formula implies that any multilinear identity $f = 0$ is equivalent to the set of identities $\{f_{\mathbf{D}'} = 0\}$, where $f_{\mathbf{D}'} = \mathcal{P}(\mathbf{D}')\mathcal{Q}(\mathbf{D}')f$ and $\mathbf{D}'$ ranges over all diagrams corresponding to all Young tables $\mathbf{D}$ with l cells. If the first column of the table $\mathbf{D}$ contains t cells and $t \geq m+1$, then for any polynomial $g \in H_\lambda$ the polynomial $\mathcal{Q}(\mathbf{D}')g$ is skew-symmetric with respect to the variables $x_{i_1}, \dots, x_{i_t}$, where $i_1, \dots, i_t$ are the numbers standing in the first column of the diagram $\mathbf{D}'$, i.e., $\mathcal{Q}(\mathbf{D}') \in \mathfrak{K}_{m+1}$. Thus, any multilinear polynomial $f \in H_\lambda$ generates, $\mod \mathfrak{K}_{m+1}$, the same ideal of weak identities as the set $\{f_{\mathbf{D}'}\}$ does, where $\mathbf{D}'$ ranges over all Young diagrams with the number of rows $\leq m$. Since $\sigma\mathcal{P}(\mathbf{D}') = \mathcal{P}(\mathbf{D}')$ for any permutation $\sigma \in R(\mathbf{D}')$, the polynomial $f_{\mathbf{D}'}$ is symmetric with respect to the variables $x_{j_1}, \dots, x_{j_{n_i}}$, where $j_1, \dots, j_{n_i}$ are the elements of ith row of the diagram $\mathbf{D}'$. After the substitution of the variable y_i instead of any of variables $x_{j_1}, \dots, x_{j_{n_i}}$ in $f_{\mathbf{D}'}$ we obtain a polynomial depending on variables whose number does not exceed m. The total linearization of this polynomial is equal to $n_1! \cdots n_t!\, f_{\mathbf{D}'}$, because $f_{\mathbf{D}'}$ is symmetric with respect to the variables $x_{j_1}, \dots, x_{j_{n_i}}$. Therefore, any identity $f_{\mathbf{D}'} = 0$, where $\mathbf{D}'$ is a Young diagram whose number of rows does not exceed k, is equivalent to some identity in k variables, and this proves Proposition 37.1.

Since in the pair $(U(\mathfrak{g}), \mathfrak{g})$ the Capelli identities of order 4 are satisfied, we see that, by Proposition 37.1, in order to prove the theorem it suffices to establish the stabilization of the chain of ideals (37.1) for the case in which the pair $(F, \mathfrak{L})$ and the triple $\{\mathcal{F}_0, \mathfrak{L}_1, F_1\}$ have a finite number of free generators $x_1, \dots, x_k$. We consider $\mathfrak{V}$ in this case as a left module over a subalgebra with unit R in $\operatorname{End}_K \mathfrak{V}$ generated by multiplications by elements of $\mathcal{F}_0$ and left multiplications of the algebra F.

LEMMA 37.2. *If the pair $(F, \mathfrak{L})$ possesses a finite number of free generators, then the ideal $\mathfrak{V}$ is a finitely generated R-module.*

PROOF. Any element of the ideal $\mathfrak{V}$ is representable in the form of a K-linear combination of elements of the form av, where v is a left-normed commutator of degree not less than two and a is a word in $x_1, \dots, x_k$. Hence, by formula (35.1), the R-module $\mathfrak{V}$ is generated by left-normed commutators of degrees 2 and 3 in $x_1, \dots, x_k$. The lemma is proved.

LEMMA 37.3. *R is a left (right) Noetherian K-algebra.*

PROOF. The K-algebra R is a $\mathcal{F}_0$-algebra. From identity (35.1) it follows that $\mathcal{F}_0$ is generated by elements (u, v), where u and v are commutators whose degree does not exceed two, i.e., $\mathcal{F}_0$ is a finitely generated commutative K-algebra. Denote the Lie subalgebra in $R^{(-)}$ generated by left multiplications by elements of $\mathfrak{L}$ by the same letter $\mathfrak{L}$. Then $\widetilde{\mathfrak{L}} = \mathfrak{L} + \mathcal{F}_0\mathfrak{L}$ is a finitely generated $\mathcal{F}_0$-submodule in R generated by commutators of degree ≤ 2 by identities (35.1) and (35.11). The assertion of the lemma now follows from the fact that if a Lie Φ-algebra is finitely generated over a Noetherian commutative algebra Φ, then any enveloping Φ-algebra of it is left (right) Noetherian (see Theorem 54.2). The lemma is proved.

Lemmas 35.1–35.3 imply that the R-module $\mathfrak{V}$ is Noetherian, and the ascending chain of ideals of weak identities (37.1) is an ascending chain of R-submodules in $\mathfrak{V}$. Hence, starting from some $N \in \mathbb{N}$ we have $\mathfrak{V}_{N+j} = \mathfrak{V}_N$ for any $j > 0$. The theorem is proved.

37.2. Finite basis property for the identities of the pair $(U(\mathfrak{g}), \mathfrak{g})$.

THEOREM 37.2. *The ideal of weak identities $\mathfrak{V}_{\mathfrak{g}}$ of the pair $(U(\mathfrak{g}), \mathfrak{g})$ possesses a finite set of generators.*

The proof of this theorem will be based on the fact that we can explicitly indicate a finite number of identities that are valid in the pair $(U(\mathfrak{g}), \mathfrak{g})$, which provides the decisive step of the arguments expounded in the previous subsection. In essence, we shall have to repeat once more the proof of Theorem 37.1.

Denote by $(\widetilde{F}, \mathfrak{L})$ the relatively free pair with the free generators $x_1, x_2, \dots$ determined by the weak identities (35.11) and (35.14)–(35.16). From Lemma 35.3 and Theorem 36.1 it follows that the Lie algebra $\mathfrak{L}$ is the free Lie algebra of the variety $\operatorname{var}\mathfrak{g}$.

REMARK 37.1. Let $\{\mathcal{F}_0, \mathfrak{L}_1, F_1\}$ have the same meaning as in the previous subsection. As we know, the commutator $[\mathfrak{L}, \mathfrak{L}]$ of the algebra $\mathfrak{L}$ is a Lie $\mathcal{F}_0$-algebra; this follows from the embedding of $\mathfrak{L}$ in $\mathfrak{L}_1$ and identity (35.3). Moreover, this action of the algebra $\mathcal{F}_0$ commutes with the restriction to $[\mathfrak{L}, \mathfrak{L}]$ of any operator $\operatorname{ad} v$, $v \in \mathfrak{L}$. In particular, this means that for any $w = \sum_i [v_i, u_i] \in [\mathfrak{L}, \mathfrak{L}]$ and $f = (v, u)$ the element $\sum_i [v_i, u_i] * f$ is equal to fw (see (35.5) and (35.3)) and does not depend on the chosen notation of the element w.

The proof of the theorem is decomposed into several steps.

LEMMA 37.4. *Let $u, v \in [\mathfrak{L}, \mathfrak{L}]$, $y \in \mathfrak{L}$, $a \in \widetilde{F}$, and $f = (y, y) \in \mathcal{F}_0$; then in $(\widetilde{F}, \mathfrak{L})$ we have the weak identity*

$$(u * f)av = ua(v * f). \tag{37.3}$$

PROOF. It suffices to consider the case $u = [x_1, x_2]$, $v = [z_1, z_2]$, a some word in free generators. If a is a nonempty word, then $a = a'x_i$, where a' is a word of lesser length. Then

$$(u * f)a'x_i v = (u * f)a'[x_i, v] + (u * f)a'vx_i.$$

Using inductive agruments and Remark 37.1 we see that the right-hand side of this equality is equal to

$$ua'[x_i, v * f] + ua'(v * f)x_i = ua'x_i(v * f).$$

Therefore, it is sufficient to prove (37.3) for the case in which a is the empty word. Then

$$([x_1, x_2] * f)[z_1, z_2] = \{[x_1, y, y, x_2] + [y, x_2, x_1, y]\}[z_1, z_2] = [[x_1, y, y][z_1, z_2], x_2] \\ - [x_1, y, y][z_1, z_2, x_2] + [[y, x_2, x_1][z_1, z_2], y] - [y, x_2, x_1][z_1, z_2, y].$$

From equality (35.11) and Remark 37.1 it follows that the right-hand side of this relation is equal to

$$[x_1, x_2]([z_1, z_2] * (y, y)) - [y, x_2]([z_1, z_2] * (x_1, y)) \\ + [y, y]([z_1, z_2] * (x_2, x_1)) - [x_2, y]([z_1, z_2] * (y, x_1)).$$

The third term in this expression is zero, the second and the fourth ones have opposite signs. The lemma is proved.

LEMMA 37.5. *Let a multilinear polynomial $g(x_1, \dots, x_l) \in \widetilde{F}$ have the form $g = \sum_k a_k v_k$, where $v_k \in [\mathfrak{L}, \mathfrak{L}]$, $a_k \in \widetilde{F}$, and g is skew-symmetric with respect to two variables x_i and x_j. Then in the pair $(\widetilde{F}, \mathfrak{L})$ we have the identity*

$$\tfrac{1}{2}\{g|_{x_i=[x_i,y,y]} + g|_{x_j=[x_j,y,y]} + g|_{x_i=[x_i,x_j,y];x_j=y}\} = \sum_k a_k(v_k * (y,y)). \tag{37.4}$$

PROOF. It suffices to consider the case of $g = (1-\tau_{ij})av$, where v is a left-normed commutator of degree ≥ 2, a is a word in $x_1, x_2, \dots$, and $\tau_{ij} \in S_l$ is a transposition. If the letters x_i and x_j occur in the commutator v, then equality (37.4) follows in this case from Lemma 35.2 and Remark 37.1. If the letters x_i and x_j occur in the word a, then $(1-\tau_{ij})a$ can be represented in the form $\sum_k a'_k v'_k$, where $v'_k \in [\mathfrak{L}, \mathfrak{L}]$. Since the degree of a is less than l, we see from the inductive argument that in this case the left-hand side of (37.4) is equal to $\sum_k a'_k(v'_k * (y,y))v$, and, by Lemma 37.4, this expression equals

$$\sum_k a'_k v'_k (v * (y,y)) = (1-\tau_{ij})\{a(v * (y,y))\}.$$

It remains to consider the case in which $g = x_i a[x_j, \dots] - x_j a[x_i, \dots]$. As in the proof of Lemma 37.4, we can assume that a is the empty word, using Remark 37.1 and inductive arguments. In this case the proof of (37.4) is carried out in the same way as the proof of identity (35.4), by using identity (35.11) instead of identity (35.1) and applying Remark 37.1 in transforming of commutators $[x_i, y, y]$, $[x_j, y, y]$, and $[x_i, x_j, y]$. The lemma is proved.

LEMMA 37.6. *Let g be an arbitrary polynomial belonging to the ideal of weak identities $\mathfrak{V}$ determined in the algebra $\widetilde{F}$ by the commutativity identity and let $\mathfrak{V}(g)$ be the least ideal of weak identities in $\widetilde{F}$ containing g. Suppose that g is represented in the form $\sum_k a_k v_k$, where $v_k \in [\mathfrak{L}, \mathfrak{L}]$. Then for any $f \in \mathcal{F}_0$ the polynomial $\sum_k a_k(v_k * f)$ belongs to $\mathfrak{V}(g)$.*

In fact, the proof repeats the proof of Lemma 37.1. Since the characteristic of the ground field K equals zero, it suffices to prove the assertion of the lemma for any multilinear polynomial $g(x_1, \dots, x_l) \in \mathfrak{V}$, for any element $f(x_{l+1}, x_{l+2})$ and for arbitrary representation of the polynomial g in the algebra $\widetilde{F}$ in the form $\sum_k a_k v_k$. By Lemma 37.5, for any transposition τ_{ij} and any permutation $\sigma \in S_l$ we have the following equality in $\widetilde{F}$:

$$(1-\tau_{ij})\sigma\left(\sum_k a_k(v_k * f)\right) = r_{\sigma.i.j},$$

where $r_{\sigma.i.j} \in \mathfrak{V}(g)$. Then for any permutation $\sigma \in S_l$ we have

$$(1-\sigma)\left(\sum_k a_k(v_k * f)\right) = r_\sigma,$$

where $r_\sigma \in \mathfrak{V}(g)$. Summing these equalities for all $\sigma \in S_l$, we obtain

$$l! \sum_k a_k(v_k * f) - \sum_{\sigma \in S_l} \sigma(a_k(v_k * f)) = \sum_{\sigma \in S_l} r_\sigma.$$

The second term on the left-hand side of this formula equals zero in $\widetilde{F}$, because the Lie polynomial $\sum_{\sigma \in S_m}([x_{\sigma(1)}, \dots, x_{\sigma(m)}] * f)$, where the sum is extended over all $\sigma \in S_m$, is equal to zero in the Lie algebra $\mathfrak{L}$ (see Remark 37.1). The lemma is proved.

LEMMA 37.7. *The ideal of weak identities $\mathfrak{V}$ in $\widetilde{F}$ determined by the commutativity identity is an $\mathcal{F}_0$-module with respect to the action*

$$f\left(\sum_k a_k v_k\right) \rightleftharpoons \sum_k a_k(v_k * f), \tag{37.5}$$

where $a_k \in \widetilde{F}$, $v_k \in [\mathfrak{L}, \mathfrak{L}]$, and $f \in \mathcal{F}_0$. Moreover, any ideal of weak identities contained in $\mathfrak{V}$ is an $\mathcal{F}_0$-submodule.

PROOF. We need to verify that the action of the algebra $\mathcal{F}_0$ on $\mathfrak{V}$ determined by formula (37.5) does not depend on the choice of representation of the element $g \in \mathfrak{V}$ in the form $\sum_k a_k v_k$. But this follows from Lemma 37.6 and Remark 37.1. The lemma is proved.

COROLLARY 1. *Any set in k variables $x_1, \dots, x_k$ determines in the algebra $\widetilde{F}$ a finitely generated ideal of weak identities.*

The proof repeats the argument of Lemmas 37.2 and 37.3 *verbatim*.

COROLLARY 2. *If the weak identities* (35.11) *and* (35.14)–(35.16) *and all Capelli identities of some order are satisfied in a variety of pairs, then the identities for this variety possess a finite basis.*

Since modulo the ideal of weak Capelli identities $\mathfrak{K}_{m+1}$ any identity is equivalent to identities in m variables (see Proposition 37.1), the proof of Corollary 2 follows from Corollary 1 and Proposition 37.2.

PROPOSITION 37.2. *If the weak identities* (35.11) *and* (35.14)–(35.16) *are satisfied in a variety of pairs, then the Capelli identities of order $m+1$ possess a finite basis.*

PROOF. Since the pair $(F, \mathfrak{L})$ is free in the variety under consideration, it suffices to prove that the ideal of weak Capelli identities $\mathfrak{K}_{m+1}$ is generated by a finite number of Capelli polynomials. It is clear that the ideal $\mathfrak{K}_{m+1}$, as an ideal of weak identities, is generated by arbitrary polynomials of the form

$$\sum_{\sigma \in S_{m+1}} \operatorname{sign}(\sigma) \left[x_{\sigma(1)}, y_{11}, \dots, y_{1l_1}\right] \cdots \left[x_{\sigma(m+1)}, y_{m+1,1}, \dots, y_{m+1,l_{m+1}}\right],$$

where $l_1, \dots, l_{m+1}$ are arbitrary positive integers. Equality (36.4) and inductive arguments show that if we replace each of these polynomials by the polynomial

$$\sum_{\sigma \in S_{m+1}} \operatorname{sign}(\sigma) \left\{(\operatorname{ad} y_1)^{l_1} x_{\sigma(1)}\right\} \cdots \left\{(\operatorname{ad} y_{m+1})^{l_{m+1}} x_{\sigma(m+1)}\right\}, \tag{37.6}$$

then we obtain another system of generators of the ideal $\mathfrak{K}_{m+1}$ as an ideal of weak identities. But this system of generators depends on $2m+2$ variables and, by Corollary 1, $\mathfrak{K}_{m+1}$ has a finite number of generators. (Identity (35.1), together with Lemmas 37.6 and 37.7, shows that we can take for these generators all the polynomials of the form (37.6) for which $0 \leq l_1, \dots, l_{m+1} \leq 2$.) The proposition is proved.

In connection with this proposition we note that the finite basis property of the Capelli identities of order $m+1$ holds without any additional assumptions. This fact was established in the papers [Raz11, Sto].

PROOF OF THEOREM 37.2. From Lemmas 35.3–35.5 it follows that in the pair $(U(\mathfrak{g}), \mathfrak{g})$ all Capelli identities of order 4 and identities (35.11) and (35.14)–(35.16) are satisfied. Therefore, the assertion of the theorem follows from Corollary 2 of Lemma 37.7. We note also that from the results of the next section (see the proof of Lemma 38.3 and of the main lemma) and from Lemmas 37.6 and 37.7 it follows rather easily that all weak Capelli identities of order 4 are consequences of identities (35.11)–(35.16), which are satisfied in the pair $(U(\mathfrak{g}), \mathfrak{g})$ (see Lemma 35.3). This permits us to prove the finite basis property for identities of any associative-Lie pair for which the identities (35.11)–(35.16) are satisfied, without using the finite basis properety of the ideal of weak Capelli identities $\mathfrak{K}_{m+1}$.

§38. Bases of identities for irreducible representations of the Lie algebra $\mathfrak{g}$

Corollary 2 of Theorem 37.1 implies the finite basis property of any irreducible representation of the Lie algebra $\mathfrak{g}$ over a field of characteristic zero. In this section we shall find an explicit form of this basis for the case in which the ground field K is algebraically closed.

THEOREM 38.1. *Let $\rho\colon \mathfrak{g} \to \mathrm{End}_K V$ be an irreducible representation of the simple three-dimensional Lie algebra $\mathfrak{g}$ over an algebraically closed field K of characteristic zero, let $(U, \mathfrak{g}) \rightleftharpoons (\rho(U(\mathfrak{g})), \rho(\mathfrak{g}))$, and let $\delta \in K$ be the value of the Casimir element γ (see (34.5) and (34.6)) in this representation. Then:*

1) *if V is infinite-dimensional and $\delta \neq 0$, then the identities*

$$(38.1) \qquad \langle x_1, x_2, x_3\rangle y = y\langle x_1, x_2, x_3\rangle,$$

$$(38.2) \qquad \delta z * \langle x_1, x_2, x_3\rangle = 2z\langle x_1, x_2, x_3\rangle$$

form a basis of identities for the pair $(U, \mathfrak{g})$;

2) *if V is infinite-dimensional and $\delta = 0$, then the identities*

$$(38.3) \qquad [z * \langle x_1, x_2, x_3\rangle]y = z[y * \langle x_1, x_2, x_3\rangle],$$

$$(38.4) \qquad \langle x_1, x_2, x_3\rangle = 0;$$

form a basis of identities for the pair $(U, \mathfrak{g})$;

3) *if V is finite-dimensional and $\dim_K V = n$, then $\delta = (n^2-1)/4$ and a basis of identities of the pair $(U, \mathfrak{g})$ consists of identities (38.1) and (38.2) plus one more identity $\mathrm{ART}_n(x, y) = 0$ of degree n in 2 variables.*

The fact that identities (38.1)–(38.4) hold in the representation ρ follows from the results of §35 (see identities (35.13), (35.12), and (35.9)). Let us clarify the meaning of the identity $\mathrm{ART}_n(x, y) = 0$.

38.1. Identities of the pair $(U, \mathfrak{g})$ in two variables. Denote by the letter P the set of associative polynomials in two variables x and y, each of which is linear with respect to the variable x.

PROPOSITION 38.1. *An identity of the form*

$$f(x, y) = 0, \tag{38.5}$$

where $f \in P$ is a nonzero polynomial, is satisfied in an irreducible representation V of the simple three-dimensional Lie algebra $\mathfrak{g}$ over an algebraically closed field K (char $K = 0$) only if V is finite-dimensional. Moreover, if $\dim_K V = n$, then all identities of the form (38.5) satisfied in $(U, \mathfrak{g})$ are equivalent to the single multihomogeneous identity $\mathrm{ART}_n(x, y) = 0$ of degree $n + 1$.

PROOF. Two operators $l_y\colon P \to yP$ and $\operatorname{ad} y\colon P \to [y, P]$ act on the space P. Obviously, l_y and $\operatorname{ad} y$ commute. Denote by $\mathcal{T}$ the subalgebra with unit generated by these operators in $\operatorname{End}_K P$. Then any multihomogeneous polynomial $f \in P$ of degree $k + 1$ can be written in the form

$$f(x, y) = r(l_y, \operatorname{ad} y)x, \tag{38.6}$$

where

$$r(l_y, \operatorname{ad} y) = \sum_{i=1}^{k} \varepsilon_i l_y^{k-i} (\operatorname{ad} y)^i \in \mathcal{T} \qquad (\varepsilon_i \in K).$$

Since the commutative polynomial r is homogeneous and the field K is algebraically closed, we have

$$r(l_y, \operatorname{ad} y) = \beta_0 (\operatorname{ad} y)^{k-t} \prod_{i=1}^{t} (l_y - \beta_i \operatorname{ad} y) \qquad (t \in \mathbb{N},\ \beta_i \in K,\ \beta_0 \neq 0). \tag{38.7}$$

Let the identity $f = 0$ be valid in the $\mathfrak{g}$-module V.

For an arbitrary regular element $g \in \mathfrak{g}$ we construct the elements h, e_+, e_- satisfying relations (34.5). If $t = k$ in (38.7), then $0 = f(h, h)V = \beta_0 h^{k+1} V$ and $g^{k+1} V = 0$ for any regular element g. Since the set of regular elements is dense in $\mathfrak{g}$, the equality $g^{k+1} V = 0$ holds for any $g \in \mathfrak{g}$. However, then $\mathfrak{g} V = 0$ and V is a trivial $\mathfrak{g}$-module. Putting in this case $\mathrm{ART}_0(x, y) = x$, we see that all identities of the $\mathfrak{g}$-module V are equivalent to the identity $\mathrm{ART}_0 = 0$. This shows that in subsequent discussions we may consider $t < k$ and suppose that the polynomial r is divisible by $\operatorname{ad} y$ in the algebra $\mathcal{T}$.

Substitute $x = e_+$ and $y = h$ in the identity $f = 0$. Then formulas (38.6) and (38.7) imply

$$\beta_0 \prod_{i=1}^{t} (h - \beta_i) e_+ V = 0 \qquad (\beta_0 \neq 0). \tag{38.8}$$

Since the subspace $V_1 = e_+ V$ is invariant under the action of the operator h and $\dim_K V > 1$, (38.8) implies:

a) all eigenvalues of the operator $h|_{V_1}$ belong to a finite set $\beta = \{\beta_1, \dots, \beta_t\}$;
b) the operator $h|_{V_1}$ has at least one nonzero eigenvector $v \in V_1$.

Let $hv = \beta_i v$. Then, since $he_+^j v = (\beta_i + j)e_+^j v$, property a) implies that, starting from some $N \in \mathbb{N}$, we have $0 = e_+^{N+1}v = e_+^{N+2}v = \cdots$, and in the $\mathfrak{g}$-module V there exists a highest vector $v_0 = e_+^N v$. Following a similar line of reasoning with the replacement of e_+ by e_-, we see that the $\mathfrak{g}$-module V possesses a lowest vector. Hence, by Proposition 34.1, the existence of a nontrivial identity of the form (38.5) implies the finite-dimensionality of the irreducible module V.

Let $\dim_K V = n > 1$. Then Corollary of Proposition 34.1 implies that a basis of the subspace $V_1 = e_+V$ can be formed by the vectors $v_j = e_-^j v_0$ ($j = 0, 1, \dots, n-2$) that are eigenvectors of the operator h, and $(n-1)/2 - j$ are their eigenvalues. Property a) of the operator $h|_{V_1}$ shows that all these eigenvalues belong to the set β. Therefore, if $f(x, y) = r(l_y, \operatorname{ad} y)x = 0$ is an identity of the module V, then

$$r(l_y, \operatorname{ad} y) = \operatorname{ad} y \cdot \prod_{i=1}^{n-1} \left(l_y - \left(j + \frac{1-n}{2} \right) \operatorname{ad} y \right) \cdot r_0(l_y, \operatorname{ad} y). \tag{38.9}$$

Conversely, if (38.9) holds, then a direct verification shows that

$$f(h, h) = f(e_+, h) = f(e_-, h) = 0$$

and, hence, $f(\mathfrak{g}, g) = 0$ for any regular element $g \in \mathfrak{g}$. Since the set of all regular elements is dense in $\mathfrak{g}$ in the Zariski topology, we have $f(\mathfrak{g}, g) = 0$ for any $g \in \mathfrak{g}$. We put by definition

$$\mathrm{ART}_n(x, y) = \operatorname{ad} y \prod_{i=1}^{n-1} \left(l_y - \left(j + \frac{1-n}{2} \right) \operatorname{ad} y \right) x \qquad (n = 1, 2, \dots).$$

It follows from what was just said above that the identity $\mathrm{ART}_n = 0$ holds in V ($\dim_K V = n$), and, by formula (38.9), any other identity of the form (38.5) of the module V is a consequence of this one. The proposition is proved.

38.2. The main lemma. Lemma 35.3 implies that in the pair $(U(\mathfrak{g}), \mathfrak{g})$ we have the identities (38.1), (38.3), and

$$[x, y, z][u, v] = x([u, v] * (y, z)) = y([u, v] * (x, z)) \tag{38.10}$$

(see (35.11)). Moreover, from equalities (35.9) for any $\delta \in K$ and from the invariance of the bilinear form $(\ ,\ )$ it follows that another identity is satisfied in $(U(\mathfrak{g}), \mathfrak{g})$:

$$\langle x, y, [u, v] \rangle = \langle [x, y], u, v \rangle. \tag{38.11}$$

Hence, these identities hold also in the irreducible representation ρ of the Lie algebra $\mathfrak{g}$.

Denote by $(F_\delta, \mathfrak{L})$ the free associative-Lie pair of the variety given by identities (38.1), (38.3), (38.10), (38.11), and either identity (38.2) for $\delta \neq 0$ or identity (38.4) for $\delta = 0$.

MAIN LEMMA. *Let $f(x_1,\dots,x_l)$ be a multilinear associative polynomial and let τ_{ij} and τ_{km} be two transpositions acting on the set of indices $\{1,\dots,l\}$ for which i,j,k,l are pairwise distinct. Then for an appropriate multilinear associative polynomial $g(x_1,\dots,\widehat{x}_j,\dots,x_l)$ of degree $l-1$ in the pair $(F_\delta,\mathfrak{L})$ we have the identity*

$$(1-\tau_{ij})(1-\tau_{km})f = g|_{x_i=[x_i,x_j]}.$$

Denote by the letter $\mathfrak{V}$ the ideal of weak identities in F_δ corresponding to the commutativity identity. Let $\mathcal{F}_0$ denote the commutative K-algebra from §37. Since in the pair $(F_\delta,\mathfrak{L})$ we have identities (38.10) and (38.3), and the Lie identities (35.14)–(35.16) are consequences of identity (38.3) (see Remark 36.2), by Lemma 37.7 the ideal of weak identities $\mathfrak{V}$ is a $\mathcal{F}_0$-module with respect to the action (37.5) and any ideal of weak identities in $\mathfrak{V}$ is an $\mathcal{F}_0$-submodule. This permits us to use the $\mathcal{F}_0$-module structure on $\mathfrak{V}$ in deducing some identities from others in $\mathrm{var}(F_\delta,\mathfrak{L})$. In particular, this implies that in the pair $\mathrm{var}(F_\delta,\mathfrak{L})$ the following identities hold:

$$2[u,v]*([x,y],z) = [u,v]*\langle x,y,z\rangle, \tag{38.12}$$

$$\delta p*([x,y],z) = p\langle x,y,z\rangle \qquad (p\in\mathfrak{V}) \tag{38.13}$$

Indeed, the first identity is a Lie identity (see (35.5)). From formulas (35.9) and (35.10) it follows that it is satisfied in the Lie algebra $\mathfrak{g}$. Therefore, by Theorem 36.1, equality (38.12) is a consequence of identities (35.14)–(35.16), which follow, as it has already been mentioned above, from identity (38.3), which holds for the pair $(F_\delta,\mathfrak{L})$. This proves that (38.12) is valid for the pair $(F_\delta,\mathfrak{L})$.

Any polynomial $r\in\mathfrak{V}$ is representable in the form $\sum_i a_i[u_i,v_i]$, where $a_i\in F_\delta$ and $u_i,v_i\in\mathfrak{L}$. Therefore, identity (38.13) follows from (38.12), (38.2), and the definition of the polynomial $r*([x,y],z)$ (see formulas (37.5) and (35.5)). This proves that identity (38.13) holds for the pair $(F_\delta,\mathfrak{L})$.

This kind of discussion will be typical also in the proof of the main lemma given below.

LEMMA 38.1. *Suppose an associative polynomial $g\in F_\delta$ is linear with respect to a variable x_i, does not depend on a variable x_j, and belongs to the ideal $\mathfrak{V}$. Then in the pair $(F_\delta,\mathfrak{L})$ we have the identity*

$$(1-\tau_{ij})(g*(x_j,x_k)) = g|_{x_i=[x_i,x_j,x_k]}. \tag{38.14}$$

PROOF. It suffices to consider the case $g = av$, where a is a monomial and v is a left-normed commutator of weight ≥ 2.

If the letter x_i appears in v, then equality (38.14) follows from identity (35.1), Lemma 37.7, and the definition of $*$ (see (37.5) and (35.5)). If $a = a'x_i$, then (38.14) follows from (38.10). If $a = a'x_ia''x_t$, then

$$(1-\tau_{ij})\{a(v*(x_j,x_k))\} = (1-\tau_{ij})\{a'x_ia''[x_t,v]*(x_j,x_k)+a'x_ia''(v*(x_j,x_k))x_t\},$$

and, since the length of the word a'' is lesser than the length of the word $a''x_t$, the assertion of the lemma follows from inductive arguments. The lemma is proved.

LEMMA 38.2. *In the absolutely free associative-Lie pair $(H, \mathfrak{M})$ over a field K of characteristic zero any element of H is representable in the form of a linear combination of elements of the form v^t, where $v \in \mathfrak{M}$, $t \in \mathbb{N}$ & $t > 0$.*

PROOF. Let $K[t_1, \dots, t_m]$ be the commutative polynomial algebra with unit. Consider an infinite matrix whose columns are indexed by various monomials $t^r \rightleftharpoons t_1^{r_1} \cdots t_m^{r_m}$ and rows—by all possible homomorphisms $\varkappa$ of the polynomial algebra into the field K, and consider that at the intersection of $\varkappa$th row and t^rth column there stands $\varkappa(t^p) \in K$. Since over an infinite field K any polynomial is uniquely determined by its values, we see that arbitrary columns of the matrix under consideration are linearly independent. Let $v_1, \dots, v_m$ be arbitrary elements of the Lie algebra $\mathfrak{M}$. Then in the algebra $K[t_1, \dots, t_m] \otimes_K H$ we have the equality

$$(t_1 v_1 + \cdots + t_m v_m)^m = \sum t^r v^{(r)} + a, \tag{38.15}$$

where the sum is taken over all collections $r = \langle r_1, \dots, r_m \rangle \in \mathbb{N}^m$ for which $r_1 + \cdots + r_m = m$, $v^{(r)} = m!\,(v_1^{r_1}/r_1!) \cdots (v_m^{r_m}/r_m!)$, and the element a is representable in the form

$$\sum f_{w_1, \dots, w_l} w_1 \cdots w_l, \tag{38.16}$$

where $l < m$, $f_{w_1, \dots, w_l} \in K[t_1, \dots, t_m]$, w_i are commutators in $v_1, \dots, v_m$. Applying to (38.15) arbitrary homomorphisms $\varkappa\colon K[t_1, \dots, t_m] \to K$, we obtain a system of linear equations

$$\sum \varkappa(t^r) v^{(r)} = (\varkappa(t_1) v_1 + \cdots + \varkappa(t_m) v_m)^m - \varkappa(a)$$

relative to $v^{(r)}$. It follows from what was said above that the column rank of the matrix of this system coincides with the number of undeterminates. Therefore, all $v^{(r)}$'s can be expressed as linear combinations of elements standing on the right-hand sides of these equalities. In particular, this applies to $v^{(1)}$. However, then (38.15) and (38.16), taking into account $\operatorname{char} K = 0$, imply that

$$v_1 \cdots v_m = \sum \beta(v) v^m + \sum \beta(v_1', \dots, v_l') v_1' \cdots v_l', \tag{38.17}$$

where $\beta(v), \beta(v_1', \dots, v_l') \in K$, $v, v_1' \cdots v_l' \in \mathfrak{M}$, $l < m$. This equality permits us to prove the assertion of the lemma by induction on the number m. Indeed, the base of induction $m = 1$ means that any Lie element is equal to itself, and the inductive step follows immediately from (38.17). The lemma is proved.

LEMMA 38.3. *Let $f(x_1, \dots, x_l)$ be an arbitrary multilinear polynomial and τ_{ij} $(i, j \in \{1, \dots, l\})$ an arbitrary transposition. Then in the algebra F_δ of the pair $(F_\delta, \mathfrak{L})$ the polynomial $(1 - \tau_{ij}) f$ can be expressed as a linear combination of elements of two types:*

$$a(\langle x_i, x_j, x_t \rangle * f_0) b, \tag{38.18}$$

$$a([x_i, x_j] * f_0) b, \tag{38.19}$$

where $a, b \in F_\delta$, $f_0 \in F_0$.

PROOF. From Lemma 38.2 it follows that it suffices to prove our assertion for the polynomial $f = x_i v^k x_j$, where $v \in \mathfrak{L}$. Then

$$f = \sum_{m=0}^{k} (-1)^m \binom{k}{m} v^{k-m} \{(\operatorname{ad} v)^m x_i\} x_j.$$

Identity (35.1) and Lemma 37.7 imply that for $m \geq 3$

$$(\operatorname{ad} v)^m x_i = \{(\operatorname{ad} v)^{m-2} x_i\} * (v, v),$$

and to complete the proof it suffices to consider two cases:

1. $f = x_i v x_j$, where v is a commutator in free generators of the pair $(F_\delta, \mathfrak{L})$. Then

$$(1 - \tau_{ij}) f = [x_i, x_j] v + v [x_i, x_j] + \langle x_i, v, x_j \rangle.$$

The first and the second terms have the form we need. If $v = x_q$, then the third summand has the form (38.18). If $v = [v_1, v_2]$, then, using the identity (38.11), we see that this term equals $-\langle [x_i, x_j], v_1, v_2 \rangle$ and it can be expressed via the elements (38.19).

2. $f = x_i v v x_j$. Then

$$(1 - \tau_{ij}) f = \tfrac{1}{2}(1 - \tau_{ij})[x_i, v, v, x_j] + (1 - \tau_{ij}) f',$$

where f' is a sum of monomials for which the distance between x_i and x_j is equal to zero or one. Therefore, $(1 - \tau_{ij}) f'$ is reducible to the form we need by the first case. Applying subsequently equality (35.1) and Lemma 38.1, we obtain the Lie identity

$$\begin{aligned} \tfrac{1}{2}(1 - \tau_{ij})[x_i, v, v, x_j] &= [x_i, x_j] * (v, v) - \tfrac{1}{2}(1 - \tau_{ij})\{[v, x_j] * (x_i, v)\} \\ &= [x_i, x_j] * (v, v) - \tfrac{1}{2}[v, [[x_j, x_i], v]]. \end{aligned}$$

Both summands on the right-hand side of this equality can be expressed via elements of the form (38.19).

As an addition to this lemma we note that from identity (35.1) it follows that in the algebra $\mathcal{F}_0$ we have the equality

$$(38.20) \qquad (1 - \tau_{ij})\{(x_i, u)(x_j, v)\} = ([x_i, x_j, v], u) \qquad (u, v \in \mathfrak{L}).$$

PROOF OF THE MAIN LEMMA. Denote by the letter d the commutator $[x_i, x_j]$, by the letter c the element of the form $\langle x_i, x_j, x_t \rangle \in F_\delta$, where $x_t \neq x_i, x_j$ but it may coincide with either of x_k and x_m, and by the letters r and e the polynomials $(1 - \tau_{km})(1 - \tau_{ij}) f, (1 - \tau_{ij}) f \in F_\delta$, respectively. In order to prove the lemma it suffices to construct a polynomial $g \in F_\delta$ such that

$$(1 - \tau_{km}) e = g|_{x_i = [x_i, x_j]}.$$

Lemma 38.3 shows that it suffices to realize this construction for polynomials e of the form (38.18) and (38.19).

Let e have the form (38.19). Then Lemma 35.6 implies that it is sufficient to consider the following cases.

1. $r = d * ([x_k, x_m], v) * f_0$, where $f_0 \in \mathcal{F}_0$, $v \in \mathfrak{L}$.

Then, applying identity (38.12) to r, we see that in F_δ we have

$$r = \tfrac{1}{2} d * \langle x_k, x_m, v\rangle * f_0,$$

and for the polynomial g we can take

$$\tfrac{1}{2} x_i * \langle x_k, x_m, v\rangle * f_0.$$

2. $r = (1 - \tau_{km})\{d * (x_k, u) * (x_m, v) * f_0\} \quad (u, v \in \mathfrak{L})$.

Applying equality (38.20) to r, we see that in F_δ we have

$$r = d * ([x_k, x_m, v], u) * f_0 = d * ([x_k, x_m], [v, u]) * f_0.$$

Therefore, this case is reduced to the first one.

3. $r = (1 - \tau_{km}) p|_{x_i = d*(x_m, v)*f_0}$,

where $v \in \mathfrak{L}$, $f_0 \in \mathcal{F}_0$, and p is a multilinear polynomial essentially depending on the variables x_k and x_i. Since $p|_{x_i=d} \in \mathfrak{V}$, Lemma 38.1 implies

$$r = p|_{x_k = [x_k, x_m, v], x_i = d*f_0}.$$

By Lemmas 37.7 and 37.4, for the polynomial g we can take the polynomial $p|_{x_k=[x_k,x_m,v]*f_0}$.

4. $r = (1 - \tau_{km})\{p|_{x_i=d*f_0}\}$,

where $f_0 \in \mathcal{F}_0$ and p is a multilinear polynomial essentially depending on x_k, x_m, and x_i. Then $(1 - \tau_{km})p \in \mathfrak{V}$, and Lemmas 37.7 and 37.4 permit us to take for g the polynomial $\{(1 - \tau_{km})p\} * f_0$.

Let e have the form (38.18). If $\delta = 0$, then from identity (38.4) it follows that the elements of this form are equal to zero in the algebra F_δ, and we can put g equal to zero.

If $\delta \neq 0$ and the element x_t in c is equal to neither x_k nor x_m, then another four cases arise, which are obtained by the replacement the element d by the element c in the cases just considered. Taking into account that c and $c * f_0$ ($f_0 \in \mathcal{F}_0$) belong to the center of the algebra F_δ and using the agrument applied above, where formula (38.12) is to be replaced by (38.13), we see that all these cases are reduced to the following single case:

5. $r = p \cdot (c * f_0)$, where $f_0 \in \mathcal{F}_0$ and $p \in \mathfrak{V}$.

However, in this case it follows from Lemmas 37.7 and 37.4 that $r = p_1 c$, where $p_1 = p * f_0 \in \mathfrak{V}$. Applying (38.13), we obtain $r = p_1 c = (1/dt) p_1 * ([x_i, x_j], x_t)$, and we can take for g the polynomial $p_1 * (x_i, x_t)$.

If $\delta \neq 0$ and x_t coincides with x_k in the element c, then it remains to consider two more cases.

6. $r = (1 - \tau_{km})\{\langle x_i, x_j, x_k\rangle * (x_m, v) * f_0\}$ $(v \in \mathfrak{L},\ f_0 \in \mathcal{F}_0)$.
Then from Lemma 38.1 and identity (38.11) we obtain

$$r = \langle x_i, x_j, [[x_k, x_m], v]\rangle * f_0 = \langle [x_i, x_j], [x_k, x_m], v\rangle * f_0,$$

and for g we can take the polynomial $\langle x_i, [x_k, x_m], v\rangle * f_0$.
7. $r = (1 - \tau_{km})\{\langle x_i, x_j, x_k\rangle x_m\}$.
Then, applying identity (38.2), we find that r is expressed as a Lie multilinear polynomial of degree 4 that is skew-symmetric with respect to the variables x_i, x_j and the variables x_k, x_m. But any Lie polynomial of this kind has the form $\beta[[x_i, x_j], [x_k, x_m]]$ $(\beta \in K)$. Therefore, $g = \beta[x_i, [x_k, x_m]]$.

The main lemma is proved.

38.3. Certain identities in three variables can be removed. Let Q be the set of associative polynomials in four variables x_1, x_2, x_3, y that are multilinear and skew-symmetric with respect to x_1, x_2, x_3. Denote by $P_{1,1}$ the set of polynomials in three variables x_1, x_2, y that are obtained from the set Q under the formal replacement of x_3 by y.

PROPOSITION 38.2. *Suppose that in the pair* $(U, \mathfrak{g})$ *constructed for a faithful irreducible representation of the Lie algebra* $\mathfrak{g}$ *in the linear space* V *we have the identity*

$$f(x_1, x_2, y) = 0, \tag{38.21}$$

where f *is a multihomogeneous polynomial from* $P_{1,1}$. *If* V *is infinite-dimensional, then the identity* (38.21) *for* $\delta \neq 0$ *is a consequence of identities* (38.1)–(38.3), (38.10), *and* (38.11), *and for* $\delta = 0$ *it follows from identities* (38.3), (38.4), (38.10), *and* (38.11). *If the space* V *is finite-dimensional and* $\dim_K V = n$, *then identity* (38.21), *modulo identities* (38.1)–(38.3), (38.10), *and* (38.11), *for* $\delta = (n^2 - 1)/4$ *follows from some identity of lesser degree satisfied in the pair* $(U, \mathfrak{g})$.

This proposition shows that any identity (38.21) of the pair $(U, \mathfrak{g})$ can be excluded from a basis of identities of this pair if this basis contains the identities (38.1)–(38.3), (38.10), and (38.11), and, for $\delta = 0$, the identity (38.4).

LEMMA 38.4. *If in an arbitrary associative-Lie pair identity* (38.1) *is satisfied then the following identities hold in this pair*:

$$\sum_{\sigma \in S_4} \operatorname{sign}(\sigma) x_{\sigma(1)} x_{\sigma(2)} x_{\sigma(3)} x_{\sigma(4)} = 0, \tag{38.22}$$

$$\sum_{\sigma \in S_3} \operatorname{sign}(\sigma) x_{\sigma(1)} y x_{\sigma(2)} x_{\sigma(3)} = \sum_{\sigma \in S_3} \operatorname{sign}(\sigma) x_{\sigma(1)} x_{\sigma(2)} y x_{\sigma(3)}, \tag{38.23}$$

$$2 \sum_{\sigma \in S_3} \operatorname{sign}(\sigma) x_{\sigma(1)} y x_{\sigma(2)} x_{\sigma(3)} = y * \langle x_1, x_2, x_3\rangle. \tag{38.24}$$

PROOF. Identity (38.22) is obtained by skew-symmetrization of identity (38.1) with respect to all its arguments. Moreover, (38.23) is the sum of identities (38.22) and (38.1). This proves (38.23). If we subtract (38.23) and (38.1) from identity (38.24), we obtain a true equality. This proves that (38.24) is a consequence of (38.1). The lemma is proved.

LEMMA 38.5. *In a relatively free pair $(F_\delta, \mathfrak{L})$ any element $f \in Q$ is representable as a linear combination of elements of two types:*

$$a\langle x_1, x_2, x_3\rangle * f_0, \tag{38.25}$$

$$a[y * \langle x_1, x_2, x_3\rangle] * f_0, \tag{38.26}$$

where $a \in F_\delta$, $f_0 \in \mathcal{F}_0$, and y is a free generator of the pair $(F_\delta, \mathfrak{L})$ not equal to x_1, x_2, x_3.

PROOF. From the definition of the set Q it follows that the polynomial f is multilinear and skew-symmetric with respect to x_1, x_2. Therefore, by Lemma 38.3, f is representable as a linear combination of elements of the two types (38.18) and (38.19), where $x_i = x_1$ and $x_j = x_2$.

If f can be expressed via the elements (38.19), then it is required to consider two cases depending on the occurrence of the element x_3.

1. $f = y^p\{\sum_{\sigma\in S_3}\operatorname{sign}(\sigma)[x_{\sigma(1)}, x_{\sigma(2)}] * (x_{\sigma(3)}, y)\}y^q * (y, y)^m$.

In this case, by (35.5), the element in curly braces equals $\beta y * \langle x_1, x_2, x_3\rangle$ ($\beta \in K$), and f can be reduced to the form (38.26) with the help of identity (38.3).

2. $f = y^r\{\sum_{\sigma\in S_3}\operatorname{sign}(\sigma)x_{\sigma(1)}x_{\sigma(2)}\{(\operatorname{ad} y)^t x_{\sigma(3)}\}\}y^q * (y, y)^m$.

If $t = 0$, then f can be represented in the form (38.26) by using (38.1). If $t = 1$, then the case $t = 0$ and equalities (38.23) and (38.24) imply that f can be expressed via elements of the form (38.25) and (38.26). If $t \geq 2$, then equality (38.14) and Lemma 38.1 imply

$$\begin{aligned} f = y^r\Big\{&\sum_{\sigma\in S_3}\operatorname{sign}(\sigma)x_{\sigma(1)}x_{\sigma(2)}\{(\operatorname{ad} y)^{t-2}x_{\sigma(3)}\} * (y, y) \\ &- \sum_{\sigma\in S_3}\operatorname{sign}(\sigma)x_{\sigma(1)}x_{\sigma(2)}\{(\operatorname{ad} y)^{t-2}y\} * (x_{\sigma(3)}, y)\Big\}y^q * (y, y)^m. \end{aligned}$$

Case 1 considered above and identity (38.3) show that the second term is an element of the form (38.26). The first summand can be reduced to the form we need by inductive arguments.

If the polynomial f is expressed via elements (38.18), then for $\delta \neq 0$ two cases arise again, depending on the occurrence of x_3:

3. $f = y^r\{\sum_{\sigma\in S_3}\operatorname{sign}(\sigma)x_{\sigma(1)}y^t\langle x_{\sigma(2)}, x_{\sigma(3)}, y\rangle\}y^q * (y, y)^m$.

Applying to f subsequently identities (38.1)–(38.3), we see that f can be reduced to the form (38.25).

4. $f = y^r\{\sum_{\sigma\in S_3}\operatorname{sign}(\sigma)\langle x_{\sigma(1)}, x_{\sigma(3)}, y\rangle * (x_{\sigma(2)}, y)\}(y, y)^m$.

Applying Lemma 38.1 and identity (38.14), we obtain

$$f = \tfrac{1}{2}y^r\Big\{\sum_{\sigma\in S_3}\operatorname{sign}(\sigma)\langle[x_{\sigma(1)}, x_{\sigma(2)}, y], x_{\sigma(3)}, y\rangle\Big\} * (y, y)^m,$$

and f is expressed via elements (38.19) and, hence, f is expressed via elements (38.25) and (38.26) by Cases 1 and 2 considered above. The lemma is proved.

PROOF OF PROPOSITION 38.2. From Lemma 38.5 and identity 38.2 it follows that if a multihomogeneous polynomial $f(x_1, x_2, y)$ belongs to $P_{1,1}$ and is of degree $l+3$, then

$$f(x_1, x_2, y) = \beta\langle x_1, x_2, y\rangle * (y, y)^{l/2} + \sum_{i=0}^{(l-1)/2} \beta_i y^{l-2i-1} y * \langle x_1, x_2, y\rangle \qquad (\beta_i \in K),$$

where $\beta = 0$ for odd l. Therefore, if in the pair $(U, \mathfrak{g})$ identity (38.21) holds, then, commuting it with a variable x and applying (38.13), we obtain

$$\left\{2\sum_{i=0}^{l/2} \beta_i [y^{l-2i}, x] * (y, y)^i\right\}([x_1, x_2], y) = 0.$$

Since the bilinear form (,) is nondegenerate and the commutator subalgebra of the Lie algebra $\mathfrak{g}$ coincides with this algebra itself, we see that the expression in curly braces is zero. Applying identity (38.14) to this expression, we find

$$\sum_{i=0}^{l/2} \beta_i [y^{l-2i}, [(\operatorname{ad} y)^{2i} x]] = 0.$$

It is clear that if some $\beta_i \neq 0$, then the polynomial on the left-hand side of this equality is nonzero. Therefore, if V is infinite-dimensional, then Proposition 38.1 gives that all β_i must be zero. However, in this case β is also equal to zero. This proves the first part of Proposition 38.2.

Let the space V be finite-dimensional, then $\delta \neq 0$ and, applying identity (38.2), we see that

$$f(x_1, x_2, y) = \sum_{i=0}^{l/2} \beta_i' \{\langle x_1, x_2, y\rangle * (y, y)^i\} y^{l-2i} = 0. \tag{38.27}$$

However, in this case from equality (35.9) and the fact that the form (,) is nondegenerate on $\mathfrak{g}$ it follows that

$$\sum_{i=0}^{l/2} \beta_i' y^{l-2i} (y, y)^i = 0 \qquad (y^0 \rightleftharpoons 1 \in \operatorname{End}_K V)$$

is an identity of the algebra with three supports $(K, \mathfrak{g}, U)$. Therefore, in the pair $(U, \mathfrak{g})$ we have the identity

$$\sum_{i=0}^{l/2} \beta_i' \{[x_1, x_2] * (y, y)^i\} y^{l-2i} = 0, \tag{38.28}$$

whose degree is $l+2$. It is clear that the identity

$$\sum_{\sigma \in S_3} \operatorname{sign}(\sigma) x_{\sigma(3)} \sum_{i=0}^{l/2} \beta_i' \{[x_{\sigma(1)}, x_{\sigma(2)}] * (y, y)^i\} y^{l-2i} = 0$$

is a consequence of (38.28). Changing the order of summation in this relation, we obtain another notation of it:

$$2\sum_{i=0}^{l/2} \beta_i'\{\langle x_3, x_1, x_2\rangle * (y, y)^i\} y^{l-2i} = 0.$$

Putting $x_3 = y$ in it, we conclude that identity (38.27) follows from (38.28). The proposition is proved.

38.4. The construction of a finite basis of identities for the pair $(U, \mathfrak{g})$**.** Let us consider the relatively free pair $(F_\delta, \mathfrak{L})$ introduced above and denote by $\mathfrak{V}(\rho)$ the ideal of weak identities in F_δ corresponding to the identities of the pair $(U, \mathfrak{g})$ constructed via a representation $\rho\colon \mathfrak{g} \to \operatorname{End}_K V$. Let T_l be a linear subspace in F_δ consisting of all multilinear polynomials depending on $x_1, \dots, x_l$. It is clear that T_l is a module over the group algebra $K[S_l]$ of the symmetric group S_l acting on the set of indices $\{1, 2, \dots, l\}$ and $W_l \rightleftharpoons \mathfrak{V}(\rho) \cap T_l$ is a $K[S_l]$-submodule of T_l. Denote by I_l the two-sided ideal in $K[S_l]$ generated by the elements $(1 - \tau_{ij})(1 - \tau_{km})$, where τ_{ij} and τ_{km} are arbitrary transpositions of the group S_l with mutually distinct i, j, k, m and by $I_{\mathbf{D}}$ the minimal two-sided ideal in $K[S_l]$ corresponding to the Young table $\mathbf{D}$.

PROPOSITION 38.3. *Over fields of characteristic zero the ideal I_l contains all ideals $I_{\mathbf{D}}$ for which the Young table $\mathbf{D}$ either contains not less than four cells in the first column or contains two columns, each of which has at least two cells.*

PROOF. We fill the table $\mathbf{D}$ by the numbers $1, \dots, l$ in such a way that i, j, k, l should either stand in the first column, if it contains not less than four cells, or i, j and k, m stand in two distinct columns, each containing at least two cells. We construct for the diagram $\mathbf{D}'$ thus obtained the elements $\mathcal{P}(\mathbf{D}')$, $\mathcal{Q}(\mathbf{D}')$, and $e(\mathbf{D}')$ of the group algebra $K[S_l]$ (see the proof of Proposition 37.1). Then $(1 - \tau_{ij})(1 - \tau_{km})\mathcal{Q}(\mathbf{D}') = \mathcal{Q}(\mathbf{D}')$ and $I_l \ni e(\mathbf{D}') = \mathcal{P}(\mathbf{D}') \circ \mathcal{Q}(\mathbf{D}')$. Since $0 \neq e(\mathbf{D}') \in I_{\mathbf{D}}$, we have $I_l \supset I_{\mathbf{D}}$. The proposition is proved.

Over a field of characteristic zero, $K[S_l]$ is decomposed into the direct sum of the ideals $I_{\mathbf{D}}$, where $\mathbf{D}$ ranges over all Young tables with l cells. Therefore, Proposition 38.3 implies that $W_l = W_{l.1} \oplus W_{l.2} \oplus W_{l.3} + W_{l.4}$ for which $W_{l.1} = I_{\mathbf{D}_1} W_l$, $W_{l.2} = I_{\mathbf{D}_2} W_l$, $W_{l.3} = I_{\mathbf{D}_3} W_l$, and $W_{l.4} = I_l W_l$, where

$\mathbf{D}_1 =$ 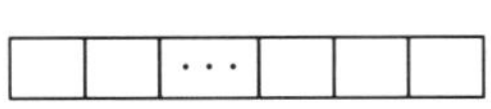, $\mathbf{D}_2 =$ 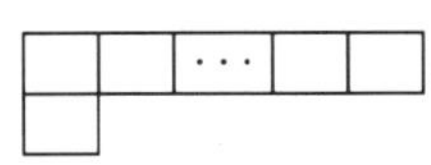, $\mathbf{D}_3 =$ 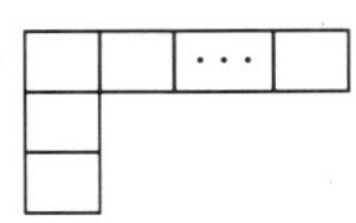.

It follows from the equality $W_{l.4} = I_l W_l$ that any polynomial from $W_{l.4}$ is representable as a linear combination of elements $r = (1 - \tau_{ij})(1 - \tau_{km})f$ $(f \in W_{l.4})$ and, therefore, by the main lemma, the identity $r = 0$ for the pair $(U, \mathfrak{g})$ is equivalent to the identity $g|_{x_i = [x_i, x_j]} = 0$ in the pair $(F_\delta, \mathfrak{L})$, where g is a polynomial of degree $l - 1$. Since $[\mathfrak{g}, \mathfrak{g}] = \mathfrak{g}$, the identity $g = 0$ holds for $(U, \mathfrak{g})$, and $r = 0$ follows from the identity $g = 0$ modulo the identities of the relatively free pair $(F_\delta, \mathfrak{L})$.

If $r \in W_{l.3} = I_{\mathbf{D}_3} W_l$, then the agrument used in the proof of Proposition 37.1 shows that the identity $r = 0$ is equivalent to the identities $f(x_1, x_2, y) = 0$, where $f \in P_{1.1}$.

From Proposition 38.2 it follows that identities of this kind can be excluded from a basis of identities for the pair $(U, \mathfrak{g})$.

Finally, if $r \in W_{l,1} \oplus W_{l,2}$, then the identity $r = 0$ is equivalent to the identities of the form $f(x, y) = 0$, where $f \in P$. From Proposition 38.1 it follows that if the dimension of the space V is infinite, then the polynomials f are zero, and if $\dim_K V = n < \infty$, then the identities follow from the identity $\mathrm{ART}_n(x, y) = 0$, where $\mathrm{ART}_n \in \mathfrak{V}(\rho)$.

Since over a field of characteristic zero any identity is equivalent to some system of multilinear identities, the above argument shows that if we join identity (38.4) for $\delta = 0$ and the identity $\mathrm{ART}_n = 0$ for $\dim_K V = n < \infty$ to the identities (38.1)–(38.3), (38.10), and (38.11), then we obtain a finite basis of identities $\mathfrak{R}_\delta$ for the pair $(U, \mathfrak{g})$. Comparing the basis of identities $\mathfrak{R}_\delta$ with the bases mentioned in Theorem 38.1, we see that in order to prove the theorem, it remains to delete the identities (38.10) and (38.11) from the basis $\mathfrak{R}_\delta$. We leave this verification to the reader.

§39. Examples of minimal nonsolvable Engel varieties of pairs over fields of characteristic $p \geq 5$

In §§6 and 33 we already come across the fact that $\boldsymbol{\alpha}$-functions on 2-words constructed with respect to an irreducible representation of a semisimple Lie algebra over a field of characteristic zero define nontrivial varieties of associative-Lie pairs over fields of positive characteristic; in these varieties the Engel identity holds, but the solvability identity fails. In this section we shall enrich the list of examples of varieties of this kind.

Let the ground field K be algebraically closed and $\operatorname{char} K = p > 2$, and let the bilinear form $(,)$ on the simple three-dimensional Lie K-algebra $\mathfrak{g}$ be given by formulas (34.1) and (34.2) and equal one half of the Killing form. Let us consider arbitrary irreducible representations $\rho_s \colon \mathfrak{g} \to \operatorname{End}_K V_s$ $(s \in S)$ of the Lie algebra $\mathfrak{g}$ in linear spaces V_s. For any $\delta \in K$ put

$$S(\delta) = \{ s \in S \mid \rho_s(\gamma) = \delta, \text{ where } \gamma \in U(\mathfrak{g}) \text{ is the Casimir element} \}.$$

Let the functions $\alpha_s \colon B \to K$, $\widetilde{\alpha}_s \colon \widetilde{B} \to K$, $\alpha \colon B \to \mathcal{E}$, and $\widetilde{\alpha} \colon \widetilde{B} \to \mathcal{E}$ be constructed with respect to the irreducible representations ρ_s and the set S in the same way as in §§34 and 35. Then we can introduce the algebras F and $\overline{F}$, the pairs $(\widetilde{U}_s, \widetilde{\mathfrak{g}})$, $(\overline{U}_s, \overline{\mathfrak{g}})$, $(\widetilde{U}(\mathfrak{g}), \widetilde{\mathfrak{g}})$, and $(\overline{U}(\mathfrak{g}), \overline{\mathfrak{g}})$, the triples $\{K, \mathfrak{g}, U_s\}$, $\{\mathcal{F}, \widetilde{\mathfrak{g}}, \widetilde{U}_s)$, $\{\overline{\mathcal{F}}, \overline{\mathfrak{g}}, \overline{U}(\mathfrak{g})\}$, and $\{\mathcal{F}, \widetilde{\mathfrak{g}}, \widetilde{U}(\mathfrak{g})\}$, the auxiliary associative algebras $\overline{A}$, $\overline{A}(\gamma)$, and $A(\gamma)$, the bilinear pairing $\widetilde{\mathfrak{b}} \colon {}_\mu A_\lambda(\gamma) \otimes {}_\nu A(\gamma)_\lambda \to K$, the mappings Φ_α, Ψ_α, $\Phi_{\widetilde{\alpha}}$, and $\Psi_{\widetilde{\alpha}}$, and the closures θ_α, η_α, $\theta_{\widetilde{\alpha}}$ $\eta_{\widetilde{\alpha}}$, as it was done in §35.

From Lemmas 35.7 and 35.8 it follows that $\widetilde{\alpha}_{s_1} = \widetilde{\alpha}_{s_2}$ if and only if $s_1 \in S(\delta_1)$ & $s_2 \in S(\delta_2)$ & $\delta_1 = \delta_2$. It is also clear that if $s \in S(\delta)$, then the ideal of weak identities $\Psi_{\alpha_s}(\{0\})$ given by the $\boldsymbol{\alpha}$-function α_s coincides with the ideal of weak identities $\Psi_\alpha(\{\gamma(\gamma - \delta)\})$, where $\{E(\gamma)\}$ is the ideal of the commutative algebra $\mathcal{E}$ generated by the polynomial $E(\gamma) \in \mathcal{E}$. Since for the two-dimensional irreducible representation we have $\delta = 3/4$, one of the main results of §6, Proposition 6.2, can be stated now as follows: if $\operatorname{char} K = p > 5$, then the ideal of weak identities $\Psi_\alpha(\{\gamma(\gamma - 3/4)\})$ contains the polynomials $[y^2, x]$, y^{p-2}, and $(\operatorname{ad} y)^{p-2}x$, and does not contain polynomials $[x_1, \dots, x_N]$ $(N = 1, 2, \dots)$. The result given below generalizes this assertion.

PROPOSITION 39.1. *Let* $E_\delta(\gamma) = \gamma(\gamma - \delta)$, *where* $\delta \in \{(n^2-1)/4 \mid n = 2, \dots, (p-1)/2\}$, *and let* $\operatorname{char} K = p > 5$. *Then for the* $\boldsymbol{\alpha}$*-function* $\alpha\colon B \to \mathcal{E}$ *constructed with respect to the set S of "all" irreducible representations of the simple three-dimensional Lie algebra* $\mathfrak{g}$ *the ideal of weak identities* $\Psi_\alpha(\{E_\delta(\gamma)\})$ *contains the polynomials* x_1^{p-2} *and* $(\operatorname{ad} x_2)^{p-2}x_1$, *but does not contain polynomials* $[x_1, \dots, x_N]$ $(N = 1, 2, \dots)$. *Moreover, if* $\delta = (n^2-1)/4$ $(n \in \{2, \dots, (p-1)/2\})$, *then* $\Psi_\alpha(\{E_\delta(\gamma)\})$ *contains the ideal of weak identities of the n-dimensional irreducible representation of the Lie algebra* $\mathfrak{g}$.

PROOF. We denote by $(F, \mathfrak{L})$ and $\{\mathcal{F}_0, \mathfrak{L}_1, F_1\}$ the free pair and the free algebra with three supports of the varieties $\operatorname{var}(U(\mathfrak{g}), \mathfrak{g})$ and $\operatorname{var}(K, \mathfrak{g}, U(\mathfrak{g}))$ with the set of free generators $X = \{x_1, x_2, \dots\}$ and put $\mathfrak{V}_\delta = \Psi_\alpha(\{E_\delta(\gamma)\})$ and $\widetilde{\mathfrak{V}}_\delta = \Psi_{\widetilde{\alpha}}(\{E_\delta(\gamma)\})$. From Lemmas 35.7 and 35.8 it follows that $\mathfrak{V}_\delta = \widetilde{\mathfrak{V}}_\delta \cap F$ and for relatively free quotient objects $(F_\delta, \mathfrak{L}_\delta)$ and $\{\mathcal{F}_{0,\delta}, \mathfrak{L}_{1,\delta}, F_{1,\delta}\}$ of the objects $(F, \mathfrak{L})$ and $\{\mathcal{F}_0, \mathfrak{L}_1, F_1\}$ by the ideals corresponding to $\mathfrak{V}_\delta$ and $\widetilde{\mathfrak{V}}_\delta$ we have the inclusions

$$\begin{gathered} \operatorname{var}(F_\delta, \mathfrak{L}_\delta) \subset \operatorname{var}(U_s, \mathfrak{g}), \\ \operatorname{var}(\mathcal{F}_{0,\delta}, \mathfrak{L}_{1,\delta}, F_{1,\delta}) \subset \operatorname{var}(K, \mathfrak{g}, U_s), \end{gathered} \tag{39.1}$$

where s is an arbitrary element of $S(\delta)$ and $U_s = \rho_s(U(\mathfrak{g}))$. The results of §35 and these inclusions show that when calculating the functions α_s and $\widetilde{\alpha}_s$ we can use identities of any representation for which the value of the Casimir element coincides with $\rho_s(\gamma)$.

Since $\mathfrak{V}_\delta = \Psi_{\alpha_s}(\{0\})$ $(s \in S_\delta)$, we see, applying (35.20) and (35.17), that

$$\begin{aligned} \mathfrak{b}_{E_\delta}([x_1, \dots, x_N], x_N \cdots x_1) &= \alpha_s\big(\tfrac{1}{2}[\overline{x}_1, \dots, \overline{x}_N, \overline{x}_N]\overline{x}_{N-1} \cdots \overline{x}_1\big) \\ &= \alpha_s([\overline{x}_1, \dots, \overline{x}_{N-1}]\overline{x}_{N-1} \cdots \overline{x}_1) = \alpha_s(\overline{x}_1\overline{x}_1) = \delta, \end{aligned}$$

and the ideal $\mathfrak{V}_\delta$ does not contain polynomials $[x_1, \dots, x_N]$ $(N = 1, 2, \dots)$ for any nonzero element δ.

Let us fix an integer $n \in \{2, \dots, (p-1)/2\}$. Since $n < p$ and the field K is algebraically closed, the results of §34 imply that there exists a unique, up to isomorphism, irreducible representation $\rho_n\colon \mathfrak{g} \to \operatorname{End}_K V_n$ of the algebra $\mathfrak{g}$ in an n-dimensional space V_n and the value of the Casimir element in this representation is equal to $(n^2-1)/4$. Therefore, from inclusions (39.1) it follows that for $\delta = (n^2-1)/4$ the ideal $\mathfrak{V}_\delta$ contains the ideal of identities of the pair $(\rho_n(U(\mathfrak{g})), \mathfrak{g})$ and the ideal $\widetilde{\mathfrak{V}}_\delta$ contains all polynomials $f \in F_1$ for which the identity $f = 0$ holds in $\{K, \mathfrak{g}, \rho_n(U(\mathfrak{g}))\}$. This proves the last assertion of Proposition 39.1 and permits us to use the identities of the representation ρ_n and the identities for $\{K, \mathfrak{g}, \rho_n(U(\mathfrak{g}))\}$ side by side with relations (35.17)–(35.21) in computing the functions α_n and $\widetilde{\alpha}_n$. Since ρ_n is irreducible, the algebra $\rho_n(U(\mathfrak{g}))$ coincides with the full matrix algebra $M_n = \operatorname{End}_K V_n$ of order n over the field K. Therefore, from the Hamilton–Cayley theorem and Corollary of Proposition 34.1 it follows that in $\{K, \mathfrak{g}, M_n\}$ we have the identity of the form

$$y^n = \sum_{i=1}^{[n/2]} \beta_i (y, y)^i y^{n-2i} \qquad (y \in \mathfrak{g},\ \beta_i \in K) \tag{39.2}$$

and this identity can also be used when computing α_n and $\widetilde{\alpha}_n$.

LEMMA 39.1. *The ideal* $\widetilde{\mathfrak{V}}_\delta$ *contains* $(x_2, x_2)^{(p-3)/2}x_2$.

PROOF. Denote by $\mathrm{Sm}_l(x_1, \dots, x_l)$ the element of F_1 equal to

$$\sum_{\sigma\in S_l}(x_{\sigma(l)}, x_{\sigma(l-1)})\cdots(x_{\sigma(3)}, x_{\sigma(2)})x_{\sigma(1)},$$

where the sum is extended over all permutations $\sigma \in S_l$. It is clear that the identity $(x_2, x_2)^{(p-3)/2}x_2 = 0$ is equivalent to the identity $\mathrm{Sm}_{p-2} = 0$. Therefore, by Corollary of Proposition 35.1, in order to prove the lemma it suffices to verify that $\mathrm{Sm}_{p-2} \in \mathrm{Ann}_{T_{p-2}}\widetilde{\mathfrak{b}}_{E_\delta}$. Using the fact that the polynomial Sm_{p-2} is symmetric with respect to all its variables, we find that we must check the equalities $\widetilde{\alpha}(\mathrm{Sm}_{p-2}f) = 0$, where $f \in T_{p-2}$ and f is symmetric with respect to $x_1, \dots, x_{p-2}$. Since $n < p$, we conclude from (39.2) that when verifying these equalities we can consider that any monomial participating in f contains at least one round bracket. Using again the fact that Sm_{p-2} is symmetric, we get that to prove the lemma we must verify the equality

$$\widetilde{\mathfrak{b}}_{E_\delta}(\mathrm{Sm}_{p-2}, g(x_{p-2}, x_{p-3})) = \widetilde{\alpha}_n(\mathrm{Sm}_{p-2}(\overline{x}_1, \dots, \overline{x}_{p-2})\overline{g}(\overline{x}_{p-2}, \overline{x}_{p-3})) = 0,$$

where g is an arbitrary element of T_{p-4}. However, from (35.17) and (35.18) for $\delta = (n^2-1)/4$ it follows that

$$\begin{aligned}\pi_\delta(\mathrm{Sm}_{p-2}(\overline{x}_1, \dots, \overline{x}_{p-2})(\overline{x}_{p-2}, \overline{x}_{p-3})) &= \pi_\delta(\mathrm{Sm}_{p-2}(\overline{x}_1, \dots, \overline{x}_{p-3}, \overline{x}_{p-3}))\\ &= \pi_\delta(\mathrm{Sm}_{p-4}(\overline{x}_1, \dots, \overline{x}_{p-4}))\Big\{2\cdot\frac{p-3}{2}(\overline{x}_{p-3}, \overline{x}_{p-3}) + 2\Big(\frac{p-3}{2}\Big)(p-3)\frac{\overline{x}_{p-3}^2}{\delta}\Big\}\\ &= (p-3)(3+p-3)\,\mathrm{Sm}_{p-4} = 0 \pmod p.\end{aligned}$$

The lemma is proved.

From identity (35.1) it follows that the Lie polynomial $(\mathrm{ad}\,x_2)^{p-2}x_1$ is equal in the algebra $F_{1.\delta}$ to the element $[(x_2, x_2)^{(p-3)/2}x_2, x_1]$. By Lemma 39.1, we have $(x_2, x_2)^{(p-3)/2}x_2 = 0$ in $F_{1.\delta}$. Therefore, $(\mathrm{ad}\,x_2)^{p-2}x_1 \in F \cap \widetilde{\mathfrak{V}}_\delta = \mathfrak{V}_\delta$.

It remains to show that $x_1^{p-2} \in \mathfrak{V}_\delta$. This follows from the next lemma.

LEMMA 39.2. *For any nonzero integers i and j such that $i + 2j = p - 2$ the ideal* $\widetilde{\mathfrak{V}}_\delta$ *contains* $(x_1, x_1)^j x_1^i$.

THE PROOF will be given by induction on the number i. The base of induction $i = 1$ is valid by Lemma 39.1. Let us prove the possibility of the inductive step. We put by definition

$$\mathrm{Sm}_{i.j}(x_1, \dots, x_{p-2}) = \sum_{\sigma\in S_l}(x_{\sigma(1)}, x_{\sigma(2)})\cdots(x_{\sigma(2j-1)}, x_{\sigma(2j)})x_{\sigma(2j+1)}\cdots x_{\sigma(p-2)}.$$

It is clear that in the algebra $F_{1.\delta}$ the equality $(x_1, x_1)^j x_1^i = 0$ is possible if and only if the equality $\mathrm{Sm}_{i.j} = 0$ holds in $F_{1.\delta}$. Since $F_{1.\delta} = F_1/\widetilde{\mathfrak{V}}_\delta$, we see that in order to prove the lemma it suffices to establish the fact $\mathrm{Sm}_{i.j} \in \mathrm{Ann}_{T_{p-2}}\widetilde{\mathfrak{b}}_{E_\delta}$. Suppose that this inclusion holds for all $i \leq l$ (l is odd). Then by Corollary of Proposition 35.1 we have the inclusion

$$\pi(\mathrm{Sm}_{l,(p-2-l)/2}(x_1, \dots, x_{p-2})x_{p-2})\,|_{\gamma=\delta} \in \mathrm{Ann}_{T_{p-2}} \widetilde{\mathfrak{b}}_{E_\delta}. \tag{39.3}$$

A direct computation by using formulas (35.17)–(35.21) shows that for some $\beta_i \in K$ we have

$$\begin{aligned} x_1\pi(\mathrm{Sm}_{l,(p-2-l)/2}(x_1, \dots, x_{p-2})x_{p-2})\,&|_{\gamma=\delta, x_1=x_1, x_2=x_1, \dots, x_{p-3}=x_1} \\ &= (p-3)!\,(p-2-l)(x_1, x_1)^{(p-4-l)/2}x_1^{l+2} + \sum_{i \leq l} \beta_i (x_1, x_1)^{(p-2-i)/2} x_1^i. \end{aligned}$$

Since $l < p-2$, we see from this equality, inductive arguments, and inclusion (39.3), that

$$x_1^{l+2}(x_1, x_1)^{(p-4-l)/2} = 0$$

in the algebra $F_{1,\delta}$ and, hence, $\mathrm{Sm}_{l+2,(p-4-l)/2} \in \widetilde{\mathfrak{V}}_\delta$. The lemma is proved.

Putting $i = p-2$ in Lemma 39.2, we obtain $x_1^{p-2} \in \widetilde{\mathfrak{V}}_\delta \cap F = \mathfrak{V}_\delta$. Proposition 39.1 is completely proved.

From Proposition 39.1 it follows that the $\boldsymbol{\alpha}$-functions α_s, where $s \in S(\delta)$ ($\delta = (n^2-1)/4$, $2 \leq n \leq p-2$), determine the Lie nonnilpotent $(p-2)$th Engel varieties of associative-Lie pairs $\overline{\mathfrak{B}}_{\alpha_s}$. The Higgins theorem (see §6) implies that then the identity of Lie solvability does not hold in these varieties. We shall clarify what general properties the varieties $\overline{\mathfrak{B}}_{\alpha_s}$ have, where $s \in S(\delta)$ and δ is an arbitrary element of the field K.

PROPOSITION 39.2. *Suppose that* $\operatorname{char} K = p \geq 3$ *and the* $\boldsymbol{\alpha}$*-function* α_δ *is constructed with respect to an arbitrary irreducible representation* $\rho\colon \mathfrak{g} \to \mathrm{End}_K V$ *for which the value of the Casimir element* $\gamma|_V$ *is equal to* $\delta \cdot \mathbf{1}$ ($\delta \in K$). *Then for* $\delta \neq 0$ *in the variety of associative-Lie pairs* $\overline{\mathfrak{B}}_{\alpha_\delta}$ *given by the* $\boldsymbol{\alpha}$*-function* α_δ, *constructed with respect to the representation* ρ, *the Engel identity of order* $2p+1$ *holds, the variety* $\overline{\mathfrak{B}}_{\alpha_\delta}$ *is Lie nonsolvable, but in any proper subvariety contained in* $\overline{\mathfrak{B}}_{\alpha_\delta}$ *the identities*

$$\langle x_1, y_1, z_1\rangle \cdots \langle x_l, y_l, z_l\rangle = 0, \tag{39.4}$$

$$[[x_1, y_1], [z_1, u_1]] \cdots [[x_l, y_l], [z_l, u_l]] = 0 \tag{39.5}$$

hold for some sufficiently large $l \in \mathbb{N}$, *and this subvariety is Lie solvable.*

PROOF. We preserve the notation used in the proof of Proposition 39.1. Let us show first that for $\delta \neq 0$ the ideal $\overline{\mathfrak{V}}_\delta$ contains the Lie polynomial $(\operatorname{ad} x_2)^{2p+1}x_1$, but does not contain any Lie commutator r_l corresponding to the solvability identity of order l. By definition, $\overline{\mathfrak{V}}_\delta$ is the largest ideal of weak identities in F for which $\mathfrak{V}_\delta \cap F_\lambda = \mathrm{Ann}_{F_\lambda} \mathfrak{b}_{\alpha_\delta}$. This means that the $(2p+1)$th Engel identity is satisfied in the variety $\overline{\mathfrak{B}}_{\alpha_\delta}$ if and only if the total linearization of this identity

$$\sum_{\sigma \in S_{2p+1}} [y, x_{\sigma(1)}, x_{\sigma(2)}, \dots, x_{\sigma(2p+1)}] = 0 \tag{39.6}$$

holds in it. However, applying the identity (35.1) repeatedly, we see that in the algebra F_1 the left-hand side of this identity equals

$$\sum_{\sigma\in S_{2p+1}} [y, x_{\sigma(1)}](x_{\sigma(2)}, x_{\sigma(3)}) \cdots (x_{\sigma(2p)}, x_{\sigma(2p+1)}).$$

Reducing similar terms in this expression, we see that any summand of the form

$$[y, x_t](x_{i_1}, x_{j_1}) \cdots (x_{i_p}, x_{j_p}),$$

where $i_1 < i_2 < i_3 < \cdots < i_p$ and $i_k < j_k$, occurs $2^p p!$ times. Hence, over fields of characteristic $p > 0$ identity (39.6) is satisfied in the pair $(U(\mathfrak{g}), \mathfrak{g})$. Since $\operatorname{var}(U(\mathfrak{g}), \mathfrak{g}) \supset \overline{\mathfrak{B}}_{\alpha_\delta}$, identity (39.6) and the $(2p+1)$th Engel identity hold in $\overline{\mathfrak{B}}_{\alpha_\delta}$.

Since the multilinear commutator r_1 can be defined by induction by the following formulas:

$$r_1 = [x_1, x_2], \qquad r_{l+1} = r_l|_{z_1=[x_1,y_1],\dots,z_{2^l}=[x_{2^l},y_{2^l}]},$$

where z_i, x_j, y_t are distinct variables from the set X, we succesively conclude, using the first property of $\boldsymbol{\alpha}$-functions and equality (35.20), that

$$\mathfrak{b}_{\alpha_\delta}(r_{l+1}, r_{l+1}) = (-2)^{2^l}\, \mathfrak{b}_{\alpha_\delta}(r_l, r_l), \qquad \mathfrak{b}_{\alpha_\delta}(r_1, r_1) = -2\delta.$$

This implies that if $\operatorname{char} K \neq 2$, then $r_l \notin \overline{\mathfrak{B}}_\delta$ $(l = 1, 2, \dots)$, and the solvability identity fails in the variety $\overline{\mathfrak{B}}_{\alpha_\delta}$.

Let $\mathfrak{B}$ be a proper subvariety of $\overline{\mathfrak{B}}_{\alpha_\delta}$ $(\delta \neq 0)$. Then in $\mathfrak{B}$ we have a certain identity $f = 0$, which does not hold in $\overline{\mathfrak{B}}_{\alpha_\delta}$. If the total linearization of the polynomial f belonged to the ideal $\mathfrak{V}_\delta$, then, as it has already been mentioned above, the polynomial f would also be contained in $\mathfrak{V}_\delta$, and this contradicts the fact that the identity $f = 0$ does not hold in $\overline{\mathfrak{B}}_{\alpha_\delta}$. Therefore, we can consider that $f \in F_\lambda$ is a multilinear polynomial, $f \notin \overline{\mathfrak{V}}_\delta$, and the identity $f = 0$ holds in $\mathfrak{B}$. Then $f \notin \operatorname{Ann}_{F_\lambda} \mathfrak{b}_{\alpha_\delta}$ and $\mathfrak{b}_{\alpha_\delta}(f, f_1) \neq 0$ for some $f_1 \in {}_\mu F_\lambda$.

LEMMA 39.3. *Suppose that for an irreducible representation $\rho\colon \mathfrak{g} \to \operatorname{End}_K V$ we have $\rho(\gamma) = \delta \cdot \mathbf{1}$ $(0 \neq \delta \in K)$. Then for an arbitrary multilinear polynomial $g(x_1, y_1, \dots, x_l, y_l)$ the following identity holds for the representation ρ:*

$$\alpha_\delta(g(\overline{x}_1, \overline{x}_1, \overline{x}_2, \overline{x}_2, \dots, \overline{x}_l, \overline{x}_l))\langle x_{11}, x_{12}, x_{13}\rangle \cdots \langle x_{l1}, x_{l2}, x_{l3}\rangle$$
$$= \left(\frac{\delta}{2}\right)^l \sum_{\sigma_1\in S_3} \cdots \sum_{\sigma_l\in S_3} \operatorname{sign}\left(\prod_{i=1}^l \sigma_i\right) g\Big|_{\substack{x_1=x_{1\sigma_1(1)},\, y_1=[x_{1\sigma_1(2)},\, x_{1\sigma_1(3)}] \\ \cdots \\ x_l=x_{l\sigma_l(1)},\, y_l=[x_{l\sigma_l(2)},\, x_{l\sigma_l(3)}]}}.$$

PROOF. Since identity (39.7) is multilinear, it suffices to check this identity on an arbitrary basis of the Lie algebra $\mathfrak{g}$. Since the Lie algebra $\mathfrak{g}$ is three-dimensional and identity (39.7) is skew-symmetric with respect to any group of variables x_{i1}, x_{i2}, x_{i3}, identity (39.7) becomes a true equality if for some i two of the variables x_{i1}, x_{i2}, x_{i3} take the same value. Therefore, it remains to verify that equality (39.7) holds in $\operatorname{End}_K V$ under the substitution $x_{i1} = \rho(e_1)$, $x_{i2} = \rho(e_2)$, and $x_{i3} = \rho(e_3)$, where $i = 1, \dots, l$,

$\{e_1, e_2, e_3\}$ is a basis of the Lie algebra $\mathfrak{g}$ satisfying relations (34.1) and (34.2). But relations (34.1) show that

$$\langle \rho(e_1), \rho(e_2), \rho(e_3)\rangle = \sqrt{-1}\rho(e_1^2 + e_2^2 + e_3^2) = \sqrt{-1}\rho(\gamma) = \sqrt{-1}\delta \cdot \mathbf{1},$$

and after the substitution mentioned above the expression on the right-hand side of (39.7) equals

$$\delta^l(\sqrt{-1})^l \sum_{i_1=1}^{3} \cdots \sum_{i_l=1}^{3} g(\rho(e_{i_1}), \rho(e_{i_1}), \rho(e_{i_2}), \rho(e_{i_2}), \ldots, \rho(e_{i_l}), \rho(e_{i_l}))$$
$$= (\sqrt{-1}\delta)^l \alpha_\delta(g(\overline{x}_1, \overline{x}_1, \overline{x}_2, \overline{x}_2, \ldots, \overline{x}_l, \overline{x}_l)).$$

The lemma is proved.

We shall consider that $\lambda = \{x_1, \ldots, x_t\}$ and $\mu = \{x_{t+1}, \ldots, x_l\}$. Choose a multilinear polynomial

$$f_2(x_{t+1}, y_{t+1}, \ldots, x_l, y_l, y_1, \ldots, y_t)$$

such that $f_1 = f_2|_{y_1 = x_1, \ldots, y_l = x_l}$ and put $g = ff_2$. It is clear that the identity $g = 0$ holds in the variety $\mathfrak{B}$, being a consequence of the identity $f = 0$. Let us construct identity (39.7) from the polynomial g. By Lemma 39.3, this identity holds in the representation ρ, and, therefore, it holds in the varieties $\overline{\mathfrak{B}}_{\alpha_\delta}$ and $\mathfrak{B}$. However, the identity $g = 0$ implies that the polynomial on the right-hand side of relation (39.7) is equal to zero in a free pair of the variety $\mathfrak{B}$. On the other hand, by construction,

$$\alpha_\delta(g(\overline{x}_1, \overline{x}_1, \overline{x}_2, \overline{x}_2, \ldots, \overline{x}_l, \overline{x}_l)) = \mathfrak{b}_{\alpha_\delta}(f, f_1) \neq 0.$$

This shows that the identity $f = 0$ in the variety $\overline{\mathfrak{B}}_{\alpha_\delta}$ implies the identities (39.7) and (39.4). Thus, identity (39.4) holds in the variety $\mathfrak{B}$.

Multiplying both sides of (39.4) by the monomial $u_1 u_2 \cdots u_l$ and applying identities (38.1) and (38.2) to the relation thus obtained, we see that the following identity holds in the variety $\mathfrak{B}$:

$$\left(\frac{\delta}{2}\right)^l [u_1 * \langle x_1, y_1, z_1\rangle] \cdots [u_l * \langle x_l, y_l, z_l\rangle] = 0. \tag{39.8}$$

LEMMA 39.4. *If the characteristic of the ground field K is not equal to* 2, *then the Lie identity* $[u * \langle x, y, z\rangle] = 0$ *is equivalent to the metabelianness identity* $r_2 = 0$.

PROOF. Since $r_2 = [[u, x], [y, z]]$ and

$$[u * \langle x, y, z\rangle] = [[u, x], [y, z]] + [[u, z], [x, y]] + [[u, y], [z, x]], \tag{39.9}$$

the identity $[u * \langle x, y, z\rangle] = 0$ is a consequence of the identity $r_2 = 0$. On the other hand, the right-hand side of (39.9) is equal to

$$-[[x, u], [y, z]] - [[x, y], [u, z]] + [[x, z], [u, y]].$$

Therefore,

$$[u * \langle x, y, z\rangle] - [x * \langle u, y, z\rangle] = 2[[u, x], [y, z]].$$

This implies that the identity $2r_2 = 0$ is a consequence of the identity $[u * \langle x, y, z\rangle] = 0$. The lemma is proved.

Applying Lemma 32.4 to each square bracket in relation (39.8) and using the fact that $\delta/2 \neq 0$, we find that in the variety of pairs $\mathfrak{B}$ we have the identity $r_2(u_1, x_1, y_1, z_1) \cdots r_2(u_l, x_l, y_l, z_l) = 0$, which is equivalent to identity (39.5). Identity (39.5) shows that in any associative-Lie pair (R, L) of the variety $\mathfrak{B}$ the bicommutator subalgebra $[[L, L], [L, L]]$ of the Lie algebra L is nilpotent of class $l - 1$. Hence, in the variety $\mathfrak{B}$ we have the solvability identity $r_t = 0$ for $t > 2l$. Proposition 39.2 is proved.

DEFINITION 39.1. A variety of associative-Lie pairs $\mathfrak{B}$ is called Lie *just nonsolvable* (*just nonnilpotent*) if in a relatively free pair (R, L) of countable rank from this variety the Lie algebra L is solvable (respectively, nilpotent) if and only if $\mathrm{var}(R, L) \neq \mathfrak{B}$.

Propositions 39.1 and 39.2 and this definition imply the following

THEOREM 39.1. *In the variety of associative-Lie pairs* $\mathrm{var}(U(\mathfrak{g}), \mathfrak{g})$ *over a field K of characteristic $p > 2$ all subvarieties $\overline{\mathfrak{B}}_{\alpha_\delta}$ $(0 \neq \delta \in K)$ are $(2p+1)$th Engel, Lie just nonsolvable, and pairwise distinct. Moreover, for $\delta \in \{(n^2-1)/4 \mid n = 2, \dots, (p-1)/2\}$ and $p \geq 5$ these subvarieties are Lie just nonnilpotent and satisfy the Engel identity of order $p - 2$.*

Indeed, the fact that the varieties $\overline{\mathfrak{B}}_{\alpha_\delta}$ are pairwise distinct follows from identity (38.2), and the just nonnilpotency of these varieties follows from Theorem 6.1.

To conclude this section, we suggest to the reader to prove that any proper subvariety of Lie algebras in the variety of 3 that algebras over a field of characteristic 5 is nilpotent, and that in any proper subvariety of the variety $\overline{\mathfrak{B}}_{\alpha_\delta}$ $\big($where $\delta \in \{(n^2-1)/4 \mid n = 2, \dots, (p-1)/2\}\big)$ of associative-Lie pairs over a field of characteristic $p \geq 5$ the identity $x_1 \cdots x_N = 0$ holds for some $N \in \mathbb{N}$ depending on the subvariety.

§40. Nonsolvability of varieties of locally finite groups of exponent 4 and of prime exponent p for $p \geq 5$

THEOREM 40.1. *The variety of locally finite groups of prime exponent p is non-solvable for $p \geq 5$.*

PROOF. The fact that locally finite groups of prime exponent form a variety follows from the positive solution of the weakened Burnside problem for groups of prime exponent (see [Kos4, KoSh]). It remains to give an example of a nonsolvable locally finite group for $p \geq 5$. We know two ways of constructing such an example. The first one is based on Proposition 39.1. It turns out that in a relatively free nonsolvable $(p-2)$th Engel Lie algebra of countable rank L of a free pair (A, L), given by an ideal of weak identities $\Psi_\alpha\{\gamma(\gamma - \frac{3}{4})\}$, the Campbell–Hausdorff formula can be realized, and this formula defines on L a binary operation $*$, which endows L with the structure of a group (see [Raz2]). This operation has the following properties:

a) $\underbrace{x * \cdots * x}_{k \text{ times}} = kx$, b) $x * y * (-x) * (-y) = [x, y] + \Sigma$, where Σ is a finite linear combination of Lie commutators in x and y of weight ≥ 3. Property a) shows that this group has a prime exponent. Property b) implies that the group thus constructed cannot be solvable, since the Lie algebra L is nonsolvable.

The second method of construction of a nonsolvable locally finite relatively free group of exponent $p \geq 5$ is based on Theorem 33.2. By this theorem, over a field of characteristic $p \geq 5$ the relatively free associative algebra of countable rank A given by the identity $y^p = 0$ is nonsolvable as a Lie algebra. By the Levitzky theorem (see [Her]), the algebra A is locally finite. We ajoin a unit to A externally. Obviously, the set of all elements of the form $1 + a$ $(a \in A)$ is a group of exponent p with respect to the operation of multiplication, and this group is nonsolvable. The theorem is proved.

THEOREM 40.2. *The variety of groups of period* 4 *is nonsolvable.*

PROOF. Local finiteness of groups of period ($\equiv$ exponent) 4 was proved by Sanov (see [San]). Therefore, all groups of this kind are locally solvable. The construction of a nonsolvable group of exponent 4 is decomposed into several steps.

First step. A group of exponent 4 will be given with the help of the following construction, which permits us to build locally finite groups of arbitrary exponent $m = p^k$, where p is prime. Let an associative algebra with unit R be given over a field of characteristic $p > 0$ by generators $z_0, z_1, z_2, \dots$ and defining relations

$$z_i a z_i = 0 \qquad (i = 0, 1, 2, \dots) \tag{40.1}$$

for any word a depending on $z_0, z_1, z_2, \dots$. We denote by the letter G the group generated by the elements $g_i = 1 + z_i$ $(i = 0, 1, \dots)$. From the defining relations of the algebra R it follows that $g_i^{-1} = 1 - z_i$, and, therefore, the group G, as a semigroup, is generated by the elements $1 \pm z_i$ $(i = 0, 1, 2, \dots)$. In order to turn G into a group of exponent p^k, it suffices to take the quotient of the algebra R by all relations of the form $g^{p^k} = 1$, where $g \in G$, or, what is the same, by the ideal W containing the elements of the form $(g - 1)^m$, where g ranges over all elements of the group G. In particular, the ideal W must contain the element $\widetilde{\Delta}_{m,l} = ((1 + z_1) \cdots (1 + z_l) - 1)^m$, which is equal, by (40.1), to the expression $\sum_{a_1, \dots, a_m} a_1 \cdots a_m$, where each $a_i = z_{i_1} z_{i_2} \cdots z_{i_t}$ $(i_1 < i_2 < \cdots < i_t)$ is a nonempty word, and the sum is extended over all distinct words $a_1 \cdots a_m$ not containing two identical letters. Denote by $\Delta_{m,l}(z_1, \dots, z_l)$ the homogeneous component of the element $\widetilde{\Delta}_{m,l}$ of highest degree. Then

$$\Delta_{m,l}(z_1, \dots, z_l) = \sum_{a_1, \dots, a_m} a_1 \cdots a_m, \tag{40.2}$$

where each nonempty word a_i is of the form $z_{i_1} z_{i_2} \cdots z_{i_t}$ $(i_1 < i_2 < \cdots < i_t)$, and the sum is extended over all $a_1, \dots, a_m$ for which $a_1 \cdots a_m$ has exactly one occurrence of each letter z_i $(i = 1, 2, \dots, l)$. It is clear that any other homogeneous components of the element $\widetilde{\Delta}_{m,l}$ are (for appropriately chosen l and $z_{j_1}, z_{j_2}, \dots, z_{j_l}$) the sums of elements of the form

$$\Delta_{m,l}(z_1, \dots, z_l)|_{z_1 = z_{j_1}, \dots, z_l = z_{j_l}}. \tag{40.3}$$

Therefore, if the ideal W contains all elements of the form (40.3), then $\widetilde{\Delta}_{m,l} \in W$. Moreover, since in the algebra R any mapping ψ of the generators z_i to the elements

$\pm z_{i_k}$ is extendable to a homomorphism, and the ideal W is stable under all these homomorphisms, we have the equality $(g-1)^m = 0$ in the quotient algebra R/W for any $g \in G$. Hence, if we add relations (40.3) to the defining relations (40.1) of the algebra R considering them also as defining relations, then the group generated by $1 \pm z_i$ $(i = 0, 1, \dots)$ will have the exponent $m = p^k$. In the algebra R we have the equality

$$c(g_1, \dots, g_k) = 1 + c(z_1, \dots, z_k), \tag{40.4}$$

where $c(g_1, \dots, g_k)$ is a certain group commutator of the elements $g_1, \dots, g_k$ and $c(z_1, \dots, z_k)$ is the corresponding Lie commutator. Therefore, the solvability problem for the group G is reduced to the solvability problem for the Lie algebra L generated by $z_0, z_1, \dots$ with respect to the commutation operation $[x, y] = xy - yx$.

Let us consider $\Delta_{m,l}(z_1, \dots, z_l)$ as a polynomial in noncommuting variables $z_1, \dots, z_l$ defined by equality (40.2). The following lemma reflects the main property of this polynomial.

LEMMA 40.1. *For any polynomials a and b in the free associative algebra we have the equality*

$$\begin{aligned}\Delta_{m,l}(z_1, \dots, z_l)|_{z_i=a, z_{i+1}=b} &- \Delta_{m,l}|_{z_i=b, z_{i+1}=a} \\ &= \Delta_{m,l-1}(z_1, \dots, z_{l-1})|_{z_i=[a,b], z_{i+1}=z_{i+2}, \dots, z_{l-1}=z_l}.\end{aligned} \tag{40.5}$$

PROOF. Let us represent the sum on the right-hand side of equality (40.2) in the form of two summands f_1 and f_2. Let f_1 contain all monomials $a_1 \cdots a_m$ in which the variables z_i and z_{i+1} occur in distinct words a_j and f_2 contain all monomials in which the variables z_i and z_{i+1} occur in the same word a_j. From the definition of $\Delta_{m,l}(z_1, \dots, z_l)$ it follows that the polynomial f_1 is symmetric with respect to z_i and z_{i+1}. Therefore, the left-hand side of equality (40.5) coincides with the polynomial $f = f_2|_{z_i=a, z_{i+1}=b} - f_2|_{z_i=b, z_{i+1}=a}$. Equality (40.2) implies

$$f_2(z_1, \dots, z_l) = \Delta_{m,l-1}(z_1, \dots, z_{l-1})|_{z_i=z_i z_{i+1}, z_{i+1}=z_{i+2}, \dots, z_{l-1}=z_l}.$$

The lemma is proved.

This lemma implies that if the algebra R is given by the defining relations (40.1) and (40.3), then the identity $\Delta_{m,l} = 0$ holds in the pair (R, L). Moreover, if $W_{m,k}$ is the ideal of weak identities generated by all $\Delta_{m,k}$ for m fixed and $l < k$, then

$$\Delta_{m,k}(z_1, \dots, z_k) = \Delta_{m,k}(z_{i_1}, \dots, z_{i_k}) \pmod{W_{m,k}}, \tag{40.6}$$

where $z_{i_1}, \dots, z_{i_k}$ is a permutation of $z_1, \dots, z_k$.

We note another property of the polynomial $\Delta_{m,l}$: in the free associative algebra for some multilinear polynomial $h_{m,l}$ with integer coefficients we have the equality

$$\Delta_{m,l}(z_1, \dots, z_{l-2}, t, t) = \Delta_{m,l-1}(z_1, \dots, z_{l-1})|_{z_{l-1}=t^2} + 2h_{m,l}|_{z_{l-1}=t, z_l=t}, \tag{40.7}$$

where t is an arbitrary free variable.

Second step. Let K be an algebraically closed field of characteristic zero and let $\mathbb{Z}$ be the ring of integers. Using the commutative K-algebras $\mathcal{F}$ and $\overline{\mathcal{F}}$, we construct, from the pair $(U(\mathfrak{g}), \mathfrak{g})$ and the triple $\{K, \mathfrak{g}, U(\mathfrak{g})\}$, the free pair $(F, \mathfrak{L})$ and the free algebra with three supports $\{\mathcal{F}_0, \mathfrak{L}_1, F_1\}$ with the set of free generators $X = \{x_i = \sum_{q=1}^{3} x_i^{(q)} e_q \in \widetilde{\mathfrak{g}} \mid i = 0, 1, 2, \dots\}$ and also their quotient objects $(\overline{F}, \overline{\mathfrak{L}})$ and $\{\overline{\mathcal{F}}_0, \overline{\mathfrak{L}}_1, F_1\}$ (see 35.2 and 35.3 in §35). Let $A(\gamma)$, $\overline{A}(\gamma)$, $\pi\colon {}_{\mu}A_{\lambda}(\gamma) \to A_{\lambda}(\gamma)$, and $\alpha\colon B \to \mathcal{E}$ denote the same objects as in Subsection 35.3, and let the bilinear pairing and the bilinear form $\widetilde{\mathfrak{b}}_0$ be defined in the same way as in Subsection 35.4.

Denote by L the $\mathbb{Z}$-submodule of $\mathfrak{L}$ generated by all commutators in $x_0, x_1, \dots$ and by $\widetilde{L}$ the $\mathbb{Z}$-submodule in $\mathfrak{L}_1$ generated by the elements of the form

$$(v_1, w_1) \cdots (v_k, w_k) v, \tag{40.8}$$

where $v_i, w_i, v \in L$, $k = 0, 1, 2, \dots$. From identity (35.1) it follows that $\widetilde{L}$ and L are Lie rings. We put ${}_{\mu}\widetilde{L}_{\lambda} \rightleftharpoons \widetilde{L} \cap {}_{\mu}A_{\lambda}(\gamma)$.

LEMMA 40.2. *The elements of two types:*

$$\begin{aligned} &\textit{the first type } (x_{i_1}, x_{j_1}) \cdots (x_{i_k}, x_{j_k}) x_n, \\ &\textit{the second type } (x_{i_1}, x_{j_1}) \cdots (x_{i_k}, x_{j_k})[x_m, x_n], \end{aligned} \tag{40.9}$$

are the generators of the $\mathbb{Z}$-module $\widetilde{L}$. Moreover, for any element $u \in {}_{\mu}\widetilde{L}_{\lambda}$ we have $\pi(u) \in {}_{\varnothing}\widetilde{L}_{\lambda}$.

PROOF. By the discussion in the proof of Lemma 35.6 and identity (35.1), for the proof of the first assertion of the lemma it suffices to prove that the element $([x_i, x_j], x_t)v$, where v is equal to x_n or $[x_m, x_n]$, can be expressed as a linear combination of elements of the form (40.9). For $v = x_n$ this follows from identity (35.2), and for $v = [x_m, x_n]$ we are first to use the fact that the operation $[\,,\,]$ is $\mathcal{F}_0$-bilinear and then apply (35.2) and (35.1).

In order to prove the second part of the assertion of the lemma, it suffices to consider the case when u is an element of the form (40.9). We shall give the proof by induction on the number k of round brackets present in the notation of the element u, using relations (35.17)–(35.21) for $\Delta = \gamma$.

The base of induction is $k = 0$, $\pi([x_i, x_j]) = 0$. Let $k > 0$. If $u \in \widetilde{L}_{\lambda}$, then $\pi(u) = u$, and everything is proved. If in u there is a generator x_i occurring twice, then the following three cases are possible.

1. Both occurrences of x_i are placed into square brackets. Then $u = 0$ and $\pi(u) = 0$.

2. $u = (x_i, x_i) \cdot u'$. Then $\pi(u) = \pi(3u')$ (see formula (35.17)).

3. $u = (x_i, x_j)u'$. Then, applying (35.18), we obtain $\pi(u) = \pi(u'|_{x_i = x_j})$.

In the latter two cases we have $\pi(u) \in \widetilde{L}$ by the induction assumption. The lemma is proved.

In the following statements we denote by $|\lambda|$ the cardinality of the finite set λ.

LEMMA 40.3. *If* $|\mu| + |\lambda| > 1$, $f \in L \cap {}_{\mu}A_{\lambda}(\gamma)$, *and* $g \in \widetilde{L}_{\lambda}$, *then* $\frac{1}{2\gamma}\widetilde{\alpha}(fg)$ *is an integer.*

PROOF. It suffices to consider the case in which f is a commutator. If $|\lambda| = 0$, then, by Lemma 40.2, $\pi(f) \in \widetilde{L}$ and we have $\pi(f) = 0$ and $\frac{1}{2\gamma}\widetilde{\alpha}(fg) = \frac{1}{2\gamma}(\widetilde{\alpha}(f))\widetilde{\alpha}(g) = 0$. Let $x_i \in \lambda$, then $f = [x_i, x_j, \dots, x_l, x_k]$, and from the first determining property of the $\boldsymbol{\alpha}$-function, which is valid for $\widetilde{\alpha}$, too, it follows that for $g' = g|_{x_i=[x_j,[\dots,[x_l,x_i]]]}$ we have

$$\frac{1}{2\gamma}\widetilde{\alpha}(fg) = \frac{1}{2\gamma}\widetilde{\alpha}([x_i, x_k]g') = \frac{1}{2\gamma}\widetilde{\alpha}([x_i, x_k]\pi(g'));$$

applying Lemma 37.2 and formula (35.20) to the right-hand side, we see that for some integer β

$$\frac{1}{2\gamma}\widetilde{\alpha}(fg) = \frac{\beta}{2\gamma}\widetilde{\alpha}([x_i, x_k][x_k, x_i]) = \frac{\beta}{2\gamma}\widetilde{\alpha}([x_i, x_k, x_k]x_i) = \beta.$$

The lemma is proved.

Let us introduce a bilinear pairing of $\mathbb{Z}$-modules L_λ and $\widetilde{L}_\lambda$ by setting $\langle f, g\rangle \rightleftharpoons \frac{1}{2\gamma}\widetilde{\mathfrak{b}}_0(f, g)$.

COROLLARY. *For* $|\lambda| > 1$ *the pairing* $\langle\, ,\, \rangle$ *takes integer values.*

It is clear that L is the relatively free Lie ring with free generators $x_0, x_1, x_2, \dots$ and L is a free module over $\mathbb{Z}$. Let p be a prime. We put

$$\operatorname{Ann}_{L_\lambda}\langle\, ,\, \rangle \rightleftharpoons \{ f \in L_\lambda \mid \langle f, g\rangle = 0 \pmod p \text{ for any } g \in \widetilde{L}_\lambda\}.$$

LEMMA 40.4. *Let* $|\lambda| > 1$ *and let* $f \in L_\lambda$ *be some multilinear Lie polynomial; let* $V(f)$ *be the ideal of Lie identities generated in* L *by this polynomial. Then for* $f \in \operatorname{Ann}_{L_\lambda}\langle\, ,\, \rangle$ *we have* $V(f) \cap L_\mu \subseteq \operatorname{Ann}_{L_\mu}\langle\, ,\, \rangle$ *for any* μ.

The proof is the same as the one of Proposition 8.1. Here we are to use Lemma 40.2, Proposition 35.1, and Corollary of Lemma 40.3.

Third step. Lemma 40.4 permits us to define, in accordance with the general conception described in Chapter 2, the ideal of Lie identities $\mathfrak{V}_p$ in the ring L for which $\mathfrak{V}_p \cap L_\lambda = \operatorname{Ann}_{L_\lambda}\langle\, ,\, \rangle$. The quotient ring $L_p \rightleftharpoons L/\mathfrak{V}_p$ is then a relatively free Lie algebra over a field with p elements.

LEMMA 40.5. *The Lie algebra* L_p *is nonsolvable.*

PROOF. Denote by r_n the multilinear Lie commutator in 2^n variables corresponding to the solvability identity of order n. Let $n = 1$. Then, as it was shown in the proof of Lemma 40.3, we have $\langle [x_i, x_k], [x_k, x_i]\rangle = 1$. Let $n > 1$. We represent r_n in the form

$$r_{n-2}\big|_{x_1=[[x_1,x_2],[x_3,x_4]],\dots,x_{2^n-2}=[[x_{2^n-3},x_{2^n-2}],[x_{2^n-1},x_{2^n}]]}.$$

From inductive arguments it follows that for

$$f_1 = r_{n-2}\big|_{x_1=[x_1,x_4],\dots,x_{2^n-2}=[x_{2^n-3},x_{2^n}]}$$

there exists an element $g_1 \in \widetilde{L}_\lambda$ such that $\langle f_1, g_1\rangle \neq 0 \pmod p$. From relations (35.17)–(35.21) it follows that

$$\begin{aligned}[][[\overline{x}_i, \overline{x}_j], [\overline{x}_k, \overline{x}_l]](\overline{x}_j, \overline{x}_k) &= \frac{\overline{x}_k^2}{\delta}[[\overline{x}_i, \overline{x}_j], [\overline{x}_j, \overline{x}_l]] \\ &= \frac{\overline{x}_k^2}{\delta} \cdot \frac{1}{2}[\overline{x}_i, [\overline{x}_j, [\overline{x}_j, \overline{x}_l]]] = \frac{\overline{x}_k^2}{\delta}\frac{\overline{x}_j^2}{\delta}[\overline{x}_i, \overline{x}_l].\end{aligned}$$

Applying this equality 2^{n-2} times, we obtain

$$\langle r_n, (x_2, x_3)\cdots(x_{2^n-2}, x_{2^n-1})g_1\rangle = \langle f_1, g_1\rangle \neq 0 \pmod p.$$

The lemma is proved.

Denote by $f_{m,l}(x_0, x_1, \dots, x_l)$ the Lie polynomial

$$x_0\Delta_{m,l}(\operatorname{ad} x_1, \dots, \operatorname{ad} x_l).$$

LEMMA 40.6. *Let $p = 2$ and $m > 2$. Then in the Lie algebra L_2 we have the identities $f_{m,l} = 0$.*

PROOF. We must verify that $\langle f_{m,l}, g\rangle = 0 \pmod 2$ for any generator g of the form (40.9) of the $\mathbb{Z}$-module $\widetilde{L}_\lambda$. We shall prove this by induction on l.

We prove first the possibility of the inductive step. Let $l > 3$; then $|\lambda| > 4$, and in the element $g \in \widetilde{L}_\lambda$ there are at least two pairs of round brackets. We take the pair not containing x_0. Then we can consider that $g = (x_i, x_j)g_1$, where $i, j \neq 0$. From Lemma 40.1 and equality (40.6) it follows that for any permutation σ of the set $\{1, 2, \dots, l\}$ the Lie identity

$$f_{m,l}(x_0, x_1, \dots, x_l) - f_{m,l}(x_0, x_{\sigma(1)}, \dots, x_{\sigma(l)}) = 0$$

is a consequence of the Lie identities $f_{m,k} = 0$, where $k < l$, and, therefore, holds in L_2, by induction assumption and Lemma 40.4. Hence, if $\langle f_{m,l}(x_0, x_1, \dots, x_l), g\rangle = 0 \pmod p$, then the equality remains valid for any permutation of $x_1, \dots, x_l$ in $f_{m,l}$. Therefore, we can assume that x_i and x_j are standing on the last place and, by the second property of the $\boldsymbol{\alpha}$-function, it suffices to check that for any $g_1 \in \widetilde{L}_{\lambda\setminus\{x_{l-1}, x_l\}}$ we have the equality

$$\langle f_{m,l}(x_0, x_1, \dots, x_l), (x_{l-1}, x_l)g_1\rangle = 0 \pmod 2.$$

Applying successively relations (35.18) and (40.7), we obtain

$$\begin{aligned}\langle f_{m,l}, g\rangle &= \frac{1}{2\gamma}\widetilde{\alpha}(f_{m,l}(\overline{x}_0, \overline{x}_1, \dots, \overline{x}_{l-2}, \overline{x}_{l-1})\overline{g}_1) \\ &= \frac{1}{2\gamma}\widetilde{\alpha}(\overline{x}_0\Delta_{m,l-1}(\operatorname{ad}\overline{x}_1, \dots, \operatorname{ad}\overline{x}_{l-2}, (\operatorname{ad}\overline{x}_{l-1})^2)\overline{g}_1) \\ &\quad + \frac{1}{2\gamma}\widetilde{\alpha}(\overline{x}_0 h_{m,l}(\operatorname{ad}\overline{x}_1, \dots, \operatorname{ad}\overline{x}_{l-2}, \operatorname{ad}\overline{x}_{l-1}, \operatorname{ad}\overline{x}_{l-1})\overline{g}_1)\cdot 2.\end{aligned} \tag{40.10}$$

The second term on the right-hand side of this equality is divided by 2 by Lemma 40.3. In the first term any left-normed commutator has the form

$$[v, \overline{x}_{l-1}, \overline{x}_{l-1}, \dots]$$

and is equal to $2(\overline{x}_{l-1}^2/\gamma)[v, \dots]$ by equality (35.20). Therefore, by Lemma 40.3, the first term is also divided by 2. This proves the possibility of an inductive step.

The base of induction is $l = 3 = m$. It can be checked directly that over a field of characteristic 2 the Lie polynomial $f_{3.3}$ is symmetric with respect to all its variables. Therefore, by the second property of the $\boldsymbol{\alpha}$-function, it suffices to make a single check of the equality $\langle f_{3.3}, g\rangle = 0 \pmod 2$ for $g = [x_0, x_1](x_2, x_3)$, which can be done by using (40.10) for $m = l = 3$. The lemma is proved.

Fourth step. Since the Lie algebra L_2 has a natural gradation with respect to multi-homogeneous Lie polynomials, the associative subalgebra R generated in $\operatorname{End}_{\mathbb{Z}_2} L_2$ by elements $z_i = \operatorname{ad} x_i$ has the gradation induced by the Lie algebra L_2. Therefore, the ideal W generated in the algebra R by the elements on the left-hand side of (40.1) has the zero intersection with the direct sum $\sum_\lambda R_\lambda$, where R_λ is the space of multilinear polynomials of the algebra R in the set of generators $\lambda = \{z_{i_1}, \dots, z_{i_k}\}$. Let us join a unit to the quotient algebra R/W and generate by the elements $1 \pm z_i$ $(i = 1, 2, \dots)$ the group G. By Lemma 40.6, in R/W we have the equalities $\Delta_{3.l}(z_{i_1}, \dots, z_{i_l}) = 0$. Therefore, as it was shown at the first step, the identity $g^4 = 1$ holds in G. From Lemma 40.5 and (40.4) it follows that the group G is nonsolvable. The theorem is proved.

§41. A basis of identities for the full matrix algebra of order two

THEOREM 41.1. *If the ground field K is of characteristic zero, then the set of all identities of the algebra M_2 of matrices of order two is equivalent to the identities of degrees* 4, 5, *and* 6, *that are valid in M_2.*

REMARK 41.1. In essence, we shall prove that if the characteristic of the field K is distinct from two and all multilinear Lie identities of the algebra $\mathfrak{g}$ are equivalent to a finite set of Lie identities, then all multilinear identities of the associative algebra M_2 are equivalent to a finite set of identities. Therefore, the theorem will follow from Theorem 36.1 and from the fact that over a field of characteristic zero any set of identities is equivalent to some set of multilinear identities.

Let F be the absolutely free associative algebra with free generators $x_1, x_2, \dots$. Our aim is to find a finite set of identities P satisfying in M_2 such that all multilinear identities of the algebra M_2 are consequences of P. This set will be completely determined in the course of the proof of Theorem 41.1. Running ahead, we note that P will consist of Lie identities (35.14)–(35.16), multilinear identities

$$\sum_{\sigma \in S_4} \operatorname{sign}(\sigma) x_{\sigma(1)} x_{\sigma(2)} x_{\sigma(3)} x_{\sigma(4)} = 0, \tag{41.1}$$

$$[(v_1 \circ v_2), z] = 0, \tag{41.2}$$

(41.3)
$$4[z, x](v_1 \circ v_2) = [z, v_1, v_2, x] + [z, v_2, v_1, x] - [x, v_1, z, v_2] - [x, v_2, z, v_1],$$

where $(v_1 \circ v_2) = v_1v_2 + v_2v_1$, v_1 and v_2 are commutators of weight two, and of another three multilinear identities of degree 6, which are dual, in a certain sense, to the Lie identities (35.14)–(35.16).

REMARK 41.2. An analysis of identities of degrees 4, 5, and 6, developed by W. Drensky, shows that all these identities are consequences of identities (41.1) and (41.2) (see [Dre2]). Therefore, we can take for P the set of two identities (41.1) and (41.2).

LEMMA 41.1. *Identities* (35.14)–(35.16), (41.1), *and* (41.2) *hold in the algebra* M_2.

PROOF. Since over a field whose characteristic is distinct from two any matrix of second order can be represented as a sum of a scalar matrix and a matrix with zero trace, and the identities under consideration are multilinear and turn into true equalities under the substitution of a unit instead of any one of its variables, it suffices to verify that all these identities are satisfied in the pair $(M_2, \mathfrak{g})$, where $\mathfrak{g} = \mathrm{sl}(2, K)$ is the simple three-dimensional Lie algebra of second order matrices with zero trace. However, this follows from the identities $2(x \circ y) = (x, y)$, (35.13), (6.27), and (35.3) satisfied in $\{K, \mathfrak{g}, M_2\}$ and from equality (35.5) and Remark 35.1. The lemma is proved.

LEMMA 41.2. *Modulo identities* (41.2) *and* (41.3), *any polynomial of the form* $a = u_1 \cdots u_n$, *where* u_i *are commutators in the generators* $x_1, x_2, \ldots$ *of weight* ≥ 2, *is representable in the form*

$$a = c + v; \tag{41.4}$$

here v *is a Lie polynomial and* $c = \sum_i \beta_i \left(u_i' \circ [x_i, x_t]\right)$, *where* u_i' *are commutators in the generators* $x_1, x_2, \ldots$ *of weight* ≥ 2, *and we can take for* x_t *any generator on which* a *depends.*

The proof will be given by induction on the number n. The base of induction is $n = 1$. Let $n > 1$. Then

$$u_1 \cdots u_{n-1}u_n = u_1 \cdots u_{n-2}\left\{\tfrac{1}{2}(u_{n-1} \circ u_n) + \tfrac{1}{2}[u_{n-1}, u_n]\right\}.$$

If $n > 2$, then the second summand is reducible to the form we need by the inductive assumption, and identity (41.3) implies that $u_{n-2}(u_{n-1} \circ u_n)$ is a Lie polynomial and the first term is also reducible to the form (41.4). Let $n = 2$. Then

$$a = \tfrac{1}{2}\{(u_1 \circ u_2) + [u_1, u_2]\}.$$

From (41.3) we obtain the identity

$$([v_1, z] \circ v_2) = (v_1 \circ [z, v_2]). \tag{41.5}$$

Since any commutator depending on x_t can be represented as a linear combination of left-normed commutators of the form $[x_t, \ldots]$, we see that $(u_1 \circ u_2)$ (and therefore a, too) can be reduced to the form (41.4) with the help of (41.5). The lemma is proved.

LEMMA 41.3. *For any associative algebra A with unit the set of all multilinear identities satisfied in it is equivalent to a subset of multilinear identities of the form $f(x_1, \dots, x_l) = 0$, where $f|_{x_1=1}, \dots, f|_{x_l=1}$ are zero polynomials.*

PROOF. Let $g(x_1, \dots, x_l) = 0$ be a multilinear identity of the algebra A, where $g|_{x_1=1}, \dots, g|_{x_l=1}$ are zero polynomials. Then $g|_{x_{l+1}=1} = 0$ is also an identity of the algebra A and, modulo identities of lesser degree, $g = 0$ is equivalent to the identity $f = 0$, where $f = g - g|_{x_{l+1}=1}x_{l+1}$. However, $f|_{x_1=1}, \dots, f|_{x_{l+1}=1}$ are zero polynomials, and the assertion of the lemma follows from the inductive arguments.The lemma is proved.

Since any associative multilinear polynomial $f(x_1, \dots, x_l)$ for which $f|_{x_1=1}, \dots, f|_{x_l=1}$ are zero polynomials can be represented as a linear combination of elements of the form $u_1 \cdots u_t$, where $u_1, \dots, u_t$ are commutators of weight > 1, Lemmas 41.3 and 41.2 imply that, modulo identities (41.2) and (41.3), a basis of identities of the algebra M_2 can be chosen among the identities of the form $c + v = 0$ in which $v(x_1, \dots, x_l)$ is a multilinear Lie polynomial and $c = \sum_i \beta_i \left(u_i' \circ [x_i, x_l]\right)$, where u_i' is a multilinear Lie polynomial in $x_1, \dots, \widehat{x_i}, \dots, x_{l-1}$ and $\beta_i \in K$. From identity (41.2) it follows that all values of the polynomial c on M_2 are scalar matrices. Obviously, the values of the Lie polynomial v have zero trace. Therefore, if char $K \neq 2$, then an identity of the form $c + v = 0$ holds in M_2 if and only if the identities $c = 0$ and $v = 0$ hold in M_2. By Theorem 36.1, the identity $v = 0$ is a consequence of identities (35.14)–(35.16). Therefore, if P contains identities (35.14)–(35.16), (41.2), and (41.3), then, modulo P, we can choose a basis of identities for the algebra M_2 among the identities $c = 0$.

LEMMA 41.4. *Let $c = \sum_i \beta_i \left(u_i \circ [v_i, x_l]\right)$ be a multilinear polynomial, where $\beta_i \in K$, u_i and v_i are commutators in $x_1, \dots, x_{l-1}$, and the weight of u_i is ≥ 2. Then the identity $c = 0$ holds in the algebra M_2 if and only if $\sum_i \beta_i[u_i, v_i] = 0$ is a Lie identity for the Lie algebra $\mathfrak{g} = \mathrm{sl}(2, K)$.*

PROOF. Let $c = 0$ be an identity for M_2. Then $c = 0$ is an identity for the pair $(M_2, \mathfrak{g})$, and from the identity $2(x \circ y) = (x, y)$ and from the invariance of the form $(\,,\,)$ we conclude that $\left(\sum_i \beta_i[u_i, v_i], x_l\right) = 0$ in $\mathcal{F}_0$. Since the Killing form on $\mathfrak{g}$ is nondegenerate, this implies that $\sum_i \beta_i[u_i, v_i] = 0$ is an identity for the Lie algebra $\mathfrak{g}$.

Conversely, let $\sum_i \beta_i[u_i, v_i] = 0$ be a Lie identity for the Lie algebra $\mathfrak{g}$. Then, using the argument just produced in the opposite direction, we find that $c = 0$ is an identity for the pair $(M_2, \mathfrak{g})$. Since the polynomial c is multilinear and it vanishes under the substitution of a unit instead of any one of its variables, $c = 0$ is an identity for the algebra M_2. The lemma is proved.

In the variety of the Lie algebras we have the Jacobi identity $[x_1, x_2, x_3] + [x_3, x_1, x_2] + [x_2, x_3, x_1] = 0$; therefore, by Lemma 41.4, in M_2 we have the identity

$$([x_1, x_2] \circ [x_3, x_l]) + ([x_3, x_1] \circ [x_2, x_l]) + ([x_2, x_3] \circ [x_1, x_l]) = 0, \tag{41.6}$$

which is, as can easily be seen, a standard identity of degree 4; see (41.1).

In the next lemma by basic commutators we shall mean the basic commutators of F. Hall. We recall their definition. Let all commutators in the generators $x_1, \dots, x_{l-1}$ be totally ordered in such a way that if the weight of u is less that the weight of v, then $u < v$. Then the basic commutators are defined as follows:

1) the basic commutators of weight 1 are $x_1, \dots, x_{l-1}$;

2) if the basic commutators of weights less that n are defined, then the commutator $[u, v]$ of weight n is a basic one if and only if: a) u and v are basic commutators and $u > v$, b) if $u = [u_1, u_2]$, then $u_2 \leq v$. From Theorem 52.2 it follows that the basic commutators form a basis of the absolutely free Lie algebra with free generators $x_1, \dots, x_{l-1}$.

LEMMA 41.5. *Let u and v be Lie polynomials in the generators $x_1, \dots, x_{l-1}$ and the weight of u is ≥ 2 and, moreover, let $[u, v] = \sum_i \delta_i [u_i, v_i]$, where $[u_i, v_i]$ are basic commutators in $x_1, \dots, x_{l-1}$; then the equality*

$$(u \circ [v, x_l]) = \sum_i \delta_i (u_i \circ [v_i, x_l]) \tag{41.7}$$

is a consequence of identities (41.6) *and* (41.2).

The proof will repeat one of the possible proofs of the representability of any Lie polynomial via basic commutators, and we shall use identity (41.6) instead of the Jacobi identity and the identity (41.5) in the form

$$(z_1 \circ [z_2, x_l]) = -(z_2 \circ [z_1, x_l]),$$

where z_1 and z_2 are commutators with weight greater than one, instead of the identity $[x, x] = 0$. (As it was mentioned in the proof of Lemma 41.2, identity (41.5) is a consequence of identity (41.2).)

From (41.8) and the fact that any commutator of weight n can be represented in the form of a linear combination of basic commutators of the same weight it follows that it suffices to prove the lemma for the case in which u and v are basic commutators, $u > v$, and the weight of u is ≥ 2.

We shall prove equality (41.7) by induction on the basic commutator v. The base of induction is the case in which $[u, v]$ is a basic commutator.

Suppose that for commutators of the form $[w_1, w_2]$, where the weight of $[w_1, w_2]$ is equal to the weight of $[u, v]$, the weight of w_1 is ≥ 2, $w_1 > w_2 > v$, and w_1 and w_2 are basic commutators, the assertion of the lemma has already been proved. Let us prove our assertion for the commutator $[u, v]$. Since the weight of u is ≥ 2, we have $u = [u_1, u_2]$ and, by the definition of basic commutators, u_1 and u_2 are basic commutators and $u_1 > u_2$. If $u_2 \leq v$, then $[u, v]$ is a basic commutator, and everything is proved. Suppose that $u_2 > v$. Then $[[u_1, u_2], v] = [[u_1, v], u_2] - [[u_2, v], u_1]$, and identity (41.6) implies

$$([u_1, u_2] \circ [v, x_l]) = ([u_1, v] \circ [u_2, x_l]) - ([u_2, v] \circ [u_1, x_l]),$$

therefore, it is sufficient to prove the assertion of the lemma for the commutators $[[u_1, v], u_2]$ and $[[u_2, v], u_1]$. Consider each of these cases.

Let $[u_1, v] = \sum_i \delta_i w_i$, where w_i are basic commutators. Then from the relation $u_1 > u_2$ we see that the weight of w_i is equal to the weight of $[u_1, v]$, which is greater than that of u_2, and, hence, $w_i > u_2$. However, since $u_2 > v$, the lemma is valid for commutators $[w_i, u_2]$ by the induction assumption, and, hence, for the commutator $[[u_1, v], u_2]$, too.

Let $[u_2, v] = \sum_i \varepsilon_i w_i'$, where w_i' are basic commutators. If $w_i' = u_1$, then $[w_i', u_1] = 0$, and the assertion of the lemma is valid in this case by identity (41.8). If $w_i' > u_1$, then from the relation $u_1 > u_2 > v$ it follows that the assertion of the lemma holds for the commutator $[w_i', u_1] = 0$ by the induction assumption. If $w_i' < u_1$, then the weight of w_i' is equal to the weight of $[u_2, v]$, which is greater than that of v, and, therefore, $w_i' > v$ and, by the induction assumption, the assertion of the lemma holds for the commutator $[u_1, w_i']$ and, by (41.8), also for $[w_i', u_1]$. Hence, the assertion of the lemma is proved also for the commutator $[[u_2, v], u_1]$. The lemma is proved.

COROLLARY. *Suppose that in the free Lie algebra with free generators* $x_1, \dots, x_{l-1}$ *the equality* $\sum_i \delta_i[u_i, v_i] = 0$ *holds, where* u_i *and* v_i *are commutators and the weight of* u_i *is* ≥ 2; *then the identity* $\sum_i \delta_i(u_i \circ [v_i, x_l]) = 0$ *is a consequence of identities* (41.6) *and* (41.2).

PROOF. Since the basic commutators form a basis in the free Lie algebra, the element $\sum_i \delta_i[u_i, v_i] = 0$ can be represented as a linear combination of basic commutators with zero coefficients. However, in this case our assertion follows from Lemma 41.5. The corollary is proved.

Let us transfer all the terms of Lie identities (35.14)–(35.16) to the left-hand side and write them in the form $f_2 = 0$, $f_3 = 0$, and $f_4 = 0$. Let us write the Lie polynomials f_i themselves in the form $f_i = \sum_j \beta_{ij}[u_{ij}, v_{ij}]$, where u_{ij} and v_{ij} are commutators and the weight of u_{ij} is ≥ 2. Then the identities

$$f_i' \rightleftharpoons \sum_j \beta_{ij}(u_{ij} \circ [v_{ij}, x_6]) = 0 \qquad (i = 2, 3, 4)$$

are identities for the algebra M_2 by Lemma 41.4.

It was shown above that, modulo the identities (41.2), (41.3), and (35.14)–(35.16), a basis of identities of the algebra M_2 can be chosen among the identities of the form $c = 0$, where $c = \sum_i \beta_i(u_i \circ [v_i, x_l])$ and u_i are commutators of weight ≥ 2. From Lemma 41.4 it follows that $\sum_i \beta_i[u_i, v_i] = 0$ is a Lie identity in $\mathfrak{g}$. From Theorem 36.1 we conclude that in the free Lie algebra we have the equality

$$\sum_i \beta_i[u_i, x_i] = \sum_{ij} \delta_{ij} f_{ij},$$

where the Lie polynomial f_{ij} is obtained from f_i by the substitution of some Lie polynomials instead of its agruments. If f_{ij}' is the associative polynomial obtained from $f_i'|_{x_6 = x_l}$ by the same substitution as in f_i, then from Corollary of Lemma 41.5 we see that the equality

$$\sum_i \beta_i(u_i \circ [x_i, x_l]) = \sum_{i,j} \delta_{ij} f_{ij}'$$

is a consequence of identities (41.1) and (41.2). The identities $f_{ij}' = 0$ are consequences of the identities $f_2' = 0$, $f_3' = 0$, and $f_4' = 0$. Therefore, we have proved that all identities for the algebra M_2 follow from the set of identities $P = \{(41.1)\text{–}(41.3), (35.14)\text{–}(35.16), f_i' = 0 \ (i = 2, 3, 4)\}$. The theorem is proved.

Comments

The results expounded in this chapter form a development of the ideas laid in §6. Theorems 36.1, 37.1, and 37.2 were first published in author's papers [Raz4, Raz15]. In this connection, the solution of the following problem is of interest.

PROBLEM 5.1. Clarify, whether the variety of associative-Lie pairs $(U(\mathfrak{h}), \mathfrak{h})$, where the Lie algebra $\mathfrak{h}$ is the direct sum of a finite set of isomorphic copies of the simple three-dimensional Lie algebra $\mathfrak{g}$ over a field of characteristic zero, has the Specht property. Find explicitly identities providing that this variety has the Specht property.

Theorem 38.1 is due to Trishin [Tri1]. The proof of it given in §38 is based on the papers [Raz15, Raz19]. These results were developed later in the papers of Trishin and Il′tyakov (see [Tri2, Il]), where the Specht property of varieties of representations of Lie triple systems and varieties of alternative and Jordan algebras was studied. Theorem 39.1 was announced by Zyrichev in 1983 (see [Zyr]). Theorems 39.1 and 39.2 suggest an idea to express the following conjecture:

PROBLEM 5.2. In any variety of associative-Lie pairs $\mathfrak{V}_\alpha$ (see §8), where the $\boldsymbol{\alpha}$-function $\alpha\colon B \to K$ ($\operatorname{char} K = p > 3$) is constructed for any irreducible p-representation of an arbitrary simple classical Lie K-algebra $\mathfrak{g}$, the Engel identity of order $(p-1)$ is valid.

The proof of Theorem 40.1 was first given in the author's paper [Raz1]. An example of a just nonsolvable variety of $(p-2)$th Engel Lie algebras, for which an analog of the Campbell–Hausdorff formula holds, is given in the paper [Raz9]. Theorem 41.1 is contained in the paper [Raz4]. The finite basis property of identities for an arbitrary algebra A with unit from $\operatorname{var} M_2$ ($\operatorname{char} K = 0$) was proved in [Raz5], where it was also shown that any such subvariety $\operatorname{var} A$ is generated by the algebra A', which is a finitely generated module over its center. The Specht property of the variety $\operatorname{var} M_2$ ($\operatorname{char} K = 0$) was proved later by Latyshev [Lat2].

PROBLEM 5.3. Do the identities of the associative matrix algebra of second order M_2 over an infinite field of characteristic 2 have the finite basis property?

CHAPTER VI

VARIETIES GENERATED BY LIE ALGEBRAS OF CARTAN TYPE

In this chapter we shall continue the investigation started in §§2 and 21 and concentrate our main attention on the problem of description of simple algebras in varieties of Lie algebras of exponential type.

§42. Summary of results

DEFINITION 42.1. A simple Lie algebra $\mathfrak{g}$ over an arbitrary field K is called a *Lie algebra of Cartan type* if there exists an extension $\widehat{K}$ of the centroid C of the algebra $\mathfrak{g}$ for which the $\widehat{K}$-algebra $\widehat{\mathfrak{g}} = \widehat{K} \otimes_K \mathfrak{g}$ possesses a proper $\widehat{K}$-subalgebra of finite codimension.

In §43 it will be shown that any simple Lie K-algebra of Cartan type satisfies a nontrivial Lie identity of the form

$$\sum_{\sigma \in S_N} \beta_\sigma [z, x_{\sigma(1)}, \dots, x_{\sigma(N)}] = 0, \tag{42.1}$$

where not all $\beta_\sigma \in K$ are zero and σ ranges over all permutations of the set $\{1, \dots, N\}$. At present no other simple Lie algebras satisfying some Lie identity of the form (42.1) are known. This motivates us to state the following conjecture.

CONJECTURE. Any simple Lie algebra satisfying a nontrivial Lie identity is a Lie algebra of Cartan type.

In §44 we shall prove that over a field K of characteristic zero an arbitrary simple Lie algebra $\mathfrak{g}$ such that $\widehat{\mathfrak{g}} = \widehat{K} \otimes_K \mathfrak{g}$ possesses a proper $\widehat{K}$-subalgebra of finite codimension n can be embedded in the Lie algebra $\widetilde{W}_n(\widehat{K})$ of all derivations of the power series algebra $\widehat{K}[[t_1, \dots, t_n]]$, where the Lie $\widehat{K}$-algebra $\widetilde{W}_n(\widehat{K})$ is simple and contains a unique Lie subalgebra of codimension n. Therefore, the Lie algebras $\widetilde{W}_n(\widehat{K})$ are universal simple objects containing an arbitrary simple Lie algebra of Cartan type. Similar universal simple objects can be defined also over fields of positive characteristic (see §44).

In §46 for fields of characteristic different from two we shall describe the properties of simple Lie algebras satisfying the standard Lie identity of degree 5:

$$\sum_{\sigma \in S_4} \operatorname{sign}(\sigma)[z, x_{\sigma(1)}, x_{\sigma(2)}, x_{\sigma(3)}, x_{\sigma(4)}] = 0. \tag{42.2}$$

It will turn out that all simple Lie K-algebras of this kind are precisely all the Lie algebras of Cartan type contained in $\widetilde{W}_1(\widehat{K})$. If for any Lie algebra $\mathfrak{h}$ we denote by var $\mathfrak{h}$ the variety of Lie algebras generated by $\mathfrak{h}$, i.e., the class of all Lie algebras satisfying all identities satisfied in $\mathfrak{h}$, then the result on simple Lie algebras with identity (42.2) can be stated as follows: any simple Lie algebra $\mathfrak{g}$ of the variety var $\widetilde{W}_1(K)$ is a Lie algebra of Cartan type and it can be embedded in the Lie algebra $\widetilde{W}_1(\widehat{K})$ for a suitable extension $\widehat{K}$ of the ground field K.

In §48 we prove an analog of this result for an arbitrary integer n in the case of Lie algebras over fields of characteristic zero. Therefore, the conjecture stated above is true for simple Lie algebras in the varieties var $\widetilde{W}_n(K)$ (char $K = 0$). We now give precise statements of the main results proved in §48.

THEOREM 42.1. *Let $\mathfrak{B}$ be an arbitrary variety of Lie algebras, of exponential type, over a field K of characteristic zero. Then any simple Lie K-algebra in the variety $\mathfrak{B}$ is a Lie algebra of Cartan type.*

(We recall (see §2) that a variety of Lie algebras is called a variety of exponential type if for the dimensions of the subspaces of multilinear polynomials T_l in variables $x_1, \dots, x_l$ in the free Lie algebra of this variety with free generators $x_1, x_2, \dots$ the following asymptotic bound holds: $\dim_K T_l < d^l$, where d is an integer; by Theorem 2.3, any variety var $\widetilde{W}_n(\widehat{K})$ is a variety of exponential type.)

THEOREM 42.2. *Any simple Lie K-algebra $\mathfrak{g}$ in the variety* var $\widetilde{W}_n(K)$ *(where* char $K = 0$*) is a Lie algebra of Cartan type. Moreover, if* $\dim_C \mathfrak{g} > n^2 + 2n$, *then* $\dim \mathfrak{g} = \infty$ *and there exists a monomorphism of* $\mathfrak{g}$ *into* $\widetilde{W}_n(\widehat{K})$.

COROLLARY. *Let K be an algebraically closed field of characteristic zero and let $\mathfrak{g}$, $0 \neq \mathfrak{g} \in$* var $\widetilde{W}_n(K)$, *be a finitely generated Lie algebra. Then there exists a nonzero homomorphism of the algebra $\mathfrak{g}$ either into a finite-dimensional Lie algebra* $\mathfrak{h}$ $(n^2 + 2n \geq \dim_K \mathfrak{h})$ *or into the Lie algebra $\widetilde{W}_n(K)$.*

(This corollary is, in essence, an analog of Hilbert's Nullstellensatz (see §51). It shows that the simple Lie algebra $\widetilde{W}_n(K)$ plays the same role for finitely generated Lie algebras in the variety var $\widetilde{W}_n(K)$ as an algebraically closed field plays for finitely generated commutative algebras.)

REMARK 42.1. In §46 we shall prove that the validity of identity (42.2) in a simple Lie K-algebra $\mathfrak{g}$ is a necessary and sufficient condition for the Lie $\widehat{K}$-algebra $\widehat{\mathfrak{g}} = \widehat{K} \otimes_C \mathfrak{g}$ to have a proper $\widehat{K}$-subalgebra of codimension 1. In the proof of Theorems 42.1 and 42.2 we shall also explicitly describe systems of identities which are responsible for the existence of a proper $\widehat{K}$-subalgebra of finite codimension in a simple Lie algebra $\widehat{\mathfrak{g}}$ (see Proposition 43.1 and Lemma 48.2); however, these systems of identities are infinite. The criteria for the existence of a subalgebra of finite codimension in a Lie algebra and the very line of reasoning in the proofs of Theorems 42.1 and 42.2 show that there are two types of identities that aspire to play the same role for the varieties var $\widetilde{W}_n(K)$ as identity (42.2) plays for the variety var $\widetilde{W}_1(\widehat{K})$. In the case of fields of characteristic zero this is the standard Lie identity

$$\sum_{\sigma \in S_N} \operatorname{sign}(\sigma)[z, x_{\sigma(1)}, \dots, x_{\sigma(N)}] = 0, \tag{42.3}$$

and for fields of positive characteristic this is the identity

$$(42.4) \qquad \sum_{\sigma \in S_N} \operatorname{sign}(\sigma)[z, [x_{\sigma(1)}, y_1], \dots, [x_{\sigma(N)}, y_N]] = 0$$

(the sums are extended over all permutations σ of the set $\{1, \dots, N\}$). The following problems arise in this connection.

PROBLEM 1. Is the variety of Lie algebras determined by identity (42.3) (by identity (42.4)) over a field of characteristic zero (respectively, of positive characteristic) ($N = 1, 2, \dots$) a variety of exponential type?

PROBLEM 2. Is the above conjecture true for a variety of Lie algebras over a field of zero (positive) characteristic if identity (42.3) (identity (42.4)) holds in this variety?

Section 45 deals with the study of the possibility of constructive recovery of the algebra of all differentiable functions on a smooth n-dimensional variety from the Lie algebra of its vector fields. In §21 for the Lie algebra $W_n(K)$ ($\operatorname{char} K = 0$) of all derivations of $\mathcal{F}_n = K[t_1, \dots, t_n]$ we proved the existence of an associative multilinear polynomial $f(x_1, \dots, x_l)$ for which the image of the linear mapping

$$(42.5) \qquad f\colon \operatorname{ad} W_n \otimes \operatorname{ad} W_n \otimes \cdots \otimes \operatorname{ad} W_n \to \operatorname{End}_K W_n$$

coincides with the image of the monomorphism $\varphi_n\colon \mathcal{F}_n \to \operatorname{End}_K W_n$ induced on W_n by its left $\mathcal{F}_n$-module structure. In §45 we shall show that any associative polynomial f of this kind is an important invariant of an arbitrary smooth n-dimensional irreducible affine variety X and it enables us to recover the algebra $K[X]$ of all its regular functions from the Lie algebra $\operatorname{Der}_K K[X]$ of all regular vector fields of X. Moreover, the polynomial f recovers the algebra of regular functions of an m-dimensional affine variety from any regular involutory n-dimensional ($n \leq m$) distribution without singularities (all necessary definitions can be found in the books [BiK, Shaf]). These results are consequences of the following assertion.

THEOREM 42.3. *Let* $\operatorname{char} K = 0$, *let* $\mathcal{E}$ *be a commutative associative* K*-algebra with unit and without nil-elements, let* $\mathfrak{g}$ *be a Lie subalgebra in* $\operatorname{Der}_K \mathcal{E}$ *for which* $\mathcal{E}\mathfrak{g} \subseteq \mathfrak{g}$, *and let* I_m *be the ideal in* $\mathcal{E}$ *generated by all elements of the form*

$$\begin{vmatrix} {}^{g_1}c_1 & \dots & {}^{g_1}c_m \\ \dots & \dots & \dots \\ {}^{g_m}c_1 & \dots & {}^{g_m}c_m \end{vmatrix} = \sum_{\sigma \in S_m} \operatorname{sign}(\sigma) \cdot {}^{g_1}c_{\sigma(1)} \cdots {}^{g_m}c_{\sigma(m)},$$

where $g_1, \dots, g_m \in \mathfrak{g}$, $c_1, \dots, c_m \in \mathcal{E}$, *and* ${}^{g}c$ *is the result of the application of the derivation* g *to the element* c. *Moreover, suppose that* $I_{n+1} = 0$ & $I_n \neq 0$ *for some* $n \in \mathbb{N}$, *and let* $f(x_1, \dots, x_l)$ *be an arbitrary associative multilinear polynomial for which the nonzero mapping* (42.5) *takes values in* $\varphi_n(\mathcal{F}_n)$. *Then* $\mathfrak{g} \in \operatorname{var} \widetilde{W_n}$, *and for the homomorphism* $\varphi\colon \mathcal{E} \to \operatorname{End}_K \mathfrak{g}$ *induced by the left* $\mathcal{E}$*-module structure on* $\mathfrak{g}$ *and for the image* $\mathcal{E}'$ *of the mapping* $f\colon (\operatorname{ad} \mathfrak{g})^l \to \operatorname{End}_K \mathfrak{g}$ *we have the following properties.*

1. $\mathcal{E}' \neq 0$, $\mathcal{E}'$ *is contained in the full ring of quotients of the algebra* $\varphi(\mathcal{E})$ *(in the central closure of* $\varphi(\mathcal{E})$*), and we have*

$$\mathcal{E}'\varphi({}^{\mathfrak{g}}\mathcal{E}) \subset \varphi({}^{\mathfrak{g}}\mathcal{E}) \ \& \ [\mathcal{E}', \operatorname{ad} \mathfrak{g}]\varphi(\mathcal{E}) \subset [\mathcal{E}', \operatorname{ad} \mathfrak{g}].$$

2. *If* ${}^{\mathfrak{g}}\mathcal{E} = \mathcal{E}$, *then* φ *is a monomorphism*, $\mathcal{E}' \subset \varphi(\mathcal{E})$, *and* $\mathcal{E}'$ *is contained in a prime ideal of the algebra* $\varphi(\mathcal{E})$ *if and only if this ideal contains* $\varphi(I_n)$.

3. *If* $I_n = \mathcal{E}$ $(n \geq 1)$, *then* $\varphi(\mathcal{E}) = \mathcal{E}'$.

REMARK 42.2. Let us endow the set of all prime (simple) ideals $\operatorname{Spec}\mathcal{E}$ of the algebra $\mathcal{E}$ with the Zariski topology. Then Theorem 42.3 possesses a natural interpretation in algebraic geometry. Every derivation of the algebra $\mathcal{E}$ determines a vector in the tangent space θ_x at any point $x \in \operatorname{Spec}\mathcal{E}$; therefore, if a Lie subalgebra $\mathfrak{g}$ satisfies the relation $\mathcal{E}\mathfrak{g} \subset \mathfrak{g}$, then it determines an involutory distribution on $\operatorname{Spec}\mathcal{E}$. The condition $I_{n+1} = 0$ is equivalent to the fact that at each point $x \in \operatorname{Spec}\mathcal{E}$ the subspace generated in θ_x by the derivations from $\mathfrak{g}$ is of dimension at most n, i.e., the distribution given by $\mathfrak{g}$ is of dimension $\leq n$ at all points of the spectrum. The condition $I_{n-j} \neq 0$ is equivalent to the fact that at all points of the spectrum the involutory distribution of $\mathfrak{g}$ has the dimension $\geq n - j$. Therefore, Property 3 means that if the involutory distribution has no singularities, then the mapping f recovers the whole algebra $\mathcal{E}$ of regular functions of the variety $\operatorname{Spec}\mathcal{E}$ from this distribution.

§43. Identities in Lie algebras of Cartan type

Let $\mathfrak{g}$ be a simple Lie algebra over a field K of arbitrary characteristic, let $\mathfrak{g}_0$ be its proper subalgebra of finite codimension, let $\mathbb{N}$ be the set of positive integers, and let $\overline{\mathbb{N}} = \mathbb{N} \cup \infty$ be the set of positive integers completed with a maximal element ∞. We define inductively the Cartan prolongation of the subalgebra $\mathfrak{g}_0$ by setting $\mathfrak{g}_{i+1} = \{g \in \mathfrak{g}_i \mid [g, \mathfrak{g}] \subset \mathfrak{g}_i\}$ and $\mathfrak{g}_{-1} = \mathfrak{g}$. This definition implies that $\dim_K \mathfrak{g}_i/\mathfrak{g}_{i+1} < \infty$, $[\mathfrak{g}_i, \mathfrak{g}_j] \subset \mathfrak{g}_{i+j}$, and $\cap_{i=0}^{\infty}\mathfrak{g}_i$ is a proper ideal of the algebra $\mathfrak{g}$. Therefore, $\mathfrak{g} = \mathfrak{g}_{-1} \supset \mathfrak{g}_0 \supset \mathfrak{g}_1 \supset \cdots$ is an exhaustive filtration of the simple Lie algebra $\mathfrak{g}$ in which all quotients $\mathfrak{g}_i/\mathfrak{g}_{i+1}$ $(i = -1, 0, 1, \dots)$ are finite-dimensional.

For each element $w \in \mathfrak{g}$ we denote by $|w|$ the integer i such that $w \in \mathfrak{g}_i$ & $w \notin \mathfrak{g}_{i+1}$. Choose in $\mathfrak{g}$ linearly independent elements $e_1, e_2, e_3, \dots$ and order them in such a way that the following properties hold:

a) if $i > j$, then $|e_j|$ is at most $|e_i|$,

b) for any integer $i \geq 0$ the set of elements e_j for which $|e_j| < i$ is a basis in the quotient space $\mathfrak{g}/\mathfrak{g}_i$.

For any simple Lie algebra $\mathfrak{g}$ and for the filtration constructed for a subalgebra $\mathfrak{g}_0$ of finite codimension we define a function $m_{\mathfrak{g}}^-: \mathbb{N} \to \overline{\mathbb{N}}$ by setting for each positive integer N

$$m_{\mathfrak{g}}^-(N) \rightleftharpoons \begin{cases} \sum_{i=1}^{n} |e_i| & \text{for } N \leq \dim_K \mathfrak{g}, \\ \infty & \text{for } N > \dim_K \mathfrak{g}. \end{cases} \tag{43.1}$$

PROPOSITION 43.1. *In an arbitrary simple Lie K-algebra $\mathfrak{g}$ possessing a proper Lie subalgebra $\mathfrak{g}_0$ of finite codimension, for any multilinear Lie polynomial of the form $r(x_1, \dots, x_N, y_1, \dots, y_t)$ that is skew-symmetric with respect to $x_1, \dots, x_N$ we have the identity*

$$r(x_1, \dots, x_N, y_1, \dots, y_t) = 0 \tag{43.2}$$

for any $t \leq m_{\mathfrak{g}}^-(N)$. In particular, for $m_{\mathfrak{g}}^-(N) \geq 1$ the standard Lie identity of the form (42.2) *holds in* $\mathfrak{g}$.

PROOF. It is clear that if $N > \dim_K \mathfrak{g}$, then identity (43.2) holds in the Lie algebra $\mathfrak{g}$ for any $t \in \mathbb{N}$. Therefore, we shall restrict ourselves to the case in which $N \leq \dim_K \mathfrak{g}$.

We first verify that for $t \leq m_{\mathfrak{g}}^{-}(N)$ all values of the polynomial r belong to the subalgebra $\mathfrak{g}_0$. Since under the substitution of an element of $\mathfrak{g}_{t+N-1}$ for an arbitrary argument of r and arbitrary elements of the Lie algebra $\mathfrak{g}$ for other arguments of r the value of this polynomial belongs to $\mathfrak{g}_0$, by property b) of the "basis" $\{e_i\}$ of the algebra $\mathfrak{g}$ it is enough to execute this verification for the elements e_j for which $|e_j| < N+t-1$, because r is multilinear. It is clear that $r(e_{i_1}, \dots, e_{i_N}, e_{j_1}, \dots, e_{j_t}) \in \mathfrak{g}_k$, where $k = |e_{i_1}| + \dots + |e_{i_N}| + \dots + |e_{j_1}| + \dots + |e_{j_t}|$. If two of the elements $e_{i_1}, \dots, e_{i_N}$ are equal to each other, then the corresponding value of r is equal to zero and, hence, it belongs to $\mathfrak{g}_0$. In the contrary case we have the inequalities $|e_{i_1}| + \dots + |e_{i_N}| \geq m_{\mathfrak{g}}^{-}(N)$ and $k \geq m_{\mathfrak{g}}^{-}(N) - t \geq 0$ by property a) of the "basis" $\{e_i\}$, and again the value of the polynomial r belongs to $\mathfrak{g}_0$.

However, for any multilinear Lie polynomial $g(z_1, \dots, z_l)$ and for any elements $u, u_1, \dots, u_l \in \mathfrak{g}$ the equality

$$[u, g(u_1, \dots, u_l)] = \sum_{i=1}^{l} g(u_1, \dots, u_{i-1}, [u, u_i], u_{i+1}, \dots, u_l),$$

is satisfied in $\mathfrak{g}$, and it implies that the linear space spanned by the values of the polynomial g forms an ideal of the Lie algebra $\mathfrak{g}$. Therefore, the linear subspace spanned by the values of the polynomial r is a proper ideal of the simple Lie algebra $\mathfrak{g}$. Hence, all values of r on $\mathfrak{g}$ are zero for $t \leq m_{\mathfrak{g}}^{-}(N)$. The proposition is proved.

Similarly, we can prove the following

LEMMA 43.1. *In an arbitrary simple Lie K-algebra $\mathfrak{g}$ containing a proper subalgebra $\mathfrak{g}_0$ of finite codimension over K we have an identity of the form*

$$f(z, y_{11}, \dots, y_{1m}, y_{21}, \dots, y_{2m}, \dots, y_{k1}, \dots, y_{km}) = 0,$$

where $m = 2\dim_K \mathfrak{g}/\mathfrak{g}_0 + \dim_K \mathfrak{g}_0/\mathfrak{g}_1 + 1$, $k = 1, 2, \dots$, and f is an arbitrary multilinear Lie polynomial, skew-symmetric with respect to each group of variables $y_{i1}, \dots, y_{im}$ $(i = 1, \dots, k)$.

Let $C = C(\mathfrak{g})$ be the centroid of the algebra $\mathfrak{g}$. From the density theorem (see §53) it follows that for any positive integer N not exceeding $\dim_C \mathfrak{g}$ we can choose a positive integer $t = t(N)$ for which there exists a multilinear identity of the form (43.2) not satisfied in $\mathfrak{g}$. We point out a rough method for constructing some function $m_{\mathfrak{g}}^{+}\colon \mathbb{N} \to \mathbb{N}$ such that for any $N \in \mathbb{N}$ and any $t \geq m_{\mathfrak{g}}^{+}(N)$ there exists a multilinear identity of the form (43.2) not satisfied in $\mathfrak{g}$. Toward this end, we assume that $\mathfrak{g}$ is a simple centrally simple infinite-dimensional Lie algebra, $\dim_K \mathfrak{g}/\mathfrak{g}_0 = n$, and in $\mathfrak{g}$ there are linearly independent elements $e_1, e_2, \dots$ which satisfy the following two conditions, in addition to a) and b):

c) for any collection of elements $e_{i_1}, \dots, e_{i_N} \in \{e_1, \dots, e_n\}$ there exists an element $w \in \mathfrak{g}$ such that in the algebra $\mathfrak{g}$ we have

$$[w, e_{i_1}, \dots, e_{i_n}] \neq 0; \tag{43.3}$$

d) for every $i = 1, 2, \dots$ there exists an integer $s = s_i$ and a collection of elements $v_i \rightleftharpoons (v_{i_1}, \dots, v_{i_s})$ of the algebra $\mathfrak{g}$ such that for the linear function $h_i(x, v_i) \rightleftharpoons [x, v_{i_1}, \dots, v_{i_s}]$ of the argument x on the Lie algebra $\mathfrak{g}$ we have the following equality in $\mathfrak{g}$:

$$h_i(e_j, v_i) = \begin{cases} 0 & \text{for } j < i, \\ \beta_i e_l, & \text{where } 0 \neq \beta_i \in K \ \& \ l \in \{1, \dots, n\}, \text{ for } j = i. \end{cases} \tag{43.4}$$

We put

$$m_{\mathfrak{g}}^{+}(N) = 1 + \sum_{i=1}^{N} s_i. \tag{43.5}$$

PROPOSITION 43.2. *For any positive integer N and $t \geq m_{\mathfrak{g}}^{+}(N)$ there exists an identity of the form* (43.2) *which is not satisfied in the algebra* $\mathfrak{g}$.

PROOF. For any element e_i we choose a corresponding collection of its own noncommuting variables $x = x(e_i)$, $y_1 = y_1(e_i), \dots, y_s = y_s(e_i)$, and denote by $h_{e_i}(x, y)$ the multilinear left-normed Lie commutator $[x, y_1, \dots, y_s]$. The function $h_i(x, v_i)$ is obtained from $h_{e_i}(x, y)$ as a result of the substitution of the elements $v_{i_1}, \dots, v_{i_s}$ instead of the variables $y_1, \dots, y_s$.

Consider the multilinear Lie identity

$$\sum_{\sigma \in S_N} \operatorname{sign}(\sigma)[y_t, h_{e_1}(x_{\sigma(1)}, y(e_1)), \dots, h_{e_N}(x_{\sigma(N)}, y(e_N))] = 0, \tag{43.6}$$

where $n \in \mathbb{N}$, $t = m_{\mathfrak{g}}^{+}(N)$, and the alternating sum is extended over all permutations σ on the set $\{1, \dots, N\}$. The polynomial h on the left-hand side of this identity is skew-symmetric with respect to $x_1, \dots, x_N$, and from the definition of the polynomials h_{e_i} it follows that in h there are $t = m_{\mathfrak{g}}^{+}(N)$ occurrences of variables of the type y. Properties c) and d) of the "basis" $\{e_i\}$ imply the existence of some $w \in \mathfrak{g}$ for which

$$[w, h_1(e_1, v_1), \dots, h_N(e_N, v_N)] \neq 0.$$

But then, by property d), the expression

$$\sum_{\sigma \in S_N} \operatorname{sign}(\sigma)[w, h_1(e_{\sigma(1)}, v_1), \dots, h_N(e_{\sigma(N)}, v_N)]$$

is nonzero, because all of its summands, except for the one corresponding to $\sigma = 1$, are equal to zero. Therefore, (43.6) is not satisfied in $\mathfrak{g}$. Proposition 43.2 is proved.

For any integer l and any infinite collection $d = (d_1, d_2, \dots)$ of positive numbers we denote by $m_{l,d}(z)$ a continuous piecewise linear function on the nonegative reals, defined by induction for $k = 0, 1, \dots$ on the intervals $(0 + d_1 + \dots + d_k, d_1 + \dots + d_k + d_{k+1}]$ via the formulas

$$m_{l,d}(0) = 0,$$
$$m_{l,d}(z) = m_{l,d}(0 + d_1 + \dots + d_k) + (l + k)(z - d_1 - \dots - d_k).$$

In the remaining part of the section we shall assume that the ground field K is of characteristic zero and the simple Lie algebra $\mathfrak{g}$ is a centrally simple ($K = C(\mathfrak{g})$) infinite-dimensional algebra. For any such algebra $\mathfrak{g}$ and its maximal subalgebra $\mathfrak{g}_0$ of finite codimension $n \geq 1$ we will present a concrete "basis" $e_1, e_2, \ldots$ satisfying conditions a)–d) and will give an explicit description of the functions $m_{\mathfrak{g}}^-(N)$ and $m_{\mathfrak{g}}^+(N)$. From the definition (43.1) of the function $m_{\mathfrak{g}}^-(N)$ it follows that this function coincides with the restriction to $\mathbb{N}$ of the piecewise linear function $m_{-1,d}(z)$, where $d_i = \dim_K \mathfrak{g}_{i-2}/\mathfrak{g}_{i-1}$. We shall indicate below possible values of the dimensions of the quotients $\mathfrak{g}_{i-2}/\mathfrak{g}_{i-1}$ and we shall prove that $m_{\mathfrak{g}}^+(N)$ coincides with the restriction to $\mathbb{N}$ of the piecewise linear function $1 + m_{\varepsilon,p}(z)$, where $\varepsilon = 0, 1$ and $p = \{p_1, p_2, \ldots\}$ is a certain definite collection of positive integers for which $p_i \leq d_i$. In particular, this description will imply $\lim_{N\to\infty} m_{\mathfrak{g}}^+(N)/m_{\mathfrak{g}}^-(N) = 1$.

In order to construct a "basis" $\{e_i\}$ satisfying conditions a)–d) and the functions $m_{\mathfrak{g}}^+(N)$ and $m_{\mathfrak{g}}^-(N)$, we shall use the classification, going back to È. Cartan, of completions $\widetilde{\mathfrak{g}}$ of the Lie algebra $\mathfrak{g}$ by the filtration $\mathfrak{g} \supset \mathfrak{g}_0 \supset \mathfrak{g}_1 \supset \cdots$ (see [Car, Vei, Kats]).

THEOREM 43.1. *Suppose that the ground field K is algebraically closed and of characteristic zero, $\mathfrak{g}$ is a centrally simple infinite-dimensional Lie K-algebra, and $\mathfrak{g}_0$ is a maximal K-subalgebra of it for which $0 < n = \dim_K \mathfrak{g}/\mathfrak{g}_0 < \infty$. Then the completion $\widetilde{\mathfrak{g}}$ of the algebra $\mathfrak{g}$ by the filtration defined by the Cartan prolongation $\{\mathfrak{g}_i\}$ of $\mathfrak{g}_0$ is isomorphic, as a filtered Lie algebra, to one of the following Lie algebras:*

1) *$\widetilde{W}_n$, the Lie algebra of all derivations of the algebra $\widetilde{\mathcal{F}} = K[[t_1, \ldots, t_n]]$ of power series in n commuting variables for which the ith term of the filtration ($i \geq 0$) consists of the derivations of the form $f_1\partial_1 + \cdots + f_n\partial_n$, where $f_1, \ldots, f_n$ belong to the $(i+1)$th power of the unique maximal ideal of the algebra $\widetilde{\mathcal{F}}$.*

2) *$\widetilde{S}_n$, the Lie subalgebra of $\widetilde{W}_n$ of all derivations $f_1\partial_1 + \cdots + f_n\partial_n$ satisfying the equality $(\partial_1 f_1) + \cdots + (\partial_n f_n) = 0$, endowed with the filtration induced by that of $\widetilde{W}_n$.*

3) *$\widetilde{H}_q$ ($n = 2q$), the Lie subalgebra in $\widetilde{W}_n$ of the derivations of the form*

$$\sum_{i=1}^{q} \left((\partial f/\partial t_i)\partial_{q+i} - (\partial f/\partial t_{q+i})\partial_i\right),$$

where f ranges over the whole algebra $\widetilde{\mathcal{F}}$, endowed with the filtration induced by that of $\widetilde{W}_n$.

4) *$\widetilde{K}_q$ ($n = 2q+1$), the Lie subalgebra in $\widetilde{W}_n$ of the derivations of the form $\{(2-\delta)f\}\partial_n + (\partial f/\partial t_n)\delta + \sum_{i=1}^{q}\left((\partial f/\partial t_i)\partial_{q+i} - (\partial f/\partial t_{q+i})\partial_i\right)$, where $\delta = t_1\partial_1 + \cdots + t_{2q}\partial_{2q}$ and f ranges over the whole algebra $\widetilde{\mathcal{F}}$, endowed with the filtration induced by that of $\widetilde{W}_n$.*

Since the Lie algebras $\mathfrak{g}$ and $\widetilde{\mathfrak{g}}$ possess the same identities, we see that in order to satisfy the conditions of Propositions 43.1 and 43.2 it suffices to construct a "basis" $\{e_i\}$ with properties a)–d) and to describe the functions $m_{\mathfrak{g}}^-(N)$ and $m_{\mathfrak{g}}^+(N)$ only for the cases $\mathfrak{g} = \widetilde{W}_n, \widetilde{S}_n, \widetilde{K}_q$ ($n = 2q+1$), and $\widetilde{H}_q$ ($n = 2q$) only.

We consider each of these cases separately.

Case 1): general Lie algebra ($\mathfrak{g} = \widetilde{W}_n$). For each monomial $e = t_1^{m_1} \cdots t_n^{m_n} \in \widetilde{\mathcal{F}} = K[[t_1, \dots, t_n]]$ and any positive integer $l \in \{1, \dots, n\}$ we put

$$R_{e,l} \rightleftharpoons e\partial_l, \qquad s(R_{e,l}) \rightleftharpoons m_1 + \cdots + m_n + 1.$$

We index all such elements by the numbers $1, 2, \dots$ in such a way that the greater index corresponds to the monomial with greater degree. It is clear that the derivations $R_{e,l}$, where $l = 1, \dots, n$, and e ranges over all monomials of degree at most i ($i = 1, 2, \dots$), form a basis of the quotient space $\mathfrak{g}/\mathfrak{g}_i$. Hence, if we denote the elements $R_{e,l}$ with the indexing chosen above by $e_1, e_2, \dots$, then they will satisfy conditions a) and b). Therefore, for $\mathfrak{g} = \widetilde{W}_n$ we have

$$\begin{aligned} d_i &= \dim_K \mathfrak{g}_{i-2}/\mathfrak{g}_{i-1} = n\binom{n+i-2}{i-1}, \\ m_{\mathfrak{g}}^-(N) &= m_{-1,d} \qquad (d = (d_1, d_2, \dots)). \end{aligned} \tag{43.7}$$

Since the first n derivations among the $R_{e,l}$'s coincide with $\partial_1, \dots, \partial_n$, property c) holds for the "basis" $\{e_i\}$. For each $e_j = R_{u,l}$, where $u = t_1^{m_1} \cdots t_n^{m_n}$, we put $s = s(R_{u,l})$ and consider the collection of elements in $\widetilde{W}_n$ of the form

$$v = (v_1, \dots, v_s) = (\underbrace{\partial_1, \dots, \partial_1}_{m_1 \text{ times}}, \dots, \underbrace{\partial_n, \dots, \partial_n}_{m_n \text{ times}}, t_l\partial_l).$$

Then

$$h_j(R_{e,k}, v) = \begin{cases} 0 & \text{if (degree of } e \leq \text{degree of } u) \ \& \ (e \neq u \vee k \neq l), \\ m_1! \cdots m_n!\partial_l & \text{for } e = u \ \& \ k = l, \end{cases}$$

and, hence, condition d) is satisfied for the basis $\{e_i\}$ just constructed. Therefore, for $\mathfrak{g} = \widetilde{W}_n$ we have

$$\begin{aligned} m_{\mathfrak{g}}^+(N) = 1 + m_{1,d}(N), \qquad m_{\mathfrak{g}}^+(N) - m_{\mathfrak{g}}^-(N) = 2N + 1, \\ \lim_{N\to\infty} m_{\mathfrak{g}}^+(N)/m_{\mathfrak{g}}^-(N) = 1. \end{aligned} \tag{43.8}$$

Case 2): special Lie algebra ($\mathfrak{g} = \widetilde{S}_n$). For each monomial $e = t_1^{m_1} \cdots t_n^{m_n} \in \widetilde{\mathcal{F}}$ and any positive integer $l \in \{1, \dots, n\}$ we put

$$\begin{aligned} D_{e,l} &\rightleftharpoons \begin{cases} e\partial_l & \text{for } m_l = 0, \\ e\partial_l - (m_l/(m_n+1))(e \cdot t_n/t_l)\partial_n & \text{for } m_l > 0 \ \& \ l \neq n, \\ 0 & \text{for } l = n \ \& \ m_l > 0, \end{cases} \\ s(D_{e,l}) &= 1 + m_1 + \cdots + m_n. \end{aligned} \tag{43.9}$$

From relations (43.9) it follows that $D_{e,l} \in \widetilde{S}_n$ and the set of nonzero elements of the form $D_{e,l}$, where $l = 1, \dots, n$ and e ranges over all monomials of degree at most i, forms a basis of the quotient space $\widetilde{S}_n/\mathfrak{g}_i$. We index all nonzero elements $D_{e,l}$ by the numbers $1, 2, \ldots$ in such a way that the greater index corresponds to the monomial with greater degree and denote them by $e_1, e_2, \ldots$. It is clear that the "basis" $\{e_i\}$ satisfies conditions a) and b). Therefore, for $\mathfrak{g} = \widetilde{S}_n$ we have

$$(43.10)\qquad \begin{aligned} d_i &= \dim_K \mathfrak{g}_{i-2}/\mathfrak{g}_{i-1} = (n-1)\binom{n+i-2}{i-1} + \binom{n-1+i-2}{i-1}, \\ m_{\mathfrak{g}}^{-}(N) &= m_{-1,d}(N) \qquad (d = (d_1, d_2, \ldots, d_i, \ldots)). \end{aligned}$$

Since the first n derivations among $D_{e,l}$'s coincide with $\partial_1, \ldots, \partial_n$, condition c) holds for the "basis" $\{e_i\}$.

Let j be the index of a nonzero element $D_{u,l}$, where $u = t_1^{q_1} \cdots t_n^{q_n}$. We set $s = s(D_{u,l})$ and $l' = l + 1$ for $l < n$, $l' = 1$ for $l = n$, and consider the set of elements in $\widetilde{S}_n$ of the form

$$v = (v_1, \ldots, v_s) = (\underbrace{\partial_1, \ldots, \partial_1}_{q_1 \text{ times}}, \ldots, \underbrace{\partial_n, \ldots, \partial_n}_{q_n \text{ times}}, t_l\partial_{l'}).$$

Then

$$h_j(D_{e,k}, v) = \begin{cases} 0 & \text{if (degree of } e \leq \text{degree of } u) \;\&\; (e \neq u \vee l \neq k), \\ q_1! \cdots q_n!\partial_{l'} & \text{for } e = u \;\&\; k = l, \end{cases}$$

and, hence, condition d) is satisfied for the basis just constructed. Therefore, equalities (43.8) are satisfied also for $\mathfrak{g} = \widetilde{S}_n$.

Case 3): contact Lie algebra ($\mathfrak{g} = \widetilde{K}_q$, $n = 2q + 1$). We realize the contact Lie algebra $\widetilde{K}_q$ on the space of the K-algebra $\widetilde{\mathcal{F}}$ of power series in $2q + 1$ commuting variables by defining the multiplication operation by the formula

$$(43.11)\quad [f, g] \rightleftharpoons ((2-\delta)f)\frac{\partial}{\partial t_n}g - ((2-\delta)g)\frac{\partial}{\partial t_n}f + \sum_{i=1}^{q}\left(\frac{\partial f}{\partial t_i}\frac{\partial g}{\partial t_{i+q}} - \frac{\partial f}{\partial t_{i+q}}\frac{\partial g}{\partial t_i}\right),$$

where $f, g \in \widetilde{\mathcal{F}}$ and $\delta = t_1\partial_1 + \cdots + t_{n-1}\partial_{n-1}$. For arbitrary positive integers k and l denote by $\widetilde{\mathcal{F}}_{k,l}$ the subspace of $\widetilde{\mathcal{F}}$ consisting of the elements $h \in \widetilde{\mathcal{F}}$ in the natural notation of which the monomials of the form $t_1^{m_1} \cdots t_n^{m_n}$, where $m_n < l$ or $m_1 + \cdots + m_{n-1} < k$, occur with zero coefficients only. Formula (43.11) implies

$$[\widetilde{\mathcal{F}}_{k,l}, \widetilde{\mathcal{F}}_{0,t}] \subseteq \widetilde{\mathcal{F}}_{k,t+l-1}, \qquad [\widetilde{\mathcal{F}}_{k_1,l_1}, \widetilde{\mathcal{F}}_{k_2,l_2}] \subseteq \widetilde{\mathcal{F}}_{k_1+k_2-2,l_1+l_2} + \widetilde{\mathcal{F}}_{k_1+k_2,l_1+l_2-1}.$$

Hence, $\mathfrak{g}_0 = \sum_{k+2l\geq 2} \widetilde{\mathcal{F}}_{k,l}$ is a subalgebra in $\widetilde{K}_q$ of codimension n. It can be directly verified that $\mathfrak{g}_0$ is a maximal subalgebra in $\widetilde{K}_q$ and the ith member $\mathfrak{g}_i$ of the Cartan prolongation coincides with $\widetilde{\mathcal{F}}_{0,i+1} + \sum_{k+l\geq i+2} \widetilde{\mathcal{F}}_{k,l}$. We index all monomials of the algebra $\widetilde{\mathcal{F}}$ in such a way that the greater index corresponds to the monomial with greater degree and the monomial t_n^l has the greatest index among all the monomials of degree l. Let us denote the monomials thus indexed by $e_1, e_2, \ldots$. It is clear that the monomials of the form $t_1^{m_1} \cdots t_n^{m_n}$, where $m_1 + \cdots + m_n = i+2$ and $m_n \neq i+2$, and the monomial t_n^{i+1} form a basis of the quotient space $\mathfrak{g}_i/\mathfrak{g}_{i+1}$. Therefore, the "basis" $\{e_i\}$ satisfies conditions a) and b). Since the first n monomials coincide with $1, t_1, \dots, t_{n-1}$, it follows from formula (43.11) that $\{e_i\}$ satisfies condition c).

Let $e_j = u = t_1^{m_1} \cdots t_n^{m_n}$. We set $s = s(e_j) = m_1 + \cdots + m_n$ and consider the set of elements in $\widetilde{K}_q$ of the form

$$\begin{aligned} v &= (v_1, \dots, v_s) \\ &\rightleftharpoons (\underbrace{t_{q+1}, \dots, t_{q+1}}_{m_1 \text{ times}}, \dots, \underbrace{t_{2q}, \dots, t_{2q}}_{m_q \text{ times}}, \underbrace{t_1, \dots, t_1}_{m_{q+1} \text{ times}}, \dots, \underbrace{t_q, \dots, t_q}_{m_{2q} \text{ times}}, \underbrace{1, \dots, 1}_{m_n \text{ times}}). \end{aligned}$$

From (43.11) it follows that

$$h_j(e, v) = \begin{cases} 0 & \text{if } e \neq u \ \& \ (\text{the degree of } e \text{ is at most } s), \\ \pm m_1! \cdots m_n! & \text{for } e = u, \end{cases}$$

and, hence, condition d) is satisfied for the basis just constructed.

Therefore, from formulas (43.1) and (43.5) defining the functions $m_{\mathfrak{g}}^-(N)$ and $m_{\mathfrak{g}}^+(N)$ for the Lie algebra $\mathfrak{g} = \widetilde{K}_q$ we have

$$d_i = \dim_K \mathfrak{g}_{i-2}/\mathfrak{g}_{i-1} = \binom{n+i-1}{i}, \tag{43.12}$$

$$m_{\mathfrak{g}}^-(N) = m_{-1,d}(N) \qquad (d = (d_1, d_2, \dots, d_i, \dots));$$

$$m_{\mathfrak{g}}^+(N) = 1 + m_{0,p}(N), \qquad p = (p_1, \dots, p_i, \dots),$$

$$p_i = \binom{n+i-2}{i-1}, \tag{43.13}$$

$$m_{\mathfrak{g}}^+(N) - m_{\mathfrak{g}}^-(N) \leq 2N + 1, \qquad \lim_{N \to \infty} m_{\mathfrak{g}}^+(N)/m_{\mathfrak{g}}^-(N) = 1.$$

Case 4): Hamiltonian Lie algebra ($\mathfrak{g} = \widetilde{H}_q$, $n = 2q$). Defining on the K-algebra $\widetilde{\mathcal{F}}$ of power series in $2q$ commuting variables an operation via the Poisson bracket

$$[f, g] \rightleftharpoons \sum_{i=1}^{q} \left(\frac{\partial f}{\partial t_i} \cdot \frac{\partial g}{\partial t_{i+q}} - \frac{\partial g}{\partial t_i} \cdot \frac{\partial f}{\partial t_{i+q}} \right), \tag{43.14}$$

we equip the space $\widetilde{\mathcal{F}}$ with the structure of a Lie algebra with filtration determined by the powers of the unique maximal ideal of the algebra $\widetilde{\mathcal{F}}$. This Lie algebra possesses a unique nontrivial ideal consisting of the polynomials of degree zero. The quotient algebra by this ideal is isomorphic to the Hamiltonian Lie algebra $\widetilde{H}_q$, and the ith term $\mathfrak{g}_i$ of the filtration in this quotient algebra consists of the elements of $\widetilde{\mathcal{F}}$ such that in their natural notation all the monomials of degree at most $i+1$ occur with zero coefficients only.

We index all monomials of degree greater than zero in $\widetilde{\mathcal{F}}$ in such a way that the greater index corresponds to the monomial with greater degree and denote them by $e_1, e_2, \dots$. It is clear that the monomials of degree i form a basis of the quotient space $\mathfrak{g}_{i-2}/\mathfrak{g}_{i-1}$. Since the first n monomials coincide with $x_1, \dots, x_n$, formula (43.14) implies that $\{e_i\}$ satisfies condition c), in addition to properties a) and b). Therefore,

$$
\begin{aligned}
d_i &= \dim_K \mathfrak{g}_{i-2}/\mathfrak{g}_{i-1} = \binom{n+i-1}{i}, \\
m_{\mathfrak{g}}^-(N) &= m_{-1.d}(N) \qquad (d = (d_1, d_2, \dots, d_i, \dots)).
\end{aligned}
\tag{43.15}
$$

Let $e_j = u = t_1^{m_1} \cdots t_n^{m_n}$. We fix an integer l for which $m_l > 0$ in the monomial u and set $s = s(e_j) = m_1 + \dots + m_n$. Let $m_l' = m_l - 1$ and $m_i' = m_i$ for $i \neq l$, and let $k = l + q$ for $l \leq q$ and $k = l - q$ for $l > q$. We assign to the monomial u a collection of elements of $\widetilde{H}_q$ as follows:

$$
v = (v_1, \dots, v_s) \rightleftharpoons (\underbrace{t_{1+q}, \dots, t_{1+q}}_{m_1' \text{ times}}, \dots, \underbrace{t_{2q}, \dots, t_{2q}}_{m_q' \text{ times}}, \underbrace{t_1, \dots, t_1}_{m_{1+q}' \text{ times}}, \dots, \underbrace{t_q, \dots, t_q}_{m_{2q}' \text{ times}}, t_k^2).
$$

From formula (43.14) it follows that

$$
h_j(e, v) = \begin{cases} 0 & \text{if } e \neq u \ \& \ (\text{the degree of the monomial } e \text{ is at most } s), \\ \pm 2m_1'! \cdots m_n'! t_k & \text{for } e = u, \end{cases}
$$

and, hence, the "basis" under consideration satisfies condition d). Therefore, equalities (43.8) are satisfied also for $\mathfrak{g} = \widetilde{H}_q$.

Now we sum up the above results.

THEOREM 43.2. *Let the ground field K be algebraically closed and of characteristic zero, let $\mathfrak{g}$ be a centrally simple infinite-dimensional Lie K-algebra, let $\mathfrak{g}_0$ be its proper maximal subalgebra of finite codimension n, and let $\tilde{\mathfrak{g}}$ be the completion of the algebra $\mathfrak{g}$ with respect to the filtration given by the Cartan prolongation $\{\mathfrak{g}_i\}$ $(i > 0)$ of the subalgebra $\mathfrak{g}_0$. Then for the number sequence $d = (d_1, d_2, \dots, d_i, \dots)$, where $d_i = \dim_K \mathfrak{g}_{i-2}/\mathfrak{g}_{i-1}$, and for any positive integer N an arbitrary identity of the form* (43.2) *holds in the algebra $\mathfrak{g}$ for any $t \leq m_{-1.d}(N)$; however, for $t \geq m_{-1.d}(N) + 2N + 1$ there exists an identity of the form* (43.2) *which is not satisfied in $\mathfrak{g}$. The values of d_i $(i = 1, 2, \dots)$ are given in the following table.*

$\tilde{\mathfrak{g}}$	$d_i = \dim_K \mathfrak{g}_{i-2}/\mathfrak{g}_{i-1}$
$\widetilde{W}_n$, $n > 0$	$i(i+1)\cdots(i+n-2)n/(n-1)!$
$\widetilde{S}_n$, $n > 1$	$i(i+1)\cdots(i+n-3)(i+n-1)/(n-2)!$
$\widetilde{K}_q$, $n = 2q+1 > 1$	$(i+1)(i+2)\cdots(i+n-1)/(n-1)!$
$\widetilde{H}_q$, $n = 2q > 0$	$(i+1)(i+2)\cdots(i+n-1)/(n-1)!$

COROLLARY 1. *Let K, $\mathfrak{g}$, $\mathfrak{g}_0$, and n have the same meaning as in Theorem* 43.2. *Then for $n > l$*

$$\lim_{N\to\infty} m^-_{\widetilde{W}_l}(N)/m^-_{\mathfrak{g}}(N) = \infty, \tag{43.16}$$

and the variety $\operatorname{var}\widetilde{W}_l$ *does not contain the variety* $\operatorname{var}\mathfrak{g}$.

COROLLARY 2. *Let $\mathfrak{g}$ be a centrally simple infinite-dimensional Lie algebra of Cartan type over a field K of characteristic zero. Then* $\operatorname{var}\mathfrak{g} \subseteq \operatorname{var}\widetilde{W}_l$ *if and only if $\mathfrak{g}$ can be embedded in the Lie algebra $\widetilde{W}_l(\widehat{K})$ of all derivations of the algebra $\widehat{K}[[t_1, \dots, t_l]]$ of power series for some extension $\widehat{K}$ of the field K.*

PROOF OF COROLLARY 1. Theorem 43.1 and formulas (43.7), (43.10), (43.12), and (43.15) imply that $m^-_{\mathfrak{g}}(N) = m^-_{\widetilde{\mathfrak{g}}}(N) = m_{-1,d}(N)$. From the table in Theorem 43.2 it follows that d_i is a polynomial function of degree $n-1$ in the parameter i with nonzero coefficient of the leading term. Similarly, $m^-_{\widetilde{W}_l}(N) = m_{-1,d'}(N)$, where d'_i is a polynomial function of degree $l-1$ in the parameter i with coefficient $l/(l-1)!$ of the leading term. Therefore, (43.16) is implied by the following lemma.

LEMMA 43.2. *If each term d_i of a sequence of positive numbers $d = \{d_i\}$ is equal to $g(i)$, where g is a polynomial of degree $q-1$, then*

$$\lim_{N\to\infty} m_{t,d}(N)/N^{(1+1/q)} = c > 0.$$

Since $\lim_{N\to\infty}(1+2N)/m^-_{\widetilde{W}_l}(N) = 0$ for $l \geq 1$, equality (43.16) implies the inequality $m^-_{\widetilde{W}_l}(N) \geq m^-_{\mathfrak{g}}(N) + 1 + 2N$ for a sufficiently large positive integer N. Therefore, by Theorem 43.2, all identities of the form (43.2) are satisfied in the Lie algebra $\widetilde{W}_l$ for $t = m^-_{\mathfrak{g}}(N) + 1 + 2N$, but among them there are identities which does not hold in $\mathfrak{g}$. Thus, $\operatorname{var}\widetilde{W}_l \not\supset \operatorname{var}\mathfrak{g}$, and Corollary 1 is proved.

PROOF OF COROLLARY 2. By the definition of a Lie algebra of Cartan type, for some algebraically closed extension $\widehat{K}$ of the field K the $\widehat{K}$-algebra $\widehat{\mathfrak{g}} = \widehat{K} \otimes_K \mathfrak{g}$ contains a proper $\widehat{K}$-subalgebra of finite codimension n.

If $\operatorname{var}\mathfrak{g} \subseteq \operatorname{var}\widetilde{W}_l$, then all the identities of the form (43.2) hold in $\widehat{\mathfrak{g}}$ for any N and $t \leq m^-_{\widetilde{W}_l}(N)$, and some of these identities would not be satisfied in $\mathfrak{g}$ if n were greater than l, as the argument in the proof of Corollary 1 shows. Therefore, $l \geq n$, and by Theorem 43.1 we have the following chain of inclusions:

$$\mathfrak{g} \subset \widehat{\mathfrak{g}} \subset \widetilde{\widehat{\mathfrak{g}}} \subset \widetilde{W}_n(\widehat{K}) \subset \widetilde{W}_l(\widehat{K}).$$

This proves the "only if" part of the assertion of Corollary 1.

The converse assertion is trivial. The corollary is proved.

§44. Embedding of algebras of Cartan type in $\widetilde{W}_n(K)$

44.1. Completed divided power algebra.

DEFINITION 44.1. For any linear K-space V by a *divided power algebra over V* we mean a commutative K-algebra $O(V)$ with unit given by a set of formal generators

$\{v^{(i)} \mid v \in V,\ i = 1, 2, \dots\}$ and defining relations of the form

$$(44.1)\qquad \begin{aligned} v^{(i)}v^{(j)} &= \binom{i+j}{i} v^{(i+j)}, \qquad v^{(0)} = 1, \\ (u+v)^{(n)} &= \sum_{i=0}^{n} u^{(i)} v^{(n-i)}, \qquad (\beta v)^{(n)} = \beta^n v^{(n)} \qquad (\beta \in K). \end{aligned}$$

It can be directly verified that for any basis $\{v_i\}$ of the space V a basis of $O(V)$ is formed by the monomials of the form $v_{i_1}^{(n_1)} \cdots v_{i_l}^{(n_l)}$ $(i_1 < \cdots < i_l,\ n_1, \dots, n_l \geq 0)$. We denote by O_i $(i = 0, 1, 2, \dots)$ the subspace of $O(V)$ spanned by all homogeneous monomials of degree i, i.e., monomials for which $n_1 + \cdots + n_l = i$. From relations (44.1) it follows that $O_i O_j \subset O_{i+j}$. Hence, $O(V) = O_0 + O_1 + O_2 + \cdots$ is a gradation on $O(V)$. The completion of $O(V)$ with respect to the descending filtration induced by this gradation is called the *completed divided power algebra*. We shall denote it by $\widetilde{O}(V)$. The canonical filtration on this algebra is given by the subspaces

$$\widetilde{O}_i \rightleftharpoons \big\{ \sum \beta_{n_1,\dots,n_l}^{i_1,\dots,i_l} v_{i_1}^{(n_1)} \cdots v_{i_l}^{(n_l)} \mid n_1 + \cdots + n_l < i \Rightarrow \beta_{n_1,\dots,n_l}^{i_1,\dots,i_l} = 0 \big\}.$$

If the ground field K is of characteristic zero, then the algebras $O(V)$ and $\widetilde{O}(V)$ are isomorphic to the polynomial algebra $K[V]$ and to the power series algebra over the space V, respectively. The isomorphism is determined by a transformation mapping an element $v_{i_1}^{(n_1)} \cdots v_{i_l}^{(n_l)} \in O(V)$ to the element $(v_{i_1}^{(n_1)}/n_1!) \cdots (v_{i_l}^{(n_l)}/n_l!) \in K(V)$.

If $\operatorname{char} K = p > 0$, then the defining relations (44.1) show that the algebras $O(V)$ and $\widetilde{O}(V)$ are local algebras whose unique maximal ideals $O_1 + O_2 + \cdots$ and $\widetilde{O}_1$, respectively, consist of those and only those elements a for which $a^p = 0$.

In the case of a finite-dimensional space V equipped with the structure of a Lie algebra, the completed divided power algebra can be obtained by another method.

Let $\mathfrak{g}$ be an arbitrary Lie K-algebra. We adjoin an external unit to its universal enveloping algebra $U(\mathfrak{g})$ and denote the algebra thus obtained by $U_{\mathrm{id}}(\mathfrak{g})$. We shall denote the extension of the diagonal homomorphism of Lie algebras $\Delta\colon \mathfrak{g} \to \mathfrak{g} \oplus \mathfrak{g}$ to a homomorphism of associative algebras with unit $U_{\mathrm{id}}(\mathfrak{g}) \to U_{\mathrm{id}}(\mathfrak{g} \oplus \mathfrak{g}) \simeq U_{\mathrm{id}}(\mathfrak{g}) \otimes_K U_{\mathrm{id}}(\mathfrak{g})$ by the same letter Δ. It is well known (see [Bou]) that this homomorphism $\Delta\colon U_{\mathrm{id}} \to U_{\mathrm{id}} \otimes_K U_{\mathrm{id}}$ $(\Delta g \rightleftharpoons g \otimes 1 + 1 \otimes g,\ g \in \mathfrak{g})$ endows $U_{\mathrm{id}}(\mathfrak{g})$ with the structure of a commutative associative coalgebra and, hence, the set of all linear functionals $\widetilde{\widetilde{O}}(\mathfrak{g}) = \operatorname{Hom}_K(U_{\mathrm{id}}(\mathfrak{g}), K)$ can be equipped by the structure of a commutative associative algebra by setting

$$(44.2)\qquad \langle f_1 f_2, u \rangle \rightleftharpoons \langle f_1 \otimes f_2, \Delta u \rangle,$$

where $f_1, f_2 \in \widetilde{\widetilde{O}}(\mathfrak{g})$ and $u \in U_{\mathrm{id}}(\mathfrak{g})$. For any $f \in \widetilde{\widetilde{O}}(\mathfrak{g})$ and $g \in \mathfrak{g}$ we define the functionals ${}^g f$ and f^g by setting for each $u \in U_{\mathrm{id}}(\mathfrak{g})$

$$(44.3)\qquad \langle f^g, u \rangle \rightleftharpoons \langle f, gu \rangle, \qquad \langle {}^g f, u \rangle \rightleftharpoons \langle f, ug \rangle.$$

Formula (44.2) implies that the mappings $f \to f^g$ and $f \to {}^g f$ are derivations of the algebra $\widetilde{\widetilde{O}}(\mathfrak{g})$ and endow $\widetilde{\widetilde{O}}(\mathfrak{g})$ with the structures of a right and a left $\mathfrak{g}$-module, respectively. It will be our convention that any space V is a left module over the operators $\mathrm{Hom}_K(V, V)$, and, therefore, the mappings $f \to {}^g f$ and $f \to f^{-g}$ determine homomorphisms of the Lie algebra $\mathfrak{g}$ into the Lie algebra $\mathrm{Der}_K \widetilde{\widetilde{O}}(\mathfrak{g})$ of all K-derivations of the algebra $\widetilde{\widetilde{O}}(\mathfrak{g})$.

Let us order totally some basis of the Lie algebra $\mathfrak{g}$ and denote it by $\{g_i\}$. By the Birkhoff–Witt theorem a basis of the algebra $U_{\mathrm{id}}(\mathfrak{g})$ is formed by all monomials of the form $g_{i_1}^{n_1} \cdots g_{i_l}^{n_l}$ ($i_1 < \cdots < i_l$, $n_1, \dots, n_l \geq 0$). We shall denote this basis by the letter E. Denote by $\left(g_{i_1}^{n_1} \cdots g_{i_l}^{n_l}\right)^*$ the functional on $U_{\mathrm{id}}(\mathfrak{g})$ that vanishes on all monomials distinct from $g_{i_1}^{n_1} \cdots g_{i_l}^{n_l}$ and equals one on this monomial. Denote the set of these functionals by E^*. Then any element f of $\widetilde{\widetilde{O}}(\mathfrak{g})$ can be considered as a formal infinite linear combination of different elements of E^* with coefficients in the field K.

It follows immediately from the formula (44.2) that in the algebra $\widetilde{\widetilde{O}}(\mathfrak{g})$ we have the equality

$$(44.4) \qquad \left(g_{i_1}^{n_1} \cdots g_{i_l}^{n_l}\right)^* \left(g_{i_1}^{m_1} \cdots g_{i_l}^{m_l}\right)^* = \prod_{i=1}^{l} \binom{n_i + m_i}{n_i} \left(g_{i_1}^{m_1+n_1} \cdots g_{i_l}^{m_l+n_l}\right)^*.$$

Moreover, for any $f_1, f_2 \in \widetilde{\widetilde{O}}(\mathfrak{g})$

$$(44.5) \quad \langle f_1 f_2, g_{i_1}^{k_1} \cdots g_{i_l}^{k_l} \rangle = \sum_{\substack{m_i+n_i=k_i \\ i=1,\dots,l}} \prod_{i=1}^{l} \binom{m_i+n_i}{n_i} \langle f_1, g_{i_1}^{n_1} \cdots g_{i_l}^{n_l} \rangle \langle f_2, g_{i_1}^{m_1} \cdots g_{i_l}^{m_l} \rangle.$$

These formulas show that if Lie algebras $\mathfrak{g}_1$ and $\mathfrak{g}_2$ have the same dimension over K then $\widetilde{\widetilde{O}}(\mathfrak{g}_1) \simeq \widetilde{\widetilde{O}}(\mathfrak{g}_2)$.

We define a mapping $\varkappa \colon \widetilde{O}(\mathfrak{g}) \to \widetilde{\widetilde{O}}(\mathfrak{g})$ by the formula

$$\varkappa \sum_{\substack{i_1,\dots,i_l \\ n_1,\dots,n_l}} \beta_{n_1,\dots,n_l}^{i_1,\dots,i_l} g_{i_1}^{(n_1)} \cdots g_{i_l}^{(n_l)} = \sum_{\substack{i_1,\dots,i_l \\ n_1,\dots,n_l}} \beta_{n_1,\dots,n_l}^{i_1,\dots,i_l} \left(g_{i_1}^{n_1} \cdots g_{i_l}^{n_l}\right)^*,$$

where $\beta_{n_1,\dots,n_l}^{i_1,\dots,i_l} \in K$. Comparing the defining relations (44.1) of the algebra $\widetilde{O}(\mathfrak{g})$ with the multiplication formulas (44.4) and (44.5) in $\widetilde{\widetilde{O}}(\mathfrak{g})$, we see that $\varkappa$ is a monomorphism of K-algebras. Moreover, if $\mathfrak{g}$ is finite-dimensional, then $\varkappa$ is an epimorphism and $\widetilde{O}(\mathfrak{g}) \simeq \widetilde{\widetilde{O}}(\mathfrak{g})$.

It is easy to see that the subspaces

$$\widetilde{\widetilde{O}}_i(\mathfrak{g}) \rightleftharpoons \left\{ f \in \widetilde{\widetilde{O}}(\mathfrak{g}) \mid n_1 + \cdots + n_l < i \Rightarrow \langle f, g_{i_1}^{n_1} \cdots g_{i_l}^{n_l} \rangle = 0 \right\} \qquad (i = 0, 1, \dots)$$

determine a descending filtration on $\widetilde{\widetilde{O}}(\mathfrak{g})$ and that $\widetilde{\widetilde{O}}_1(\mathfrak{g}) = \{f \in \widetilde{\widetilde{O}}(\mathfrak{g}) \mid \langle f, 1 \rangle = 0\}$ is the unique maximal ideal in $\widetilde{\widetilde{O}}(\mathfrak{g})$.

REMARK 44.1. It follows from what was said above that the structure of $\tilde{\tilde{O}}(\mathfrak{g})$ depends on the structure of the linear space of $\mathfrak{g}$ itself rather than on the particular Lie algebra structure of $\mathfrak{g}$. At the same time, the derivations $f \to {}^g f$ and $f \to f^g$ defined by formulas (44.3) are solely dependent on the Lie algebra structure. Therefore, introducing different binary multilinear operations equipping $\mathfrak{g}$ with the structure of a Lie algebra, we can obtain different derivations of $\tilde{\tilde{O}}(\mathfrak{g})$. In particular, if $\mathfrak{g}$ is an abelian Lie algebra, then formulas (44.2) show that the linear mappings $\partial_j = \partial(g_j)$ for which

$$(44.6)\quad \partial_j\Big(\beta + \sum_{i_1<\cdots<i_l} \beta^{i_1,\dots,i_l}_{n_1,\dots,n_l}(g_{i_1}^{n_1}\cdots g_{i_l}^{n_l})^*\Big) = \sum_{i_1<\cdots<i_l} \beta^{i_1,\dots,j,\dots,i_l}_{n_1,\dots,m+1,\dots,n_l}(g_{i_1}^{n_1}\cdots g_j^{m}\cdots g_{i_l}^{n_l})^*$$

are derivations of $\tilde{\tilde{O}}(\mathfrak{g})$.

44.2. A representation of a Lie algebra by derivations of the divided power algebra.

DEFINITION 44.2. Suppose that for any basis element $g_i \in \mathfrak{g} \cap E$ we are given a functional $f_i \in \tilde{\tilde{O}}(\mathfrak{g})$ such that for each monomial $g_{i_1}^{n_1}\cdots g_{i_l}^{n_l} \in E$ there is only a finite set of functionals f_i that are nonzero at this monomial. Then the mapping $\sum_i f_i\partial_i \colon \tilde{\tilde{O}}(\mathfrak{g}) \to \tilde{\tilde{O}}(\mathfrak{g})$ for which

$$(44.7)\qquad \sum_i f_i\partial_i(f) \rightleftharpoons \sum_i f_i(\partial_i f) \qquad (f \in \tilde{\tilde{O}}(\mathfrak{g}))$$

is well defined, and it is a derivation of the algebra $\tilde{\tilde{O}}(\mathfrak{g})$. Any such derivation is called a *special derivation of the algebra* $\tilde{\tilde{O}}(\mathfrak{g})$.

It can be verified directly that the definition of a special derivation does not depend either on the choice of a basis in $\mathfrak{g}$ or on its particular total ordering.

THEOREM 44.1. *For any Lie algebra $\mathfrak{g}$ over a field K of arbitrary characteristic the derivations $f \to {}^g f$ and $f \to f^g$ given by formulas* (44.3) *are special derivations of the algebra $\tilde{\tilde{O}}(\mathfrak{g})$, and for any $f \in \tilde{\tilde{O}}(\mathfrak{g})$ and $g \in \mathfrak{g}$ we have the equalities*

$$(44.8)\qquad {}^g f = \sum_i {}^g(g_i^*)\partial_i f, \qquad f^g = \sum_i (g_i^*)^g\partial_i f,$$

where the sums are extended over all elements g_i of a fixed basis $\{g_i\} \subset \mathfrak{g}$ endowed with a total ordering.

PROOF. We denote by $E(u)$ the set of basis monomials of E which occur with nonzero coefficients in the decomposition of the element $u \in U_{\mathrm{id}}(\mathfrak{g})$ with respect to the basis E. If $u = 0$, then $E(u)$ is empty. For $u \neq 0$ the set $E(u)$ is finite. Therefore, for any elements $u \in U_{\mathrm{id}}(\mathfrak{g})$ and $g \in \mathfrak{g}$ the sets $E(ug)$ and $E(gu)$ are finite. Hence, $0 \neq \langle {}^g(g_i^*), u\rangle$ if and only if $g_i \in E(ug)$. Therefore, the set $M(g,u)$ of all indices i for which the functionals ${}^g(g_i^*)$ are nonzero on u is finite. This shows that the expression $\sum_i {}^g(g_i^*)\partial_i$ is well defined and that it is a special derivation of the algebra $\tilde{\tilde{O}}(\mathfrak{g})$.

Formula (44.6) defining ∂_i implies

$$\langle \partial_i f, u\rangle = \langle f, \partial_i^* u\rangle, \tag{44.9}$$

where $f \in \widetilde{\widetilde{O}}(\mathfrak{g})$, $u \in U_{\text{id}}(\mathfrak{g})$, and ∂_i^* is a linear operator on $U_{\text{id}}(\mathfrak{g})$ transforming a basis monomial $ag_i^r b$ into the basis monomial $ag_i^{r+1} \cdot b$.

We first show that it suffices to produce the verification of equalities (44.8) only for functionals $f \in \widetilde{\widetilde{O}}(\mathfrak{g})$ for which only a finite number of values on basis monomials in E differs from zero. For fixed $g \in \mathfrak{g}$ and $u \in U_{\text{id}}(\mathfrak{g})$ and for a decomposition

$$\Delta u = \sum_{j=1}^{k} w_j \otimes v_j$$

we put

$$M \rightleftharpoons \bigcup_{j=1}^{k} M(g, w_j), \qquad L \rightleftharpoons \bigcup_{j=1}^{k} \bigcup_{i\in M} E(\partial_i^* v_j) \cup E(ug).$$

It is clear that M and L are finite sets. Let h be an arbitrary functional from $\widetilde{\widetilde{O}}(\mathfrak{g})$ vanishing on L. Then it follows directly from the definitions of the sets M and L and from formulas (44.2), (44.3), and (44.9) that

$$\begin{gathered} \langle {}^g h, u\rangle = \langle h, ug\rangle = 0, \\ \left\langle \sum_i {}^g g_i^* \cdot \partial_i h, u \right\rangle = \sum_i \sum_{j=1}^{k} \langle g_i^*, w_j g\rangle \cdot \langle h, \partial_i^* v_j\rangle = 0. \end{gathered} \tag{44.10}$$

Since any functional $f \in \widetilde{\widetilde{O}}(\mathfrak{g})$ is representatable in the form

$$f = \sum_{v\in L} \beta(v) v^* + h,$$

where $\beta(v) \in K$, $h|_L = 0$, and L is a finite set, equalities (44.10) show that if we prove equalities (44.8) for all functionals of the form $\sum_{v\in L} \beta(v)v^*$, where L is an arbitrary finite subset of E, then the values of the functionals on the left-hand side and on the right-hand side of equalities (44.8) will coincide on an arbitrary $u \in U_{\text{id}}(\mathfrak{g})$ and, hence, equalities (44.8) will be satisfied for every $f \in \widetilde{\widetilde{O}}(\mathfrak{g})$.

The functionals of the form $\sum_{v\in L} \beta(v)v^*$, where L is an arbitrary finite subset, form a subalgebra of $\widetilde{\widetilde{O}}(\mathfrak{g})$ isomorphic to the divided power algebra whose generators are the elements $(g_i^m)^*$, where $g_i \in \mathfrak{g} \cap E$ and $m = 0, 1, 2, \ldots$. Since the mappings $f \to {}^g f$ and $\sum_i {}^g(g_i^*)\partial_i$ are derivations of $\widetilde{\widetilde{O}}(\mathfrak{g})$, it remains to verify equalities (44.8) for functionals of the form $(g_i^n)^*$.

If char $K = 0$, then $(g_i^*)^n = n!\,(g_i^n)^*$ in the algebra $\widetilde{\widetilde{O}}(\mathfrak{g})$, and the verification of equalities (44.8) can be reduced to the case in which f ranges over the functionals of the form g_i^*. In this case (44.8) follows immediately from the definition (44.6) of the operator ∂_i.

If $\operatorname{char} K = p > 0$, then this reduction is impossible for $n = p^l$, but the validity of the theorem for fields of characteristic zero suggests finding a direct proof of the system of equalities

$$\langle {}^g(g_i^m)^*, u\rangle = \langle (g_i^{m-1})^* \cdot {}^g(g_i^*), u\rangle, \qquad (u \in E,\ m = 1, 2, \dots),$$

which is equivalent to the first equality of (44.8). If $u = ag_i^n b \in E$, then this system is equivalent to the system

$$\text{(44.11)} \qquad \langle (g_i^m)^*, ag_i^n bg\rangle = \binom{n}{m-1}\langle g_i^*, ag_i^{n-m+1}bg\rangle,$$

where $\binom{n}{m-1} \rightleftharpoons 0$ for $n < m-1$. Relations (44.11) mean that in the decompositions of the elements $ag_i^n bg$ and $a(g_i)^{n-m+1}bg$ ($ag_i^n b \in E$) with respect to the basis E the coefficient of g_i^m in the first decomposition differs from that in the second decomposition by the multiple of $\binom{n}{m-1}$. Therefore, the assertion of the theorem follows immediately from the next lemma.

LEMMA 44.1. *Let $g_i \in E \cap \mathfrak{g}$ and let δ_i be the linear operator on $U_{\mathrm{id}}(\mathfrak{g})$ of deleting one occurrence of the letter g_i in each basis monomial $u = g_{i_1} \cdots g_{i_l}$ ($i_1 \le i_2 \le \cdots \le i_l$) ($\delta_i u = 0$ if g_i does not occur in u). If g_i occurs in u precisely n times, then for an arbitrary element $g \in \mathfrak{g}$ and any decomposition $u = ab$, where $a, b \in E$ and either b does not begin with g_i or a does not end with g_i, we have the equality*

$$\text{(44.12)} \quad \langle (g_i^m)^*, agb\rangle = \binom{n}{m-1}\langle g_i^*, \delta_i^{m-1}(a)gb + ag\delta_i^{m-1}(b)\rangle \qquad (m = 2, 3, \dots).$$

PROOF. Since equality (44.12) is linear with respect to g, it suffices to prove it for any $g = g_i \in E \cap \mathfrak{g}$. For any monomial $ag_j b$, where $ab = u \in E$, we define an integer $\operatorname{ind}(ag_j b)$ by induction as follows:

$$\operatorname{ind}(ag_j b) = \begin{cases} 0 & \text{for } ag_j b \in E, \\ \operatorname{ind}(a'g_j g_t b) + 1 & \text{for } a = a'g_t \text{ and } t > j, \\ \operatorname{ind}(ag_t g_j b') + 1 & \text{for } b = g_t b' \text{ and } t < j. \end{cases}$$

We shall prove equality (44.12) for $g = g_j$ by induction on the degree l of the monomial u and on $\operatorname{ind}(ag_j b)$.

The base of induction: l is an arbitrary positive integer and $\operatorname{ind}(ag_j b) = 0$. Then $ag_j b$ is a basis monomial; for $ag_j b \ne g_i^m$ both sides of equality (44.12) are equal to zero, and for $ag_j b = g_i^m$ the numbers n and $m-1$ are equal to each other and both sides of equality (44.12) are equal to one.

The inductive step. Let $\operatorname{ind}(ag_j b) > 0$ and let the degree of the monomial $u = ab$ equal l. Then the following cases are posssible:

1) $a = a'g_t$ & $g_t > g_j$ & $g_t \ne g_i$,

2) $b = g_t b'$ & $g_t < g_j$ & $g_t \ne g_i$,

3) $a = a'g_i^n$ & $i > j$,

4) $b = g_i^n b'$ & $i < j$.

In the first and second cases we have, respectively,

$$ag_j b = a'[g_t, g_j]b + a'g_j g_t b, \qquad ag_j b = a[g_j, g_t]b' + ag_t g_j b', \tag{44.13}$$

and relation (44.12) follows from inductive arguments, since in the first summand on the right-hand side of both equalities in (44.13) the basis monomials $a'b$ and ab' are of degree $l-1$ and the second summands have lesser indices.

In the third case we have

$$ag_j b = a' g_i^n g_j b = \sum_{t=0}^{n} \binom{n}{t} a'((\operatorname{ad} g_i)^{n-t} g_j) g_i^t b.$$

For $t > 0$ the basis monomials $a'(g_i)^{n-t}b$ in the summands on the right-hand side are of degree $\leq l-1$, and for $t = 0$ the index of the summand $a'g_j g_i b$ is equal to $\operatorname{ind}(ag_j b) - n$. Therefore, from inductive arguments we have

$$\begin{aligned}\langle (g_i^m)^*, ag_j b\rangle &= \left\langle g_i^*, \sum_{t=m-1}^{n} \binom{n}{t'} \binom{t}{m-1} a'((\operatorname{ad} g_i)^{n-t} g_j) g_i^{t-m+1} b \right\rangle \\ &= \binom{n}{m-1} \langle g_i^*, a' g_i^{t-m+1} g_j b\rangle = \binom{n}{m-1} \langle g_i^*, \delta_i^{m-1}(a) g_j b\rangle.\end{aligned}$$

This equality proves (44.12), because $\delta_i b = 0$.

The fourth case is treated similarly by means of the formula

$$g_j g_i^n = \sum_{t=0}^{n} \binom{n}{t} g_i^t ((-\operatorname{ad} g_i)^{n-t} g_j).$$

The lemma is proved.

THEOREM 44.2. *Let $\mathfrak{g}_0$ be a proper subalgebra of a Lie K-algebra $\mathfrak{g}$ and let $\dim_K(\mathfrak{g}/\mathfrak{g}_0) = n < \infty$. Then the completed divided power algebra $\widetilde{O}(\mathfrak{g}/\mathfrak{g}_0)$ can be equipped with an action of the Lie algebra $\mathfrak{g}$ by special derivations for which the algebra $\widetilde{O}(\mathfrak{g}/\mathfrak{g}_0)$ possesses only trivial $\mathfrak{g}$-invariant ideals and $\mathfrak{g}_0$ consists of those and only those elements of $\mathfrak{g}$ that leave invariant the unique maximal ideal of $\widetilde{O}(\mathfrak{g}/\mathfrak{g}_0)$.*

PROOF. Denote by R the right ideal in $U_{\mathrm{id}}(\mathfrak{g})$ generated by the subspace $\mathfrak{g}_0$. Choose a basis $\{v_i\}$ in $\mathfrak{g}_0$ and extend this basis to a basis of the algebra $\mathfrak{g}$ by certain elements $e_1, \dots, e_n$. We introduce a total ordering on this basis in such a way that the elements $e_1, \dots, e_n$ be the last n elements and $e_1 < \cdots < e_n$. Then the elements $v_{i_1}^{t_1} \cdots v_{i_k}^{t_k} e_1^{m_1} \cdots e_n^{m_n}$, where $t_1 + \cdots + t_k > 0$, $t_i, m_i \geq 0$, and $i_1 < \cdots < i_k$, form a basis of the ideal R, and the elements of the form $e_1^{m_1} \cdots e_n^{m_n}$ ($m_1 \geq 0, \dots, m_n \geq 0$) complete this basis of R to a basis of $U_{\mathrm{id}}(\mathfrak{g})$. We denote by $R^{\perp}$ the subset of $\widetilde{\widetilde{O}}(\mathfrak{g})$ consisting of those and only those functionals that vanish on R. Since $\Delta R \subseteq R \otimes_K U_{\mathrm{id}}(\mathfrak{g}) + U_{\mathrm{id}}(\mathfrak{g}) \otimes_K R$, $R^{\perp}$ is a subalgebra of $\widetilde{\widetilde{O}}(\mathfrak{g})$ by formula (44.2). From the definition in (44.3) it follows that if $f \in R^\perp$, $u \in R$, and $g \in \mathfrak{g}$, then $ug \in R$ and $\langle {}^g f, u\rangle = \langle f, ug\rangle = 0$. This shows that $R^\perp$ is a $\mathfrak{g}$-invariant subalgebra of $\widetilde{\widetilde{O}}(\mathfrak{g})$. Since the elements of the subalgebra $R^\perp$ have the form

$$\sum \beta^{i_1,\dots,i_n}_{m_1,\dots,m_n} \left(e_1^{m_1}\cdots e_n^{m_n}\right)^*,$$

it follows from formulas (44.4) and (44.5) and the defining relations of the divided power algebra (44.1) that the subalgebra $R^\perp$ is isomorphic to the completed divided power algebra $\widetilde{O}(\mathfrak{g}/\mathfrak{g}_0)$. The first of the equalities (44.8) has now the following form for $f \in R^\perp \simeq \widetilde{O}(\mathfrak{g}/\mathfrak{g}_0)$:

$${}^g f = \sum_i {}^g(e_i^*)\cdot \partial_{e_i} f,$$

where, by the $\mathfrak{g}$-invariance of $R^\perp$, the elements ${}^g(e_i^*)$ belong to $R^\perp$, and from the definition of ∂_{e_i} (see formula (44.6)) we have $\partial_{e_i}\colon R^\perp \to R^\perp$. This shows that the given action of the element $g \in \mathfrak{g}$ on the algebra $\widetilde{O}(\mathfrak{g}/\mathfrak{g}_0)$ is a special derivation.

It was mentioned above that $\widetilde{O}(\mathfrak{g}/\mathfrak{g}_0) = R^\perp$ is a local algebra and its unique maximal ideal consists of the functionals $f \in R^\perp$ for which $\langle f, 1\rangle = 0$. Suppose that $0 \neq f \in R^\perp$. Then, for some $u = e_1^{r_1}\cdots e_n^{r_n}$, we have $0 \neq \langle f, u\rangle = \langle {}^u f, 1\rangle$. Therefore, ${}^u f$ does not belong to the maximal ideal of the algebra $R^\perp$. This proves that any nonzero $\mathfrak{g}$-invariant ideal contains an invertible element of the algebra $R^\perp$ and, hence, coincides with $R^\perp$.

It is clear that if $g \in \mathfrak{g}_0$ and $f \in R^\perp$, then $0 = \langle {}^g f, 1\rangle = \langle f, g\rangle$ and ${}^g f$ belongs to the maximal ideal of the algebra $R^\perp$. Thus, ${}^{\mathfrak{g}_0}\operatorname{Rad} R^\perp \subset R^\perp$. Conversely, if for some $g \in \mathfrak{g}$ we have the inclusion ${}^g\operatorname{Rad} R^\perp \subset R^\perp$, then ${}^g R^\perp \subset R^\perp$. This means that $\langle f, g\rangle = \langle {}^g f, 1\rangle = 0$ for any $f \in R^\perp$, however, this is possible only if $g \in \mathfrak{g}_0$. The theorem is proved.

DEFINITION 44.3. Let V be a finite-dimensional space over a field K of arbitrary characteristic and $\dim_K V = n$. The Lie algebra of all special derivations of the completed divided power algebra $\widetilde{O}(V)$ is called the *full Cartan Lie algebra of general type* and is denoted by $\widetilde{W}_n(K)$.

The elements of the Lie algebra $\widetilde{W}_n(K)$ have the form $f_1\partial_1 + \cdots + f_n\partial_n$, where $f_1, \dots, f_n \in \widetilde{O}(V)$. Therefore, the filtration $\widetilde{O}_i(V)$ $(i = 0, 1, \dots)$ induces a filtration on $\widetilde{W}_n(K)$ for which

$$(\widetilde{W}_n(K))_j \rightleftharpoons \widetilde{O}_{j+1}(V) + \cdots + \widetilde{O}_{j+1}(V)\partial_n \qquad (j = -1, 0, 1, \dots).$$

From Theorem 44.2 we immediately obtain the following

COROLLARY 1. *Suppose that for a certain extension $\widehat{K}$ of the ground field K there exists a proper $\widehat{K}$-subalgebra $\widehat{\mathfrak{g}}_0$ of codimension n in the Lie $\widehat{K}$-algebra $\widehat{\mathfrak{g}} = \widehat{K}\otimes_K \mathfrak{g}$. Then there exists a $\widehat{K}$-homomorphism $\rho_{\widehat{\mathfrak{g}}_0}\colon \widehat{\mathfrak{g}} \to \widetilde{W}_n(\widehat{K})$ such that*

a) *the kernel of the homomorphism $\rho_{\widehat{\mathfrak{g}}_0}$ coincides with the largest ideal contained in $\widehat{\mathfrak{g}}_0$,*

b) $\rho_{\widehat{\mathfrak{g}}_0} + (\widetilde{W}_n(\widehat{K}))_0 = \widetilde{W}_n(\widehat{K})$.

PROOF. It suffices to take for $\rho_{\widehat{\mathfrak{g}}_0}$ the representation of the Lie algebra $\widehat{\mathfrak{g}}$ in the Lie algebra of all special derivations of the algebra $\widetilde{O}(\widehat{\mathfrak{g}}/\widehat{\mathfrak{g}}_0)$; this representation was constructed in the proof of Theorem 44.2. Then properties a) and b) of the homomorphism $\rho_{\widehat{\mathfrak{g}}_0}$ are equivalent to the subalgebra $\widehat{\mathfrak{g}}_0$ consisting of those and only those elements that preserve the unique maximal ideal of the algebra $\widetilde{O}(\widehat{\mathfrak{g}}/\widehat{\mathfrak{g}}_0)$, which is true by Theorem 44.2. The corollary is proved.

REMARK 44.2. Since the Lie subalgebra $(\widetilde{W}_n(K))_1$ is approximated by nilpotent algebras and any Lie subalgebra of the quotient algebra $(\widetilde{W}_n)_0/(\widetilde{W}_n)_1$ possesses a proper subalgebra of codimension $\leq n$, an arbitrary Lie subalgebra of $\widetilde{W}_n(K)$ possesses a proper subalgebra of codimension $\leq n$. Therefore, if a Lie K-algebra $\mathfrak{g}$ admits a nonzero K-homomorphism into $\widetilde{W}_n(\widehat{K})$, then the Lie K-algebra $\widehat{\mathfrak{g}} = \widehat{K} \otimes_K \mathfrak{g}$ possesses a $\widehat{K}$-subalgebra $\widehat{\mathfrak{g}}_0$ for which

$$1 \leq \dim_{\widehat{K}} \widehat{\mathfrak{g}}/\widehat{\mathfrak{g}}_0 \leq n.$$

COROLLARY 2. *Let $\mathfrak{g}$ be a finitely generated Lie algebra over an algebraically closed field K of characteristic zero. Suppose that for a certain extension $\widehat{K}$ of the ground field K there exists a proper $\widehat{K}$-subalgebra $\widehat{\mathfrak{g}}_0$ of the Lie $\widehat{K}$-algebra $\widehat{\mathfrak{g}} = \widehat{K} \otimes_K \mathfrak{g}$ for which* $\dim_{\widehat{K}} \widehat{\mathfrak{g}}/\widehat{\mathfrak{g}}_0 = n < \infty$. *Then in the Lie algebra $\mathfrak{g}$ there exists a proper K-subalgebra $\mathfrak{g}_0$ for which* $\dim_K \mathfrak{g}/\mathfrak{g}_0 \leq n$.

PROOF. Since $\operatorname{char} K = 0$, the $\widehat{K}$-algebra $\widetilde{O}(\widehat{\mathfrak{g}}/\widehat{\mathfrak{g}}_0)$ is isomorphic to the algebra $\widetilde{\mathcal{F}} = K[[t_1, \dots, t_n]]$ of power series and we have $\widetilde{W}_n(K) = \operatorname{Der}_K \widetilde{\mathcal{F}}$. Let $g_1, \dots, g_k$ be the generators of the K-algebra $\mathfrak{g}$. By Theorem 44.2 and its Corollary 1, there exists a $\widehat{K}$-homomorphism $\rho_{\widehat{\mathfrak{g}}_0}$ of the Lie algebra $\widehat{\mathfrak{g}}$ into the Lie algebra $\operatorname{Der}_{\widehat{K}} \widetilde{\mathcal{F}}$ such that

$$\rho_{\widehat{\mathfrak{g}}_0}(g_i) = \sum_{j=1}^{n} f_{ij} \frac{\partial}{\partial t_j} \qquad (f_{ij} \in \widetilde{\mathcal{F}},\ i = 1, \dots, k) \tag{44.14}$$

and the algebra $\widetilde{\mathcal{F}}$ has only trivial $\rho_{\widehat{\mathfrak{g}}_0}$-invariant ideals. We denote by the letter $\mathcal{E}$ the differential K-subalgebra with unit in $\widetilde{\mathcal{F}}$ (see §51) generated by the f_{ij} $(i = 1, \dots, k;$ $j = 1, \dots, n)$. Let $\mathfrak{h} = \mathcal{E}\frac{\partial}{\partial t_1} + \dots + \mathcal{E}\frac{\partial}{\partial t_n}$. It is clear that $\mathfrak{h}$ is a Lie K-subalgebra of $\operatorname{Der}_K \widetilde{\mathcal{F}}$ and $\rho_{\widehat{\mathfrak{g}}_0}(\mathfrak{g}) \subset \mathfrak{h}$. Since $\mathcal{E}$ has no zero divisors and $\mathcal{E}$ is a finitely generated differential K-algebra, by Theorem 51.2 there exists a homomorphism $\varphi\colon \mathcal{E} \to K$ mapping the nonzero elements $f_{ij} \in \mathcal{E}$ into nonzero elements of the field K. Then $\mathfrak{h}_0 \rightleftharpoons (\operatorname{Ker} \varphi)\mathfrak{h}$ is a Lie K-subalgebra of $\mathfrak{h}$ and $\dim_K \mathfrak{h}/\mathfrak{h}_0 = n$. From equalities (44.14) it follows that the Lie K-subalgebra $\mathfrak{g}_0 = \left(\rho_{\widehat{\mathfrak{g}}_0}\right)^{-1}\mathfrak{h}_0$ is a proper Lie K-subalgebra in $\mathfrak{g}$ and $\dim_K \mathfrak{g}/\mathfrak{g}_0 \leq n$. The corollary is proved.

44.3. Taylor homomorphisms into the algebra $\widetilde{\widetilde{O}}(\mathfrak{g})$. In §45 we shall be forced to use the fact that any homomorphism ψ of a commutative K-algebra $\mathcal{E}$, on which a Lie algebra $\mathfrak{g}$ is acting by derivations, into the field K can be canonically lifted to a homomorphism of $\mathfrak{g}$-differential K-algebras $\widetilde{\psi}\colon \mathcal{E} \to \widetilde{\widetilde{O}}(\mathfrak{g})$ for which the following diagram is commutative:

$$\begin{array}{ccc} \mathcal{E} & \xrightarrow{\widetilde{\psi}} & \widetilde{\widetilde{O}}(\mathfrak{g}) \\ {\scriptstyle\psi}\big\downarrow & & \big\downarrow{\scriptstyle\varepsilon} \\ K & \xrightarrow{\simeq} & \widetilde{\widetilde{O}}(\mathfrak{g})/\operatorname{Rad}\widetilde{\widetilde{O}}(\mathfrak{g}) \end{array} \qquad \left(\operatorname{Rad}\widetilde{\widetilde{O}}(\mathfrak{g}) = \{f \in \widetilde{\widetilde{O}}(\mathfrak{g}) \mid \langle f, \mathbf{1}\rangle = 0\}\right).$$

This homomorphism $\widetilde{\psi}$ is a natural generalization of the Taylor series expansion of infinitely differentiable functions at some point.

We denote by ${}^{g}e$ the result of differentiation of the element $e \in \mathcal{E}$ by the element $g \in \mathfrak{g}$. The $\mathfrak{g}$-module structure on $\mathcal{E}$ endows $\mathcal{E}$ by the structure of a $U_{\mathrm{id}}(\mathfrak{g})$-module; therefore, ${}^{u}e$ will denote the result of applying the operator $u \in U_{\mathrm{id}}(\mathfrak{g})$ to the element $e \in \mathcal{E}$.

PROPOSITION 44.1. *For any homomorphism $\psi\colon \mathcal{E} \to K$ of the K-algebra $\mathcal{E}$ on which the Lie K-algebra $\mathfrak{g}$ acts by derivations, the mapping $\widetilde{\psi}\colon \mathcal{E} \to \widetilde{\widetilde{O}}(\mathfrak{g})$ given by the formula*

$$\langle \widetilde{\psi}(e), u\rangle \rightleftharpoons \psi({}^{u}(e)), \qquad (e \in \mathcal{E},\ u \in U_{\mathrm{id}}(\mathfrak{g})), \tag{44.15}$$

is a homomorphism of K-algebras, and we have

$$\begin{aligned} \widetilde{\psi}({}^{g}e) &= {}^{g}\widetilde{\psi}(e) \qquad (g \in \mathfrak{g},\ e \in \mathcal{E}) \\ \psi(e) &= \varepsilon\widetilde{\psi}(e) \qquad (e \in \mathcal{E}), \end{aligned} \tag{44.16}$$

where ε is the unique nonzero K-homomorphism of the local algebra $\widetilde{\widetilde{O}}(\mathfrak{g})$ into the field K.

The proof of the proposition is obtained immediately from the definition (44.2) of the multiplication in the algebra $\widetilde{\widetilde{O}}(\mathfrak{g})$ and from formulas (44.3).

§45. Recovery of the algebra of regular functions of a smooth irreducible affine variety from the Lie algebra of its vector fields

In this section we shall prove Theorem 42.3, which generalizes Theorem 12.2 to the case of arbitrary smooth algebraic varieties.

Let f be an arbitrary multilinear polynomial mapping of the form (42.5) recovering the polynomial K-algebra $\mathcal{F}_n = K[t_1, \dots, t_n]$ ($\operatorname{char} K = 0$) from the Lie algebra of all its derivations $W_n(K)$ and let φ_n have the same meaning as in Theorem 42.3. The existence of an associative multilinear polynomial f of this kind follows from Theorem 12.2.

LEMMA 45.1. *In the Lie algebras $W_n(K)$ and $\widetilde{W}_n(K)$ we have the following identities:*

$$[f(\operatorname{ad} y_1, \dots, \operatorname{ad} y_l), f(\operatorname{ad} z_1, \dots, \operatorname{ad} z_l)] = 0, \tag{45.1}$$

$$\begin{aligned} &[\operatorname{ad}\{f(\operatorname{ad} y_1, \dots, \operatorname{ad} y_l) \times x\}, f(\operatorname{ad} z_1, \dots, \operatorname{ad} z_l)] \\ &\quad = f(\operatorname{ad} y_1, \dots, \operatorname{ad} y_l)[\operatorname{ad} x, f(\operatorname{ad} z_1, \dots, \operatorname{ad} z_l)], \end{aligned} \tag{45.2}$$

$$f\big|_{\operatorname{ad} \widetilde{W}_{n-1}} = 0. \tag{45.3}$$

PROOF. Since the algebras W_n and $\widetilde{W}_n$ have the same identities and the linear space of $f|_{\operatorname{ad} W_n}$ coincides with $\varphi_n(\mathcal{F}_n)$, identities (45.1) and (45.2) follow from the obvious equalities:

$$(45.4) \qquad \begin{aligned} &[a,b] = 0, \qquad &&[\operatorname{ad} g, \varphi_n(a)] = \varphi_n({}^g a), \\ &{}^{(bg)}a = b({}^g a) \qquad &&(a, b \in \mathcal{F}_n,\ g \in W_n(K)). \end{aligned}$$

We identify the Lie algebra $W_{n-1}(K)$ with the subalgebra of $W_n(K)$ consisting of the derivations of the form $a_i \frac{\partial}{\partial t_1} + \cdots + a_{n-1} \frac{\partial}{\partial t_{n-1}}$, where $a_i \in K[t_1, \ldots, t_{n-1}]$. Then from the definition of the polynomial f it follows that $f|_{\operatorname{ad} W_n} \subset \varphi_n(\mathcal{F}_n)$. Since $bg \neq 0$ for any nonzero elements $b \in \varphi_n(\mathcal{F}_n)$ and $g \in W_n$, equality (45.3) is equivalent to the equality $f|_{\operatorname{ad} W_{n-1}} \frac{\partial}{\partial t_n} = 0$, which follows from the relation $\left[a_i \frac{\partial}{\partial t_i}, \frac{\partial}{\partial t_n}\right] = 0$, where $i < n$ and $a_i \in K[t_1, \ldots, t_{n-1}]$. The lemma is proved.

COROLLARY. *The image of the linear mapping f of $(\otimes \operatorname{ad} \widetilde{W}_n(K))^l$ coincides with the algebra $\varphi_n(\widetilde{\mathcal{F}}_n)$ ($\widetilde{\mathcal{F}}_n = K[[t_1, \ldots, t_n]]$) of power series.*

PROOF. Let us denote by the letter $\mathcal{F}$ the image of the mapping f. It is clear that $\widetilde{W}_n(K) \supset W_n(K)$ & $\mathcal{F} \supset \varphi_n(\mathcal{F}_n)$. Moreover, $\mathcal{F} \supset [\mathcal{F}, \operatorname{ad} \widetilde{W}_n(K)]$. Therefore, $\mathcal{F} \supset [\varphi_n(\mathcal{F}_n), \operatorname{ad} \widetilde{W}_n] \supset \varphi_n(\widetilde{\mathcal{F}})$. However, then for any $e \in \widetilde{\mathcal{F}}_n$ it follows from formula (45.2) that for $f(\operatorname{ad} z_1, \ldots, \operatorname{ad} z_l) \in \varphi_n(\mathcal{F}_n)$ we have

$$(45.5) \qquad f(\operatorname{ad} g_1, \ldots, \operatorname{ad} g_l)\varphi_n({}^g e) = \varphi_n\left({}^{f(\operatorname{ad} g_1, \ldots, \operatorname{ad} g_l) \times g} e\right),$$

where $g, g_i \in \widetilde{W}_n(K)$. Since the right-hand side of this equality belongs to $\varphi_n(\widehat{\mathcal{F}})$, the left-hand side of it also belongs to $\varphi_n(\widehat{\mathcal{F}})$. In particular, for $e = t_1$ and $g = \frac{\partial}{\partial t_1}$ we obtain that all values of f belong to $\varphi_n(\widehat{\mathcal{F}})$. The corollary is proved.

Let $K, \mathcal{E}, \mathcal{E}', \mathfrak{g}, I_m, n$, and φ have the same meaning as in Theorem 42.3. Denote by $\widehat{K}$ an algebraically closed field for which $\widehat{K} \supset K$ & $\dim_K \mathcal{E} < |\widehat{K}|$. We set $\widehat{\mathcal{E}} = \widehat{K} \otimes_K \mathcal{E}$, $\widehat{\mathfrak{g}} = \widehat{K} \otimes_K \mathfrak{g}$, and $\widehat{\mathcal{E}}' = \widehat{K} \otimes_K \mathcal{E}'$.

LEMMA 45.2. *For any elements $e \in \widehat{\mathcal{E}}$, $e' \in \widehat{\mathcal{E}}'$, and $g \in \widehat{\mathfrak{g}}$ the following equalities hold in $\operatorname{End}_{\widehat{K}} \widehat{\mathfrak{g}}$:*

$$(45.6) \qquad \varphi({}^{e' \times g} e) = \varphi({}^g e)e',$$

$$(45.7) \qquad [\operatorname{ad}(e \times g), e'] = \varphi(e)[\operatorname{ad} g, e'].$$

PROOF. It is enough to prove (45.6) and (45.7) for $e' = f(\operatorname{ad} g_1, \ldots, \operatorname{ad} g_l)$ ($g_i \in \mathfrak{g}$). We note that for the case $\widehat{\mathcal{E}} = \widehat{K}[[t_1, \ldots, t_n]]$, $\widehat{\mathfrak{g}} = \widetilde{W}_n(\widehat{K})$, and $\varphi = \varphi_n$ equality (45.6) has already been proved (see formula (45.5)). The validity of (45.7) in this special case is established in a similar way. Now we shall show how we can derive (45.6) from (45.5) by using Proposition 44.1 and Theorem 44.2.

Let ψ be an arbitrary homomorphism $\widehat{\mathcal{E}} \to \widehat{K}$, let $\widetilde{\psi}\colon \widehat{\mathcal{E}} \to \widetilde{\widetilde{O}}(\widehat{\mathfrak{g}})$ be the Taylor $\widehat{K}$-homomorphism of $\widehat{\mathfrak{g}}$-differential algebras constructed via the homomorphism ψ (see Proposition 43.1), let

$$\widehat{\mathfrak{g}}_0 = \widehat{\mathfrak{g}}_0(\psi) = \{g \in \widehat{\mathfrak{g}} \mid {}^g \operatorname{Ker} \psi \subset \operatorname{Ker} \psi\}$$

be a Lie $\widehat{K}$-subalgebra of $\widehat{\mathfrak{g}}$, let $R = R(\psi) = \widehat{\mathfrak{g}}_0 U_{\mathrm{id}}(\widetilde{\mathfrak{g}})$ be a right ideal of the universal enveloping algebra $U_{\mathrm{id}}(\widetilde{\mathfrak{g}})$ with unit, let

$$R^{\perp} = R^{\perp}(\psi) = \{h \in \widetilde{\widetilde{O}}(\widehat{\mathfrak{g}}) \mid \langle h, R\rangle = 0\},$$

and let $m = m(\psi)$ be an integer such that $\psi(I_m) \neq 0$ & $\psi(I_{m+1}) = 0$. By the assumption of Theorem 42.3 we have $I_{n+1} = 0$. Therefore, $m \leq n$ and, as it was mentioned in Remark 42.2, we have $\dim_{\widehat{K}} \widehat{\mathfrak{g}}/\widehat{\mathfrak{g}}_0 = m < \infty$. Since $\widehat{\mathcal{E}} = \widehat{K}\cdot \mathbf{1} + \operatorname{Ker}\psi$, we have ${}^{\mathfrak{g}_0}\widehat{\mathcal{E}} \subset \operatorname{Ker}\psi$. Therefore, $\widetilde{\psi}(\widehat{\mathcal{E}}) \subset R^{\perp}$. As proved in Theorem 44.2, $R^{\perp}$ is a $\widehat{\mathfrak{g}}$-invariant $\widehat{K}$-subalgebra of $\widetilde{\widetilde{O}}(\widehat{\mathfrak{g}})$, isomorphic to the algebra $\widetilde{\mathcal{F}}_m(\widehat{K})$ of power series in m variables. Hence, Proposition 44.1 and Theorem 44.2 determine homomorphisms

$$\widetilde{\psi}\colon \widehat{\mathcal{E}} \to R^{\perp} \simeq \widetilde{\mathcal{F}}_m(\widehat{K}),$$
$$\rho_{\psi}\colon \widehat{\mathfrak{g}} \to \operatorname{Der}_{\widehat{K}} R^{\perp} \simeq \widetilde{W}_n(\widehat{K}),$$

with the relationship

$$\psi({}^{g}e) = {}^{\rho_{\psi}(g)}\widetilde{\psi}(e) \qquad (e \in \widehat{\mathcal{E}},\ g \in \widehat{\mathfrak{g}}). \tag{45.8}$$

From the definition of φ and from the faithfulness of the action of $\widehat{\mathfrak{g}}$ on $\widehat{\mathcal{E}}$ it follows that equality (45.6) is equivalent to the equalities

$$ {}^{((e'\times g_e)\times v)}c - {}^{(g_e\times(e'\times v))}c = 0 \qquad (v \in \widehat{\mathfrak{g}},\ c \in \widehat{\mathcal{E}}). \tag{45.9}$$

Applying the homomorphism $\widetilde{\psi}$ to (45.9) and using (for $m = n$) formulas (45.8), (45.4), and (45.5) and (for $m < n$) formula (45.3), we see that under the action of any homomorphism $\widetilde{\psi}$ the left-hand side of (45.9) is mapped into zero. But then the same is also true for any homomorphism $\psi\colon \widehat{\mathcal{E}} \to \widehat{K}$ (see (44.16)). Since there are no nilpotent elements in the algebra $\mathcal{E}$, neither there are in the algebra $\widehat{\mathcal{E}}$. Therefore, by Hilbert's Nullstellensatz (see Theorem 51.1), $\bigcap_{\psi} \ker\psi = 0$ and the element on the left-hand side of relation (45.9) is equal to zero in the algebra $\widehat{\mathcal{E}}$.

Equality (45.7) is proved similarly. The lemma is proved.

Note that in the course of proving this lemma we have constructed, for each $\psi\colon \widehat{\mathcal{E}} \to \widehat{K}$, a homomorphism $\rho_{\psi}\colon \widehat{\mathfrak{g}} \to \widetilde{W}_m(\widehat{K})$ $(m \leq n)$ which coincides with the homomorphism $\rho_{\widehat{\mathfrak{g}}_0}$ of Corollary 1 of Theorem 44.2. Therefore, $\operatorname{Ker}\rho_{\psi} \subset \widehat{\mathfrak{g}}_0(\psi) = \{g \in \widehat{\mathfrak{g}} \mid {}^{g}\mathcal{E} \subset \operatorname{Ker}\psi\}$. Since $\widehat{\mathfrak{g}}$ acts faithfully on $\widehat{\mathcal{E}}$, for any $g \neq 0$ $(g \in \widehat{\mathfrak{g}})$ there exists an element $e \in \mathcal{E}$ such that $0 \neq {}^{g}e$, and, by Theorem 51.1, we have $0 \neq \psi({}^{g}e)$ for some homomorphism $\psi\colon \widehat{\mathcal{E}} \to \widehat{K}$. This means that $g \notin \widehat{\mathfrak{g}}_0(\psi)$. Hence, $\bigcap_{\psi} \operatorname{Ker}\rho_{\psi} = \bigcap_{\psi} \widehat{\mathfrak{g}}_0(\psi) = 0$ and $\widehat{\mathfrak{g}}$ can be embedded in the Cartesian product of the Lie algebras $\widetilde{W}_n(\widehat{K})$. This proves that $\mathfrak{g} \in \operatorname{var} \widetilde{W}_n(K)$.

LEMMA 45.3. *For any homomorphism $\psi\colon \widehat{\mathcal{E}} \to \widehat{K}$ the algebras $\widetilde{\psi}(\widehat{\mathcal{E}})$ and $\rho_{\psi}(\widehat{\mathfrak{g}})$ are dense in $\widetilde{\mathcal{F}}_m(\widehat{K})$ and $\widetilde{W}_m(\widehat{K})$ with respect to the topologies determined by the canonical filtrations on $\widetilde{\mathcal{F}}_m$ and $\widetilde{W}_m$, respectively.*

PROOF. Since $\psi(I_m) \neq 0$ & $\psi(I_{m+1}) = 0$, the definition of the ideal I_m and the inclusion $\mathcal{E}\mathfrak{g} \subset \mathfrak{g}$ imply the existence of elements $g_1, \dots, g_m \in \mathfrak{g}$ and $e_1, \dots, e_m \in \mathcal{E}$

such that $\psi({}^{g_i}e_j) = \delta_{ij}\beta$ $(0 \neq \beta \in \widehat{K})$. We put $t_i' = \widetilde{\psi}(e_i) - \psi(e_i)$ $(i = 1, \ldots, m)$. It is clear that then $\rho_\psi(g_1), \ldots, \rho_\psi(g_m)$ form a basis of $\widehat{\mathfrak{g}}/\widehat{\mathfrak{g}}_0(\psi)$ and all monomials of the form $(t_1')^{r_1} \cdots (t_m')^{r_m}$ form a dense basis in $R^\perp(\psi)$. However, then the elements of the form $\rho_\psi\left(\widetilde{\psi}^{-1}((t_1')^{r_1} \cdots (t_m')^{r_m})g_j\right)$ $(j = 1, \ldots, m)$ form a dense basis in $\mathrm{Der}_{\widehat{K}} R^\perp$. The lemma is proved.

COROLLARY. *For any homomorphism $\psi\colon \widehat{\mathcal{E}} \to \widehat{K}$ and any elements $e \in \widehat{\mathcal{E}}$ and $g \in \widehat{\mathfrak{g}}$ we have*

$$\rho_\psi(e \times g) = \widetilde{\psi}(e) \times \rho_\psi(g).$$

PROOF OF THEOREM 42.3. Since $I_n \neq 0$ and $\mathcal{E}$ has no nil-elements by the assumption of Theorem 42.3, there exists a homomorphism $\psi\colon \widehat{\mathcal{E}} \to \widehat{K}$ such that $m(\psi) = n$, by Theorem 51.1. Then, by Lemma 45.3, $\rho_\psi(\mathfrak{g})$ contains a dense basis of the algebra $\widetilde{W}_n(\widehat{K})$ and, hence, $\mathrm{var}\,\rho_\psi(\mathfrak{g}) = \mathrm{var}\,\widetilde{W}_n(K)$. This proves that $\mathrm{var}\,\mathfrak{g} \supseteq \mathrm{var}\,\widetilde{W}_n(K)$. The converse inclusion $\mathrm{var}\,\mathfrak{g} \subseteq \mathrm{var}\,\widetilde{W}_n(K)$ has been obtained above. Therefore, $\mathrm{var}\,\mathfrak{g} = \mathrm{var}\,\widetilde{W}_n(K)$.

Since the algebra $\mathcal{E}$ is commutative and has no nil-elements, $\mathcal{E}$ is a semiprime algebra and the Martindale central closure $Q(\mathcal{E})$ is well defined (see §3). The algebra $\varphi(\mathcal{E})$ does not contain nil-elements either. Indeed, the equality $\varphi(e)^k = 0$ implies $({}^{(eg)}b)^k = 0$ (see (45.4)) for arbitrary $b \in \mathcal{E}$ and $g \in \mathfrak{g}$, and this means that $\varphi(e)\mathfrak{g} = 0$. Hence, the algebra $\varphi(\mathcal{E})$ is also semiprime and $\varphi({}^{\mathfrak{g}}e)$ is a large ideal in $\varphi(\mathcal{E})$. Therefore, from properties of the central closure and from equality (45.6) we conclude that $\mathcal{E}' \subset Q(\varphi(\mathcal{E}))$. The first property of the image $\mathcal{E}'$ of the mapping f in Theorem 42.3 now follows from formulas (45.6) and (45.7).

Let us prove the second property of the image $\mathcal{E}'$ of the mapping f. Let ${}^{\mathfrak{g}}\mathcal{E} = \mathcal{E}$. Then $1 = {}^{g_1}c_1 + \cdots + {}^{g_t}c_t$ for some elements $c_i \in \mathcal{E}$ and $g_i \in \mathfrak{g}$. Since for any $a, b \in \mathcal{E}$ and $g \in \mathfrak{g}$ we have equality (45.4) in $\mathcal{E}$, for any $a \in \mathcal{E}$ we have

$$a = {}^{(ag_1)}c_1 + \cdots + {}^{(ag_t)}c_t$$

and, hence, φ is a monomorphism. Moreover, relation (45.6) implies that for any $e' \in \mathcal{E}'$ we have

$${}^{(e'g_1)}\varphi(c_1) + \cdots + {}^{(e'g_t)}\varphi(c_t) = e' \cdot \varphi(1) = e'.$$

This proves that $\mathcal{E}'$ is contained in $\varphi(\mathcal{E})$.

Let ψ be an arbitrary homomorphism $\mathcal{E} \to \widehat{K}$ and let $m = m(\psi)$. From Corollary of Lemma 45.3 we have

$$\varphi_m\widetilde{\psi}\varphi^{-1}(f(\mathrm{ad}\,g_1, \ldots, \mathrm{ad}\,g_l)) = f(\mathrm{ad}\,\rho_\psi(g_1), \ldots, \mathrm{ad}\,\rho_\psi(g_l)). \tag{45.10}$$

Then for $\psi(I_n) = 0$ we have $m < n$ and formula (45.3) implies $\psi(\varphi^{-1}\mathcal{E}') = 0$. If $\psi(I_n) \neq 0$, then $m = n$ and, by Lemma 45.3, $\rho_\psi(\mathfrak{g})$ contains a dense basis of $\widetilde{W}_n(\widehat{K})$, therefore, $f|_{\rho_\psi(\mathfrak{g})}$ contains a dense basis of $\widetilde{\mathcal{F}}_n(\widehat{K})$, and, hence, $\psi(\varphi^{-1}\mathcal{E}') \neq 0$ by formula (45.10). Since $\mathrm{Ker}\,\psi$ ranges over all simple (prime) ideals in $\mathcal{E}$ when $\widehat{K}$ runs through all fields containing K and ψ runs through all homomorphisms $\mathcal{E} \to \widehat{K}$, this proves the second property of the image of $\mathcal{E}'$.

We note that if we put $f'(t_1, \dots, t_{l+1}) = [f, t_{l+1}]$, then the image of the mapping

$$f : (\otimes \operatorname{ad} W_n)^{l+1} \to \operatorname{End}_K W_n$$

also coincides with $\varphi_n(\mathcal{F}_n)$. Therefore, if $\mathcal{E} = {}^{\mathfrak{g}}\mathcal{E}$, then $[\mathcal{E}', \operatorname{ad} \mathfrak{g}] = f'|_{\operatorname{ad} \mathfrak{g}}$ is an ideal of $\varphi(\mathcal{E})$. Applying Property 2 of Theorem 42.3 for the image of the mapping f', we see that this ideal is contained in those and only those simple ideals of the algebra $\varphi(\mathcal{E})$ which contain $\varphi(I_n)$. Since there is a unit in the algebra $\mathcal{E}$, this implies the third property of the mapping f. Theorem 42.3 is proved.

§46. Simple Lie algebras satisfying the standard Lie identity of degree 5

In §44 for an arbitrary field K and for any positive integer $n \in \mathbb{N}$ a simple Lie algebra $\widetilde{W}_n(K)$ was defined (see Definition 44.3). In the following two sections general results on simple Lie algebras of the variety var $\widetilde{W}_n(K)$ will be obtained, and these imply Theorems 42.1 and 42.2. In this section we shall give a complete description of algebras of this kind for the case $n = 1$ and $\operatorname{char} K \neq 2$. This special case is very important for applications of Lie algebra theory, since this class of simple Lie algebras, as we shall see below, possesses an interesting geometric interpretation and is closely related to the class of Lie algebras which arises from smooth one-dimensional involutory distributions on affine algebraic varieties, and, essentially, the study of these objects of algebraic geometry is equivalent to the study of Lie algebras of the variety given by identity (42.2).

The class of simple Lie algebras investigated in this section is described by the following

PROPOSITION 46.1. *Let $\mathfrak{g}$ be a simple Lie algebra over a field K whose characteristic differs from two and let C be the centroid of the algebra $\mathfrak{g}$. Then the following properties of the algebra $\mathfrak{g}$ are equivalent.*

1. *There exists an extension $\widehat{K}$ of the field C such that the $\widehat{K}$-algebra $\widehat{\mathfrak{g}} = \widehat{K} \otimes_C \mathfrak{g}$ contains a $\widehat{K}$-subalgebra of codimension* 1.

2. *All Lie identities valid in $\widetilde{W}_1(K)$ are also valid in the Lie algebra $\mathfrak{g}$, i.e., $\mathfrak{g} \in$ var $\widetilde{W}_1(K)$.*

3. *In the Lie algebra $\mathfrak{g}$ the standard Lie identity of degree 5 is valid*:

$$\sum_{\sigma \in S_4} \operatorname{sign}(\sigma)[z, y_{\sigma(1)}, y_{\sigma(2)}, y_{\sigma(3)}, y_{\sigma(4)}] = 0, \tag{46.1}$$

where the sum is taken over all permutations of the set $\{1, 2, 3, 4\}$ and $\operatorname{sign}(\sigma)$ *is the sign of the permutation σ.*

4. *The Lie algebra $\mathfrak{g}$ can be embedded, as a C-algebra, in a Lie C-algebra* $\operatorname{Diff}_\partial \mathcal{E} = \{f\partial \mid f \in \mathcal{E}\}$ *of all special derivations of some commutative differentially simple C-algebra $\mathcal{E}$ with unit element and signature derivation ∂.*

First we prove the implications $1 \Rightarrow 2, 4$ and $4 \Rightarrow 2$.

Suppose that Property 1 holds for a simple Lie algebra $\mathfrak{g}$. Then the Lie $\widehat{K}$-algebra $\widehat{\mathfrak{g}} = \widehat{K} \otimes_C \mathfrak{g}$ contains a $\widehat{K}$-subalgebra $\widehat{\mathfrak{g}}_0$ for which $\dim_{\widehat{K}} \widehat{\mathfrak{g}}/\widehat{\mathfrak{g}}_0 = 1$, and from Corollary

1 of Theorem 44.2 it follows that there exists a nonzero homomorphism $\rho_{\widehat{\mathfrak{g}}_0}\colon \widehat{\mathfrak{g}} \to \widetilde{W}_1(\widehat{K})$. Since the algebra $\mathfrak{g}$ is simple, $\rho_{\widehat{\mathfrak{g}}_0}$ is a monomorphism and $\operatorname{var}\mathfrak{g} \subset \operatorname{var}\widetilde{W}_1(\widehat{K})$. Moreover, from Definition 44.3 of the algebra $\widetilde{W}_1(\widehat{K})$ and from Theorem 44.2 it follows that the $\widehat{K}$-algebra $\widehat{\mathfrak{g}}$ can be embedded in the Lie algebra $\operatorname{Diff}_\partial \tilde{\tilde{O}}(M)$ of all special derivations of the completed divided power algebra $\tilde{\tilde{O}}(M)$ constructed from the one-dimensional Lie algebra M $(0 \neq \partial \in M)$. Since the algebra $\tilde{\tilde{O}}(M)$ contains only trivial ∂-invariant ideals, this proves the implications $1 \Rightarrow 2$ and $1 \Rightarrow 4$.

The validity of the implication $4 \Rightarrow 2$ is almost obvious and follows from the next lemma.

LEMMA 46.1. *Let $\widehat{\mathcal{E}}$ be a commutative differential $\widehat{K}$-algebra with a derivation ∂ and let $\psi\colon \widehat{\mathcal{E}} \to \widehat{K}$ be an arbitrary nonzero $\widehat{K}$-homomorphism. Then the Taylor homomorphism $\tilde{\psi}\colon \widehat{\mathcal{E}} \to \tilde{\tilde{O}}(M)$ (see §44), where M is a one-dimensional Lie algebra containing ∂ as a basis element, is a homomorphism of ∂-differential $\widehat{K}$-algebras and the mapping $\rho_\psi\colon \operatorname{Diff}_\partial \widehat{\mathcal{E}} \to \operatorname{Diff}_\partial \tilde{\tilde{O}}(M)$ for which $\rho_\psi(f\partial) = \tilde{\psi}(f)\partial$ $(f \in \mathcal{E})$ is a homomorphism of Lie $\widehat{K}$-algebras.*

The proof of this lemma can be obtained by a direct computation. The lemma itself is a special case of Proposition 44.1 and of Corollary of Lemma 45.3.

Suppose that Property 4 mentioned in Proposition 46.1 holds for a simple Lie $\widehat{K}$-algebra $\mathfrak{g}$. Then, extending (if necessary) the field C to a field $\widehat{K}$ for which there exists a nonzero $\widehat{K}$-homomorphism $\psi\colon \widehat{\mathcal{E}} = \widehat{K} \otimes_C \mathcal{E} \to \widehat{K}$ (see Theorem 51.1), we see that ρ_ψ defines a C-homomorphism of the simple C-algebra $\mathfrak{g}$ into the Lie $\widehat{K}$-algebra $\widetilde{W}_1(\widehat{K}) = \operatorname{Diff}_\partial \tilde{\tilde{O}}(M)$. It is clear that any Lie identity for the Lie algebra $\widetilde{W}_1(K)$ is a Lie identity for the algebra $\widetilde{W}_1(\widehat{K})$. Therefore, if the homomorphism $\rho_\psi\colon \mathfrak{g} \to \widetilde{W}_1(\widehat{K})$ be injective, then property 2 would hold in the algebra $\mathfrak{g}$. However, the C-algebra $\mathcal{E}$ is differentially simple, therefore, $\operatorname{Ker}\tilde{\psi} \cap \mathcal{E} = 0$ and the homomorphism ρ_ψ induces an injective homomorphism of the algebra $\operatorname{Diff}_\partial \mathcal{E}$ into $\widetilde{W}_1(\widehat{K})$. This proves the validity of the implication $4 \Rightarrow 2$ in Proposition 46.1.

To complete the proof of the proposition, it remains to establish the validity of the implications $2 \Rightarrow 3$ and $3 \Rightarrow 1$. To this end, we shall study the identities of the algebra $\widetilde{W}_1(K)$.

46.1. Some identities of the Lie algebra $\widetilde{W}_1(K)$. From Definition 44.3 it follows that $\widetilde{W}_1(K) = \operatorname{Diff}_\partial \tilde{\tilde{O}}(M)$, where ∂ is an arbitrary nonzero element of the one-dimensional Lie algebra M. Since the algebra $\tilde{\tilde{O}}(M)$ contains the divided power algebra $O(M)$ as a ∂-invariant subalgebra, the Lie algebra $\widetilde{W}_1(K)$ contains the Lie subalgebra $\operatorname{Diff}_\partial O(M)$ of special derivations of the algebra $O(M)$. We denote this Lie subalgebra by $W_1(K)$. It is clear that if $t^{(i)}$ $(i = 0, 1, 2, \dots)$ is a standard basis of the algebra $O(M)$ for which

$$t^{(0)} = 1, \qquad t^{(i)}t^{(j)} = \binom{i+j}{i} t^{(i+j)}, \qquad \partial t^{(i+1)} = t^{(i)}, \tag{46.2}$$

then for the subspaces L_i in $W_1(K)$ spanned by the elements $t^{(i+1)}\partial$ (where $i =$

$-1, 0, 1, 2, \ldots$) we have the following properties:

1) $[L_i, L_j] \subset L_{i+j}$;

2) $\dim_K L_i = \begin{cases} 0 & \text{for } i < -1, \\ 1 & \text{for } i \geq -1; \end{cases}$

3) $[L_{-1}, L_j] \neq 0$ for $j \geq 0$ (left transitivity of the algebra W_1);

4) the subspace $L = L_{-1} + L_0 + L_1$ is a simple three-dimensional Lie algebra.

Hence, $\sum_{i=-\infty}^{\infty} L_i$ is a $\mathbb{Z}$-gradation of the algebra W_1. Obviously, the completion of the algebra $W_1(K)$ with respect to the descending filtration $\sum_{i=k}^{\infty} L_i$ ($k = -1, 0, 1, 2, \ldots$) induced by this gradation coincides with the algebra $\widetilde{W_1}(K)$. Therefore, in the algebras $W_1(K)$ and $\widetilde{W_1}(K)$ we have the same identities. Hence, in our search for the identities of the algebra $\widetilde{W_1}(K)$, we can use Proposition 21.1 for the algebra $W_1(K)$ for an arbitrary field K.

For any Lie polynomials v, v_1, v_2, and v_3 we introduce the notation

$$[v * \langle v_1, v_2, v_3 \rangle] \rightleftharpoons \sum_{\sigma \in S_3} \operatorname{sign}(\sigma)[v, v_{\sigma(1)}, v_{\sigma(2)}, v_{\sigma(3)}].$$

The following assertion establishes the validity of the implication $2 \Rightarrow 3$.

PROPOSITION 46.2. *In the Lie algebra $W_1(K)$ we have the identities*

$$[[z * \langle x_1, x_2, x_3 \rangle] * \langle y_1, y_2, y_3 \rangle] = [[z * \langle y_1, y_2, y_3 \rangle] * \langle x_1, x_2, x_3 \rangle], \tag{46.3}$$

$$[z * [\langle x_1, x_2, x_3 \rangle, \operatorname{ad} y]] = [y * [\langle x_1, x_2, x_3 \rangle, \operatorname{ad} z]], \tag{46.4}$$

and the identity (46.1).

PROOF. We shall say that a nonzero element $w \in W_1$ is homogeneous if $w \in L_i$ for some integer i. In this case we write $|w| = i$. Denote by the letter E a certain fixed homogeneous basis of the algebra $W_1(K)$. Since the identities (46.1), (46.3), and (46.4) are multilinear, it suffices to verify that they transform into true equalities in W_1 for any replacement of the variables z, y, x_i, y_j by homogeneous elements w_1, w_2, v_i, u_j from the set E.

Since identity (46.1) is skew-symmetric with respect to y_1, y_2, y_3, y_4, it transforms into a true equality if there are equal elements among u_1, u_2, u_3, u_4. Therefore, if identity (46.1) is not satisfied in W_1, then the left-hand side of it can be nonzero only on elements w_1 and u_i for which $|w_1| \geq 1$, $-1 < |u_1| < |u_2| < |u_3| < |u_4|$. However, then $|w_1| + \sum_{j=1}^{4} |u_j| \geq 1$, which contradicts Remark 21.1 and Proposition 21.1.

Remark 21.1 shows that in the course of verifying identity (46.3) we may assume that the following relations hold:

$$|w_1| + \sum_{i=1}^{3} (|u_i| + |v_i|) = -1, \qquad |w_1|, |u_i|, |v_i| \geq -1. \tag{46.5}$$

Moreover, since identity (46.3) is skew-symmetric with respect to $\{y_1, y_2, y_3\}$ (and $\{x_1, x_2, x_3\}$), we can restrict ourselves to considering the case in which there are no equal basis elements among u_1, u_2, u_3 (respectively, v_1, v_2, v_3). However, in this case it follows from relations (46.5) that $w_1, u_j, v_j \in L_{-1} \oplus L_0 \oplus L_1 = L$ $(j = 1, 2, 3)$. The space L is a simple three-dimensional Lie algebra; therefore, the validity of identity (46.3) follows from Lemma 35.5 (see identity (35.14)).

In the same way as in the proof of identity (46.3), we may assume in the verification of identity (46.4) that

$$|w_1| \geq -1, \qquad |w_2| \geq -1, \qquad |v_1| + |v_2| + |v_3| \geq 0,$$
$$|w_1| + |w_2| + |v_1| + |v_2| + |v_3| = -1.$$

These relations give two possibilities:

$$|w_1| = |w_2| = -1, \qquad |v_1| + |v_2| + |v_3| = 1;$$
$$w_1, w_2, v_1, v_2, v_3 \in L = L_{-1} \oplus L_0 \oplus L_1.$$

In the first case $w_1 = w_2$, and everything is clear. In the second case identity (35.14) implies that both parts of equality (46.4) are equal to zero in the simple three-dimensional Lie algebra L. Proposition 46.2 is proved.

The argument exposed above shows how we can guess at many important identities in the algebras $\widetilde{W}_n(K)$. The following assertion strengthens Proposition 46.2 and gives another (and a more formal) proof of identities (46.3) and (46.4).

PROPOSITION 46.3. *Over a field of arbitrary characteristic identities* (46.3) *and* (46.4) *are consequences of identity* (46.1).

LEMMA 46.2. *The identities*

$$2\sum_{\sigma\in S_3} \operatorname{sign}(\sigma)[z, y_{\sigma(1)}, x, y_{\sigma(2)}, y_{\sigma(3)}] = \sum_{\sigma\in S_3} \operatorname{sign}(\sigma)[z, y_{\sigma(1)}, y_{\sigma(2)}, y_{\sigma(3)}, x] \tag{46.6}$$

and

$$\sum_{\sigma\in S_3}\sum_{\delta\in S_3} \operatorname{sign}(\sigma)\operatorname{sign}(\delta)[x_{\delta(1)}, y_{\sigma(1)}, y_{\sigma(2)}, y_{\sigma(3)}, x_{\delta(2)}, x_{\delta(3)}] = 0$$

are consequences of identity (46.1).

PROOF. The Jacobi identity and the identity $[x, x] = 0$ permit us to write an arbitrary multilinear Lie polynomial in the form of a linear combination of left-normed commutators in which any previously chosen variable stands at the first place from the left. Therefore, we can write down the polynomial St_4 on the left-hand side of identity (46.1) as a linear combination of left-normed commutators in which the variable y_4 stands at the first place. An immediate verification shows that

$$\begin{aligned}\mathrm{St}_4(z, y_1, y_2, y_3, y_4) = 2\sum_{\sigma\in S_3} &\operatorname{sign}(\sigma)[y_4, y_{\sigma(1)}, z, y_{\sigma(2)}, y_{\sigma(3)}]\\ &- \sum_{\sigma\in S_3} \operatorname{sign}(\sigma)[y_4, y_{\sigma(1)}, y_{\sigma(2)}, y_{\sigma(3)}, z].\end{aligned}$$

By the replacement $z \to x$ and $y_4 \to z$ in this equality we obtain the assertion of the lemma for identity (46.6).

If the characteristic of the ground field equals two, then (46.7) is a consequence of identity (46.6).

Let the characteristic of the field differ from two. Note that any two juxtaposed skew-symmetric variables in a left-normed commutator can be considered as a commutator of weight 2. Therefore, applying identity (46.6) several times, we have

$$\begin{aligned}
&\sum_{\sigma\in S_3}\sum_{\delta\in S_3} \operatorname{sign}(\sigma)\operatorname{sign}(\delta)[x_{\delta(1)}, y_{\sigma(1)}, y_{\sigma(2)}, y_{\sigma(3)}, x_{\delta(2)}, x_{\delta(3)}]\\
&\quad = -\frac{1}{2}\sum_{\sigma\in S_3}\sum_{\delta\in S_3} \operatorname{sign}(\sigma)\operatorname{sign}(\delta)[y_{\sigma(1)}, y_{\sigma(2)}, x_{\delta(1)}, y_{\sigma(3)}, x_{\delta(2)}, x_{\delta(3)}]\\
&\quad = -\frac{1}{4}\sum_{\sigma\in S_3}\sum_{\delta\in S_3} \operatorname{sign}(\sigma)\operatorname{sign}(\delta)[y_{\sigma(1)}, y_{\sigma(2)}, x_{\delta(1)}, x_{\delta(2)}, x_{\delta(3)}, y_{\sigma(3)}]\\
&\quad = \frac{1}{2}\sum_{\sigma\in S_3}\sum_{\delta\in S_3} \operatorname{sign}(\sigma)\operatorname{sign}(\delta)[x_{\delta(1)}, y_{\sigma(1)}, y_{\sigma(2)}, x_{\delta(2)}, x_{\delta(3)}, y_{\sigma(3)}]\\
&\quad = -\frac{1}{2}\sum_{\sigma\in S_3}\sum_{\delta\in S_3} \operatorname{sign}(\sigma)\operatorname{sign}(\delta)[y_{\sigma(1)}, x_{\delta(1)}, y_{\sigma(2)}, x_{\delta(2)}, x_{\delta(3)}, y_{\sigma(3)}]\\
&\quad = -\frac{1}{4}\sum_{\sigma\in S_3}\sum_{\delta\in S_3} \operatorname{sign}(\sigma)\operatorname{sign}(\delta)[y_{\sigma(1)}, x_{\delta(1)}, x_{\delta(2)}, x_{\delta(3)}, y_{\sigma(2)}, y_{\sigma(3)}]\\
&\quad = \frac{1}{4}\sum_{\sigma\in S_3}\sum_{\delta\in S_3} \operatorname{sign}(\sigma)\operatorname{sign}(\delta)[x_{\delta(1)}, y_{\sigma(1)}, x_{\delta(2)}, x_{\delta(3)}, y_{\sigma(2)}, y_{\sigma(3)}]\\
&\quad = \frac{1}{8}\sum_{\sigma\in S_3}\sum_{\delta\in S_3} \operatorname{sign}(\sigma)\operatorname{sign}(\delta)[x_{\sigma(1)}, y_{\delta(1)}, y_{\delta(2)}, y_{\delta(3)}, x_{\sigma(2)}, x_{\sigma(3)}]
\end{aligned}$$

(identity (46.6) has been used in the second, the fifth, and the seventh equalities). By comparing in these equalities the first and the last lines, we see that (46.7) is a consequence of identity (46.6) over fields whose characteristic differs from seven. By comparing the sixth and the last lines, we see that the same is true over fields whose characteristic is not three. The lemma is proved.

Proof of Proposition 46.3. The Jacobi identity applied to the subcommutator $[z, y_{\sigma(1)}, x]$ on the left-hand side of equality (46.6) permits one to write (46.6) in the form

$$\sum_{\sigma\in S_3} \operatorname{sign}(\sigma)\{2[z, x, y_{\sigma(1)}, y_{\sigma(2)}, y_{\sigma(3)}] + 2[x, y_{\sigma(1)}, z, y_{\sigma(2)}, y_{\sigma(3)}] - [z, y_{\sigma(1)}, y_{\sigma(2)}, y_{\sigma(3)}, x]\} = 0.$$

Applying identity (46.6) to the second summand in this equality, we obtain another form of identity (46.4). By Lemma 46.2, identity (46.6) is a consequence of identity (46.1).

In order to prove the fact that identity (46.3) follows from (46.1), we shall join juxtaposed skew-symmetric variables in a commutator and apply identity (46.6), as in Lemma 46.2. Then the right-hand side of equality (46.3) is equal to

$$\begin{aligned}
&2\sum_{\sigma\in S_3}\sum_{\delta\in S_3}\operatorname{sign}(\sigma)\operatorname{sign}(\delta)[z, x_{\sigma(1)}, y_{\delta(1)}, x_{\sigma(2)}, x_{\sigma(3)}, y_{\delta(2)}, y_{\delta(3)}]\\
&\quad=\sum_{\sigma\in S_3}\sum_{\delta\in S_3}\operatorname{sign}(\sigma)\operatorname{sign}(\delta)[z, x_{\sigma(1)}, y_{\delta(1)}, y_{\delta(2)}, y_{\delta(3)}, x_{\sigma(2)}, x_{\sigma(3)}]\\
&\quad=[[z * \langle y_1, y_2, y_3\rangle] * \langle x_1, x_2, x_3\rangle]\\
&\quad-\sum_{\sigma\in S_3}\operatorname{sign}(\sigma)[z * [\langle y_1, y_2, y_3\rangle, \operatorname{ad} x_{\sigma(1)}]]\operatorname{ad} x_{\sigma(2)}\operatorname{ad} x_{\sigma(3)}.
\end{aligned}$$

Identities (46.4) and (46.7) show that the last summand is equal to zero. The proposition is proved.

Propositions 46.2 and 46.3 and the implication $1 \Rightarrow 2$ of Proposition 46.1 imply the following

COROLLARY. *In an arbitrary simple Lie algebra satisfying condition* 1 *of Proposition* 46.1 *we have identities* (46.1), (46.3), *and* (46.4).

To finally complete the proof of Proposition 46.1, it remains to establish the validity of the implication $3 \Rightarrow 1$. This requires a detailed study of simple Lie algebras satisfying identity (46.1).

46.2. Characteristic properties of simple algebras with identity (46.1). Let us introduce the following notation and definitions:

U is the associative subalgebra in $\operatorname{End}_K \mathfrak{g}$ generated by the subspace $\operatorname{ad}\mathfrak{g}$;

C is the centroid of the Lie algebra $\mathfrak{g}$;

$\langle v_1, v_2, v_3\rangle$ is a notation of the element

$$\sum_{\sigma\in S_3}\operatorname{sign}(\sigma)\operatorname{ad} v_{\sigma(1)}\operatorname{ad} v_{\sigma(2)}\operatorname{ad} v_{\sigma(3)};$$

${}^g u$ is a notation of the element $[\operatorname{ad} g, u]$, where $g \in \mathfrak{g}$ and $u \in \operatorname{End}_K \mathfrak{g}$;

${}^{\mathfrak{g}} I$ is the set of linear combinations of elements of the form ${}^g i$, where i ranges over all elements of the subset I of the algebra U and g runs through the elements of the Lie algebra $\mathfrak{g}$.

DEFINITION 46.1. An A-module V is said to be *multiplicative* if any A-submodule in V has the form $I \cdot V$ for some ideal I of the algebra A. A module V is said to be *strictly multiplicative* if V is multiplicative and the mapping that assigns to each ideal I the submodule $I \cdot V$ is one-to-one.

The following assertion shows that any simple Lie algebra $\mathfrak{g}$ with identity (46.1) uniquely determines a certain commutative $\mathfrak{g}$-invariant subalgebra A in $\operatorname{End}_K \mathfrak{g}$ for which the natural representation of $\mathfrak{g}$ into $\operatorname{Der}_K A$ is faithful for $\dim_C \mathfrak{g} \neq 3$ and defines a one-dimensional distribution on the variety $\operatorname{Spec} A$.

THEOREM 46.1. *Suppose that identity* (46.1) *holds in a simple Lie algebra* $\mathfrak{g}$ *over a field* K *whose characteristic differs from two. Let* A *be an associative* K*-subalgebra of* $U \subset \operatorname{End}_K \mathfrak{g}$ *generated by the elements* $\langle v_1, v_2, v_3\rangle$, *where* v_1, v_2, v_3 *range over the whole Lie algebra* $\mathfrak{g}$. *Then we have the following.*

1. *The algebra* A *is commutative, it contains* ${}^{\mathfrak{g}}A$ *and* C (*in particular, the unit element* $\mathbf{1}$ *of the algebra* $\operatorname{End}_K \mathfrak{g}$, *too*) *and possesses only trivial* $\mathfrak{g}$*-invariant ideals* I, *i.e.*, $I = {}^{\mathfrak{g}}I \Rightarrow (I = A) \vee (I = 0)$. *Moreover, for any elements* $a, b \in A$ *and* $g, h \in \mathfrak{g}$ *we have the following equality in the* A*-module* $\mathfrak{g}$:

$$({}^{g}a)h = ({}^{h}a)g, \tag{46.8}$$

and the following relations in the algebra A:

$$\begin{aligned} {}^{(bg)}a &= b({}^{g}a), \\ {}^{g}(ab) &= ({}^{g}a)b + a({}^{g}b). \end{aligned} \tag{46.9}$$

2. *If* $\dim_C \mathfrak{g} \neq 3$, *then* $A = {}^{\mathfrak{g}}A$ *and the* A*-module* $\mathfrak{g}$ *is finitely generated and strictly multiplicative; moreover, any* A*-submodule of* $\mathfrak{g}$ *is a Lie subalgebra of* $\mathfrak{g}$ *and any Lie subalgebra of* $\mathfrak{g}$ *whose codimension equals one is an* A*-submodule of* $\mathfrak{g}$.
3. *If* $\dim_C \mathfrak{g} = 3$, *then* ${}^{\mathfrak{g}}A = 0$ *and* $A = C$.

PROOF. The commutativity of the algebra A follows from Proposition 46.3 and identity (46.3). The inclusion $[\operatorname{ad} g, A] \subset A$ follows from the formula

$$[\operatorname{ad} g, \langle v_1, v_2, v_3\rangle] = \langle [g, v_1], v_2, v_3\rangle + \langle v_1, [g, v_2], v_3\rangle + \langle v_1, v_2, [g, v_3]\rangle.$$

Therefore, identity (46.8), which holds for the generators $a = \langle v_1, v_2, v_3\rangle$ of the algebra A by identity (46.4) and Proposition 46.3, also holds for any element $a \in A$. In the same way, with the help of identities (46.4) and (46.3), we obtain the equality

$$({}^{(bg)}a)h = ({}^{h}a)bg = b({}^{h}a)g = b({}^{g}a)h \qquad (g, h \in \mathfrak{g},\ a, b \in A),$$

whence, by the faithfulness of the action of the algebra A on the space of $\mathfrak{g}$, the first one of the assertions (46.9) follows. The second equality of (46.9) is now obvious.

LEMMA 46.3. *In the Lie algebra* $\mathfrak{g}$ *there exists a finite set of elements* $w_1, \dots, w_l$ *such that*

$$\mathfrak{g} = Aw_1 + \cdots + Aw_l. \tag{46.10}$$

PROOF. First we show that $A \neq 0$. Suppose that $A = 0$. Then in the Lie algebra $\mathfrak{g}$ the identity $[z * \langle x_1, x_2, x_3\rangle] = 0$ would be satisfied, which is equivalent to the metabelianness identity $[[x, y], [z, u]] = 0$ over a field of characteristic different from two (see Lemma 39.4). This would mean that the Lie algebra $\mathfrak{g}$ is solvable, which contradicts the simplicity of the Lie algebra $\mathfrak{g}$. Hence, for some element $c = \langle v_1, v_2, v_3\rangle$ ($v_i \in \mathfrak{g}$) and some $v \in \mathfrak{g}$ we have the inequality $cv \neq 0$ in the Lie algebra $\mathfrak{g}$, and $A \neq 0$.

Since the Lie algebra $\mathfrak{g}$ is simple, there is an element $u \in U$ such that

$$ucv = v \neq 0. \tag{46.11}$$

The element u can be expressed via a finite set of elements $\operatorname{ad} g_1, \dots, \operatorname{ad} g_t$. We shall show that

$$\mathfrak{g} = Av_1 + Av_2 + Av_3 + Ag_1 + \cdots + Ag_t. \tag{46.12}$$

It is clear that for any $a \in A$ and $g \in \mathfrak{g}$ the following relation holds:

$$\operatorname{ad} g\, a = a \operatorname{ad} g + [\operatorname{ad} g, a] = a \operatorname{ad} g + {}^g a \qquad ({}^g a \in A). \tag{46.13}$$

Applying this relation to an element a of the form ${}^{(h_1 \cdots h_l)}c$ (for c from equality (46.11)) several times, we obtain the inclusion

$$(\operatorname{ad} \mathfrak{g})^k v = (\operatorname{ad} \mathfrak{g})^k (uc)^{k+1} v \subset c\mathfrak{g} + \sum_{i=1}^{t} \sum_j {}^{g_i} a_j \mathfrak{g},$$

where $a_j \in A$. From the equality

$$\langle v_1, v_2, v_3 \rangle g = \langle g, v_2, v_3 \rangle v_1 + \langle v_1, g, v_3 \rangle v_2 + \langle v_1, v_2, g \rangle v_3 \tag{46.14}$$

we conclude that $c\mathfrak{g} \subset Av_1 + Av_2 + Av_3$, and from relation (46.8) it follows that ${}^{g_i} a_j \mathfrak{g} \subset Ag_i$. Therefore,

$$(\operatorname{ad} \mathfrak{g})^k v \subseteq Av_1 + Av_2 + Av_3 + Ag_1 + \cdots + Ag_t \qquad (k = 1, 2, \dots).$$

Since $v \neq 0$ and the Lie algebra $\mathfrak{g}$ is simple, this inclusion implies (46.12), and we can take $v_1, v_2, v_3, g_1, \dots, g_t$ for the elements $w_1, \dots, w_l$. The lemma is proved.

COROLLARY 1. *The algebra A contains the identity operator and possesses only trivial $\mathfrak{g}$-invariant ideals.*

PROOF. From equality (46.10) it follows that

$$w_i = c_{i1} w_1 + \cdots + c_{il} w_l \qquad (i = 1, \dots, l)$$

for some $c_{ij} \in A$. From this system of equalities and from the commutativity of the algebra A it follows that

$$\operatorname{Det}(E - d) \cdot w = 0, \tag{46.15}$$

where E is the unit matrix, d is a square matrix of order l with the elements c_{ij} $(i, j = 1, \dots, l)$, and $w = w_1, \dots, w_l$. Since the w_i generate the module $\mathfrak{g}$ over the commutative algebra A, formula (46.15) holds for any $w \in \mathfrak{g}$. But then $e = \mathbf{1} - \operatorname{Det}(E - d) \in A$, and the operator e determines the identity action on the space of $\mathfrak{g}$. Hence, $e = \mathbf{1}$.

Let I be a $\mathfrak{g}$-invariant ideal of the algebra A. The inclusion

$$\operatorname{ad} \mathfrak{g} I\, \mathfrak{g} = I \operatorname{ad} \mathfrak{g}\, \mathfrak{g} + {}^{\mathfrak{g}} I\, \mathfrak{g} \subseteq I\mathfrak{g}$$

shows that $I\mathfrak{g}$ is an ideal of the simple Lie algebra $\mathfrak{g}$. Then we have either $I\mathfrak{g} = 0$ or $I\mathfrak{g} = \mathfrak{g}$. If $I\mathfrak{g} = 0$, then $I = 0$, since $I \subset \operatorname{End}_K \mathfrak{g}$. If $I\mathfrak{g} = \mathfrak{g}$, then equality (46.10) implies

$$\mathfrak{g} = Iw_1 + \cdots + Iw_l$$

for some $w_1, \dots, w_l \in \mathfrak{g}$. Using the discussion developed in the beginning of the proof of the corollary, we obtain from this equality that the ideal I contains the identity operator. Therefore, the equality $I\mathfrak{g} = \mathfrak{g}$ implies the inclusion $I \supseteq A$. Thus, $I = A$, and Corollary 1 is proved.

COROLLARY 2. *The centroid C of the Lie algebra $\mathfrak{g}$ is contained in A. Moreover, if* ${}^{\mathfrak{g}}A = 0$, *then* $A = C$ *and* $\dim \mathfrak{g} = 3$.

PROOF. It follows from the definition of the centroid that

$$c\langle v_1, v_2, v_3\rangle = \langle cv_1, v_2, v_3\rangle \qquad (c \in C,\ v_i \in \mathfrak{g}).$$

Therefore, $CA \subset A$. But the algebra A contains the identity operator. Therefore, $C = C \cdot \mathbf{1} \subseteq A$.

If ${}^{\mathfrak{g}}A = 0$, then $C \supseteq A$. Since $A \neq 0$, we have $0 \neq \langle v_1, v_2, v_3\rangle \in C$ for some elements $v_i \in \mathfrak{g}$, and the equality $\dim_C \mathfrak{g} = 3$ follows from relation (46.14). Corollary 2 is proved.

COROLLARY 3. *If* ${}^{\mathfrak{g}}A \neq 0$, *then* ${}^{\mathfrak{g}}A = A$, *the A-module $\mathfrak{g}$ is strictly multiplicative, and any A-submodule of $\mathfrak{g}$ is a Lie subalgebra.*

PROOF. Relation (46.9) shows that ${}^{\mathfrak{g}}A \supseteq A \cdot {}^{\mathfrak{g}}A$. Hence, ${}^{\mathfrak{g}}A$ is a nonzero $\mathfrak{g}$-invariant ideal of A. Therefore, ${}^{\mathfrak{g}}A = A$ (see Corollary 1); in particular, for some $g_i \in \mathfrak{g}$ and $c_i \in A$ $(i = 1, \dots, q)$ the unit element has the following decomposition:

$$(46.16) \qquad {}^{g_1}c_1 + \cdots + {}^{g_q}c_q = 1.$$

Let V be an arbitrary A-submodule of $\mathfrak{g}$. Denote by ${}^{V}A$ the subspace of A spanned by the elements of the form ${}^{v}a$, where $v \in V$ and $a \in A$. Relation (46.9) shows that ${}^{V}A$ is an ideal in the algebra A and relation (46.8) means that ${}^{V}A\mathfrak{g} \subset V$. Applying equalities (46.16) and (46.8), we obtain

$$v = \sum_{i=1}^{q} ({}^{g_i}c_i)v = \sum_{i=1}^{q} ({}^{v}c_i)g_i$$

for any $v \in V$. This relation shows that we have also the reverse inclusion $({}^{V}A)\mathfrak{g} \supset V$. Thus, $V = {}^{V}A\mathfrak{g}$, and the A-module $\mathfrak{g}$ is multiplicative.

Suppose that for some ideal I of the algebra A we have the equality $I\mathfrak{g} = V$. Then relations (46.9) and (46.16) imply the equalities ${}^{V}A = {}^{I\mathfrak{g}}A = I \cdot {}^{\mathfrak{g}}A = I$. This proves that the module $\mathfrak{g}$ is strictly multiplicative.

It remains to show that for any ideal I of A the submodule $I\mathfrak{g}$ is a Lie subalgebra. This immediately follows from the relation

$$(46.17) \qquad [bg, ah] = ({}^{bg}a)h + a[bg, h] = b \cdot {}^{g}a \cdot h + a[bg, h],$$

which holds, by equalities (46.8) and (46.9), for any $a, b \in I$. The corollary is proved.

LEMMA 46.4. *In an arbitrary Lie algebra $\mathfrak{g}$ any subalgebra $\mathfrak{g}_0$ of codimension one is invariant under the action of the operators $\langle u_1, u_2, u_3\rangle \in \mathrm{End}_K\, \mathfrak{g}$, where $u_1, u_2, u_3 \in \mathfrak{g}$.*

PROOF. For the case $\mathfrak{g} = \widetilde{W}_1(K)$ and $\mathfrak{g}_0 = (\widetilde{W}_1(K))_0$ the assertion of the lemma can be verified directly. The general case is reduced to this one by the homomorphism $\rho_{\mathfrak{g}_0} \colon \mathfrak{g} \to \widetilde{W}_1(K)$ (see Corollary 1 of Theorem 44.2). The lemma is proved.

Now all assertions of Theorem 46.1 follow from Corollaries 1–3 of Lemma 46.3 and from Lemma 46.4.

For the case of fields of positive characteristic the results of Theorem 46.1 can be strengthened.

THEOREM 46.2. *Let C, $\mathfrak{g}$, K, and A have the same meaning as in Theorem* 46.1 *and, moreover, let the Lie algebra $\mathfrak{g}$ be centrally simple* (*i.e.*, $C = K \cdot \mathbf{1}$), *let the characteristic of the field K be equal to $p > 2$, and* $\dim_K \mathfrak{g} \neq 3$. *Then*:

1) *for any element $a \in A$ the element a^p belongs to $K \cdot \mathbf{1}$, and in the algebra A there exists a unique maximal ideal*

$$\operatorname{Rad} A = \{a \in A \mid a^p = 0\};$$

2) *the A-module $\mathfrak{g}$ is cyclic, and for any generator ∂ of it the algebra A with unit is differentially simple with respect to the derivation $\partial \colon a \to {}^{\partial}a$ and $\mathfrak{g}$ is isomorphic to the Lie algebra of all special derivations* $\operatorname{Diff}_\partial A = \{a\partial \mid a \in A\}$ *of the K-algebra A.*

PROOF. Since $\operatorname{char} K = p > 2$, we have

$${}^{g}(a^p) = pa^{p-1} \cdot {}^{g}a = 0 \qquad (a \in A,\ g \in \mathfrak{g}).$$

Thus, all elements a^p, where $a \in A$, belong to the centroid C of the Lie algebra $\mathfrak{g}$, which coincides with $K \cdot \mathbf{1}$, because $\mathfrak{g}$ is centrally simple. In particular, for any element b belonging to a proper ideal of the algebra A we have $b^p = 0$. This proves the first assertion of the theorem.

Since, by assumption, $\dim_K \mathfrak{g} \neq 3$ and $\mathfrak{g}$ is a centrally simple K-algebra, Theorem 46.1 implies that ${}^{\mathfrak{g}}A = A$ and in the algebra A we have a decomposition of the unit element (46.16). Taking both sides of equality (46.16) to the pth power and taking into account that $\operatorname{char} K = p$, we see that for some $b \in A$ and $g \in \mathfrak{g}$ we have

$$0 \neq ({}^{g}b)^p \in K \cdot \mathbf{1}. \tag{46.18}$$

However, then from identity (46.8) for any $h \in \mathfrak{g}$ we obtain the equality

$$({}^{g}b)^p h = ({}^{g}b)^{p-1}({}^{g}b)h = ({}^{g}b)^{p-1}({}^{h}b)g,$$

which, together with (46.18), proves that g is a generator of the A-module $\mathfrak{g}$. Since ${}^{\mathfrak{g}}A = A$, this implies that the algebra A is differentially simple with respect to the derivation $\partial \colon a \to {}^{g}a$. Identity (46.9) shows that $\mathfrak{g}$ coincides with the Lie algebra of all special derivations $\operatorname{Diff}_g A = \{ag \mid a \in A\}$ of the algebra A. The theorem is proved.

PROOF OF THE IMPLICATION $3 \Rightarrow 1$ OF PROPOSITION 46.1. Suppose that $\mathfrak{g}$ is an arbitrary simple Lie K-algebra ($\operatorname{char} K \neq 2$) and in the algebra $\mathfrak{g}$ we have identity (46.1). For the field $\widehat{K}$ we take an arbitrary algebraically closed field containing the field C and of greater cardinality than $\dim_C \mathfrak{g}$. Then the Lie $\widehat{K}$-algebra $\widehat{\mathfrak{g}} = \widehat{K} \otimes_C \mathfrak{g}$ is a simple Lie algebra satisfying identity (46.1). From Lemmas 3.4 and 3.5 it follows that $\widehat{\mathfrak{g}}$ is a centrally simple $\widehat{K}$-algebra. Therefore, by Theorem 46.1 applied to the Lie $\widehat{K}$-algebra $\widehat{\mathfrak{g}}$, either $\widehat{\mathfrak{g}}$ is three-dimensional or the $\widehat{A}$-module $\widehat{\mathfrak{g}}$ is finitely generated and strictly multiplicative. For the first case the existence of a $\widehat{K}$-subalgebra of codimension 1 is obvious. For the second case we choose for the required subalgebra a maximal $\widehat{A}$-submodule V of $\widehat{\mathfrak{g}}$ (the existence of a submodule of that kind follows from Zorn's Lemma and from the fact that the $\widehat{A}$-module $\widehat{\mathfrak{g}}$ is finitely generated). Since the algebra

$\widehat{A}$ is commutative, the quotient algebra $\widehat{A}/\operatorname{Ann}(\widehat{\mathfrak{g}}/V)$ is contained in the centralizer of the irreducible quotient $\widehat{A}$-module $\widehat{\mathfrak{g}}/V$. Since $\dim_{\widehat{K}}\widehat{\mathfrak{g}}/V < |\widehat{K}|$, the centralizer of the module $\widehat{\mathfrak{g}}/V$ coincides with $\widehat{K}\cdot\mathbf{1}$ by Lemma 3.4. Therefore, $\widehat{K}\cdot\mathbf{1}\supset\widehat{A}/\operatorname{Ann}(\widehat{\mathfrak{g}}/V)$. But the algebra $\widehat{A}$ contains the identity operator; therefore, $\widehat{A}/\operatorname{Ann}(\widehat{\mathfrak{g}}/V)=\widehat{K}\cdot\mathbf{1}$ and $\dim_{\widehat{K}}\widehat{\mathfrak{g}}/V = 1$. From assertion 2 of Theorem 46.1 it follows that $\operatorname{Ann}(\widehat{\mathfrak{g}}/V)\widehat{\mathfrak{g}} = V$ and V is a Lie subalgebra of $\widehat{\mathfrak{g}}$. The implication $3\Rightarrow 1$ in Proposition 46.1 is proved.

46.3. Relationships with simple differential algebras. Theorem 46.2 shows that for the case of fields of positive characteristic $p>2$ any centrally simple non-three-dimensional Lie algebra $\mathfrak{g}$ with identity (46.1) is realized as the Lie algebra of special derivations of a suitable differentially simple commutative algebra A. Generally speaking, this fact fails over fields of characteristic zero. However, the realization of the Lie algebra $\mathfrak{g}$ by derivations of the commutative algebra A (the one constructed in Theorem 46.1) satisfies relations (46.8) and (46.9) and permits one to interpret the Lie algebra $\mathfrak{g}$ as a one-dimensional involutory distribution without singularities on the spectrum $\operatorname{Spec}A$ of the differentially simple algebra A with respect to $\mathfrak{g}$ (see Remark 42.2). It turns out that the converse of the statements of Theorems 46.1 and 46.2 are also true.

THEOREM 46.3. *Let a field K be of characteristic different from two, let $\mathcal{E}$ be a commutative K-algebra with unit and with only trivial $\mathfrak{g}$-invariant ideals, where $\mathfrak{g}$ is a nonzero Lie subalgebra of the algebra of all derivations* $\operatorname{Der}_K\mathcal{E}$ *coinciding with an $\mathcal{E}$-submodule of* $\operatorname{Der}_K\mathcal{E}$. *Suppose, moreover, that for any elements $a,b\in\mathcal{E}$ and $g,h\in\mathfrak{g}$ relations* (46.8) *and* (46.9) *hold in the $\mathcal{E}$-module $\mathfrak{g}$ and in the algebra $\mathcal{E}$, respectively. Then the Lie algebra $\mathfrak{g}$ is simple, identity* (46.1) *holds in it, and, with the exception for the case* $\operatorname{char}K=3$ & $\dim_C\mathfrak{g}=3$, *the subalgebra A of* $\operatorname{End}_K\mathfrak{g}$ *generated by $\langle v_1,v_2,v_3\rangle$ ($v_i\in\mathfrak{g}$) coincides with $\varphi(\mathcal{E})$, where φ is an injective homomorphism of the algebra $\mathcal{E}$ into* $\operatorname{End}_K\mathcal{E}$, *induced by the structure of a left $\mathcal{E}$-module on $\mathfrak{g}$ for which*

$$\varphi({}^{g}f)={}^{g}\varphi(f). \tag{46.19}$$

Moreover, any Lie subalgebra $\mathfrak{h}$ in $\operatorname{Der}_K\mathcal{E}$ *containing $\mathfrak{g}$ and for which relations* (46.8) *and* (46.9) *are satisfied for arbitrary elements $a,b\in\mathcal{E}$ and $g,h\in\mathfrak{h}$ coincides with $\mathfrak{g}$.*

COROLLARY. *For any differentially simple commutative K-algebra $\mathcal{E}$* ($\operatorname{char}K\neq 2$) *with unit and with a nonzero signature derivation ∂ the Lie algebra $\mathfrak{g}=\operatorname{Diff}_\partial\mathcal{E}$ of all special derivations of the algebra $\mathcal{E}$ is a simple algebra in which identity* (46.1) *is satisfied. Moreover, $\varphi(\mathcal{E})=A$ with the exception for the only case in which* $\dim_C\mathfrak{g}=3$ & $\operatorname{char}K=3$.

The proof of these statements is divided into several lemmas.

Let $\mathcal{E}_1$ be a commutative K-algebra and let ∂ be an arbitrary derivation of it. For arbitrary three elements $f_1,f_2,f_3\in\mathcal{E}_1$ we denote by $W_\partial(f_1,f_2,f_3)$ the Wronskian of these functions:

$$\begin{vmatrix} f_1 & {}^{\partial}f_1 & {}^{\partial^2}f_1\\ f_2 & {}^{\partial}f_2 & {}^{\partial^2}f_2\\ f_3 & {}^{\partial}f_3 & {}^{\partial^2}f_3\end{vmatrix}.$$

LEMMA 46.5. *In the Lie algebra* Diff_∂ *identity* (46.1) *holds. Moreover, for any elements* $f_1, f_2, f_3, f_0, f \in \mathcal{E}_1$ *we have the equalities*

$$\langle f_1\partial, f_2\partial, f_3\partial\rangle f_0\partial = -2W_\partial(f_1, f_2, f_3)f_0\partial, \tag{46.20}$$

$$^{f\partial}\langle f_1\partial, f_2\partial, f_3\partial\rangle \cdot f_0\partial = -2 \cdot {}^{f\partial}W_\partial(f_1, f_2, f_3)f_0\partial. \tag{46.21}$$

PROOF. It is immediate that

$$\begin{aligned}[[f_1\partial, f_2\partial], [f_3\partial, f_0\partial]] = &\{(f_1 \cdot {}^{\partial}f_2 - {}^{\partial}f_1 \cdot f_2)(f_3 \cdot {}^{\partial^2}f_0 - {}^{\partial^2}f_3 \cdot f_0) \\ &- (f_1 \cdot {}^{\partial^2}f_2 - {}^{\partial^2}f_1 \cdot f_2)(f_3 \cdot {}^{\partial}f_0 - {}^{\partial}f_3 \cdot f_0)\}\partial.\end{aligned}$$

By skew-symmetrizing both sides of this equality with respect to f_1, f_2, f_3, we see that the terms containing ${}^{\partial}f_0$ and ${}^{\partial^2}f_0$ cancel each other, and we obtain the doubled equality (46.20). Using this relation, we obtain that the left-hand side of formula (46.21) is equal to

$$\begin{aligned}&-2\{[f\partial, W_\partial(f_1, f_2, f_3)f_0\partial] - W_\partial(f_1, f_2, f_3)[f\partial, f_0\partial]\} \\ &\quad = -2 \cdot {}^{f\partial}W_\partial(f_1, f_2, f_3)f_o\partial.\end{aligned}$$

This proves (46.21). By skew-symmetrizing both sides of the equality (46.21) with respect to f_1, f_2, f_3, f, we obtain identity (46.1). The lemma is proved.

Let $\mathcal{E}$, A, $\mathfrak{g}$, and φ have the same meaning as in Theorem 46.3, and let $\operatorname{char} K \neq 2$. Denote by the letter C the set of all elements $c \in \mathcal{E}$ for which ${}^{g}c = 0$ for any $g \in \mathfrak{g}$. Since the algebra $\mathcal{E}$ contains a unit and has only trivial $\mathfrak{g}$-invariant ideals and $\mathcal{E}c$ is a $\mathfrak{g}$-invariant ideal for each $c \in C$, C is a field.

LEMMA 46.6. *There exist elements* $\partial \in \mathfrak{g}$ *and* $a \in \mathcal{E}$ *such that the element* ${}^{\partial}a$ *is not a zero divisor in* $\mathcal{E}$.

PROOF. Since $\mathfrak{g} \neq 0$ by assumption, we have ${}^{\mathfrak{g}}\mathcal{E} \neq 0$. Identity (46.9) shows that ${}^{\mathfrak{g}}\mathcal{E}$ is an ideal of $\mathcal{E}$. Obviously, this ideal is $\mathfrak{g}$-invariant. Therefore, ${}^{\mathfrak{g}}\mathcal{E} = \mathcal{E}$, and for some elements $c_i \in \mathcal{E}$ and $g_i \in \mathfrak{g}$ $(i = 1, \dots, q)$ we have the decomposition (46.16) of the unit element of the algebra $\mathcal{E}$. Then one of the elements ${}^{g_i}c_i$ is not a nil-element. We set $\partial = g_i$ and $a = c_i$. Denote by $P({}^{\partial}a)$ the set of all elements $f \in \mathcal{E}$ such that for some positive integer $i = i(f)$ we have the equality $({}^{\partial}a)^i \cdot f = 0$. However, in this case $0 = {}^{\partial}f({}^{\partial}a)^{i+1}$, and $P({}^{\partial}a)$ is a $\mathfrak{g}$-invariant ideal in $\mathcal{E}$ not containing the unit element of the algebra $\mathcal{E}$. Therefore, $P({}^{\partial}a) = 0$ and ${}^{\partial}a$ is not a zero divisor in $\mathcal{E}$. The lemma is proved.

We denote by $\mathcal{E}_1$ the localization of the algebra $\mathcal{E}$ with respect to the element ${}^{\partial}a$ just found. Then, by Lemma 46.6 and by Proposition 51.2, we can assume that the algebra $\mathcal{E}$ is embedded in $\mathcal{E}_1$. All derivations of the algebra $\mathcal{E}$ are uniquely extendable to derivations of the algebra $\mathcal{E}_1$. Since the element ${}^{\partial}a$ is invertible in the algebra $\mathcal{E}_1$, from identity (46.8) for $h = \partial$ it follows that the Lie algebra $\mathfrak{g}$ is contained in $\mathfrak{g}_1 = \mathrm{Diff}_\partial \mathcal{E}_1$.

LEMMA 46.7. *Let φ_1 be a homomorphism of the algebra $\mathcal{E}_1$ into* $\mathrm{End}_K \mathfrak{g}_1$ *induced by the left $\mathcal{E}_1$-module structure on $\mathfrak{g}_1$ and let A' be a K-subalgebra of* $\mathrm{End}_K \mathfrak{g}_1$ *generated by the elements $\langle g_1, g_2, g_3\rangle$, where g_i range over the Lie algebra $\mathfrak{g}$. Then φ_1 is an injective homomorphism, $A' \subset \varphi_1(\mathcal{E}_1)$,*

$$[\mathrm{ad}\, g, \varphi_1(f)] = \varphi_1({}^g f) \qquad (g \in \mathfrak{g},\ f \in \mathcal{E}_1), \tag{46.22}$$

and A' is a $\mathfrak{g}$-invariant subalgebra of $\varphi_1(\mathcal{E}_1)$.

PROOF. The algebra $\mathcal{E}_1$ contains only trivial ∂-invariant ideals, because in the contrary case such an ideal would be $\mathfrak{g}$-invariant and would have a nontrivial intersection with the subalgebra $\mathcal{E}$. Therefore, $\mathcal{E}_1 \cdot {}^\partial\mathcal{E}_1 = \mathcal{E}_1$ and, by formula (46.9), for any $b \in \mathrm{Ker}\,\varphi_1$ we have the equality $b\mathcal{E}_1 = \mathcal{E}_1 \cdot {}^{b\partial}\mathcal{E}_1 = 0$. Since the algebra $\mathcal{E}_1$ contains a unit, this implies that $b = 0$ and φ_1 is a monomorphism. The inclusion $A' \subset \varphi(\mathcal{E}_1)$ follows from formula (46.20) and from the fact $\mathfrak{g} \subset \mathrm{Diff}_\partial \mathcal{E}_1$. The validity of equality (46.22) and the $\mathfrak{g}$-invariance of the subalgebra A' of $\varphi_1(\mathcal{E}_1)$ are now obvious. The lemma is proved.

We denote by C_1 the set of all elements c of the algebra $\mathcal{E}_1$ for which $\partial c = 0$. It is clear that $C = C_1 \cap \mathcal{E}$ and two cases are possible: $A' \subseteq \varphi_1(C_1)$ and $A' \not\subseteq \varphi_1(C_1)$.

LEMMA 46.8. *If $A' \subseteq \varphi_1(C_1)$, then $\mathcal{E}_1 = \mathcal{E}$, $C = C_1$, $A' = A = \varphi(C)$,* $\mathrm{char}\, K = 3$, $\mathfrak{g} = \mathrm{Diff}_\partial \mathcal{E}$, *and $\mathfrak{g}$ is a simple three-dimensional algebra over the field C.*

PROOF. It is immediate that

$$W_\partial(1, a, a^2) = 2({}^\partial a)^3. \tag{46.23}$$

By formula (46.20), the element $\langle \partial, a\partial, a^2\partial\rangle$ is equal to $-4\varphi(\partial a)^3$ and, hence, it belongs to the intersection $A' \cap \varphi_1(\mathcal{E})$. By Lemma 46.6, the element $(\partial a)^3$ is not a zero divisor, and by the assumption of the lemma $A' \subseteq \varphi_1(C_1)$. Therefore, $0 \neq (\partial a)^3 \in \mathcal{E} \cap C_1 = C$ and $(\partial a)^3$ is an invertible element of the algebra $\mathcal{E}$. However, in this case it is immediate from the definition of the algebra $\mathcal{E}_1$ that $\mathcal{E}_1 = \mathcal{E}$ and, hence, $\mathfrak{g} = \mathrm{Diff}_\partial \mathcal{E}$, $C = C_1$, $A' = A \subset \varphi(C)$. Moreover, since C is a field, $\varphi(C)A \subset A$ and $0 \neq A \subset \varphi(C)$, we have $\varphi(C) = A$.

From (46.23) and the equality

$$-4(\partial a)^3 g = \langle \partial, a\partial, a^2\partial\rangle g = \langle g, a\partial, a^2\partial\rangle\partial + \langle \partial, g, a^2\partial\rangle a\partial + \langle \partial, a\partial, g\rangle a^2\partial,$$

where $g \in \mathfrak{g}$, we conclude that $\mathfrak{g}$ is a three-dimensional Lie algebra over its centroid C. By Lemma 46.7, we see that in this case $\mathcal{E}$ is a three-dimensional algebra over the field C and the elements $1, a, a^2$ form a basis of it. Hence, for some $c_1, c_2, c_3 \in C$ we have

$$a^3 + c_1 a^2 + c_2 a + c_3 = 0.$$

Differentiating this equality, we obtain

$$(3a^2 + 2c_1 a + c_2) \cdot {}^\partial a = 0.$$

Since ∂a is an invertible element of the algebra $\mathcal{E}$, the equalities $3 = 0$, $2c_1 = 0$, and $c_2 = 0$ must hold. This proves that $\mathrm{char}\, K = 3$.

From the equality

$$\langle \partial, a\partial, a^2\partial\rangle \mathfrak{g} = -4(\partial a)^3 \cdot \mathfrak{g} = \mathfrak{g}$$

it follows that the Lie algebra $\mathfrak{g}$ is not metabelian. Since any three-dimensional Lie algebra is either simple or metabelian, $\mathfrak{g}$ is simple. The lemma is proved.

Now we consider the second case in which $A' \not\subseteq \varphi_1(C_1)$.

LEMMA 46.9. *If A' is not contained in $\varphi_1(C_1)$, then $A' \supseteq \varphi_1(\mathcal{E})$ and $\mathfrak{g}$ is a simple Lie algebra.*

PROOF. By Lemma 46.7, A' is a $\mathfrak{g}$-invariant subalgebra of $\varphi_1(\mathcal{E}_1)$. As it was shown above, $\varphi_1(\partial a)^3 \in A' \cap \varphi_1(\mathcal{E}_1)$. Therefore, there must exist an element b in $A' \cap \varphi_1(\mathcal{E})$ such that ${}^g b \neq 0$ for some $g \in \mathfrak{g}$. (Indeed, in the contrary case the element $(\partial a)^3$ would belong to $C_1 \cap \mathcal{E} = C$ and would be invertible in the field C. However, in this case the element ${}^{\partial} a$ would be invertible in the algebra $\mathcal{E}$, and the definition of the algebra $\mathcal{E}_1$ would imply $\mathcal{E}_1 = \mathcal{E}$ and, hence, $A' \cap \varphi_1(\mathcal{E}) = A' \subset \varphi_1(C) = \varphi_1(C_1)$, which would contradict the assumption of the lemma.)

Let $b \in A' \cap \varphi_1(\mathcal{E})$ and ${}^{\mathfrak{g}} b \neq 0$. Then a $\mathfrak{g}$-invariant subspace $\sum_{i=1}^{\infty} {}^{\mathfrak{g}^i} b$ is nonzero, and from identity (46.9) it follows that this subspace is an ideal of $\varphi_1 \mathcal{E}$. Therefore, $A' \cap \varphi_1(\mathcal{E}) = \varphi_1(\mathcal{E})$ and $A' \supseteq \varphi_1(\mathcal{E})$.

Any ideal J of the Lie algebra $\mathfrak{g}$ is an A'-submodule of $\mathfrak{g}$, and, since $A' \supseteq \varphi_1(\mathcal{E})$, we obtain, in particular, $\mathcal{E}J \subseteq J$. Therefore, by identity (46.9), we have $\mathcal{E} \cdot {}^J\mathcal{E} \subseteq {}^{\mathcal{E}J}\mathcal{E} \subseteq {}^J\mathcal{E}$. Simultaneously, ${}^{\mathfrak{g}}({}^J\mathcal{E}) = {}^{[J,\mathfrak{g}]}\mathcal{E} + {}^J({}^{\mathfrak{g}}\mathcal{E}) \subseteq {}^J\mathcal{E}$. Hence, an arbitrary ideal in the Lie algebra $\mathfrak{g}$ either is zero or coincides with $\mathfrak{g}$. The lemma is proved.

LEMMA 46.10. *If A' is not contained in $\varphi_1(C_1)$, then in the Lie algebra $\mathfrak{g}$ identity* (46.1) *holds and $A = \varphi(\mathcal{E})$.*

PROOF. It follows from Lemma 46.9 that $\mathfrak{g}$ is a simple Lie algebra. The inclusion $\mathfrak{g} \subseteq \mathrm{Diff}_{\partial} \mathcal{E}_1$, together with Lemma 46.5, shows that identity (46.1) holds for $\mathfrak{g}$. Therefore, by Theorem 46.1, in the algebra A there are only trivial $\mathfrak{g}$-invariant ideals, and for any elements $a, b \in A$ and $g \in \mathfrak{g}$ identity (46.9) holds in A. From the definition of the algebras A and A' it follows that $A'|_{\mathfrak{g}} = A$, and, since $\varphi_1(\mathcal{E})|_{\mathfrak{g}} = \mathcal{E}$, Lemma 46.9 implies $A \supseteq \varphi(\mathcal{E})$. Equality (46.19) shows that $\varphi(\mathcal{E})$ is a $\mathfrak{g}$-invariant subalgebra of A. Furthermore, since ${}^{\mathfrak{g}}\mathcal{E} = \mathcal{E}$, we see, applying identity (46.9) in the algebra A, that $A \cdot {}^{\mathfrak{g}}\varphi(\mathcal{E}) = {}^{A\mathfrak{g}}\varphi(\mathcal{E}) = {}^{\mathfrak{g}}(\varphi(\mathcal{E})) = \varphi(\mathcal{E})$. This proves that $\varphi(\mathcal{E})$ is a nonzero ideal of A. Hence, $\varphi(\mathcal{E}) = A$. The lemma is proved.

To conclude the proof of Theorem 46.3, it remains to show that the Lie algebra $\mathfrak{g}$ cannot be contained in a larger Lie algebra $\mathfrak{h}$ satisfying the same conditions as $\mathfrak{g}$ does. We denote by $A(\mathfrak{h})$ the subalgebra of $\mathrm{End}_K \mathfrak{h}$ generated by the elements $\langle h_1, h_2, h_3 \rangle$, where the h_i range over $\mathfrak{h}$. Lemmas 46.8–46.10 show that two cases are possible: a) $A(\mathfrak{h}) = \varphi_{\mathfrak{h}}(C)$, b) $A(\mathfrak{h}) = \varphi_{\mathfrak{h}}(\mathcal{E})$. Therefore, in both cases the restriction of the algebra $A(\mathfrak{h})$ to $\mathfrak{g}$ coincides with $A = A(\mathfrak{g})$. In the first case the coincidence of the algebras $\mathfrak{g}$ and $\mathfrak{h}$ follows from the fact that they are three-dimensional over C. In the second case $\mathfrak{g}$ is an $A(\mathfrak{h})$-submodule of $\mathfrak{h}$. Since, by Lemmas 46.8 and 46.9, $\mathfrak{h}$ is a simple Lie algebra and identity (46.1) holds in it, we see by Theorem 46.1 that in $A(\mathfrak{h})$ there exists an ideal I for which $\mathfrak{g} = I\mathfrak{h}$. Let $J = \varphi_{\mathfrak{h}}^{-1}(I)$; then $\mathfrak{g} = J \cdot \mathfrak{h}$ and

$$\mathcal{E} = {}^{\mathfrak{g}}\mathcal{E} = {}^{J\mathfrak{h}}\mathcal{E} = J \cdot {}^{\mathfrak{h}}\mathcal{E} \subseteq J.$$

Hence, $J = \mathcal{E}$ and $\mathfrak{g} = \mathfrak{h}$. Theorem 46.3 is completely proved.

REMARK 46.1. Theorem 46.2 and Corollary of Theorem 46.3, together with our Theorem 46.1, give a complete description of centrally simple Lie algebras with identity (46.1) over fields of positive characteristic $p > 2$ by reducing their classification to differentially simple commutative local algebras $\mathcal{E}$ with unit and with a unique nonzero signature derivation ∂ for which $\operatorname{Rad}\mathcal{E} = \{a \in \mathcal{E} \mid a^p = 0\}$. If the field K is perfect, then all such algebras $\mathcal{E}$ allow a very clear description. Indeed, if the field K is perfect, then $\psi\colon \mathcal{E} \to \mathcal{E}/\operatorname{Rad}\mathcal{E} = K$ and, applying the Taylor homomorphism $\widetilde{\psi}\colon \mathcal{E} \to \widetilde{O}(M)$ of Lemma 46.1, we obtain that $\mathcal{E}$ is isomorphic to a ∂-invariant subalgebra of the completed divided power algebra $\widetilde{O}(M)$ over a one-dimensional space M. Since in the algebra $\widetilde{O}(M)$ for any element $f = \beta_0 + \beta_1 t^{(1)} + \beta_2 t^{(2)} + \cdots$ ($\beta_i \in K$) we have the equality $({}^{\partial^l}f)^p = (\beta_l)^p$, it is clear that, conversely, any ∂-invariant subalgebra $\mathcal{E}$ of $\widetilde{O}(M)$ is a differentially simple local algebra with unit for which $\operatorname{Rad}\mathcal{E} = \{f \in \mathcal{E} \mid f^p = 0\}$. To put this technique into practice, we recommend the reader to describe, up to an isomorphism, all simple finite-dimensional Lie algebras with identity (46.1) over an algebraically closed field.

46.4. Division algebras in the variety var $\widetilde{W_1}(K)$**.** In this subsection we assume that $\operatorname{char} K = 0$.

DEFINITION 46.2. A Lie algebra $\mathfrak{g}$ is called a *division algebra* if for any nonzero $g \in \mathfrak{g}$ and any $h \in \mathfrak{g}$ the equation

$$[g, x] = h, \tag{46.24}$$

has at least one solution in $\mathfrak{g}$.

THEOREM 46.4. *Over a field K of characteristic zero any simple Lie algebra $\mathfrak{h}$ for which identity* (46.1) *holds can be embedded in a Lie division algebra $\mathfrak{g}$ such that $\mathfrak{g} \in$* var $\widetilde{W_1}(K)$. *Any such Lie division algebra $\mathfrak{g}$ has the following properties*:

1. *The subalgebra A of* $\operatorname{End}_K \mathfrak{g}$ *generated by the elements $\langle v_1, v_2, v_3\rangle$, where v_1, v_2, v_3 range over $\mathfrak{g}$, is a field.*
2. *The A-module $\mathfrak{g}$ is cyclic.*
3. *For any nonzero element $\partial \in \mathfrak{g}$ and any $a \in A$ the equation*

$$^{\partial}y = a \tag{46.25}$$

is solvable in the field A, and the Lie algebra $\mathfrak{g}$ is the Lie algebra of all special derivations of the field A with respect to ∂.

Conversely, any Lie algebra of all special derivations of a differential field $\mathcal{E}$ for which equation (46.25) *is solvable for the signature derivation ∂ and any $a \in \mathcal{E}$ is a Lie division algebra from the variety* var $\widetilde{W_1}(K)$ *and $\varphi(\mathcal{E}) = A$.*

PROOF. First we shall show that a simple algebra $\mathfrak{h}$ with identity (46.1) can be embedded in a Lie division algebra $\mathfrak{g} \in$ var $\widetilde{W_1}(K)$. Denote by $A(\mathfrak{h})$ the associative subalgebra of $\operatorname{End}_K \mathfrak{h}$ generated by the elements $\langle h_1, h_2, h_3\rangle$, where the h_i range over $\mathfrak{h}$.

LEMMA 46.11. *The algebra $A(\mathfrak{h})$ has no zero divisors.*

PROOF. By Theorem 46.1, the algebra $A(\mathfrak{h})$ contains a unit and has only trivial $\mathfrak{h}$-invariant ideals. Since the characteristic of the ground field is equal to zero, the assertion of the lemma follows from Corollary of Proposition 51.1. The lemma is proved.

In the construction of the division algebra $\mathfrak{g}$ we can restrict ourselves to the case in which ${}^{\mathfrak{h}}A(\mathfrak{h}) = A(\mathfrak{h})$. This follows from Theorem 46.1. (Indeed, if $A(\mathfrak{h}) = 0$, then by Theorem 46.1 $A(\mathfrak{h})$ coincides with the centroid C of the Lie algebra $\mathfrak{h}$ and $\dim_C \mathfrak{h} = 3$. In this case the Lie C-algebra $\mathfrak{h}$ can be embedded in the Lie algebra $W_1(\widehat{K})$ over the algebraic closure $\widehat{K}$ of the field K, and for the Lie algebra $\mathfrak{h}' = W_1(\widehat{K})$ the equality ${}^{\mathfrak{h}'}A(\mathfrak{h}') = A(\mathfrak{h}')$ is already valid.)

Thus, suppose that ${}^{\mathfrak{h}}A(\mathfrak{h}) = A(\mathfrak{h})$. Denote by the letter Q the field of quotients of the algebra $A(\mathfrak{h})$ (the existence of this field is guaranteed by Lemma 46.11). Then $A(\mathfrak{h}) \subseteq Q$, and we can assume that the Lie algebra $\mathfrak{h}$ is contained in the Lie algebra $\operatorname{Der}_K Q$. Since $A(\mathfrak{h}) = {}^{\mathfrak{h}}A(\mathfrak{h})$, for an arbitrary nonzero element $\partial \in \mathfrak{h}$ there exists $a \in A(\mathfrak{h})$ for which ${}^{\partial}a \neq 0$. Since by Theorem 46.1 in the algebra $A(\mathfrak{h})$ relation (46.8) holds, any element $h \in \mathfrak{h}$ is representable in the Lie algebra $\operatorname{Der}_K Q$ in the form $({}^{\partial}a)^{(-1)} \cdot ({}^{h}a)\partial$. This shows that for any nonzero element $\partial \in \mathfrak{h}$ the algebra $\mathfrak{h}$ is contained in $\operatorname{Diff}_\partial Q$. From the theory of differential algebras (see [Rit]) it is well known that any differential field Q can be embedded in a differential field $\widehat{Q}$ such that equation (46.25) is solvable in $\widehat{Q}$ for any $a \in \widehat{Q}$.

LEMMA 46.12. *The Lie algebra* $\mathfrak{g} = \operatorname{Diff}_\partial \widehat{Q}$ *of special derivations of an arbitrary differential field* $\widehat{Q}$ *in which any equation* (46.25) *is solvable is a division algebra satisfying identity* (46.1).

PROOF. Corollary of Theorem 46.3 implies that identity (46.1) holds in the Lie algebra $\operatorname{Diff}_\partial \widehat{Q}$. The solvability of equation (46.24) in the Lie algebra $\mathfrak{g}$ for any $g, h \in \mathfrak{g}$, where $g = a\partial$, $h = b\partial$, $0 \neq a \in \widehat{Q}$, and $b \in \widehat{Q}$, follows from the equality $[a\partial, aq\partial] = b\partial$, where $q \in \widehat{Q}$ is a solution of the equation ${}^{\partial}y = b/a^2$. The lemma is proved.

This lemma shows that any simple Lie algebra $\mathfrak{h}$ with identity (46.1) can be embedded in a Lie division algebra $\operatorname{Diff}_\partial \widehat{Q}$ of the variety var $\widetilde{W_1}(K)$.

Since any Lie division algebra is simple, this lemma and Theorem 46.3 imply the last assertion of Theorem 46.4.

Now we show that in any Lie division algebra $\mathfrak{g}$ satisfying identity (46.1) Properties 1–3 are satisfied. From the solvability of equation (46.24) it follows that $\mathfrak{g}$ is a simple Lie algebra and $[g, \mathfrak{g}] = \mathfrak{g}$ for any nonzero element $g \in \mathfrak{g}$. Therefore, by Theorem 46.1 in the A-module $\mathfrak{g}$ identities (46.8) and (46.9) are valid, and the commutator subalgebra of the Lie algebra $\mathfrak{g}$ coincides with $\mathfrak{g}$ and, as a linear space, it is spanned by the elements $[[g, g_1], [g, g_2]]$, where g_1 and g_2 range over $\mathfrak{g}$. It can be verified directly that

$$\langle g_2, g_1, g\rangle g = 2[[g, g_1], [g, g_2]].$$

Therefore, the A-module $\mathfrak{g}$ is cyclic and any nonzero element g of it can be taken for a generator. From the strict multiplicativity of this module it follows that A is a field and $\mathfrak{g}$ is a free A-module. Relation (46.9) shows that $\mathfrak{g} = \operatorname{Diff}_g A$. Therefore, any element $h \in \mathfrak{g}$ has the form ag and a solution x of equation (46.24) is representable in the form $x = yg$, where $y \in A$. Then we have

$$[g, yg] = ({}^{g}y)g = ag,$$

whence it follows that equation (46.25) is solvable in A for any $a \in A$ and for any nonzero $g \in \mathfrak{g}$. Theorem 46.4 is completely proved.

§47. Criteria for existence of proper subalgebras of finite codimension in Lie algebras

In this section we give necessary and sufficient conditions for the existence of a proper subalgebra of finite codimension in a Lie algebra under extensions of the ground field. These criteria are such that there is certainly no need to extend the ground field if it is algebraically closed and its cardinality is greater than the dimension of the Lie algebra.

PROPOSITION 47.1. *A Lie K-algebra $\mathfrak{g}$ contains a proper subalgebra $\mathfrak{g}_0$ of finite codimension if and only if there exists a nonzero linear functional $h\colon \mathfrak{g} \to K$ for which the kernel of the bilinear form $\mathfrak{b}_h\colon \mathfrak{g}\otimes_K\mathfrak{g} \to K$ ($\mathfrak{b}_h(x,y) \rightleftharpoons h([x,y])$) is of finite codimension.*

PROOF. *Sufficiency.* Denote by $\mathfrak{h}$ the annihilator of the bilinear form $\mathfrak{b}_h$ and by $\mathfrak{g}^*$ the $\mathfrak{g}$-module dual to the adjoint representation of the Lie algebra $\mathfrak{g}$. It is clear that $\mathfrak{h} = \{g \in \mathfrak{g} \mid {}^g h = 0\}$ (here ${}^g h$ denotes the linear functional from $\mathfrak{g}^*$ for which ${}^g h(y) \rightleftharpoons -h(\operatorname{ad} g\, y)$). Hence, $\mathfrak{h}$ is a Lie subalgebra of $\mathfrak{g}$ consisting of the elements annihilating h in the dual representation $\mathfrak{g}^*$ of the Lie algebra $\mathfrak{g}$. Therefore, if the kernel of $\mathfrak{b}_h$ is of finite codimension, then either $\mathfrak{h} \neq \mathfrak{g}$ and then we can take $\mathfrak{h}$ for $\mathfrak{g}_0$, or $\mathfrak{g} = \mathfrak{h}$ and then the commutator subalgebra of the algebra $\mathfrak{g}$ is a proper ideal of $\mathfrak{g}$, and we can take any subspace of codimension 1 containing the commutator subalgebra for $\mathfrak{g}_0$.

Necessity. Suppose that $0 < \dim_K \mathfrak{g}/\mathfrak{g}_0 < \infty$. Let us consider the first member $\mathfrak{g}_1 \rightleftharpoons \{g \in \mathfrak{g}_0 \mid [g,\mathfrak{g}] \subset \mathfrak{g}_0\}$ of the Cartan prolongation of the subalgebra $\mathfrak{g}_0$. Since the representation $\operatorname{ad}\colon \mathfrak{g}_0/\mathfrak{g}_1 \to \operatorname{End}_K \mathfrak{g}/\mathfrak{g}_0$ in a finite-dimensional space $\mathfrak{g}/\mathfrak{g}_0$ is faithful, we have $\dim_K \mathfrak{g}/\mathfrak{g}_0 < \infty$. Therefore, for any linear functional $h \in \mathfrak{g}^*$ satisfying $h(\mathfrak{g}_0) = 0$ the annihilator of the bilinear form $\mathfrak{b}_h$ contains $\mathfrak{g}_1$ and is of finite codimension. The proposition is proved.

For an arbitrary Lie K-algebra $\mathfrak{g}$ denote by $S(\mathfrak{g})$ the algebra of commutative polynomials over the space of $\mathfrak{g}$ with zero constant term. The algebra $S(\mathfrak{g})$ has a natural gradation

$$S(\mathfrak{g}) = S_1(\mathfrak{g}) \oplus S_2(\mathfrak{g}) \oplus \cdots \oplus S_k(\mathfrak{g}) \oplus \cdots,$$

where $S_k(\mathfrak{g})$ is the space of homogeneous polynomials of degree k.

For any $h \in \mathfrak{g}^*$ denote by $\mathfrak{b}_h$ the matrix of values of the bilinear form $\mathfrak{b}_h(x,y)$, i.e., the matrix whose rows and columns are indexed by all elements of $\mathfrak{g}$ and whose entry at the intersection of the g_1th row and the g_2th column is $\mathfrak{b}_h(g_1,g_2) = h([g_1,g_2])$. It is clear that the codimension of the annihilator of the bilinear form $\mathfrak{b}_h$ coincides with the rank of the matrix $\mathfrak{b}_h$. Hence, the existence of a proper subalgebra of finite codimension is related to the existence of a functional $h \in \mathfrak{g}^*$ for which the matrix $\mathfrak{b}_h$ is of finite rank. Since any functional $h\colon \mathfrak{g} \to K$ can be extended to a homomorphism $h\colon S(\mathfrak{g}) \to K$, this suggests an idea of introducing the following objects:

a) a "generic" matrix $\mathfrak{b}$ whose entry at the intersection of the g_1th row and the g_2th column is the element $[g_1,g_2] \in \mathfrak{g}$ of the algebra $S(\mathfrak{g})$;
b) homogeneous ideals $J_n(\mathfrak{g})$ ($n = 0,1,2,\ldots$) of $S(\mathfrak{g})$ generated by all minors of order $n+1$ of the matrix $\mathfrak{b}$,

and leads to the following criterion for the existence of a proper subalgebra of finite codimension in a Lie algebra $\mathfrak{g}$.

PROPOSITION 47.2. *Suppose that the ground field K is algebraically closed and the dimension of a Lie algebra $\mathfrak{g}$ over K is less than the cardinality $|K|$ of the field K. Then the algebra $\mathfrak{g}$ has no proper subalgebras of finite codimension if and only if the quotient algebra $S(\mathfrak{g})/J_n(\mathfrak{g})$ is a nil-algebra for any positive integer n.*

PROOF. Suppose that the algebra $\mathfrak{g}$ contains a proper K-subalgebra $\mathfrak{g}_0$ of finite codimension. Then, by Proposition 47.1, there exists a nonzero functional $h\colon \mathfrak{g} \to K$ for which the annihilator of the bilinear form $\mathfrak{b}_h(x, y)$ is of finite codimension, say, n. The extension of this functional to a homomorphism $h\colon S(\mathfrak{g}) \to K$ transforms the matrix $\mathfrak{b}$ into the matrix $\mathfrak{b}_h$. Since the rank of the matrix $\mathfrak{b}_h$ is equal to n, we have $h(J_n(\mathfrak{g})) = 0$ and h induces a nonzero homomorphism of the quotient algebra $S(\mathfrak{g})/J_n(\mathfrak{g})$ into the field K. Therefore, for n chosen as above the algebra $S(\mathfrak{g})/J_n(\mathfrak{g})$ possesses nonnilpotent elements.

Conversely, suppose that for some positive integer n the quotient algebra of the form $S(\mathfrak{g})/J_n(\mathfrak{g})$ has a nonnilpotent element. Then, since the ground field K is algebraically closed and the set of generators of the commutative algebra $S(\mathfrak{g})/J_n(\mathfrak{g})$ does not exceed $\dim_K \mathfrak{g}$, where $\dim_K \mathfrak{g} < |K|$, it follows from Hilbert's Nullstellensatz that there exists a nonzero homomorphism of this algebra into the field K. In this case the through homomorphism $h\colon S(\mathfrak{g}) \to S(\mathfrak{g})/J_n(\mathfrak{g}) \to K$ determines a nonzero linear functional $h \in \mathfrak{g}^*$ on $\mathfrak{g}$ for which the matrix $\mathfrak{b}_h$ is of rank at most n. Therefore, the annihilator of the bilinear form $\mathfrak{b}_h(x, y)$ is of codimension less than n. The existence of a proper subalgebra $\mathfrak{g}_0$ of $\mathfrak{g}$ satisfying the condition $\dim \mathfrak{g}/\mathfrak{g}_0 \leq n$ now follows by Proposition 47.1. The proposition is proved.

Let $a = \sum_{\sigma \in S_{n+1}} \beta_\sigma \sigma$ be an element of the group algebra $K[S_{n+1}]$ of the symmetric group S_{n+1} acting on the set $\{1, \dots, n+1\}$. Let us denote by $J_{a,n}(\mathfrak{g})$ the ideal of the algebra $S(\mathfrak{g})$ generated by the elements of the form

$$\sum_{\sigma \in S_{n+1}} \beta_\sigma [u_{\sigma(1)}, v_1] \cdots [u_{\sigma(n)}, v_n][u_{\sigma(n+1)}, v_{n+1}], \tag{47.1}$$

where v_i and u_j range over the whole Lie algebra $\mathfrak{g}$, and by $\varphi_{a,n}$ the canonical homomorphism of the algebra $S(\mathfrak{g})$ onto the quotient algebra $S_{a,n}(\mathfrak{g}) \rightleftharpoons S(\mathfrak{g})/J_{a,n}(\mathfrak{g})$. The following statement is a generalization of Proposition 47.2.

THEOREM 47.1. *Let $\mathfrak{g}$ be a Lie algebra over an arbitrary field K and let $\mathfrak{g} = [\mathfrak{g}, \mathfrak{g}]$. Then the $\widehat{K}$-algebra $\widehat{\mathfrak{g}} = \widehat{K} \otimes_K \mathfrak{g}$ has no proper $\widehat{K}$-subalgebras of finite codimension for any extension $\widehat{K}$ if and only if for any Lie algebra $\mathfrak{h}$ containing $\mathfrak{g}$ as a subalgebra and for any ideal $J_{a,n}(\mathfrak{h})$, where $0 \neq a \in K[S_{n+1}]$, $n = 0, 1, 2, \dots$, the image of the Lie ideal $\mathrm{id}_{\mathfrak{h}}([\mathfrak{g}, \mathfrak{h}])$ generated in $\mathfrak{h}$ by the mutual commutator $[\mathfrak{g}, \mathfrak{h}]$ under the canonical homomorphism $\varphi_{a,n}\colon S(\mathfrak{h}) \to S_{a,n}(\mathfrak{h}) = S(\mathfrak{h})/J_{a,n}(\mathfrak{h})$ consists of nilpotent elements only.*

PROOF. *Sufficiency.* Suppose that for any Lie algebra $\mathfrak{h}$ containing $\mathfrak{g}$ the image $\varphi_{a,n}(\mathrm{id}_{\mathfrak{h}}[\mathfrak{h}, \mathfrak{g}])$ consists of nilpotent elements. Then, in particular, for $\mathfrak{h} = \mathfrak{g}$ and $a = \sum_{\sigma \in S_{n+1}} \mathrm{sign}(\sigma)\sigma$ all quotient algebras $S_n(\mathfrak{g}) + S(\mathfrak{g})/J_n(\mathfrak{g})$ $(n = 1, 2, \dots)$ consist of nilpotent elements, because in this case $J_{a,n}(\mathfrak{g}) = J_n(\mathfrak{g})$ and $\mathrm{id}_{\mathfrak{h}}([\mathfrak{h}, \mathfrak{g}]) = \mathfrak{g}$. Assume that for some extension $\widehat{K}$ of the field K the $\widehat{K}$-algebra $\widehat{\mathfrak{g}}$ contains a proper subalgebra of finite codimension over $\widehat{K}$. Then for any extension $\widetilde{K}$ of the field $\widehat{K}$ the Lie algebra

$\widetilde{\mathfrak{g}} = \widetilde{K} \otimes_K \mathfrak{g}$ also has a proper $\widetilde{K}$-subalgebra of finite codimension. Let us take for $\widetilde{K}$ an algebraically closed field for which $\dim_K \mathfrak{g} < |\widetilde{K}|$. From Proposition 47.2 it follows that for some $n \in \mathbb{N}$ the quotient algebra $S_n(\widetilde{\mathfrak{g}})$ contains nonnilpotent elements. However, $J_n(\widetilde{\mathfrak{g}}) = \widetilde{K} \otimes_K J_n(\mathfrak{g})$ and $S_n(\widetilde{\mathfrak{g}}) = \widetilde{K} \otimes_K S_n(\mathfrak{g})$, which contradicts the fact that the algebra $S_n(\mathfrak{g})$ consists of nilpotent elements.

Necessity. Suppose for any extension $\widehat{K}$ of the field K there is no proper $\widehat{K}$-subalgebra of finite codimension of the $\widehat{K}$-algebra $\widehat{\mathfrak{g}}$.

We assume that for some Lie algebra $\mathfrak{h} \supset \mathfrak{g}$ and for some nonzero element $a \in K[S_{n+1}]$ there exist nonnilpotent elements of the subspace $\varphi_{a.n}(\mathrm{id}_{\mathfrak{h}}[\mathfrak{h}, \mathfrak{g}])$. Since the linear space of $\mathrm{id}_{\mathfrak{h}}([\mathfrak{h}, \mathfrak{g}])$ is spanned by left-normed commutators of the form $[g_0, g_1, \dots, g_t]$, where $g_0 \in \mathfrak{g}$ and $g_1, \dots, g_t \in \mathfrak{h}$, $t \geq 1$, a nonnilpotent element c of the form $\varphi_{a.n}([g_0, g_1, \dots, g_t])$ must exist in $S_{a.n}(\mathfrak{h})$. Then this element c is not nilpotent in the $\widehat{K}$-algebra $\widehat{K} \otimes_K S_{a.n}(\mathfrak{h})$ for any extension $\widehat{K}$ of the field K.

Let us take for $\widehat{K}$ an algebraically closed field whose cardinality is greater than $\dim_K \mathfrak{h}$ and put $\widehat{\mathfrak{h}} = \widehat{K} \otimes_K \mathfrak{h}$. From the explicit form (47.1) of the generators of the ideal $J_{a.n}$ it follows that $\widehat{K} \otimes_K J_{a.n}(\mathfrak{h}) = J_{a.n}(\widehat{\mathfrak{h}})$. Therefore, $\widehat{K} \otimes_K S_{a.n}(\mathfrak{h}) = S_{a.n}(\widehat{\mathfrak{h}})$ and $\varphi_{a.n}(c)$ is a nonnilpotent element of $S_{a.n}(\widehat{\mathfrak{h}})$. By Hilbert's Nullstellensatz, there exists a $\widehat{K}$-homomorphism of the algebra $S_{a.n}(\widehat{\mathfrak{h}})$ into the field $\widehat{K}$ transforming $\varphi_{a.n}(c)$ into a nonzero element. Then the through $\widehat{K}$-homomorphism $h\colon S(\widehat{\mathfrak{h}}) \to S_{a.n}(\widehat{\mathfrak{h}}) \to \widehat{K}$ determines a nonzero functional on $\widehat{\mathfrak{h}}$, and we have

$$\mathfrak{b}_h([g_0, g_1, \dots, g_{t-1}], g_t) \neq 0 \qquad (g_0 \in \mathfrak{g},\ t \geq 1). \tag{47.2}$$

Hence, the element $[g_0, g_1, \dots, g_{t-1}] \in \widehat{\mathfrak{h}}$ is outside the annihilator $\widehat{\mathfrak{h}}_0$ of the bilinear form $\mathfrak{b}_h$, and $\widehat{\mathfrak{h}}_0$ is a proper $\widehat{K}$-subalgebra of $\widehat{\mathfrak{h}}$.

We shall show first that $\dim_{\widehat{K}} \widehat{\mathfrak{h}}/\widehat{\mathfrak{h}}_0 \leq n$. Since $J_{\sigma a.n} = J_{a.n}$ for any permutation $\sigma \in S_{n+1}$, we shall assume in what follows that the coefficient β_1 of the identity permutation in the decomposition $a = \sum_{\sigma \in S_{n+1}} \beta_\sigma \sigma$ is nonzero. Let us consider in the algebra $S(\widehat{\mathfrak{h}})$ a descending chain of ideals $J_{a_i.i}$ $(i = 0, 1, 2, \dots)$, where $a_i = \sum_{\sigma \in S_{i+1}} \beta_\sigma \sigma \in K[S_{i+1}]$ and the permutation groups $S_1, \dots, S_{i+1}, \dots$ are regarded as naturally embedded in one another. We choose the smallest N such that $h(J_{a_N.N}(\widehat{\mathfrak{h}})) = 0$. Since $a_n = a$ and $h(J_{a.n}(\widehat{\mathfrak{h}})) = 0$, the set $J_{a.n}(\widehat{\mathfrak{h}})$ belongs to the kernel of the homomorphism h and $0 \leq N \leq n$. From the explicit form (47.1) of the generators for the ideal $J_{a_N.N}$ it follows that

$$h\Bigl(\sum_{\sigma \in S_{n+1}} \beta_\sigma [u_{\sigma(1)}, v_1] \cdots [u_{\sigma(N)}, v_N][u_{\sigma(N+1)}, v_{N+1}]\Bigr) = 0, \tag{47.3}$$

where u_i and v_i are arbitrary elements of the algebra $\widehat{\mathfrak{h}}$. Since $\beta_1 \neq 0$ and $\mathfrak{b}_h \neq 0$, this equality can hold only for $N > 1$. However, then $h(J_{a_{N-1}.N-1}(\widehat{\mathfrak{h}})) \neq 0$, and there exist elements $u_1, v_1, \dots, u_N, v_N$ for which

$$h\Bigl(\sum_{\sigma \in S_N} \beta_\sigma [u_{\sigma(1)}, v_1] \cdots [u_{\sigma(N)}, v_N]\Bigr) \neq 0. \tag{47.4}$$

We fix these elements $u_1, v_1, \dots, u_N, v_N$ and for any $u_{N+1} \in \widehat{\mathfrak{h}}$ we put

$$b_i(u_{N+1}) \rightleftharpoons \sum_{\sigma(N+1)=i} \beta_\sigma[u_{\sigma(1)}, v_1]\cdots[u_{\sigma(N)}, v_N].$$

Then inequality (47.4) shows that $h(b_{N+1}) \neq 0$ in the field $\widehat{K}$, and equality (47.3) can be rewritten in the form

$$\mathfrak{b}_h\left(\left\{u_{N+1} - h(b_{N+1})^{-1}\sum_{i=1}^{N} h(b_i(u_{N+1}))u_i\right\}, v_{N+1}\right) = 0,$$

where $h(b_i(u_{N+1})) \in K$. Since v_{N+1} and u_{N+1} are arbitrary and $u_1, \dots, u_N$ are fixed elements of the algebra $\widehat{\mathfrak{h}}$ in this equality, we have $\dim_{\widehat{K}} \widehat{\mathfrak{h}}/\widehat{\mathfrak{h}}_0 \leq N \leq n$.

Let us construct the Cartan prolongation $\widehat{\mathfrak{h}}_1, \widehat{\mathfrak{h}}_2, \dots$ of $\widehat{\mathfrak{h}}_0$ by successively setting $\widehat{\mathfrak{h}}_{i+1} = \{g \in \widehat{\mathfrak{h}}_i \mid [g, \widehat{\mathfrak{h}}] \subset \widehat{\mathfrak{h}}_i\}$. As already mentioned in §43, the $\widehat{K}$-subspaces $\widehat{\mathfrak{h}}_i$ are Lie subalgebras, and the inequality $\dim_{\widehat{K}} \widehat{\mathfrak{h}}/\widehat{\mathfrak{h}}_0 \leq n$ implies $\infty > \dim_{\widehat{K}} \widehat{\mathfrak{h}}/\widehat{\mathfrak{h}}_i$ for each $i \in \mathbb{N}$. Hence, $\widehat{\mathfrak{g}}_0 \rightleftharpoons \widehat{\mathfrak{g}} \cap \widehat{\mathfrak{h}}_{t-1}$ is a $\widehat{K}$-subalgebra of $\widehat{\mathfrak{g}}$ of finite codimension. From inequality (47.2) it follows that $[g_0, g_1, \dots, g_{t-1}] \notin \widehat{\mathfrak{h}}_0$; however, then the element g_0 is outside $\widehat{\mathfrak{h}}_{t-1}$. Therefore, $\widehat{\mathfrak{g}}_0$ is a proper $\widehat{K}$-subalgebra of $\widehat{\mathfrak{g}}$ of finite codimension. This contradiction shows that for any Lie algebra $\mathfrak{h}$ and for any nonzero element $a \in K[S_{n+1}]$ the subspace $\varphi_{a,n}(\mathrm{id}_{\mathfrak{h}}[\mathfrak{h}, \mathfrak{g}])$ of the algebra $S_{a,n}(\mathfrak{h})$ consists of nilpotent elements. The theorem is proved.

COROLLARY 1. *Suppose that for any extension $\widehat{K}$ of the field K the $\widehat{K}$-algebra $\widehat{\mathfrak{g}} = \widehat{K} \otimes_K \mathfrak{g}$ has no proper subalgebras of finite codimension. Then for any ideal $J_{a,n}(\mathfrak{g})$ $(a \neq 0)$ the quotient algebra $S_{a,n}(\mathfrak{g}) = S(\mathfrak{g})/J_{a,n}(\mathfrak{g})$ is a nil-algebra.*

PROOF. The commutator subalgebra of the Lie algebra $\mathfrak{g}$ coincides with $\mathfrak{g}$, since otherwise any subspace of codimension 1 containing the commutator subalgebra would be an ideal of $\mathfrak{g}$. Applying Theorem 47.1 for $\mathfrak{h} = \mathfrak{g}$ and taking into account that $\mathrm{id}_{\mathfrak{h}}([\mathfrak{g}, \mathfrak{h}]) = \mathfrak{g}$ in this case, we obtain that the generators $\varphi_{a,n}(\mathfrak{g})$ of the algebra $S_{a,n}(\mathfrak{g})$ are nil-elements. The corollary is proved.

COROLLARY 2. *Suppose that for some Lie K-algebra $\mathfrak{h}$ containing $\mathfrak{g}$ as a subalgebra and for some $n \in \mathbb{N}$ and $0 \neq a \in K[S_{n+1}]$ the subspace $\varphi_{a,n}(\mathrm{id}_{\mathfrak{h}}[\mathfrak{g}, \mathfrak{h}])$ contains a nonnilpotent element of the algebra $S_{a,n}(\mathfrak{h})$. Then for some extension $\widehat{K}$ of the field K the Lie algebra $\mathfrak{g}$ admits a nonzero homomorphism into the Lie algebra $\widetilde{W}_n(\widehat{K})$.*

PROOF. The proof of Theorem 47.1 shows that for an algebraically closed field $\widehat{K}$ with sufficienly large cardinality and containing K we can choose a proper Lie $\widehat{K}$-subalgebra $\widehat{\mathfrak{h}}_0$ of the algebra $\widehat{\mathfrak{h}} = \widehat{K} \otimes_K \mathfrak{h}$ for which $\dim_{\widehat{K}} \widehat{\mathfrak{h}}/\widehat{\mathfrak{h}}_0 = N \leq n$ and $\mathrm{id}_{\widehat{\mathfrak{h}}}(\mathfrak{g}) \not\subset \widehat{\mathfrak{h}}_0$ (see inequality (47.2)). Therefore, the existence of a $\widehat{K}$-homomorphism $\varkappa \colon \widehat{\mathfrak{h}} \to \widetilde{W}_n(\widehat{K})$ for which $\varkappa(\mathfrak{g}) \neq 0$ follows from Corollary 1 of Theorem 44.2.

§48. Simple Lie algebras in varieties of exponential type

In this section we shall prove Theorems 42.1 and 42.2.

48.1. Identities in varieties of exponential type.

LEMMA 48.1. *Let $\mathfrak{g}$ be a simple Lie algebra over an arbitrary field K. Then in $\mathfrak{g}$ there exists a finite set of elements $g_1, \dots, g_k$ for which we have the equality*

$$\mathfrak{g} = [g_1, \mathfrak{g}, \mathfrak{g}] + \cdots + [g_k, \mathfrak{g}, \mathfrak{g}]. \tag{48.1}$$

PROOF. Since $\mathfrak{g}$ is simple, for an arbitrary nonzero element $g_0 \in \mathfrak{g}$ there exists an element $u \in \operatorname{End}_K \mathfrak{g}$ such that it can be expressed via a finite set of operators from $\operatorname{ad} \mathfrak{g}$ and

$$u \times g_0 = g_0. \tag{48.2}$$

We denote by $g_1, \dots, g_k$ all elements of $\mathfrak{g}$ participating in the expression of u. We shall show that for these elements equality (48.1) holds. By the simplicity of the Lie algebra $\mathfrak{g}$ we have

$$\mathfrak{g} = \sum_{t=1}^{\infty} (\operatorname{ad} g)^t \cdot g_0.$$

Since from equality 48.2 it follows that $u^{t+1} g_0 = g_0$, in order to prove the lemma, it suffices to verify the inclusion

$$(\operatorname{ad} \mathfrak{g})^t \operatorname{ad} g_{i_1} \cdots \operatorname{ad} g_{i_l} g_0 \subset \sum_{i=1}^{k} [g_i, \mathfrak{g}, \mathfrak{g}]$$

for arbitrary $i_1, \dots, i_l \in \{1, \dots, k\}$ and $l > t$. We shall verify this by induction on the number t.

The base of induction for $t = 0$ is obvious.

The induction step. Applying the equality $xy = yx + [x, y]$ for $x = \operatorname{ad} g_i$ and $y \in \operatorname{ad} \mathfrak{g}$ and then for $x \in \operatorname{ad}[\mathfrak{g}, g_i]$, we obtain the inclusions

$$\begin{aligned}(\operatorname{ad} \mathfrak{g})^t \operatorname{ad} g_{i_1} &\subset \operatorname{ad} g_{i_1} (\operatorname{ad} \mathfrak{g})^t + \sum_{j=0}^{t-1} (\operatorname{ad} \mathfrak{g})^j [\operatorname{ad} g_{i_1}, \operatorname{ad} \mathfrak{g}] (\operatorname{ad} \mathfrak{g})^{t-j-1} \\ &\subset \operatorname{ad} g_{i_1} (\operatorname{ad} \mathfrak{g})^t + \operatorname{ad}[g_{i_1}, \mathfrak{g}] \cdot (\operatorname{ad} \mathfrak{g})^{t-1} + (\operatorname{ad} \mathfrak{g})^{t-1},\end{aligned}$$

which imply that

$$(\operatorname{ad} \mathfrak{g})^t \operatorname{ad} g_{i_1} \cdots \operatorname{ad} g_{i_l} g_0$$

is contained in $[\mathfrak{g}, g_{i_1}] + [g_{i_1}, \mathfrak{g}, \mathfrak{g}] + (\operatorname{ad} \mathfrak{g})^{t-1} \operatorname{ad} g_{i_2} \cdots \operatorname{ad} g_{i_l} g_0$. Since the commutator subalgebra of the algebra $\mathfrak{g}$ is $\mathfrak{g}$ itself, the first two terms are contained in the subspace $[g_{i_1}, \mathfrak{g}, \mathfrak{g}]$. For the last summand the inclusion of it into $\sum_{j=0}^{k} [g_j, \mathfrak{g}, \mathfrak{g}]$ follows from the inductive arguments. The lemma is proved.

LEMMA 48.2. *Let $\mathfrak{B}$ be an arbitrary variety of Lie algebras of exponential type over a field K of arbitrary characteristic. Then for any positive integer m there exist positive integers $N = N(m)$ and $M = M(m)$ and elements $\beta_\sigma = \beta_\sigma(m)$ and $\beta'_\sigma = \beta'_\sigma(m)$*

($\sigma \in S_N, S_M$) *such that in the variety* $\mathfrak{B}$ *we have nontrivial multilinear identities of the form*

$$\sum_{\sigma \in S_N} \beta_\sigma[[x_{\sigma(1)}, y_{11}, \dots, y_{1m}] \cdots [x_{\sigma(N)}, y_{N1}, \dots, y_{Nm}]] = 0, \tag{48.3}$$

$$\sum_{\sigma \in S_M} \beta_\sigma[x_{\sigma(1)}, y_{11}, \dots, y_{1m}, \dots, x_{\sigma(N)}, y_{N1}, \dots, y_{Nm}] = 0, \tag{48.3$'$}$$

where not all β_σ *(respectively,* β'_σ*) are zero.*

PROOF. Let us consider the space of all multilinear polynomials P_{mN} in mN variables x_i and y_{ij} ($i = 1, \dots, N$; $j = 1, \dots, m$) in the free Lie algebra of the variety $\mathfrak{B}$. From the assumption of the lemma and the definition of a variety of exponential type it follows that $\dim_K P_{mN} \le d^{mN}$ for some integer d and for all sufficiently large numbers N. Therefore, from the Stirling formula it follows that for sufficiently large N there are linearly dependent left-normed commutators among the $N!$ ones of the form

$$[[x_{\sigma(1)}, y_{11}, \dots, y_{1m}] \cdots [x_{\sigma(N)}, y_{N1}, \dots, y_{Nm}]] \qquad (\sigma \in S_N).$$

This means that we have a nontrivial identity of the form (48.3) in the variety $\mathfrak{B}$. The validity of (48.3$'$) is proved similarly. The lemma is proved.

REMARK 48.1. Let us fix a nontrivial identity of the form (48.3) and (48.3$'$) for any positive integer m. We shall show below that any such system of identities determines a variety in which all simple algebras are Lie algebras of Cartan type. At the first step of the proof the identity of the form

$$\sum_{\sigma \in S_{N+1}} \beta_\sigma[z, [x_{\sigma(1)}, y_1], \dots, [x_{\sigma(N)}, y_N][x_{\sigma(N+1)}, y_{N+1}]] = 0, \tag{48.4}$$

plays the main role; this identity is satisfied in simple Lie algebras of the varieties under consideration.

48.2. Criteria for a simple algebra to be of Cartan type in a variety $\mathfrak{R}$ of Lie algebras given by identity (48.4). Suppose that identity (48.4) is satisfied in a simple Lie K-algebra $\mathfrak{g}$ which is not a Lie algebra of Cartan type. Let us denote by the letter a the element of the group algebra $K[S_{N+1}]$ equal to $\sum_{\sigma \in S_{N+1}} \beta_\sigma \sigma$, where β_σ are the same elements which occur in (48.4). Then, by Corollary 1 of Theorem 47.1, the quotient algebra $S(\mathfrak{g})/J_{a,N}(\mathfrak{g})$ is a nil-algebra. This means that for any element $g \in \mathfrak{g} \subset S(\mathfrak{g})$ there exists a positive integer $l = l(g)$ for which in the algebra $S(\mathfrak{g})$ we have the equality

$$g^l = \sum_{i=1}^{k} a_i b_i,$$

where b_i are generators of the ideal $J_{a,N}(\mathfrak{g})$ of the form (47.1) and a_i are homogeneous elements of the graded algebra $S(\mathfrak{g})$ of degree $l - N - 1$. Since the graded algebra $\mathrm{Gr}(U(\mathfrak{g}))$, constructed with respect to the universal enveloping algebra $U(\mathfrak{g})$ (without unit element) with its natural filtration $U_l = \sum_{i=1}^{l} \mathfrak{g}^i$, is isomorphic to $S(\mathfrak{g})$, in $U(\mathfrak{g})$ we have the inclusion

$$g^l - \sum_{i=1}^{k} \widetilde{a}_i \widetilde{b}_i \in \sum_{i=1}^{l-1} \mathfrak{g}^i,$$

where $\widetilde{a}_i$ and $\widetilde{b}_i$ are the natural preimages in $U(\mathfrak{g})$ of the homogeneous elements a_i and b_i. Comparing identity (48.4) with the element (47.1), we see that we can choose the elements $\widetilde{b}_i$ so that their images in the adjoint representation of the Lie algebra $\mathfrak{g}$ are zero. Hence, for any element $g \in \mathfrak{g}$ there exists an integer $l = l(g)$ for which in the adjoint representation of the Lie algebra $\mathfrak{g}$ we have

$$(\operatorname{ad} g)^{l(g)} \in \sum_{i=1}^{l-1} (\operatorname{ad} \mathfrak{g})^i. \tag{48.5}$$

DEFINITION 48.1. Let $\rho\colon \mathfrak{g} \to \operatorname{End}_K V$ be a representation of the Lie algebra $\mathfrak{g}$ in a linear space V. An element g of the algebra $\mathfrak{g}$ is called an *element of finite type* in this representation if for some $l = l(g) > 0$ we have the inclusion

$$\rho(g)^l \in \sum_{i=1}^{l-1} \rho(\mathfrak{g})^i. \tag{48.6}$$

The number l is called the *degree* of the element of finite type g.

LEMMA 48.3. *Let $\mathfrak{g}$ be a simple Lie algebra over an arbitrary field K and let identity* (48.4) *hold in $\mathfrak{g}$. If for any extension $\widehat{K}$ of the centroid C of the algebra $\mathfrak{g}$ the Lie $\widehat{K}$-algebra $\widehat{\mathfrak{g}} = \widehat{K} \otimes_K \mathfrak{g}$ has no proper subalgebras of codimension $\leq N(N-1)$, then all elements of $\mathfrak{g}$ are of finite type in its adjoint representation and, moreover, we can choose a basis E in $\mathfrak{g}$ and an integer l depending only on this basis such that for any element $e \in E$*

$$(\operatorname{ad} e)^l \in \sum_{i=1}^{l-1} (\operatorname{ad} \mathfrak{g})^i. \tag{48.7}$$

PROOF. The first part of the lemma was proved above (see the inclusion in (48.5)). In order to construct the desired basis E and the number l, we consider the Lie C-algebra $\mathfrak{h}$ which is the quotient algebra of the free product of the Lie C-algebra $\mathfrak{g}$ with the free Lie C-algebra $\mathfrak{M}$ with a countable set of the free generators $\{x_i, y_i \mid i = 1, 2, \dots\}$ by the ideal of identities determined by identity (48.4). Obviously, the Lie algebra $\mathfrak{g}$ is embedded in $\mathfrak{h}$; therefore, we identify it with a subalgebra of $\mathfrak{h}$. Let $a = \sum_{\sigma \in S_{N+1}} \beta_\sigma \sigma$ as before. In accordance with Lemma 48.1, we can find in the algebra $\mathfrak{g}$ a finite set of elements $g_1, \dots, g_k$ for which equality (48.1) holds. By Theorem 47.1, for some $l \in \mathbb{N}$ in the algebra $S_{a,N}(\mathfrak{h})$ we have the equality

$$([g_1, x_1, y_1] + \cdots + [g_k, x_k, y_k])^l = 0.$$

Since any mapping of the generators x_i and y_i $(i = 1, 2, \dots)$ into $\mathfrak{g}$ can be extended to a Lie algebra homomorphism $\mathfrak{h} \to \mathfrak{g}$ that is the identity on the subalgebra $\mathfrak{g}$, and any homomorphism of this kind maps the ideal $J_{a,N}(\mathfrak{h})$ into the ideal $J_{a,N}(\mathfrak{g})$, we see that in $S(\mathfrak{g})/J_{a,N}(\mathfrak{g})$ we have the equality

$$([g_1, u_1, v_1] + \cdots + [g_k, u_k, v_k])^l = 0,$$

where u_i and v_i are arbitrary elements of the Lie algebra $\mathfrak{g}$. However, by Lemma 48.1 the elements in the parentheses in the latter equality span the Lie algebra $\mathfrak{g}$ as a linear space. Hence, we can choose the desired basis among these elements. The lemma is proved.

COROLLARY. *If* char $K = p > 0$ *and all the assumptions of Lemma* 48.3 *are satisfied, then there exists an integer k such that for any $g \in \mathfrak{g}$ we have the inclusion*

$$(\operatorname{ad} g)^k \in \sum_{i=1}^{k-1} (\operatorname{ad} \mathfrak{g})^i. \tag{48.8}$$

PROOF. Let $k = p^t > l$. Then the inclusion (48.8) follows from (48.7), since

$$(u+v)^{p^t} - u^{p^t} - v^{p^t} \in \mathfrak{g} \qquad (u, v \in \mathfrak{g}).$$

The corollary is proved.

The following two lemmas show that if a simple Lie algebra $\mathfrak{g}$ of Cartan type consists of elements of finite type, then it is finite-dimensional over its centroid.

LEMMA 48.4. *Let $\widehat{\mathfrak{g}}$ be a simple Lie algebra over an arbitrary field $\widehat{K}$ containing a proper subalgebra $\widehat{\mathfrak{g}}_0$ of codimension n and the images of elements $e_1, \dots, e_n \in \widehat{\mathfrak{g}}$ in the quotient space $\widehat{\mathfrak{g}}/\widehat{\mathfrak{g}}_0$ form a basis of this space. If the elements $e_1, \dots, e_n$ are of finite type of degree l in the adjoint representation, then all multilinear Lie identities of the form*

$$f(x_1, \dots, x_N, y_1, \dots, y_t) = 0 \tag{48.9}$$

hold in $\widehat{\mathfrak{g}}$, where $N = N(n, l) \in \mathbb{N}$ and the Lie polynomial f is skew-symmetric with respect to $x_1, \dots, x_N$.

The proof of this lemma proceeds similarly to that of Proposition 43.1. We can take for N in identity (48.9) any integer satisfying the inequality $n(N-1)(l-1) \le m^-_{\widehat{\mathfrak{g}}}(N)$. Since $\widehat{\mathfrak{g}}$ can be embedded in $\widetilde{W}_n(\widehat{K})$, we have $m^-_{\widehat{\mathfrak{g}}}(N) \ge m^-_{\widetilde{W}_n}(N)$. Therefore, N can be chosen as a solution of the inequality $nN(l-1) \le m^-_{\widetilde{W}_n}(N)$. Then N will depend on n and l only. Details of the proof are left to the reader.

LEMMA 48.5. *Suppose that for some $N \in \mathbb{N}$ and for any $m \in \mathbb{N}$ in a simple Lie algebra $\mathfrak{g}$ over an arbitrary field K all identities of the form* (48.3) *hold, where β_σ are fixed elements of the field K and $\beta_1 \neq 0$. Then the algebra $\mathfrak{g}$ is finite-dimensional over its centroid C and* $\dim_C \mathfrak{g} < N$.

PROOF. The centroid of a simple algebra is a field, and it coincides with the centralizer of the adjoint representation of the Lie algebra $\mathfrak{g}$. Suppose that in $\mathfrak{g}$ we can choose N linearly independent (over C) elements $e_1, \dots, e_N$. Then, by the irreducibility of the adjoint representation of $\mathfrak{g}$ and by the density theorem (see §53), for any elements $v_1, \dots, v_N$ we can choose operators $a_1, \dots, a_N \in \operatorname{End}_K \mathfrak{g}$ belonging to a subalgebra of U, generated by $\operatorname{ad} \mathfrak{g}$, for which

$$a_i e_j = \delta_{ij} v_j \qquad (i, j = 1, \dots, N). \tag{48.10}$$

Since $\mathfrak{g}$ coincides with its commutator subalgebra, we have $(\operatorname{ad}\mathfrak{g})^{m+1} \supset (\operatorname{ad}\mathfrak{g})^m$ $(m = 1, 2, \dots)$. Therefore, from identities (48.3), where $N, \beta_1, \dots, \beta_\sigma$ are fixed and $m = 1, 2, \dots$, it follows that for any $b_1, \dots, b_N \in U \subset \operatorname{End}_K \mathfrak{g}$ the equality

$$\sum_{\sigma \in S_N} \beta_\sigma [b_1 e_{\sigma(1)}, \dots, b_N e_{\sigma(N)}] = 0$$

holds in $\mathfrak{g}$. Putting $b_i = a_i$ $(i = 1, \dots, N)$ in it and taking into account the equalities (48.10), we obtain $\beta_1[v_1, \dots, v_N] = 0$ for any $v_1, \dots, v_N \in \mathfrak{g}$. Since $\beta_1 \neq 0$ by the assumption of the lemma, this implies that the algebra $\mathfrak{g}$ is nilpotent, which contradicts its simplicity. Hence, any N elements of $\mathfrak{g}$ are linearly dependent over C. The lemma is proved.

As a result we obtain the following criterion.

THEOREM 48.1. *Let $\mathfrak{g}$ be a simple Lie algebra over an arbitrary field K in which identity* (48.4) *holds. Then $\mathfrak{g}$ is a Lie algebra of Cartan type if and only if either $\mathfrak{g}$ is finite-dimensional over its centroid C or $\mathfrak{g}$ possesses elements of nonfinite type. Moreover, if the algebra $\mathfrak{g}$ cannot be embedded in a Lie algebra of the form $\widetilde{W}_n(\widehat{K})$, where $\widehat{K}$ is an extension of the field C, then $\mathfrak{g}$ has a basis of elements of finite type whose degree is bounded, and either the Lie algebra $\mathfrak{g}$ is finite-dimensional over C or it is not a Lie algebra of Cartan type.*

48.3. The proof of Theorem 42.1.

LEMMA 48.6. *Let $\mathfrak{g}$ be a simple Lie algebra satisfying an identity of the form*

$$\sum_{\sigma \in S_M} \beta'_\sigma [y_1, x_{\sigma(1)}, y_2, x_{\sigma(2)}, \dots, y_M, x_{\sigma(M)}] = 0. \tag{48.11}$$

Then there exists a finite set of elements $e_1, \dots, e_l$ of $\mathfrak{g}$ for which we have the equality

$$\mathfrak{g} = [e_1, \mathfrak{g}] + \dots + [e_l, \mathfrak{g}]. \tag{48.12}$$

PROOF. Among the identities of the form (48.11) satisfied in the algebra $\mathfrak{g}$ we can choose an identity for which M is minimal and $\beta'_1 \neq 0$. Then this identity can be represented in the form

$$\sum_{i=1}^{N} [f_i(x_1, \dots, \widehat{x}_i, \dots, x_M, y_1, \dots, y_{M-1}), y_M, x_i] = 0, \tag{48.13}$$

where f_i are Lie multilinear polynomials and the identity $f_M = 0$ does not hold in $\mathfrak{g}$. Since the subspace of $\mathfrak{g}$ spanned by all values of any multilinear polynomial in $\mathfrak{g}$ is an ideal of $\mathfrak{g}$, we have $f_M|_{\mathfrak{g}} = \mathfrak{g}$. Therefore, a finite set of elements $g_1, \dots, g_k$ whose existence is stated in Lemma 48.1 can be chosen among the values of the polynomial $f_M|_{\mathfrak{g}}$. Let us assume that

$$g_i = f_M\big|_{x_1=e_{i1},\dots,x_{M-1}=e_{i,M-1},y_1=e'_{i1},\dots,y_{M-1}=e'_{i,M-1}}.$$

Then the substitution of the elements $e_{i1}, e'_{i1}, e_{i2}, e'_{i2}, \dots, e_{i,M-1}, e'_{i,M-1}$ instead of the variables $x_1, y_1, x_2, y_2, \dots, x_{M-1}, y_{M-1}$ and arbitrary elements of $\mathfrak{g}$ instead of y_M and x_M transforms identity (48.13) into an equality in $\mathfrak{g}$ which implies the inclusion

$$[g_i, \mathfrak{g}, \mathfrak{g}] \subset \sum_{j=1}^{M-1} [\mathfrak{g}, e_{ij}] \qquad (i = 1, \dots, k).$$

A comparison of these inclusions with equality (48.1) shows that the elements

$$e_{ij} \qquad (i = 1, 2, \dots, k;\ j = 1, 2, \dots, M-1)$$

can be taken for the desired elements $e_1, \dots, e_l$. The lemma is proved.

THEOREM 48.2. *Let $\mathfrak{g}$ be a simple Lie algebra over an arbitrary field K of characteristic zero for which identities* (48.4) *and* (48.11) *are satisfied for some $N \in \mathbb{N}$. If the algebra $\mathfrak{g}$ cannot be embedded in the Lie algebra $\widetilde{W}_N(\widehat{K})$ for any extension $\widehat{K}$ of the centroid C of the algebra $\mathfrak{g}$, then for some integer k we have the inclusion*

$$(\mathrm{ad}\,\mathfrak{g})^k \subset \sum_{i=1}^{k-1} (\mathrm{ad}\,\mathfrak{g})^i. \tag{48.14}$$

PROOF. Let $e_1, \dots, e_l$ be elements satisfying equality (48.12) whose existence is ensured by Lemma 48.6, and let C be the centroid of the Lie algebra $\mathfrak{g}$. Let the Lie algebra $\mathfrak{h}$ be the quotient algebra of the free product of the Lie C-algebra $\mathfrak{g}$ with the free Lie C-algebra $\mathfrak{m}$ on a countable set of the free generators $\{x_i \mid i = 1, 2, \dots\}$ by the ideal of identities determined by the identities (48.4) and (48.11). Obviously, the Lie algebra $\mathfrak{g}$ is embedded in $\mathfrak{h}$. Since the simple Lie algebra $\mathfrak{g}$ cannot be embedded in $\widetilde{W}_N(\widehat{K})$ for any extension $\widehat{K} \supset C$, by Corollary 2 of Theorem 47.1 all elements of $\varphi_{a,n}(\mathrm{id}_{\mathfrak{h}}[\mathfrak{h}, \mathfrak{g}])$ are nilpotent in the quotient algebra $S(\mathfrak{h})/J_{a,n}(\mathfrak{h})$, where $a = \sum_{\sigma \in S_{N+1}} \beta_\sigma \sigma$, and $\beta_\sigma \sigma$ are the coefficients participating in the notation of identity (48.4). In particular, the element $[x_1, e_1] + \dots + [x_l, e_l] \in \mathrm{id}_{\mathfrak{g}}([\mathfrak{h}, \mathfrak{g}])$ is nilpotent. Then for some integer $k \in \mathbb{N}$ we have the following identity in the algebra $S(\mathfrak{h})$:

$$([x_1, e_1] + \dots + [x_l, e_l])^k = \sum_i a_i b_i, \tag{48.15}$$

where b_i are generators of the ideal $J_{a,n}(\mathfrak{h})$ of the form (47.1) and a_i are homogeneous elements of the graded algebra $S(\mathfrak{h})$ of degree $k - N - 1$. Comparing identity (48.4) with the element (47.1) for $n = N$ we see that relation (48.15) implies the following inclusion in the adjoint representation of the Lie algebra $\mathfrak{h}$:

$$\big(\operatorname{ad}([x_1, e_1] + \cdots + [x_l, e_l])\big)^k \in \sum_{i=1}^{k-1} (\operatorname{ad}\mathfrak{h})^i.$$

Since by Lemma 48.6 any element $g \in \mathfrak{g}$ can be represented in the form $g = [g_1, e_1] + \cdots + [g_l, e_l]$ ($g_i \in \mathfrak{g}$) and any mapping $x_1 \to g_1, \ldots, x_l \to g_l$ can be extended to a homomorphism $\mathfrak{h} \to \mathfrak{g}$ that is the identity on $\mathfrak{g}$ and for which $[x_1, e_1] + \cdots + [x_l, e_l]$ is transformed into g, we have

$$(\operatorname{ad} g)^k \in \sum_{i=1}^{k-1} (\operatorname{ad}\mathfrak{g})^i$$

for any element $g \in \mathfrak{g}$. Since the ground field K is infinite, the linearization of the left-hand side of this inclusion gives

$$\sum_{\sigma \in S_k} \operatorname{ad} v_{\sigma(1)} \cdots \operatorname{ad} v_{\sigma(k)} \in \sum_{i=1}^{k-1} (\operatorname{ad}\mathfrak{g})^i,$$

whence it follows that

$$k!\, \operatorname{ad} v_1 \cdots \operatorname{ad} v_k \in \sum_{i=1}^{k-1} (\operatorname{ad}\mathfrak{g})^i$$

for any $v_1, \ldots, v_k \in \mathfrak{g}$. This proves (48.14). The theorem is proved.

LEMMA 48.7. *Suppose that in a simple Lie algebra $\mathfrak{g}$ we have the inclusion*

$$(\operatorname{ad}\mathfrak{g})^k \in \sum_{i=1}^{k-1} (\operatorname{ad}\mathfrak{g})^i, \tag{48.16}$$

and the identity

$$0 = \sum_{\sigma \in S_N} \beta_\sigma(k)[[x_{\sigma(1)}, y_{11}, \ldots, y_{1,k-1}] \cdots [x_{\sigma(N)}, y_{N1}, \ldots, y_{N,k-1}]]. \tag{48.17}$$

Then the Lie algebra $\mathfrak{g}$ is finite-dimensional over its centroid C and $\dim_C \mathfrak{g} \le N$.

PROOF. Since the commutator subalgebra of the Lie algebra $\mathfrak{g}$ coincides with $\mathfrak{g}$, we have $(\operatorname{ad}\mathfrak{g})^2 \supset \operatorname{ad}\mathfrak{g}$. Therefore, in $\mathfrak{g}$ we have the equalities

$$\sum_{\sigma \in S_N} \beta_\sigma(k)[[b_1 e_{\sigma(1)}], \ldots, [b_N e_{\sigma(N)}]] = 0, \tag{48.18}$$

where $e_1, \ldots, e_N$ are arbitrary elements of the Lie algebra $\mathfrak{g}$ and $b_1, \ldots, b_N$ range over the subspace $\sum_{i=1}^{k-1} (\operatorname{ad}\mathfrak{g})^i$. Inclusion (48.16) shows that then equality (48.18) holds in $\mathfrak{g}$ for any $b_1, \dots, b_N \in \sum_{i=1}^{\infty}(\operatorname{ad}\mathfrak{g})^i$. Therefore, the assertion of the lemma is an immediate consequence of Lemma 48.5. The lemma is proved.

Now the proof of Theorem 42.1 is obvious. Indeed, by Lemma 48.2 for a simple Lie K-algebra from a variety of exponential type for any integer m and for some $N = N(m)$, $\beta_\sigma = \beta_\sigma(m)$ and $\beta'_\sigma = \beta'_\sigma(m)$ a system of nontrivial identities of the form (48.3) and (48.3′) is satisfied. Since the Lie algebra $\mathfrak{g}$ coincides with its commutator subalgebra, it follows from identities (48.3) for $m = 2$ that in the algebra $\mathfrak{g}$ a nontrivial identity of the form (48.4) holds, where $N = N(2)$ and $\beta_\sigma = \beta_\sigma(2)$. Similarly, it follows from identities (48.3′) for $m = 2$ that in the algebra $\mathfrak{g}$ a nontrivial identity of the form (48.11) holds, where $N = N(2)$ and $\beta'_\sigma = \beta'_\sigma(2)$. Then applying Lemma 48.6 and Theorem 48.2 to the Lie algebra $\mathfrak{g}$ we find that if the Lie algebra $\mathfrak{g}$ is not a Lie algebra of Cartan type, moreover, if $\mathfrak{g}$ cannot be embedded in $\widetilde{W}_{N(2)}(\widehat{K})$ for an algebraically closed field $\widehat{K} \supset K$ of cardinality more than $\dim_c \mathfrak{g}$, then for some $k \in \mathbb{N}$ the inclusion (48.14) holds in the adjoint representation of the algebra $\mathfrak{g}$. Since in the algebra $\mathfrak{g}$ a nontrivial identity of the form (48.3) is satisfied for $m = k - 1$, we can apply Lemma 48.7 to the algebra $\mathfrak{g}$ and finally find that if a simple algebra $\mathfrak{g}$ cannot be embedded in $\widetilde{W}_{N(2)}(\widehat{K})$, then $\mathfrak{g}$ is a finite-dimensional algebra over its centroid. Therefore, the Lie $\widehat{K}$-algebra $\widehat{\mathfrak{g}} = \widehat{K} \otimes_C \mathfrak{g}$ must contain a $\widehat{K}$-subalgebra of finite codimension. Theorem 42.1 is proved.

REMARK 48.2. The only point in the proof of Theorem 42.1 where the zero characteristic of the ground field has been used was in passing from inclusion (48.8) to inclusion (48.14) (see the proof of Theorem 48.2).

48.4. The proof of Theorem 42.2. From Proposition 43.1 it follows that for the variety var $\widetilde{W}_n(K)$ identities of the form (48.3) and (48.3′) hold for each $m \in \mathbb{N}$ with $\beta'_\sigma = \beta_\sigma = \operatorname{sign}(\sigma) \in \mathbb{Z}$. Since in the proof of Theorem 42.1 we have used only identities of the form (48.3) and (48.3′), it follows that in the variety var $\widetilde{W}_n(K)$ ($\operatorname{char} K = 0$) all simple Lie algebras are Lie algebras of Cartan type. Therefore, if a simple Lie algebra $\mathfrak{g} \in$ var $\widetilde{W}_n(K)$ is infinite-dimensional over its centroid C, then, by Corollary 2 of Theorem 43.2, $\mathfrak{g}$ can be embedded in the Lie algebra $\widetilde{W}_n(\widehat{K})$, where $\widehat{K}$ is an extension of the field K. In order to prove Theorem 42.2, it remains to show that any simple Lie algebra that is finite-dimensional over its centroid C is of dimension $\leq n^2+2n$. However, this follows from Lemma 43.1 and from the next assertion proved in §18.

THEOREM 48.3. *Let $\mathfrak{g}$ be a centrally simple finite-dimensional algebra over an algebraically closed field $\widehat{K}$ of characteristic zero, and let $m = \dim_{\widehat{K}} \widehat{\mathfrak{g}}$. Then there exists a Lie polynomial*

$$f(z, y_{11}, \dots, y_{1m}, \dots, y_{k1}, \dots, y_{km}), \tag{48.19}$$

that is skew-symmetric with respect to any set of variables of the form $y_{i1}, \dots, y_{im}$ ($i = 1, \dots, k$) and takes nonzero values on $\widehat{\mathfrak{g}}$.

Indeed, if $\dim_C \mathfrak{g} = m$ and $m > n^2+2n$, then we have $\dim_{\widehat{K}} \widehat{\mathfrak{g}} = m$ for the algebraic closure $\widehat{K}$ of the field C as well, where $\widehat{\mathfrak{g}} = \widehat{K} \otimes_C \mathfrak{g}$, and $\widehat{\mathfrak{g}}$ is a centrally simple Lie algebra over $\widehat{K}$. Then, by Theorem 48.3, some Lie polynomial of the form (48.19) does not vanish on $\widehat{\mathfrak{g}}$. On the other hand, all these polynomials vanish on $\widehat{\mathfrak{g}}$ by Lemma 43.1. The theorem is proved.

The proof of Corollary of Theorem 42.2 now follows from Corollary 2 of Theorem 44.2.

Comments

The study of simple infinite-dimensional Lie algebras containing a proper subalgebra of finite codimension was originated in the papers of É. Cartan, who also proved Theorem 43.1 mentioned in §43 (see [Car]). A modern exposition of this theorem can be found in the papers [Vei, Kats, SiS]. The interest to Lie algebras of this kind rapidly increases at the end of sixties after the paper of Shafarevich and Kostrikin [KoSh] in which simple finite-dimensional Lie algebras over fields of positive characteristic were systematized. This systematization clarified relationships between these simple algebras and the Lie algebras studied by Cartan and it leads in succession to the proof of the Kostrikin–Shafarevich conjecture for simple Lie p-algebras by Block and Wilson (see [BW2]). In 1975 the author and Sumenkov [Sum] noticed that in simple Lie algebras of Cartan type nontrivial Lie identities hold. This leads to the replacement of the conjecture that any simple Lie algebra with an identity over its centroid is finite-dimensional, which existed at that time, by the conjecture stated in §42. Theorem 12.2, which has been proved in the paper [Raz18], gives a possibility to trace a way to justify this conjecture for the varieties var $W_n(K)$ and to prove Theorems 46.1–46.4 (see [Raz22]). Special cases of Theorem 42.3 were published in the author's papers [Raz21, Raz22]. Theorems 42.3 and 46.1–46.3 complete the classical Frobenius theorem (see [BiK]). The results of §45 are originated from Cartan's paper; their precise exposition including the case of fields of positive characteristic is given by Block and Wilson [BW1]. I announced Theorems 42.1 and 42.2 in the paper [Raz21].

PROBLEM 6.1. Prove the assertions of Theorems 42.1 and 42.2 for the case of fields of positive characteristic. (Consider the special case: var $W_1(K)$ with char $K = 2$.)

PROBLEM 6.2. Whether there exist nonfinite-dimensional centrally simple Lie algebras with an identity which consist of the elements of finite type with respect to the adjoint representation? Whether there exist simple Engel Lie algebras over fields of positive characteristic?

CHAPTER VII

ALGEBRAIC SUPPLEMENTS

§49. Generalities of the theory of universal algebras and algebras with several supports

49.1. Binary algebras. A linear space A over a field K with a fixed bilinear operation $*: A \otimes_K A \to A$ on it is called a *(binary) algebra*. The result $*(a \otimes b)$ of the application of this operation to elements $a, b \in A$ is usually denoted by $a * b$. A linear subspace R of A is called a *K-subalgebra* of the algebra A if R itself is a K-algebra with respect to the operation $*$, i.e., $a * b \in R$ for any $a, b \in R$. Let S be an arbitrary subset of A; the least K-subalgebra of A containing S will be denoted by $\mathrm{alg}_A S$ and will be called the *K-subalgebra generated by the set S*.

A linear subspace I of A is called an *ideal* of the K-algebra A if $i * a, a * i \in I$ for any elements $i \in I$ and $a \in A$. For any ideal I the operation $*$ on the K-algebra A induces an operation $*$ on the quotient space A/I. To this end, we must put $(a + I) * (b + I) \rightleftharpoons ab + I$. The quotient space A/I with the operation so defined is called the *quotient algebra* of the K-algebra A by the ideal I.

Let A_1 and A_2 be linear spaces over a field K, each with its own binary bilinear operation, both denoted by the same symbol $*$. Then a mapping $\varphi: A_1 \to A_2$ is called an algebra *homomorphism* if φ is K-linear and $\varphi(a * b) = \varphi(a) * \varphi(b)$ for each $a, b \in A_1$. If $\varphi(A_1) = A_2$, then the homomorphism φ is called an *epimorphism*. If φ is a one-to-one epimorphism we say that φ is an *isomorphism* and that the K-algebras A_1 and A_2 are isomorphic; notation: $A_1 \simeq A_2$.

For any homomorphism of K-algebras $\varphi: A_1 \to A_2$ the set $\operatorname{Ker}\varphi = \{a \in A_1 \mid \varphi(a) = 0\}$ forms an ideal in the K-algebra A_1; it is called the *kernel of the homomorphism φ*. The subset $\operatorname{Im}\varphi = \{\varphi(a) \mid a \in A_1\}$ is a K-subalgebra of the algebra A_2; it is called the *image of the homomorphism φ*. It is clear that φ is an isomorphism if and only if $\operatorname{Im}\varphi = A_2$ and $\operatorname{Ker}\varphi$ is the zero ideal of A_1.

Obviously, the mapping $\psi: A \to A/I$ such that $\psi(a) = a + I$ determines an epimorphism of the K-algebra A onto the quotient algebra A/I of the algebra A by the ideal I of it, and the kernel of ψ is the ideal I. This homomorphism is called *canonical*.

THEOREM 49.1 (Theorem on homomorphisms). *For any K-algebra homomorphism $\varphi: A_1 \to A_2$ and any subalgebra R of A:*

a) $\operatorname{Im}\varphi = \varphi(A_1) \simeq A_1/\operatorname{Ker}\varphi$;

b) $\varphi(R) \simeq R/R \cap \operatorname{Ker}\varphi$.

For any K-algebra A and any ideals $J \supseteq I$ in it and a K-subalgebra R:

c) *J/I is an ideal of A/I and $(A/I)/(J/I) \simeq A/J$;*

d) $\psi(R) \simeq R/(I \cap R)$ *under the canonical homomorphism* $\psi\colon A \to A/I$.

The proof of this assertion follows immediately from the definitions given above.

An algebra A is called a *simple* K-algebra if $A * A \neq 0$ and any K-ideal of A coincides either with A or with the zero ideal.

If in an algebra A there is a countable system of subspaces V_i $(i = 0, \pm 1, \dots, \pm n, \dots)$ for which $A = \oplus_{i=-\infty}^{+\infty} V_i$ and $V_i * V_j \subseteq V_{i+j}$ (i and j are arbitrary integers), then it is said that a *gradation* V_i is given on the algebra A, and the algebra A is called *graded*.

A system of pairwise embedded K-subspaces L_i $(i = 0, \pm 1, \dots)$ of the algebra A is called a *descending* (*ascending*) filtration of the algebra A if $L_{i+1} \subseteq L_i$ (respectively, $L_{i+1} \supseteq L_i$) and $A = \cup_{i=-\infty}^{+\infty} L_i$ & $L_i * L_j \subseteq L_{i+j}$ (i and j are arbitrary integers). The filtration is called *exhaustive* if $\cap_{i=-\infty}^{+\infty} L_i = \{0\}$.

For any gradation $\{V_i\}$ of an algebra A an exhaustive ascending filtration $\{L_i\}$ can be constructed by setting $L_i \rightleftharpoons \sum_{k=i}^{\infty} V_k$. In turn, for any ascending filtration $\{L_i\}$ of the algebra A we can construct an associated graded algebra $\operatorname{Gr} A = \oplus_{i=-\infty}^{+\infty} L_i/L_{i+1}$ with gradation $V_i = L_i/L_{i+1}$ and operation $*$ such that for any $v_i \in V_i$ and $v_j \in V_j$ the element $v_i * v_j$ is equal (by definition) to $\widetilde{v}_i * \widetilde{v}_j + L_{i+j+1}$, where $v_i = \widetilde{v}_i \pmod{L_{i+1}}$ and $v_j = \widetilde{v}_j \pmod{L_{j+1}}$.

Let X be an arbitrary nonempty set. Denote by $\Gamma(X)$ the *free groupoid* whose elements are defined by induction via the following rules:

a) the elements of the set X belong to $\Gamma(X)$,

b) if $g_1, g_2 \in \Gamma(X)$, then $(g_1 g_2) \in \Gamma(X)$.

Let us introduce on $\Gamma(X)$ an operation $*$ by setting $g_1 * g_2 = (g_1 g_2)$. We denote by $K\Gamma(X)$ the linear space over the field K whose basis consists of all elements of $\Gamma(X)$ and extend the operation $*$ by bilinearity from $\Gamma(X)$ to $K\Gamma(X)$:

$$\left(\sum_u \beta_u u\right) * \left(\sum_v \beta'_v v\right) = \sum_{u,v} \beta_u \beta'_v (uv) \qquad (u, v \in \Gamma(X),\ \beta_u, \beta'_v \in K),$$

where under the summation sign there is only a finite set of nonzero β_u and β'_v. The algebra $K\Gamma(X)$ with the operation thus defined is called the *algebra of nonassociative polynomials*.

DEFINITION 49.1. A K-algebra F is called an (*absolutely*) *free K-algebra* if in it there exists a system Y of generators such that any mapping $\varphi\colon Y \to A$ into an arbitrary K-algebra A can be extended (uniquely) to a homomorphism $\psi\colon F \to A$ $(\psi|_Y = \varphi)$. In this case Y is called the *set of free generators* of the K-algebra F.

PROPOSITION 49.1. *The K-algebra $K\Gamma(X)$ of nonassociative polynomials is absolutely free and X is its set of free generators.*

PROOF. Let φ be an arbitrary mapping of X into a K-algebra A. The inductive definition of the set $\Gamma(X)$ shows that there exists a unique method of extending the mapping φ to a mapping $\psi\colon \Gamma(X) \to A$ compatible with the operation $*$. This method is determined by the following inductive rules:

a) if $g \in X$, then $\psi(g) = \varphi(g)$,

b) if $g_1, g_2 \in \Gamma(X)$ and $\psi(g_1)$ and $\psi(g_2)$ have already been defined, then $\psi((g_1 g_2))$ coincides with $\psi(g_1) * \psi(g_2)$.

Thus, the K-linear mapping $\psi\colon K\Gamma(X) \to A$ is well defined on the basis $\Gamma(X)$ of the K-algebra $K\Gamma(X)$. These rules imply $\psi|_X = \varphi$ and $\psi((g_1g_2)) = \psi(g_1) * \psi(g_2)$ $(g_i \in K\Gamma(X))$. Therefore, ψ is a homomorphism of K-algebras. Proposition 49.1 is proved.

Let $f = f(x_1, \dots, x_l)$ be a nonassociative polynomial from $K\Gamma(X)$, where $x_1, \dots, x_l$ are all those variables from the set X that occur in its representation in the form $\beta_1 g_1 + \dots + \beta_l g_l$ $(\beta_i \in K, g_i \in \Gamma(X))$, and let $a_1, \dots, a_l$ be arbitrary elements of some K-algebra A. Let us denote by

$$f(a_1, \dots, a_l), \qquad f|_{x_1=a_1,\dots,x_l=a_l}$$

the element $\psi(f)$ of the K-algebra A, where ψ is the homomorphism from $K\Gamma(X)$ to A for which $\psi(x_i) = a_i$ $(i = 1, \dots, l)$. This element is called the *result of substitution* (or a *specialization*) of the elements $a_1, \dots, a_l$ into the polynomial f. It follows from Proposition 49.1 that the result of a substitution is determined uniquely by the polynomial f and $a_1, \dots, a_l \in A$. It is said that the *identity* $f(x_1, \dots, x_l) = 0$ holds in the algebra A if $f(a_1, \dots, a_l) = 0$ for any $a_1, \dots, a_l \in A$.

Let S be an arbitrary subset of nonassociative polynomials of $K\Gamma(X)$. The class of all binary algebras for which all identities $f = 0$ hold, where f ranges over S, is called a *variety*, and it is said that this variety is *given by the set of identities* $\{f = 0 \mid f \in S\}$.

49.2. Algebras of arbitrary signature. Most classes of objects treated in algebra are varieties of binary algebras. For example, the associative algebras are distinguished by the identity $(x_1 * x_2) * x_3 = x_1 * (x_2 * x_3)$, the Lie algebras by the two identities

$$x_1 * x_1 = 0, \qquad (x_1 * x_2) * x_3 + (x_2 * x_3) * x_1 + (x_3 * x_1) * x_2 = 0,$$

the alternative algebras by the identities

$$(x_1 * x_2) * x_2 - x_1 * (x_2 * x_2) = 0,$$
$$(x_1 * x_1) * x_2 - x_1 * (x_1 * x_2) = 0.$$

Similarly, identities of small degrees distinguish the classes of right alternative algebras, Jordan algebras, and so on. However, in the study of these classical varieties of algebras we have to use the language of varieties in a somewhat more general situation and to consider algebras with greater collection of operations, which cannot formally be kept within the scheme of binary algebras. For example, associative and commutative differential algebras (see §51 and Chapter 6) are objects of this kind.

Suppose that for any positive integer $n \in \mathbb{N}$ a certain subset $\Omega(n)$ (possibly empty) is chosen in the set of all K-multilinear operations $\mathrm{Hom}_K(A^{\otimes n}, A)$ on a linear K-space A and a certain name is assigned to each operation from $\Omega = \cup_{n=0}^{+\infty}\Omega(n)$. In this case the linear space A endowed with the operations $\omega \in \Omega$,

$$\omega\colon A \otimes_K \dots \otimes_K A \to A, \qquad (\omega \in \Omega(n)),$$

is called a *K-algebra of signature* Ω, and the number n is said to be the '*arity*' of the operation ω. The result $\omega(a_1 \otimes \dots \otimes a_n)$ of applying the operation ω to the elements

$a_1, \ldots, a_n \in A$ is usually denoted by $\omega(a_1, \ldots, a_n)$. A linear subspace R of A is called a *K-subalgebra* of the algebra *A of signature* Ω if $\omega(r_1, \ldots, r_n) \in R$ for any $\omega \in \Omega$ and $r_1, \ldots, r_n \in R$. A linear subspace I of A is called an *ideal* of the K-algebra A of signature Ω if for any $\omega \in \Omega$, $a_1, \ldots, a_n \in A$, and $v \in I$ all elements of the form

$$\omega(a_1, \ldots, a_{j-1}, v, a_{j+1}, \ldots, a_n) \qquad (j = 1, \ldots, n;\ \omega \in \Omega(n))$$

belong to I. It is clear that any subalgebra R and any ideal I of A are themselves K-algebras of signature Ω with respect to the operations of Ω restricted to R and I (here it is understood that the new operations preserve the old names). For any ideal I an arbitrary operation $\omega \in \Omega(n)$ on A induces the operation ω on the quotient space A/I for which

$$\omega(a_1 + I, \ldots, a_n + I) \rightleftharpoons \omega(a_1, \ldots, a_n) + I.$$

Thus, the quotient space A/I is endowed with the structure of a K-algebra of the same signature Ω. This algebra is called the *quotient algebra* of the K-algebra A of signature Ω by the ideal I.

Suppose that in linear spaces A_1 and A_2 over a field K multilinear operations with their names are fixed so that the list of operation names for A_1 coincides with the list of operation names for A_2 and the 'arities' of the operations with the same names in A_1 and A_2 coincide. In this case we can assume that the K-algebras A_1 and A_2 have the same signature Ω. In this case a K-linear mapping $\varphi\colon A_1 \to A_2$ is called a *homomorphism* of K-algebras of signature Ω if for any operation $\omega \in \Omega(n)$ $(n = 0, 1, \ldots)$ and any elements $a_1, \ldots, a_n \in A_1$ the equality

$$\varphi\omega(a_1, \ldots, a_n) = \omega(\varphi(a_1), \varphi(a_2), \ldots, \varphi(a_n))$$

holds in A_2. A homomorphism φ is called a *monomorphism* if the equality $\varphi(a) = 0$ in A_2 is possible only when a is the zero element. It is immediate that

$$\operatorname{Ker}\varphi = \{a \in A_1 \mid \varphi(a) = 0\}, \qquad \operatorname{Im}\varphi = \{\varphi(a) \mid a \in A_1\}$$

are an ideal and a K-subalgebra in A_1 and A_2, respectively, and the mapping $\psi\colon A \to A/I$ for which $\psi(a) = a + I$ is an epimorphism of the K-algebra A of signature Ω onto A/I and $\operatorname{Ker}\psi = I$.

The theorem on homomorphisms (49.1) can be extended to K-algebras of arbitrary signature Ω without any modifications.

Let us consider the class of all K-algebras of a fixed signature Ω. The fixation of a signature means that for each name from Ω we can determine the 'arity' of the operation corresponding to this name. Therefore, given a set Ω of such names, we can construct without difficulties a universal object in the class of all K-algebras of signature Ω, using the case of binary algebras discussed above.

Let X be an arbitrary nonempty set. Denote by $\Gamma_\Omega(X)$ a *free groupoid of signature* Ω whose elements are defined by induction via the following rules:

a) the elements of the set X belong to $\Gamma_\Omega(X)$,

b) if ω is an arbitrary name of an operation from Ω and $\omega \in \Omega(n)$ and $g_1, \ldots, g_n \in \Gamma_\Omega(X)$, then $\omega(g_1, \ldots, g_n) \in \Gamma_\Omega(X)$.

For each name $\omega \in \Omega$ we introduce on $\Gamma_\Omega(X)$ an operation

$$\omega\colon \underbrace{\Gamma_\Omega(X) \times \cdots \times \Gamma_\Omega(X)}_{n \text{ times}} \to \Gamma_\Omega(X),$$

which assigns an element $\omega(g_1, \dots, g_n)$ to arbitrary $g_1, \dots, g_n \in \Gamma_\Omega(X)$. Denote by $K\Gamma_\Omega(X)$ the linear space over the field K whose basis consists of all elements of the groupoid $\Gamma_\Omega(X)$. The extension of all operations $\omega \in \Omega$ from the groupoid $\Gamma_\Omega(X)$ to the space $K\Gamma_\Omega(X)$ by multilinearity endows $K\Gamma_\Omega(X)$ with the structure of a K-algebra of signature Ω. This algebra is called the *polynomial algebra* of signature Ω.

A K-algebra F of signature Ω is called an *absolutely free K-algebra* of signature Ω if in it there exists a system of generators Y such that any mapping $Y \to A$ into an arbitrary K-algebra A of the same signature Ω can be extended to a homomorphism of K-algebras $F \to A$. The set of generators Y of this kind is called the *set of free generators*.

PROPOSITION 49.2. *The polynomial K-algebra $K\Gamma_\Omega(X)$ of signature Ω is absolutely free and X is its set of free generators.*

The proof of this proposition repeats the proof of Proposition 49.1.

Let $X = \{x_1, x_2, \dots\}$ be a countable set. An expression of the form $f = 0$ ($f \in K\Gamma_\Omega(X)$) is called an *identity*. It is said that *for a K-algebra A* of signature Ω *the identity $f(x_1, \dots, x_l) = 0$ holds* if $f(a_1, \dots, a_l) = 0$ in A for any $a_1, \dots, a_l \in A$. As in the case of binary algebras, any subset S of $K\Gamma_\Omega(X)$ permits one to single out the class $\mathfrak{B} = \mathfrak{B}(S)$ of all K-algebras of signature Ω in which the identities $f = 0$, where f ranges over S, are valid. A class $\mathfrak{B}$ of this sort is called a *variety of K-algebras of signature* Ω. Let $\{A_i \mid i \in I\}$ be a set of K-algebras of signature Ω. Denote by the letter A the linear space that is the full Cartesian product $\prod_{i\in I} A_i$ of the linear spaces A_i. Its elements are arbitrary functions a on the set I for which $a(i) \in A_i$. By setting

$$(\omega(a_1, \dots, a_n))(i) \rightleftharpoons \omega(a_1(i), \dots, a_n(i)), \tag{49.1}$$

for each operation $\omega \in \Omega(n)$ and any elements $a_1, \dots, a_n \in A$, we endow the linear space A with the structure of a K-algebra of signature Ω. This algebra is called *the full Cartesian product of K-algebras A_i $(i \in I)$ of signature* Ω and is again denoted by $\prod_{i\in I} A_i$. From formula (49.1) it follows that if in all algebras A_i $(i \in I)$ an identity $f = 0$ holds, then this identity holds also in their full Cartesian product. It is clear also that the validity of an identity $f = 0$ in some algebra A' implies its validity in any subalgebra and in any homomorphic image of the algebra A'. Therefore, any variety $\mathfrak{B}$ of K-algebras of signature Ω contains the full Cartesian products, subalgebras, and homomorphic images of the algebras contained in $\mathfrak{B}$ (or, as it is usual to say, $\mathfrak{B}$ is closed with respect to the operations of taking Cartesian products, subalgebras and homomorphic images.

THEOREM 49.2 (Birkhoff). *A class $\mathfrak{B}$ of K-algebras of signature Ω is a variety if and only if it is nonvoid and closed with respect to the operations of taking subalgebras, homomorphic images, and full Cartesian products.*

PROOF. Denote by the letter S the subset of all polynomials f from $K\Gamma_\Omega(X)$ for which the identity $f = 0$ holds in all algebras of the class $\mathfrak{B}$. Let R be an arbitrary algebra of signature Ω, in which all identities of the form $f = 0$ ($f \in S$) are valid. In order to prove the theorem, it suffices to show that if the class $\mathfrak{B}$ is closed with respect to the three operations mentioned above, then R belongs to S.

Denote by the letter P the set $K\Gamma_\Omega(X) \setminus S$ and by the letter Y an arbitrary set. For any polynomial $g \in P$ there is an algebra in the class $\mathfrak{B}$ for which the identity $g = 0$ does not hold. Denote this algebra by A_g. Let $\Phi(g)$ be the set of all homomorphisms from the free algebra $K\Gamma_\Omega(X)$ into A_g, let $\Phi = \cup_{g\in P}\Phi(g)$, and let C_φ be the image of the algebra $K\Gamma_\Omega(X)$ under the homomorphism $\varphi \in \Phi$. It is clear that any algebra C_φ is contained in some algebra A_g as a subalgebra and, therefore, the algebras C_φ and their Cartesian product $C = \prod_{\varphi\in\Phi} C_\varphi$ belong to the class $\mathfrak{B}$. For any element $y \in Y$ denote by $\overline{y}$ the element of the algebra C for which $\overline{y}(\varphi) = \varphi(y)$ and generate a subalgebra F of C by the set $\overline{Y} = \{\overline{y} \in C \mid y \in Y\}$. The algebra F belongs to the calss $\mathfrak{B}$ as well.

Let $\overline{\psi}$ be an arbitrary mapping of the set Y into R and let $\psi\colon K\Gamma_\Omega(Y) \to R$ and $\psi_0\colon K\Gamma_\Omega(Y) \to F$ be the homomorphisms of K-algebras of signature Ω for which $\psi(y) = \overline{\psi}(\overline{y})$ and $\psi_0(y) = \overline{y}$ (the existence of these homomorphisms follows from Proposition 49.2). Suppose that there exists a polynomial $f(y_1, \dots, y_l) \in K\Gamma_\Omega(Y)$ for which

$$f \in \operatorname{Ker}\psi \ \ \& \ \ f \notin \operatorname{Ker}\psi_0.$$

Then $f(\overline{y}_1, \dots, \overline{y}_l) = 0$ in the algebra $F \subset C$. But then by construction

$$0 = (f(y_1, \dots, y_l))(\varphi) = f(\varphi(y_1), \dots, \varphi(y_l))$$

for any homomorphism $\varphi \in \Phi$. This shows that the polynomial $f(x_1, \dots, x_l) \in K\Gamma_\Omega(X)$ is not in P and, hence, $f \in S$. However, in this case $f = 0$ is an identity of the algebra R, which contradicts the fact that $f(\psi(y_1), \dots, \psi(y_l)) \neq 0$ in R. The contradiction thus obtained shows that $\operatorname{Ker}\psi \supset \operatorname{Ker}\psi_0$, and, by the theorem on homomorphisms, there exists a homomorphism $\overline{\psi}\colon F \to R$ for which $\overline{\psi}\psi_0 = \psi$. This shows that any mapping of the generators $\overline{Y}$ of the algebra F into R can be extended to an algebra homomorphism $F \to R$.

In the special case $Y = R$ we see that R is a homomorphic image of the algebra $F \in \mathfrak{B}$. Hence, $R \in \mathfrak{B}$. The theorem is proved.

DEFINITION 49.2. An algebra F from a variety $\mathfrak{B}$ of K-algebras of signature Ω is called a *(relatively) free algebra* of this variety if in F there exists a set of generators Y such that any mapping of it into an arbitrary algebra $A \in \mathfrak{B}$ can be extended to a homomorphism $F \to A$. In this case the set Y is called *the set of free* generators.

The discussion used in the proof of Theorem 49.2 shows that if a variety $\mathfrak{B}$ consists not of one-element algebras only, then in this variety there are relatively free algebras with an arbitrary set of free generators.

For an arbitrary class of algebras $\mathfrak{B}$ denote by $\operatorname{var}\mathfrak{B}$ the variety such that any algebra of $\operatorname{var}\mathfrak{B}$ can be obtained from the algebras of $\mathfrak{B}$ by operations of taking subalgebras, Cartesian products, and homomorphic images.

THEOREM 49.3. *Let A be a finite K-algebra of arbitrary signature Ω. Then any finitely generated K-algebra of the variety* $\operatorname{var} A$ *is finite.*

It suffices to give the proof only for a free algebra F of the variety $\operatorname{var} A$ with an arbitrary finite set Y of free generators. From the construction of this algebra it follows that F is contained as a subalgebra in the Cartesian product $\prod_{\varphi\in\Phi} A_\varphi$, where $A_\varphi \simeq A$ and φ ranges over arbitrary homomorphisms of the polynomial algebra $K\Gamma_\Omega(Y)$ into A. Since any homomorphism of this kind is determined uniquely by its values on the set Y, in the Cartesian product indicated above there is a finite number of factors, equal to the cardinality of the set of all mappings of Y into A. The theorem is proved.

49.3. Algebras with several supports. The algebraic parlance and elementary facts presented in the two previous subsections can be extended to the next level of generalization, which is very useful in the study of varieties of classical binary algebras and their representations.

Denote m linear spaces $A_1, \dots, A_m$ over the same field K by a symbol $\mathbf{A}$. Let for any m-tuple $\mathbf{n} = (n_1, \dots, n_m)$ of positive integers n_j and for any number $i \in \{1, \dots, m\}$ a subset $\Omega(\mathbf{n}; i)$ (possibly empty) be chosen in the set of all K-linear transformations

$$\mathrm{Hom}_K\left(A_1^{\otimes n_1} \otimes_K A_2^{\otimes n_2} \otimes_K \cdots \otimes_K A_m^{\otimes n_m}, A_i\right)$$

and let to any mapping from $\Omega = \cup_{\mathbf{n},i}\Omega(\mathbf{n}; i)$ a certain (own) name be assigned. In this case the m linear spaces $\mathbf{A} = (A_1, \dots, A_m)$ with operations $\omega \in \Omega$,

$$\omega\colon A_1^{\otimes n_1} \otimes_K A_2^{\otimes n_2} \otimes_K \cdots \otimes_K A_m^{\otimes n_m} \to A_i \qquad (\omega \in \Omega(\mathbf{n}; i)),$$

are called an *algebra with m supports of signature* Ω (a heterogeneous algebra). If m linear subspaces $R_i \subset A_i$ $(i = 1, \dots, m)$ have the property that for any operation $\omega \in \Omega$ and any elements $r_{ij} \in R_i$ the element

$$\omega(r_{11}, \dots, r_{1n_1}, \dots, r_{m1}, \dots, r_{mn_m}) \qquad (\omega \in \Omega(\mathbf{n}; q))$$

is contained in R_q, then $\mathbf{R} = (R_1, \dots, R_m)$ is said to be a *K-subalgebra of the algebra A with m supports*. Furthermore, if for any operation $\omega \in \Omega$ and any $a_{ij} \in A_i$ and $v \in R_t$ $(t = 1, \dots, m)$ the element

$$\omega(a_{11}, \dots, a_{1n_1}, \dots, a_{t1}, \dots, a_{t(i-1)}, v, a_{t(i+1)}, \dots, a_{tn_t}, \dots, a_{m1}, \dots, a_{mn_m}),$$

where $\omega \in \Omega(\mathbf{n}; q)$, belongs to R_q, then $\mathbf{R} = (R_1, \dots, R_m)$ is called an *ideal of the algebra A with m supports*. We denote the least K-subalgebra of the algebra with m supports A containing the subsets $T_1, \dots, T_m$ of the sets $A_1, \dots, A_m$, respectively, by $\mathrm{alg}_A\,\mathbf{T}$ and we shall say that $\mathbf{T} = (T_1, \dots, T_m)$ generates the algebras with m supports $\mathrm{alg}_A\,\mathbf{T}$. Let $\mathbf{J} = (J_1, \dots, J_m)$ be an ideal of $\mathbf{A} = (A_1, \dots, A_m)$, let $\mathbf{A}/\mathbf{J} = (A_1/J_1, \dots, A_m/J_m)$, and let $\psi_i\colon A_i \to A_i/J_i$ be the canonical epimorphisms of K-spaces. From the definition of ideal it follows that for any $\omega \in \Omega(\mathbf{n}; q)$ and any $b_{ij} \in A_i/J_i$ the element

$$\psi_q\left(\omega(\psi_1^{-1}(b_{11}), \dots, \psi_1^{-1}(b_{1n_1}), \dots, \psi_m^{-1}(b_{m1}), \dots, \psi_m^{-1}(b_{mn_m}))\right)$$

is determined uniquely in A_q/J_q and, hence, any operation $\omega \in \Omega(\mathbf{n}; q)$ of the algebra $\mathbf{A}$ induces a multilinear mapping belonging to the set

$$\mathrm{Hom}_K\left((A_1/J_1)^{\otimes n_1} \otimes_K \cdots \otimes_K (A_m/J_m)^{\otimes n_m}, (A_q/J_q)\right),$$

and $\mathbf{A}/\mathbf{J}$ becomes a K-algebra with m supports of signature $\boldsymbol{\Omega}$ with respect to all these operations. The linear mappings $\boldsymbol{\psi} = (\psi_1, \dots, \psi_m)$ are called in this case the *canonical homomorphism* of the algebra with m supports $\mathbf{A}$ *onto* its *quotient algebra* $\mathbf{A}/\mathbf{J}$. (By a *homomorphism* of arbitrary algebras with m supports $\mathbf{A} \to \mathbf{R}$ of signature $\boldsymbol{\Omega}$ we mean a set of m linear mappings $\boldsymbol{\varphi} = (\varphi_1, \dots, \varphi_m)$ for which $\varphi_i \colon A_i \to R_i$ and

$$\begin{aligned}\boldsymbol{\varphi}\omega(&a_{11}, \dots, a_{1n_1}, \dots, a_{m1}, \dots, a_{mn_m}) \\ &= \omega(\boldsymbol{\varphi}(a_{11}), \dots, \boldsymbol{\varphi}(a_{1n_1}), \dots, \boldsymbol{\varphi}(a_{m1}), \dots, \boldsymbol{\varphi}(a_{mn_m})) \qquad (\omega \in \boldsymbol{\Omega}(\mathbf{n}; q)),\end{aligned}$$

where $\boldsymbol{\varphi}(b)$ is the element $\varphi_j(b) \in R_j$ for $b \in A_j$.)

By analogy with the case of algebras of arbitrary signature (see subsections 49.1 and 49.2), for any m-tuple of sets $\mathbf{X} = (X_1, \dots, X_m)$ (where some X_i can be empty sets) we can construct inductively, with the help of names of operations from $\boldsymbol{\Omega} = \cup_{\mathbf{n};q}\boldsymbol{\Omega}(\mathbf{n}; q)$, a *free groupoid with m supports*

$$\Gamma_{\Omega}(\mathbf{X}) = (\Gamma_1(\mathbf{X}), \dots, \Gamma_m(\mathbf{X}))$$

and a *free K-algebra with m supports* of signature $\boldsymbol{\Omega}$

$$K\Gamma_{\Omega}(\mathbf{X}) = (K\Gamma_1(\mathbf{X}), \dots, K\Gamma_m(\mathbf{X})),$$

generated by $\mathbf{X}$. Then, by the theorem on homomorphisms, which remains valid for algebras with m supports as well, any K-algebra with m supports is isomorphic to some quotient algebra $K\Gamma_{\Omega}(\mathbf{X})/\mathbf{J}$, and we obtain a sufficiently clear description of the variety of all algebras with m supports of signature $\boldsymbol{\Omega}$.

Let $X_1, \dots, X_m$ be countable sets. Then expressions of the form $f = 0$, where $f \in K\Gamma_i(\mathbf{X})$, are called *identities*. It is said that for a K-algebra A with m supports of signature $\boldsymbol{\Omega}$ the identity $f = 0$ *holds* if for any homomorphisms $\boldsymbol{\varphi} \colon K\Gamma_{\Omega}(\mathbf{X}) \to \mathbf{A}$ the element $\boldsymbol{\varphi}(f)$ is equal to zero. Let $\mathbf{T} = (T_1, \dots, T_m)$ (T_i are arbitrary subsets of $K\Gamma_i(\mathbf{X})$); then the class of all algebras with m supports of signature $\boldsymbol{\Omega}$ in which all identities $f = 0$, where $f \in T_i$, are valid is called a *variety*. Obviously, any variety is closed with respect to taking subalgebras, homomorphic images, and Cartesian products. The proof of Theorem 49.2, which can be repeated for algebras with m supports word for word, shows that any class of algebras with m supports of signature $\boldsymbol{\Omega}$ closed with respect to taking subalgebras, homomorphic images, and Cartesian products is a variety and contains relatively free algebras with m supports with an arbitrary set of free generators $\mathbf{Y} = (Y_1, \dots, Y_m)$.

Certainly, if we replace the set $\{1, 2, \dots, m\}$ by a set of arbitrary cardinality, the terminology introduced above and Theorems 49.1 and 49.2 retain their validity for these algebras with any sets of supports as well. In this case the algebraic language invades the region that is nowadays intensively developed by mathematical logic and becomes so rich in content that any expression in the language of predicates can be interpreted in terms of identities of algebras with several supports. Detailed introductions to the theory of algebras with several supports can be found in the books [GTsYu, Higg3] and in the papers [BL, Higg1].

§50. Partially ordered sets

50.1. Basic concepts and Zorn's lemma. Let S be an arbitrary nonempty set on which a function in two variables with values in a two-point set $\{0, 1\}$ is fixed. The function f is said to define a *partial order* on the set S if it has the following properties:

a) $p(a, a) = 1$ for any element $a \in S$,

b) if $p(a, b) = 1 \ \& \ p(b, a) = 1$, then $a = b$,

c) if $p(a, b) = 1 \ \& \ p(b, c) = 1$, then $p(a, c) = 1$.

Elements $a, b \in S$ are said to be *incomparable* with respect to this order if $p(a, b) \neq 1 \ \& \ p(b, a) \neq 1$; otherwise, they are said to be *comparable*. Usually the notation $a \leq b$ means that the elements $a, b \in S$ are comparable and $p(a, b) = 1$; if we have $p(b, a) \neq 1$ in this case, then we wrire $a < b$. A set with a partial order on it is called a *partially ordered set*.

An element $s \in S$ is said to be *maximal* (*minimal*) element of a partially ordered set S if the inequality $s \leq a$ (respectively, $a \leq s$), where $a \in S$, implies the coincidence of a with s.

Let S' be some nonempty subset of a partially ordered set S. An element $s \in S$ is called an *upper* (*lower*) *bound* of the subset S' if $s' \leq s$ (respectively, $s \leq s'$) for any element $s' \in S'$. An element $s \in S$ is called *the least upper* (*greatest lower*) *bound* of the subset S' if s is an upper (respectively, lower) bound of the subset S' and for any upper (respectively, lower) bound $a \in S$ of the subset S' we have the inequality $s \leq a$ ($a \leq s$).

In a partially ordered set the least upper bound and the greatest lower bound of two elements $a, b \in S$ are denoted (if these exist) by $a \cup b$ and $a \cap b$, respectively. If in S any two elements have the least upper bound (greatest lower bound), then the assignment to any $a, b \in S$ their least upper bound $a \cup b$ (greatest lower bound $a \cap b$) defines on S a binary operation $\cup$ ($\cap$), called the union (intersection) operation (or the join and the meet operation, respectively). In this case the set S is called a *semilattice* with respect to the union operation $\cup$ (intersection operation $\cap$). It is immediate from the definition of least upper bound and greatest lower bound that for any a, b, c the following equalities hold:

$$a \cup a = a, \qquad a \cup b = b \cup a, \qquad a \cup (a \cup b) = a \cup b, \tag{50.1}$$

$$a \cap a = a, \qquad a \cap b = b \cap a, \qquad a \cap (a \cap b) = a \cap b. \tag{50.2}$$

It is easy to verify that if on the set S an operation $\cup$ (operation $\cap$) is given satisfying equalities (50.1) (equalities (50.2)), then on S a partial order compatible with this operation can be introduced by setting $a \leq b$ if and only if $a \cup b = b$ ($a \cap b = a$).

A semilattice S with respect to the union (intersection) operation is said to be *full* if any nonempty subset of it has a least upper bound (greatest lower bound).

If in a partially ordered set S any two elements have the least upper bound as well as the greatest lower bound, then the set S with two operations $\cup$ and $\cap$ is called a *lattice* and, besides the equalities (50.1) and (50.2), in S we have the equalities

$$a \cup (b \cap a) = a, \qquad a \cap (b \cup a) = a. \tag{50.3}$$

Relations (50.1)–(50.3) permit one to reconstruct from the operations $\cup$ and $\cap$ a partial order on the set S such that with respect to this order the operations are the operations of taking the least upper bound and the greatest lower bound; to this end, it suffices to put $a \leq b$ if and only if $a \cap b = a$.

A lattice S with respect to the union and intersection operations is said to be *full* if any nonempty subset of it has a least upper bound and a greatest lower bound.

A partially ordered set S is called *totally* (or *linearly*) *ordered* if any two elements in it are comparable. A totally ordered set is said to be *completely ordered* if any nonempty subset in it has a minimal element. A subset S' of a partially ordered set S is called a *chain* if it is nonempty and totally ordered with respect to the partial order on S.

Practically, all nonconstructive proofs acknowledged presently in algebra are based on the following completely unclear statement which can be taken, in essence, as an axiom of modern set theory.

ZORN'S LEMMA. *If S is a partially ordered set in which any chain has an upper bound, then S possesses a maximal element.*

A justification of this fact based on the axiom of choice can be found in the book [VdW]. Contemporary understanding of rigor level of proofs in algebra permits one to use this lemma in verifications of algebraic assertions.

We give a typical example of an argument using Zorn's lemma and a situation where we cannot manage without it.

PROPOSITION 50.1. *Let A be an arbitrary K-algebra and a an arbitrary nonzero element of it. Then the set S of all ideals of the K-algebra A that do not contain a possesses a maximal ideal (with respect to the natural partial order defined by inclusion).*

PROOF. It is clear that the zero ideal belongs to S. Hence, the set S is nonempty. Consider an arbitrary chain of ideals $\{J_i\}$ in a partially ordered set S. Since all ideals J_i are pairwise comparable with respect to the inclusion, the union J of all these ideals is an ideal of A containing each J_i. Since J_i does not contain a, we have $a \notin J$ so that $J \in S$. This shows that any chain in the partially ordered set S has an upper bound in S. Therefore, by Zorn's lemma, the set S possesses a maximal element, i.e., an ideal I that does not contain the element a, and any other ideal J containing I satisfies either $J = I$ or $J \ni a$. The proposition is proved.

50.2. The Dilworth theorem.

THEOREM. *A necessary and sufficient condition for an arbitrary partially ordered set S to be decomposable into the union of no more than m pairwise disjoint chains is that among any distinct $m + 1$ elements of the set S there be two comparable elements.*

Necessity. Let $S = S_1 \cup \cdots \cup S_k$, where $k \leq m$ and $S_1, \ldots, S_k$ are mutually disjoint chains, and let $a_1, \ldots, a_{m+1}$ be different elements of the set S. Then some chain S_t contains two elements of the form a_i and a_j and these elements are comparable.

Sufficiency. Let us consider the case in which S is a finite set. We proceed by joint induction on the number m and on the number of elements of the set S. Since the set S is finite, it has a minimal element. Denote it by the letter s. Then from inductive arguments it follows that the set $S' = S \setminus \{s\}$ is decomposed into the union of disjoint chains $S_1, \ldots, S_k$, where $k \leq m$. If $k < m$, then putting $S_{k+1} = \{s\}$ we obtain a desired decomposition of the set S.

Suppose that $k = m$; then we decompose each set S_i into two subsets:

$$S_i^+ = \{a \in S \mid a > s\}, \qquad S_i^- = \{a \in S \mid a \text{ is not comparable with } s\}.$$

It is clear that $S_i^+ \cap S_i^- = \varnothing$ and we have $S_i = S_i^+ \cup S_i^-$, because the element s is minimal in S. If there is at least one $i \in \{1, \dots, m\}$ for which the set S_i^- is empty, then $S_i' = S_i \cup \{s\}$ is a chain and $S_i' \cup S_1 \cup \dots \cup S_{i-1} \cup S_{i+1} \cup \dots \cup S_m$ is a desired decomposition of the set S into m disjoint chains.

Now we suppose that the sets $S_1^-, \dots, S_m^-$ are nonempty. Consider the set $P_i = S \setminus S_i^+$. Assume that for any $i = 1, \dots, m$ we can choose m mutually incomparable elements $a_{i1}, \dots, a_{im}$ in P_i. Then these belong to different chains S_j and we can suppose that $a_{ij} \in S_j$. Denote by b_j the smallest of the elements $a_{1j}, \dots, a_{mj}$. Since $a_{jj} \notin S_j^+$, we have $a_{jj} \in S_j^-$, and any element b_j belongs to S_j^-. It is clear that $b_1, \dots, b_m$ are mutually incomparable elements (indeed, if $b_i < b_j$, then $b_i = a_{ti}$ and $b_j = a_{lj}$ for some t and l and we have $b_j \le a_{tj}$ by construction), which contradicts the fact that the elements a_{ti} and a_{tj} are incomparable). However, in this case the element s is not comparable with any element $b_1, \dots, b_m$, and in S there are $1 + m$ mutually incomparable elements $b_1, \dots, b_m, s$, which condradicts the assumption of the theorem. The contradiction thus obtained shows that for some $i = \{1, \dots, m\}$ any subset with m elements of a partially ordered set P_i contains a pair of comparable elements. Then by inductive arguments it follows that P_i is decomposed into the union of $m - 1$ chains that are mutually disjoint. These chains, together with the chain $S_i^+ \cup \{s\}$, form a desired decomposition of the partially ordered set S into m mutually disjoint chains. The theorem is proved for the case of a finite set S.

We do not use the Dilworth theorem for the general case; the conclusion of its proof can be found in the book [Hal2].

§51. Theorems on homomorphisms of commutative algebras

51.1. Basic notions. A linear space $\mathcal{E}$ over a field K with a fixed bilinear operation $\cdot\colon \mathcal{E} \otimes_K \mathcal{E} \to \mathcal{E}$ on it is called an *associative and commutative K-algebra* if for any elements $a, b, c \in \mathcal{E}$ we have

$$(51.1) \qquad (a \cdot b) \cdot c = a \cdot (b \cdot c), \qquad a \cdot b = b \cdot a.$$

An element $e \in \mathcal{E}$ is called a *unit element* of the K-algebra $\mathcal{E}$ if $e \neq 0$ and $e \cdot a = a \cdot e = a$ for any $a \in \mathcal{E}$. An element $a \in \mathcal{E}$ is said to be *invertible* in the algebra $\mathcal{E}$ if for some element $b \in \mathcal{E}$ the product $b \cdot a = a \cdot b$ is a unit element of the algebra $\mathcal{E}$. Equalities (51.1) imply the uniqueness of such a unit element e of the algebra $\mathcal{E}$ and of such an element b for an invertible element a.

For any K-algebra $\mathcal{E}$ denote by $\mathcal{E}_{\mathrm{id}}$ the K-algebra whose linear space is the direct sum $K \oplus \mathcal{E}$ and the bilinear operation is given by the formula

$$(\beta_1, a_1) \cdot (\beta_2, a_2) = (\beta_1 \cdot \beta_2, \beta_1 \cdot a_2 + \beta_2 \cdot a_1 + a_1 \cdot a_2) \qquad (\beta_i \in K,\ a_i \in \mathcal{E}).$$

This definition implies that $(1, 0)$ is the unit element of the algebra $\mathcal{E}_{\mathrm{id}}$ and the algebra $\mathcal{E}$ is an ideal of $\mathcal{E}_{\mathrm{id}}$ of codimension one.

It is said that $\mathcal{E}$ is a *finitely generated* K-algebra if a finite set of elements $e_1, \dots, e_k$ can be chosen in $\mathcal{E}$ for which $\mathcal{E}$, as a linear space, is spanned by all monomials of the form $e_1^{m_1} \cdots e_k^{m_k}$ $(m_1 + \cdots + m_k > 0 \ \& \ m_i \geq 0)$.

An element a of the algebra $\mathcal{E}$ is called a *nil-element* (or a nilpotent element), if for some positive integer m we have the equality $a^m = 0$. It is immediate that the subset of all nil-elements of an associative and commutative algebra $\mathcal{E}$ forms an ideal. This ideal is called the *nil-radical* of the algebra $\mathcal{E}$ and is denoted by $\operatorname{rad} \mathcal{E}$. Obviously, $\operatorname{rad}(\mathcal{E}/\operatorname{rad} \mathcal{E})$ is the zero ideal.

For any $a \in \mathcal{E}$ we put $\operatorname{Ann} a = \{b \in \mathcal{E} \mid ab = 0\}$. An element a is called a *zero divisor* of the algebra $\mathcal{E}$ if $a \neq 0$ and the ideal $\operatorname{Ann} a$ of the algebra $\mathcal{E}$ is nonzero.

A linear operator $\partial\colon \mathcal{E} \to \mathcal{E}$ is called a K-derivation of the K-algebra $\mathcal{E}$ if for any $a, b \in \mathcal{E}$ we have the equality

$$^{\partial}(a \cdot b) = {}^{\partial}a \cdot b + a \cdot {}^{\partial}b$$

(here ${}^{\partial}c$ denotes the result of application of the operator ∂ to the element $c \in \mathcal{E}$). The set of all K-derivations of the K-algebra $\mathcal{E}$ forms a linear subspace of the linear space $\operatorname{End}_K \mathcal{E}$ of all K-linear operators of the space $\mathcal{E}$. We denote this subspace by $\operatorname{Der}_K \mathcal{E}$. Assume that $\mathcal{E}$ has a unit; denote it by $\mathbf{1}$. Then ${}^{\partial}(\mathbf{1} \cdot \mathbf{1}) = {}^{\partial}\mathbf{1} \cdot \mathbf{1} + \mathbf{1} \cdot {}^{\partial}\mathbf{1}$ and, hence, ${}^{\partial}\mathbf{1} = 0$ for any $\partial \in \operatorname{Der}_K \mathcal{E}$. This shows that any K-derivation of the K-algebra $\mathcal{E}$ can be uniquely extended to a K-derivation of the K-algebra $\mathcal{E}_{\mathrm{id}}$.

PROPOSITION 51.1. *Let $\mathcal{E}$ be an associative and commutative K-algebra. Then:*

a) *for any element $a \in \mathcal{E}$ the ideal $\bigcup_{i=1}^{\infty} \operatorname{Ann} a^i$ is invariant with respect to all derivations from $\operatorname{Der}_K \mathcal{E}$;*

b) *if the characteristic of the ground field K is zero, then the subspace $\operatorname{rad}_K \mathcal{E}$ is also invariant with respect to the derivations from $\operatorname{Der}_K \mathcal{E}$.*

PROOF. Let us differentiate the equality $a^k \cdot b = 0$ and multiply the expression thus obtained by the element a. Then we have

$$0 = a \cdot ({}^{\partial}(a^k b)) = (k-1)a^k \cdot b \cdot {}^{\partial}a + a^{k+1} \cdot {}^{\partial}b = a^{k+1} \cdot {}^{\partial}b$$

for any $\partial \in \operatorname{Der}_K \mathcal{E}$. Hence, if $b \in \operatorname{Ann} a^k$, then ${}^{\partial}b \in \operatorname{Ann} a^{k+1}$. This proves assertion a).

Let a be a nil-element and $a^t = 0$. We shall show that then for any derivation ∂ of $\mathcal{E}$ we have the equalities

$$t(t-1)\cdots(t-i+1)a^{t-i} \cdot ({}^{\partial}a)^{2i-1} = 0 \qquad (i = 1, 2, \dots, t). \tag{51.2}$$

For $i = 1$ relation (51.2) is obtained by differentiation of the equality $a^t = 0$. Suppose that relation (51.2) has already been proved for $i = l$ $(l < t)$. Then, applying the derivation ∂ to this relation and multiplying the equality thus obtained by ${}^{\partial}a$, we establish relation (51.2) for $i = l + 1$. Formula (51.2) for $i = t$ shows that if the characteristic of the field K is zero, the element ${}^{\partial}a$ is a nil-element. This proves assertion b). Proposition 51.1 is proved.

COROLLARY. *Let the characteristic of the ground field K be zero and let in the K-algebra $\mathcal{E}$ there be nonnilpotent elements. Let $\mathfrak{h}$ be some subset of* $\mathrm{Der}_K \mathcal{E}$ *and let any nonzero ideal of $\mathcal{E}$ invariant with respect to all derivations of $\mathfrak{h}$ coincide with $\mathcal{E}$. Then there are no zero divisors in the algebra $\mathcal{E}$.*

For arbitrary associative and commutative K-algebra $\mathcal{E}$ with unit we denote by $\mathcal{E}[[t]]$ the K-algebra which, as a linear space, consists of arbitrary formal infinite sums of the form

$$a_0 + a_1 t + a_2 t^2 + \cdots + a_i t^i + \cdots \qquad (a_i \mathcal{E}) \tag{51.3}$$

and on which the K-bilinear operation $\cdot$ is given by the formula

$$\left(\sum_{i=0}^{\infty} a_i t^i\right) \cdot \left(\sum_{i=0}^{\infty} b_i t^i\right) \rightleftharpoons \sum_{k=0}^{\infty} \left(\sum_{j=0}^{\infty} a_j \cdot b_{k-j}\right) t^k.$$

It is immediate that $\mathcal{E}[[t]]$ is an associative and commutative K-algebra with unit containing the algebra $\mathcal{E}$ as a K-subalgebra. The algebra $\mathcal{E}[[t]]$ is called the *algebra of formal power series* in one variable t over the algebra $\mathcal{E}$.

Denote by $\mathcal{E}[t]$ the least K-subalgebra of $\mathcal{E}[[t]]$ containing the subalgebra $\mathcal{E}$ and the element $\mathbf{t} \rightleftharpoons 0 + \mathbf{1} \cdot t + \sum_{i=2}^{\infty} 0 \cdot t^i$. The elements of $\mathcal{E}[t]$ are those and only those expressions of the form (51.3) in which only finitely many elements $a_i \in \mathcal{E}$ are nonzero. The algebra $\mathcal{E}[t]$ is called the (commutative) *polynomial algebra* over $\mathcal{E}$ in one variable t and its elements are called *polynomials* in one variable with coefficients from $\mathcal{E}$. It is immediate that any polynomial $f(t) \in \mathcal{E}[t]$ can be represented in the form

$$a_1 \mathbf{t}^{n_1} + \cdots + a_l \mathbf{t}^{n_l} \qquad (0 \le n_1 < n_2 < \cdots < n_l,\ a_1, \ldots, a_l \ne 0,\ a_i \in \mathcal{E}) \tag{51.4}$$

and this representation is unique. Therefore, any homomorphism $\varphi\colon \mathcal{E} \to \mathcal{E}'$ into some associative and commutative K-algebra $\mathcal{E}'$ can be extended to a homomorphism $\psi\colon \mathcal{E}[t] \to \mathcal{E}'$ transforming $\mathbf{t}$ into any aforegiven element $b \in \mathcal{E}'$. An integer n_l in decomposition (51.4) is called the *degree* of the polynomial f.

By induction we can define the algebra $\mathcal{E}[[t_1, \ldots, t_n]]$ of power series and the algebra $\mathcal{E}[t_1, \ldots, t_n]$ of polynomials in n commuting variables $t_1, \ldots, t_n$, by setting

$$\mathcal{E}[[t_1, \ldots, t_n]] = \mathcal{E}[[t_1, \ldots, t_{n-1}]][[t_n]],$$
$$\mathcal{E}[t_1, \ldots, t_n] = \mathcal{E}[t_1, \ldots, t_{n-1}][t_n].$$

For any $a \in \mathcal{E}$ denote by $\mathcal{E}_a$ the quotient algebra of the K-algebra $\mathcal{E}_{\mathrm{id}}[t]$ by the least ideal containing the element $at - 1$ (1 is the unit element of the algebra $\mathcal{E}_{\mathrm{id}}$) and by $\varkappa_a$ the canonical homomorphism $\mathcal{E}_{\mathrm{id}}[t] \to \mathcal{E}_a$. The element $\varkappa_a(t)$ is usually denoted by $\varkappa_a(a)^{-1}$ and the algebra $\mathcal{E}_a$ is called the *algebra of fractions* (or *quotients*) of the K-algebra $\mathcal{E}$ with respect to the element a.

PROPOSITION 51.2. *For any element $a \in \mathcal{E}$ the kernel of the homomorphism $\varkappa_a\colon \mathcal{E} \to \mathcal{E}_a$ coincides with $\bigcup_{i=1}^{\infty} \mathrm{Ann}\, a^i$. In particular, if $0 \ne a$ is a nonnilpotent element of the algebra $\mathcal{E}$, then $\mathcal{E}_a$ is an algebra with unit and $\varkappa_a(a)$ is an invertible element of $\mathcal{E}_a$.*

PROOF. Let $b \in \mathcal{E}_{\mathrm{id}}$ and $b \in \bigcup_{i=1}^{\infty} \mathrm{Ann}_{\mathcal{E}_{\mathrm{id}}} a^i$. Then $a^i \cdot b = 0$ for some $i \in \mathbb{N}$ and $0 = \varkappa_a(a^i \cdot b) \cdot \varkappa_a(t^i) = \varkappa_a(b) \cdot \varkappa_a(1) = \varkappa_a(b \cdot 1) = \varkappa_a(b)$. Hence, b belongs to the kernel of the homomorphism $\varkappa_a\colon \mathcal{E}_{\mathrm{id}} \to \mathcal{E}_a$.

Conversely, suppose that $b \in \mathcal{E}_{\text{id}}$ & $\varkappa_a(b) = 0$. By the theorem on homomorphisms this implies that b belongs to the ideal $\mathcal{E}_{\text{id}}(at-1)$ and, therefore,

$$b = (a_l + a_{l-1}t + \cdots + a_0 t^l)(at-1) \qquad (a_i \in \mathcal{E}_{\text{id}}).$$

Since the representation of a polynomial in the form (51.4) is unique, this equality is equivalent to the system of equalities:

$$a_0 \cdot a = 0, \qquad a_i \cdot a = a_{i-1} \qquad (i = 1, \dots, l-1), \qquad 0 = b + a_l.$$

Multiplying the ith equality $(i = 0, 1, \dots, l)$ by a^i and adding all relations thus obtained we find $b \cdot a^l = 0$, i.e., $b \in \operatorname{Ann} a^l$. Hence, the kernel of the homomorphism $\varkappa_a \colon \mathcal{E}_{\text{id}} \to \mathcal{E}_a$ coincides with $\cup_{i=1}^{\infty} \operatorname{Ann}_{\mathcal{E}_{\text{id}}} a^i$, and the first assertion of the proposition follows from the equality $\mathcal{E} \cap \operatorname{Ann}_{\mathcal{E}_{\text{id}}} c = \operatorname{Ann}_{\mathcal{E}} c$ which holds for any $c \in \mathcal{E}$.

If $0 \neq a$ is a nonnilpotent element, then $1 \notin \operatorname{Ann}_{\mathcal{E}_{\text{id}}} a^i$ $(i \in \mathbb{N})$ and $0 \neq \varkappa_a(1)$ is the unit element of the algebra $\mathcal{E}_a$. Since $at-1$ belongs to the kernel of the homomorphism $\varkappa_a \colon \mathcal{E}_{\text{id}}[t] \to \mathcal{E}_a$, we have $\varkappa_a(a) \cdot \varkappa_a(t) = \varkappa_a(1)$, and $\varkappa_a(a)$ is an invertible element of the algebra $\mathcal{E}_a$. Proposition 51.2 is proved.

51.2. Hilbert's Nullstellensatz.

DEFINITION 51.1. A field K is said to be *algebraically closed* if for any polynomial $f(t) = \sum_{i=0}^{n} \beta_i t^i \in K[t]$ $(\beta_n \neq 0)$ there exist elements $\gamma_1, \dots, \gamma_n \in K$ such that $f(t) = \beta_n(t-\gamma_1)\cdots(t-\gamma_n)$. The elements γ_i are called the roots of the polynomial f.

DEFINITION 51.2. It is said that an associative and commutative K-algebra $\mathcal{E}$ has *sufficiently many homomorphisms* into the field K if for any nonzero nonnilpotent element $a \in \mathcal{E}$ there exists a homomorphism $\varphi \colon \mathcal{E} \to K$ for which $\varphi(a) \neq 0$.

LEMMA 51.1. *Let K be an algebraically closed field, let $\mathcal{E}$ be an associative and commutative K-algebra with unit without zero divisors, and let $\mathcal{E}_0$ be some K-subalgebra of $\mathcal{E}$ containing the unit element of the algebra $\mathcal{E}$. Suppose that $\mathcal{E}$ is generated over $\mathcal{E}_0$ by a single element $b \in \mathcal{E}$. In this case, if $\mathcal{E}_0$ has sufficiently many homomorphisms into the field K, then the algebra $\mathcal{E}$ has this property as well. Furthermore, if $f(t) = \sum_{i=0}^{l} a_i t^{l-i} \in \mathcal{E}_0[t]$ $(l > 0)$ is a polynomial of minimal degree for which $f(b) = 0$ in the algebra $\mathcal{E}$, then an arbitrary homomorphism $\varphi \colon \mathcal{E}_0 \to K$ mapping a_0 into a nonzero element of the field K can be extended to a homomorphism $\psi \colon \mathcal{E} \to K$ $(\psi|_{\mathcal{E}_0} = \varphi)$.*

PROOF. We first consider the case in which for any nonzero polynomial $g(t) \in \mathcal{E}[t]$ we have the inequality $g(b) \neq 0$. In this case the homomorphism $\varkappa \colon \mathcal{E}_0 \to \mathcal{E}$ mapping t into b and acting as identity on $\mathcal{E}_0$ has zero kernel and, by assumption, is an epimorphism. Therefore, by the theorem on homomorphisms, $\varkappa$ is an isomorphism, and any homomorphism $\varphi \colon \mathcal{E}_0 \to K$ can be extended to a homomorphism $\psi \colon \mathcal{E} \to K$ mapping the element b into an arbitrary element $\beta \in K$. Since the algebra $\mathcal{E} = \mathcal{E}_0[t]$ has no zero divisors, we have $\operatorname{rad} \mathcal{E}_0[t] = 0$, and in order to prove the lemma in the case under consideration it suffices to construct, for an arbitrary nonzero polynomial $g(t) = \sum_{i=0}^{m} b_i t^{m-i}$ $(b_0 \neq 0)$, a homomorphism $\psi \colon \mathcal{E}_0[t] \to K$ for which $\psi(g) \neq 0$. By the assumption of the lemma, there exists a homomorphism $\varphi \colon \mathcal{E}_0 \to K$ for which $\varphi(b_0) \neq 0$. Since an algebraically closed field is infinite, we can choose $\beta \in K$ such that

β is not a root of the polynomial $\sum_{i=0}^{m}\varphi(b_i)t^{m-i}\in K[t]$. Then the homomorphism $\psi\colon \mathcal{E}_0[t]\to K$ for which $\psi|_{\mathcal{E}_0}=\varphi$ & $\psi(t)=\beta$ is a desired one.

Now we consider the case in which for some polynomial $f(t)=\sum_{i=0}^{l}a_it^{l-i}\in\mathcal{E}_0[t]$ $(a_0\neq 0,\ l>0)$ in the algebra $\mathcal{E}$ we have the equality $f(b)=0$. We assume that the degree l of the polynomial f is minimal among the polynomials with this property. Let $\varphi\colon\mathcal{E}_0\to K$ be a homomorphism and $\varphi(a_0)\neq 0$ in K. We denote the natural extension of this homomorphism to a homomorphism of the polynomial algebras $\mathcal{E}_0[t]\to K[t]$ by the same letter. For any polynomial $g(t)\in\mathcal{E}_0[t]$ we put $\psi(g(b))\rightleftharpoons\varphi(g(t))|_{t=\beta}$, where $\beta\in K$ is a fixed root of the polynomial $\varphi(g(t))\in K[t]$.

We shall show that the mapping $\psi\colon\mathcal{E}\to K$ is well defined. Toward this end, it suffices to prove the following implication:

$$(g(b)=0\text{ in }\mathcal{E})\Rightarrow(\varphi(g(t))|_{t=\beta}=0\text{ in }K).$$

If $g(b)=0$, then it follows from the assumption of the lemma that the degree m of the polynomial g is not less than the degree l of the polynomial f. Let $k>m-l$. Then in $\mathcal{E}_0[t]$ we have the equality $a_0^k\cdot g(t)=p(t)\cdot f(t)+r(t)$, where $p(t),r(t)\in\mathcal{E}_0[t]$ and the degree of $r(t)$ is less than l. However, then $r(b)=0$, and $r(t)=0$, because the number l is minimal. Therefore, $a_0^k\cdot g(t)=p(t)\cdot f(t)$ and, hence,

$$\varphi(a_0^k)\cdot\varphi(g(t))|_{t=\beta}=\varphi(p(t))|_{t=\beta}\cdot\varphi(f(t))|_{t=\beta}=0.$$

Since $\varphi(a_0)\neq 0$ in the field K, the latter equality proves implication (51.5) and the fact that the mapping ψ is well defined. It is clear that $\psi\colon\mathcal{E}\to K$ is a homomorphism and $\psi|_{\mathcal{E}_0}=\varphi$.

It remains to show that in the case under consideration the algebra $\mathcal{E}$ has sufficiently many homomorphisms into the field K.

Assume the contrary. Then $g(b)\neq 0$ in the algebra $\mathcal{E}$ for some polynomial $g(t)\in\mathcal{E}_0[t]$, but $\psi(g(b))=0$ for any homomorphism $\psi\colon\mathcal{E}\to K$. We assume that the polynomial g has the minimal degree among the polynomials with this property. Denote this degree by m.

It is clear that $m\geq 1$, because otherwise $g(b)\in\mathcal{E}_0$, and for an extension $\psi\colon\mathcal{E}\to K$ of the homomorphism $\varphi\colon\mathcal{E}_0\to K$ for which $\varphi(g(b)\cdot a_0)\neq 0$ we would have the inequality $\psi(g(b))\neq 0$.

Let $1\leq m<l$ and let $c_0\in\mathcal{E}_0$ be the nonzero coefficient of the leading term t^m of the polynomial g. Then in $\mathcal{E}_0$ we have the equality $c_0^l\cdot f(t)=p(t)\cdot g(t)+r(t)$, where the degree of $p(t)$ is less than l and the degree of $r(t)$ is less than $m<l$. This equality implies $\psi(r(b))=0$ for any homomorphism $\psi\colon\mathcal{E}\to K$. Since the degree m of the polynomial g is minimal, we have $r(b)=0$ and $0=c_0^l\cdot f(b)=p(b)\cdot g(b)$. Since the algebra $\mathcal{E}$ has no zero divisors and $g(b)\neq 0$, we have $p(b)=0$. However, the polynomial f has the smallest degree among all nonzero polynomials vanishing at the element b. Hence, $p(t)=0=r(t)$ and $c_0^l\cdot f(t)$ is the zero polynomial, which contradicts the fact that the algebra $\mathcal{E}$ has no zero divisors, because a_0 and c_0 are nonzero elements of the algebra $\mathcal{E}_0$.

Let $m\geq l$. Then $c_0^m\cdot f(t)=p(t)\cdot g(t)+r(t)$, where $p(t),r(t)\in\mathcal{E}_0[t]$ and the degree of $r(t)$ is less than $l\leq m$. This equality implies $a_0^m\cdot g(b)=r(b)$, and any homomorphism $\psi\colon\mathcal{E}\to K$ maps $r(b)$ into zero. Since the algebra $\mathcal{E}$ has no zero divisors and $a_0 \neq 0 \neq g(b)$, we have $r(b) \neq 0$ in $\mathcal{E}$. Thus, the polynomial $r(t)$ has the same property as $g(t)$ but it has lesser degree. The contradiction with the minimality of the degree of the polynomial $g(t)$ thus obtained proves the lemma.

COROLLARY. *Suppose that the field K is algebraically closed and an associative and commutative K-algebra $\mathcal{E}$ is finitely generated and has no zero divisors. Then $\mathcal{E}$ has sufficiently many homomorphisms into the field K.*

PROOF. Let $e_1, \dots, e_k$ be generators of the K-algebra $\mathcal{E}$ and let e_0 be its unit element. Then the subalgebra Ke_0 is isomorphic to the field K and by Lemma 51.1 we successively find that the K-subalgebras of $\mathcal{E}$ generated by the elements $e_0, e_1, \dots, e_i$ $(i = 0, 1, \dots, k)$ have sufficiently many homomorphisms into the field K. For $i = k$ this subalgebra coincides with $\mathcal{E}$. The corollary is proved.

THEOREM 51.1 (Hilbert's Nullstellenzatz). *Suppose that the ground field K is algebraically closed, $\mathcal{E}$ is an associative and commutative K-algebra and the cardinality $|K|$ of the field K is strictly greater than the cardinality of a set of generators of the algebra $\mathcal{E}$. Then the algebra $\mathcal{E}$ has sufficiently many homomorphisms into the field K.*

PROOF. Let $a \neq 0$ be an arbitrary nonnilpotent element of the algebra $\mathcal{E}$ and let $\varkappa_a \colon \mathcal{E} \to \mathcal{E}_a$ be the canonical homomorphism of the algebra $\mathcal{E}$ into its algebra of fractions $\mathcal{E}_a$ with respect to the element a. By Proposition 51.2, the element $\varkappa_a(a)$ is invertible in $\mathcal{E}_a$ and, by Zorn's lemma, in $\mathcal{E}_a$ there exists a maximal ideal J not containing the unit element of the algebra $\mathcal{E}_a$. Obviously, the quotient algebra $\widetilde{K} = \mathcal{E}_a/J$ is a field and contains a subfield $K \cdot 1$ isomorphic to K. Since $\varphi(a) \neq 0$ for the through homomorphism $\varphi \colon \mathcal{E} \to \mathcal{E}_a \to \mathcal{E}_a/J \to \widetilde{K}$, for the proof of the theorem it remains to show that $\widetilde{K}$ is isomorphic to K.

By the assumption of the theorem, in $\mathcal{E}$ there exists a subset G of cardinality less than $|K|$ a basis of the K-algebra $\mathcal{E}$ can be chosen among the monomials of the form $g_1^{m_1} \cdot \dots \cdot g_l^{m_l}$, where $g_i \in G$, $m_1 + \dots + m_l \geq 1$, $l = 1, 2, \dots$. Two cases are possible: G is a finite set and G is an infinite set.

If the set G is finite then the field $\widetilde{K}$ is finitely generated as a K-algebra (it suffices to take for its generators the elements of the set $\varphi(G)$ and the element $\varkappa_a(a)^{-1}$) and, by Corollary of Lemma 51.1, it admits a nonzero K-homomorphism into the field K. Hence, in the case under consideration $\widetilde{K}$ coincides with $K \cdot 1$.

If the set G is infinite, then $\dim_K \widetilde{K} \leq \dim_K \mathcal{E} < |K|$ and for any element $b \in \widetilde{K}$ there are elements in the set $\{(b - \beta)^{-1} \mid \beta \in K \cdot 1\}$ linearly dependent over K, i.e., for some $\beta_1, \dots, \beta_n \in K \cdot 1$ and for some $\gamma_1, \dots, \gamma_n \in K \cdot 1$ that are not all zero for $n \geq 2$ we have $\sum_{i=1}^{n} \gamma_i (b - \beta_i)^{-1} = 0$. Multiplying it by $(b - \beta_1) \cdot \dots \cdot (b - \beta_n)$ we see that for some nonzero polynomial $f(t) \in K[t]$ of degree $n - 1$ we have the equality $f(b) = 0$ in $\widetilde{K}$. However, then we have $(b - \delta_1 \cdot 1) \cdot \dots \cdot (b - \delta_{n-1} \cdot 1) = 0$, where $\delta_1, \dots, \delta_{n-1} \in K$ are all the roots of the polynomial $f(t)$. Since $\widetilde{K}$ is a field, this is possible only if $b = \delta_i \cdot 1$ for some $i \in \{1, 2, \dots, n - 1\}$. Hence, $b \in K \cdot 1$ and in this case $\widetilde{K} = K \cdot 1$. The theorem is proved.

51.3. An analog of Hilbert's Nullstellensatz for differential algebras. Suppose that on a linear space $\mathcal{E}$ over a field K there are given a binary bilinear operation $\cdot \colon \mathcal{E} \otimes_K \mathcal{E} \to \mathcal{E}$ (which endows $\mathcal{E}$ with the structure of an associative and commutative algebra) and n linear operators $\partial_1, \dots, \partial_n$ commuting with each other and belonging to $\operatorname{Der}_K \mathcal{E}$. In this case $\mathcal{E}$ is called an *associative and commutative differential K-algebra* and $\partial_1, \dots, \partial_n$ are called its *signature derivations.*

THEOREM 51.2. *Suppose that the ground field K is algebraically closed and of characteristic zero and $\mathcal{E}$ is an associative and commutative differential K-algebra with finite set of generators with respect to the commuting signature derivations $\partial_1, \dots, \partial_n$. Then the K-algebra $\mathcal{E}$ possesses sufficiently many homomorphisms into the field K.*

The proof of this theorem is based on the following differential analog of Lemma 51.1.

LEMMA 51.2. *Let K be an algebraically closed field of characteristic zero, let $\mathcal{E}$ be an associative and commutative differential K-algebra with respect to signature derivations $\partial_1, \dots, \partial_n$ with the unit element and without zero divisors, and let $\mathcal{E}_0$ be a differential K-subalgebra of $\mathcal{E}$ containing the unit element of the algebra $\mathcal{E}$ such that $\mathcal{E}$ as a differential algebra is generated over $\mathcal{E}_0$ by an element $b \in B$. If in this case $\mathcal{E}_0$ possesses sufficiently many homomorphisms into the field K, then the algebra $\mathcal{E}$ has the same property.*

Let us introduce the following notation and assumptions: denote by $\mathbb{N}^n$ the Cartesian product of n copies of the set $\mathbb{N}$ of positive integers; denote by $\leq$ the total ordering on $\mathbb{N}^n$ for which $r = \langle r_1, \dots, r_n \rangle < s = \langle s_1, \dots, s_n \rangle$ if $|r| = r_1 + \dots + r_n < |s| = s_1 + \dots + s_n$ and for $|r| = |s|$ the n-tuples r and s are compared with respect to the lexicographic order; and denote by $\preccurlyeq$ the partial ordering on $\mathbb{N}^n$ for which $r = \langle r_1, \dots, r_n \rangle \preccurlyeq s = \langle s_1, \dots, s_n \rangle$ if and only if $r_1 \leq s_1, \dots, r_n \leq s_n$.

LEMMA 51.3. *For any nonempty subset S of the set $\mathbb{N}^n$ the set S_0 consisting of all minimal elements of the set S with respect to the partial ordering $\preccurlyeq$ is finite and any element of S is comparable with some element of S_0.*

PROOF. It is clear that from any infinite sequence of positive integers we can choose an infinite nondecreasing subsequence. Therefore, from an arbitrary infinite sequence of elements of $\mathbb{N}^n$ we can choose an infinite subsequence in which the first coordinates are nondecreasing, after that we choose an infinite subsequence of this subsequence in which the second coordinates are nondecreasing and so on, until we obtain on the nth step an infinite subsequence in which all members are mutually comparable and are nondecreasing with respect to the ordering $\preccurlyeq$. Hence, in any infinite subset of $\mathbb{N}^n$ there are two distinct elements comparable with respect to the ordering $\preccurlyeq$. Since in S_0 all elements are mutually incomparable, this proves that S_0 is a finite set.

Obviously, $r \preccurlyeq s \Rightarrow r \leq s$ and there exists only a finite set of elements r of $\mathbb{N}^n$ for which $r \leq s$. Therefore, in $\mathbb{N}^n$ there are no strictly increasing infinite chains with respect to the ordering $\leq$. This proves the second assertion of the lemma. The lemma is proved.

PROOF OF LEMMA 51.2. For any collection of n positive integers $r = \langle r_1, \dots, r_n \rangle$ we denote by $b^{(r)}$ the element $\partial_1^{r_1} \cdots \partial_n^{r_n}(b)$ and denote by $\mathcal{E}_r^-$ a K-subalgebra of $\mathcal{E}$ generated over $\mathcal{E}_0$ by all elements of the form $b^{(r')}$, where $r' < r$. We put $\mathcal{E}_r \rightleftharpoons \mathcal{E}_r^-[b^{(r)}]$. It is clear that $\mathcal{E} = \cup_{r \in \mathbb{N}^n} \mathcal{E}_r$ and all algebras $\mathcal{E}_r$ are finitely generated over $\mathcal{E}_0$.

We choose in $\mathbb{N}^n$ a subset S by setting $r \in S$ if and only if for element $b^{(r)}$ there exists a nonzero polynomial $f_r(t) \in \mathcal{E}_r^-[t]$ for which $f_r(b^{(r)}) = 0$. By Lemma 51.3, the subset S_0 consisting of minimal elements of the set S with respect to the ordering

$\preccurlyeq$ is finite. Let us fix for any collection $r \in S_0$ a nonzero polynomial $f_r(t) \in \mathcal{E}_r^-[t]$ of minimal degree $l = l(r)$ for which $f_r(b_r) = 0$. Since char $K = 0$, we have $\frac{d}{dt} f_r(t) \neq 0$ and $\frac{d}{dt} f_r(t)|_{t=b^{(r)}} \neq 0$. Suppose that

$$f_r(t) = a_0(r)t^l + \cdots + a_{l-1}(r)t + a_l(r) \qquad (a_0(r) \neq 0).$$

Then, differentiating the equality $f_r(b^{(r)}) = 0$ by means of ∂_i, we find that for the collection $r' = \langle r_1, \ldots, r_{i_1}, r_i + 1, r_{i+1}, \ldots, r_n \rangle$ and for the polynomial

$$f_{r'}(t) \rightleftharpoons \left(\frac{d}{dt} f_r(t)\right)\bigg|_{t=b^{(r)}} \cdot t + \sum_{j=0}^{l} (\partial_i a_j) \cdot (b^{(r)})^{l-j} \in \mathcal{E}_{r'}^-[t]$$

we have the equality $f_{r'}(b^{(r')}) = 0$ in the algebra $\mathcal{E}_{r'}$. Successively differentiating the equality $f_{r'}(b^{(r')}) = 0$ with the help of ∂_k, we find that for any $s \succ r \in S_0$ $(s \neq r)$ the element $b^{(s)}$ is a solution of the first order equation of the form $t\left(\frac{d}{dt} f_r(t)\right)\big|_{t=b^{(r)}}$ $+b_s = 0$, where $\frac{d}{dt} f_r(t)|_{t=b^{(r)}}, b_s \in \mathcal{E}_s^-$.

Thus, the set S consists of those and only those collections s for which there exists a collection $r \in S_0$ such that $s \succcurlyeq r$. Moreover, for s ranging over S we can choose nonzero polynomials of smallest degree $f_s(t) \in \mathcal{E}_s^-[t]$ satisfying the condition $f_s(b^{(s)}) = 0$ for which the set of coefficients $E = \{a_0(s) \mid s \in S\}$ of the leading terms of the polynomials $f_s(t)$ is finite.

Let c be an arbitrary nonzero element of the algebra $\mathcal{E}$. Then for some collection $q \in \mathbb{N}^n$ the element $a = c \cdot \prod_{e \in E} e$ is contained in the algebra $\mathcal{E}_q^-$ finitely generated over $\mathcal{E}_0$ and, by Corollary of Lemma 51.1, $\varphi(a) \neq 0$ for some homomorphism $\varphi\colon \mathcal{E}_q^- \to K$. Since $\mathcal{E} = \cup_{r \leq t} \mathcal{E}_r^-\left[b^{(r)}\right]$, $\mathcal{E}_{r_1}^- \supseteq \mathcal{E}_{r_2}^-$ for $r_1 \geq r_2$, and $\mathcal{E}_r^- = \mathcal{E}_{r'}$, where r' is a maximal element of the subset $\{s \in \mathbb{N}^n \mid s < r\}$, we can apply Lemma 51.1 in order to extend the homomorphism $\varphi_r^-\colon \mathcal{E}_r^- \to K$ successively to a homomorphism $\varphi_r\colon \mathcal{E}_r \to K$ starting from $r = q$. Since $\varphi_r^-(a) = \varphi(a) \neq 0$, we have $\varphi_r^-(a_0(r)) \neq 0$ for $r \in S$, where $a_0(r)$ is the coefficient of the leading term of the polynomial $f_r(t)$, and the last assertion of Lemma 51.1 guarantees extendability of the homomorphism φ_r^- for any $r \in S$. Finally we obtain a homomorphism $\psi\colon \mathcal{E} \to K$ for which $\psi(a) = \varphi(a) \neq 0$. Hence, for any nonzero element c of the algebra $\mathcal{E}$ there exists a homomorphism ψ such that $\psi(c) \neq 0$ in the field K. The lemma is proved.

Proof of Theorem 51.2. Let $a \neq 0$ be an arbitrary nonnilpotent element of $\mathcal{E}$ and let $\mathcal{E}_a$ be the algebra of fractions of the algebra $\mathcal{E}$ with respect to the element a. From Propositions 51.2 and 51.1 it follows that the kernel of the canonical homomorphism $\varkappa_a\colon \mathcal{E} \to \mathcal{E}_a$ is invariant with respect to any derivation $\partial \in \mathrm{Der}_K \mathcal{E}$. Hence, we can endow $\varkappa_a(\mathcal{E})$ with the structure of a differential algebra for which $\varkappa_a\colon \mathcal{E} \to \varkappa_a(\mathcal{E})$ is a homomorphism of differential K-algebras. Since any element of the algebra $\mathcal{E}_a$ is representatable in the form $b \cdot ((\varkappa_a(a))^{-1})^i$, where $b \in \varkappa_a(\mathcal{E})$, we extend an arbitrary derivation $\partial \in \mathrm{Der}_K \varkappa_a(\mathcal{E})$ to a derivation of the algebra $\mathcal{E}_a$ by setting

$$^{\partial}(b \cdot (\varkappa_a(a)^{-1})^i) = {}^{\partial} \cdot (\varkappa_a(a)^{-1})^i - ib \cdot (\varkappa_a(a)^{-1})^{i+1} \cdot \varkappa({}^{\partial}a).$$

It is immediate that this derivation is well defined and that ∂ determines a derivation on $\mathcal{E}_a$. This permits us to consider $\mathcal{E}_a$ as a differential algebra with respect to $\partial_1, \ldots, \partial_n$.

Obviously, $\mathcal{E}_a$ is also a finitely generated differential K-algebra. Let J be a maximal differentially invariant ideal of the algebra $\mathcal{E}_a$ not containing the unit element. Its existence follows from the standard argument using Zorn's lemma. It is clear that under the through homomorphism $\varphi_1: \mathcal{E} \to \mathcal{E}_a \to \mathcal{E}_a/J$ the image of the nonnilpotent element a is nonzero. From Corollary of Proposition 51.2 it follows that in the algebra $\mathcal{E}_a/J$ there are no zero divisors. Therefore, the existence of a homomorphism $\varphi_2: \mathcal{E}_a/J \to K$ for which $\varphi_2(\varphi_1(a)) \neq 0$ follows from Lemma 51.2 by induction on the number of differential generators of the algebra $\mathcal{E}_a/J$. Thus, $\varphi = \varphi_2\varphi_1$ is a desired homomorphism of the algebra $\mathcal{E}$ into the field K. Theorem 51.2 is proved.

§52. Universal enveloping algebras

52.1. The Poincaré–Birkhoff–Witt theorem.

DEFINITION 52.1. A linear space L over a field K with a fixed bilinear operation $[\,,\,]\colon L \otimes_K L \to L$ is called a *Lie K-algebra* if for any $a, b, c \in L$ the equalities

$$[a, a] = 0, \qquad [b, c] + [c, b] = 0, \tag{52.1}$$

$$[a, [b, c]] + [b, [c, a]] + [c, [a, b]] = 0 \quad \text{(the Jacobi identity)} \tag{52.2}$$

hold in L.

DEFINITION 52.2. A linear space U over a field K with a fixed bilinear operation $\cdot\colon U \otimes_K U \to U$ is called an *associative K-algebra* if for any $a, b, c \in U$ we have

$$(a \cdot b) \cdot c = a \cdot (b \cdot c). \tag{52.3}$$

Any associative K-algebra U can be endowed with the structure of a Lie algebra by introducing on the space U a new bilinear operation $[\,,\,]$ by the formula $[a, b] \rightleftharpoons a \cdot b - b \cdot a$. We denote by $U^{(-)}$ the new K-algebra on U with respect to the operation $[\,,\,]$. From (52.3) it follows that in the algebra $U^{(-)}$ identities (52.1) and (52.2) are satisfied and that $U^{(-)}$ is a Lie algebra.

By a *representation of a Lie K-algebra L in an associative K-algebra U* we mean an arbitrary homomorphism of the Lie K-algebras $\rho\colon L \to U^{(-)}$. Let us consider the set $\operatorname{End}_K V$ of all linear transformations of some linear space V over the field K. It forms an associative algebra with respect to the natural multiplication of operators. If V is endowed with the structure of a binary algebra, then it follows directly from the definition of the set $\operatorname{Der}_K V$ (see §51) that all K-derivations of the K-algebra V form a Lie subalgebra of $(\operatorname{End}_K V)^{(-)}$.

An arbitrary homomorphism $\rho\colon L \to (\operatorname{End}_K V)^{(-)}$ is called a *representation of the Lie algebra on a linear space V*. For any element g of the Lie algebra L denote by $\operatorname{ad} g$ the linear operator on L transforming c into $[g, c]$. From identities (52.1) and (52.2) it follows that $\operatorname{ad} a \cdot \operatorname{ad} b - \operatorname{ad} b \cdot \operatorname{ad} a = \operatorname{ad}[a, b]$. Hence, $\operatorname{ad}\colon L \to (\operatorname{End}_K L)^{(-)}$ is a representation of the Lie algebra L. It is called the *adjoint representation* and its kernel is called the *center* of the Lie algebra L.

A representation ρ is said to be *faithful* if its kernel is equal to zero. Let $X = \{x_i \mid i \in I\}$ be an arbitrary (generally speaking, infinite) alphabet. On the set $W = W(X)$

of all nonempty words in the alphabet X we introduce an operation $\cdot$ by the formula $a \cdot b = ab$. For any $a, b, c \in W$ this operation satisfies relation (52.3) and endows W with the structure of a semigroup. We denote by $H = H_K(X)$ the linear space over the field K whose basis is formed by all the words from W. We extend the semigroup operation $\cdot$ by bilinearity to the space H. The algebra H is associative with respect to this operation. It is called the *algebra of associative polynomials* in variables $x_i \in X$. It is clear that the set X generates the algebra H and that any mapping $\varphi \colon X \to U$ into an arbitrary associative K-algebra U can be extended to a homomorphism $\psi \colon H \to U$ for which $\psi(x_{i_1} \cdots x_{i_t}) = \varphi(x_{i_1}) \cdot \ldots \cdot \varphi(x_{i_t})$. This shows that H is an absolutely free associative K-algebra and X is its set of free generators. We denote by $\mathfrak{M} = \mathfrak{M}(X)$ the least Lie K-subalgebra in $H^{(-)}$ containing the set X. Obviously, any mapping φ of the set of generators X of the Lie algebra $\mathfrak{M}$ into an arbitrary Lie K-subalgebra L of the algebra $U^{(-)}$ is extendable to a Lie algebra homomorphism $\mathfrak{M} \to L$. This shows that the Lie algebra $\mathfrak{M}$ would be absolutely free in the class of all Lie algebras if any Lie algebra L have a faithful representation in a certain associative algebra U. The proof of existence of a representation of this kind is based on the following construction.

Let us choose a basis $E = \{e_i \mid i \in I\}$ of a Lie K-algebra L. Then

$$[e_i, e_j] = \sum_q c_{ij}^q e_q \qquad (c_{ij}^q \in K,\ i, j, q \in I).$$

From identities (52.1) and (52.2) it follows that these are equivalent to the following equalities between the structure constants c_{ij}^q:

$$\begin{gathered} c_{ii}^t = 0, \qquad c_{ij}^t = -c_{ji}^t, \\ \sum_q (c_{ij}^q \cdot c_{ql}^t + c_{jl}^q \cdot c_{qi}^t + c_{li}^q \cdot c_{qj}^t) = 0 \qquad (i, j, q, t \in I). \end{gathered} \tag{52.4}$$

Denote by $H = H(E)$ the free associative K-algebra with the set of free generators E and denote by $J = J(L)$ the least ideal of H containing all elements of the form

$$e_i e_j - e_j e_i - \sum_q c_{ij}^q e_q. \tag{52.5}$$

We introduce a total ordering on the set E (on the set of indices I) in an arbitrary way. (The existence of an ordering of this kind follows from Zorn's lemma.) This total ordering induces a partial ordering on the set $W = W(E)$ for which $a < b$ if the word a has lesser length than the word b and the words of the same length are comparable if and only if they have equal numbers of entries for each letter $e_i \in E$; in this case

$$a < b \Leftrightarrow a = ce_j a' \ \&\ b = ce_i b' \ \&\ e_i > e_j.$$

We denote by $U = U(L)$ the linear subspace in H whose basis consists of all the words of the form $e_{i_1} e_{i_2} \cdots e_{i_t}$ $(i_1 \le i_2 \le \cdots \le i_t)$.

LEMMA 52.1. *There exists a linear operator $\pi \in \mathrm{End}_K H$ such that*

a) $\pi(u) = u$ *for any* $u \in U(L)$;

b) $\pi(e_ie_j) = e_je_i + \sum_q c_{ij}^q e_q$ *for* $i > j$;
c) $\pi(w) = \pi(a\pi(b)c)$ *for* $w = abc \in W$;
d) $\pi(H) = U(L)$.

PROOF. Using property b), we define the operator π by induction: $\pi(w) = w$ for $w \in U$, $\pi(w) = \pi(a\pi(e_ie_j)b)$ for $w = ae_ie_jb \in W$ and $i > j$ & $ae_i \in U$. The mapping π is well defined, since all the words occurring with nonzero coefficients in the decomposition of $a\pi(e_ie_j)b$ with respect to the basis W are strictly lesser than the word w, and in W there are no infinite monotone decreasing word chains with respect to the ordering under consideration. Then property d) of the operator π is evident, and the first two relations of (52.4) imply

$$\pi(e_ie_j - e_je_i) = \sum_q c_{ij}^q e_q \tag{52.6}$$

for any $e_i, e_j \in E$.

Now we show that the operator π has property c). Let $w = e_ie_je_l$ $(i > j > l)$. Then

$$\begin{aligned}
&\pi(e_ie_je_l) - \pi(e_i\pi(e_je_l)) \\
&\quad = \pi(e_je_ie_l - e_ie_le_j) + \pi\left(\sum_q (c_{ij}^q e_qe_l - c_{jl}^q e_ie_q)\right) \\
&\quad = \pi(e_je_le_i - e_le_ie_j) + \sum_q \pi(c_{ij}^q e_qe_l - c_{jl}^q e_ie_q + c_{il}^q(e_je_q - e_qe_j)) \\
&\quad = \sum_q \pi(c_{ij}^q(e_qe_l - e_le_q) + c_{jl}^q(e_qe_i - e_ie_q) + c_{il}^q(e_je_q - e_qe_j)) \\
&\quad = \sum_{t,q} (c_{ij}^q c_{ql}^t + c_{jl}^q c_{qi}^t + c_{il}^q c_{jq}^t)e_t = 0.
\end{aligned}$$

(The latter two equalities are obtained from (52.6), (52.1), and (52.2).)

Consider the general case: $w = abc$. Since with respect to the ordering $<$ introduced above on the set W there are no infinite monotone decreasing word chains in W, we can assume that property c) has already been proved for all words preceding w. If $b \in U$, then everything is clear. Let $b = b'e_pe_qb''$ and $w = w'e_ie_jw''$, where $p > q$, $i > j$, and $b'e_p, w'e_i \in U$. Then three cases are possible.

1. $p = i$ & $q = j$. Then $w' = ab'$ and by the definition of the operator π and by inductive arguments we have

$$\pi(a\pi(b)c) = \pi(a\pi(b'\pi(e_ie_j)b'')c) = \pi(ab'\pi(e_ie_j)b''c) = \pi(w).$$

2. $j = p$. Then $b = e_je_qb''$ and $w = w'e_ie_je_qb''d$. Using once more the definition of the operator π, inductive agruments, and also the equality $\pi(\pi(e_ie_j)e_q) = \pi(e_i\pi(e_je_q))$ which has already been proved, we obtain

$$\begin{aligned}
\pi(a\pi(b)c) &= \pi(a\pi(\pi(e_je_q)b'')d) = \pi(w'e_i\pi(e_je_q)b''d) \\
&= \pi(w'\pi(e_i\pi(e_je_q))b''d) = \pi(w'\pi(\pi(e_ie_j)e_q)b''d) \\
&= \pi(w'\pi(e_ie_j)e_qb''d) = \pi(w'\pi(e_ie_j)w'') = \pi(w).
\end{aligned}$$

3. $w = de_ie_jfe_pe_qg$. Then the definition of the operator π and inductive agruments lead to the equalities

$$\begin{aligned}\pi(a\pi(b)c) &= \pi(a\pi(b'\pi(e_pe_q)b'')c) = \pi(ab'\pi(e_pe_q)b''c) \\ &= \pi(d\pi(e_ie_j)f\pi(e_pe_q)g) = \pi(d\pi(e_ie_j)fe_pe_qg) = \pi(w).\end{aligned}$$

Property c) holds in all three cases. The lemma is proved.

LEMMA 52.2. *H is the direct sum of the subspaces $J(L)$ and $U(L)$. The kernel* $\operatorname{Ker}\pi$ *of the operator π coincides with $J(L)$ and $\pi^2 = \pi$.*

PROOF. The equality $\pi^2 = \pi$ follows from properties a) and d) of the operator π. From property c) it follows that $\operatorname{Ker}\pi$ is an ideal of the algebra H, and formula (52.6) shows that $J(L)$ is contained in $\operatorname{Ker}\pi$. Furthermore, since $\pi(e_ie_j) - e_ie_j \in J(L)$ $(i > j)$, it follows from the inductive definition of the operator π that any word $w \in W$ and, hence, any element $w \in H$ can be represented in the form $v + u$, where $v \in J$ and $u \in U$. Now if $w \in \operatorname{Ker}\pi$, then $\pi(w) = 0 = \pi(v) + \pi(u) = 0 + u$. Therefore, $w = v \in J$. This proves that $J = \operatorname{Ker}\pi$. Hence, for any element $w \in H$ we have a decomposition $w = (w - \pi(w)) + \pi(w)$, where $\pi(w) \in U$ and $w - \pi(w) \in J$, and this decomposition is unique. The lemma is proved.

Let us define on the linear space $U(L)$ a bilinear operation $\cdot$ by the formula $a \cdot b \rightleftharpoons \pi(ab)$. From Lemma 52.2 it follows that the K-algebra $U(L)$ equipped with this operation is isomorphic to the quotient algebra $H/J(L)$ and, hence, $U(L)$ is an associative K-algebra. Formula (52.6) shows that the linear subspace spanned by E is a Lie subalgebra of $U(L)^{(-)}$ isomorphic to the Lie algebra L.

THEOREM 52.1 (Poincaré–Birkhoff–Witt). *For each Lie K-algebra L there exists a unique* (*up to isomorphism*) *associative K-algebra $U(L)$ having the following properties*:

a) *the Lie algebra $U(L)^{(-)}$ contains L as a Lie subalgebra*;

b) *for any basis $E = \{e_i \mid i \in I\}$ with a given total ordering on it a basis of the algebra $U(L)$ is formed by the monomials of the form $e_{i_1} \cdot e_{i_2} \cdot \ldots \cdot e_{i_t}$, where $i_1 \leq i_2 \leq \cdots \leq i_t$ and $e_{i_j} \in E$*;

c) *any representation $\rho\colon L \to U^{(-)}$ in some associative K-algebra U can be extended to a homomorphism $\rho\colon U(L) \to U$ of associative algebras.*

PROOF. We shall show that the associative K-algebra $U(L)$ constructed above for the Lie algebra L with a fixed totally ordered basis E in it has the property c). By Lemma 52.2, the algebra $U(L)$ is isomorphic to $H(E)/J(L)$. Let us extend the mapping $\rho\colon L \to U$ to a homomorphism $\psi\colon H(E) \to U$ for which $\psi(e_i) = \rho(e_i)$ (this is possible, because $H(E)$ is an absolutely free associative K-algebra with the set of free generators E). Then from the definition of the representation ρ it follows that

$$\psi\Big(e_ie_j - e_je_i - \sum_q c_{ij}^q e_q\Big) = \rho\Big([e_i, e_j] - \sum_q c_{ij}^q e_q\Big) = 0.$$

Hence, $\psi(J(L)) = 0$ and, by the theorem on homomorphisms, ψ induces a homomorphism of the quotient algebra $H(E)/J(L)$ into U. This homomorphism of $U(L)$

into U restricted on L coincides with the Lie algebra homomorphism ρ. This proves that property c) holds for the algebra $U(L)$.

The remaining assertions of the theorem (property b) and uniqueness of the algebra $U(L)$ up to isomorphism) are now clear. The theorem is proved.

The associative K-algebra $U(L)$ is called the *universal enveloping algebra* of the Lie K-algebra L.

COROLLARY. *The universal enveloping algebra of the free Lie K-algebra $\mathfrak{M}$ with the set of free generators X is the absolutely free associative algebra with the same set of free generators.*

We denote by $U_n(L)$ $(n = 1, 2, \dots)$ the linear subspace of $U(L)$ spanned by all monomials whose length is at most n. It is clear that

$$U_1(L) = L, \qquad U_i(L) \subset U_{i+1}(L), \qquad U(L) = \sum_{i=1}^{\infty} U_i(L).$$

The multiplication formula in the algebra $U(L)$ shows that

$$(e_{i_1}^{m_1} \cdot \dots \cdot e_{i_l}^{m_l}) \cdot (e_{i_1}^{n_1} \cdot \dots \cdot e_{i_l}^{n_l}) = e_{i_1}^{m_1+n_1} \cdot \dots \cdot e_{i_l}^{m_l+n_l} + u, \tag{52.7}$$

where $u \in U_{i-1}(L)$ $(i = m_1 + n_1 + \dots + m_l + n_l)$. Therefore, $U_m \cdot U_n \subseteq U_{m+n}$, and the subspaces $U_i(L)$ determine an ascending filtration, which is called the canonical filtration on the universal enveloping algebra $U(L)$. Denote by $\mathrm{Gr}\, U(L)$ the algebra whose linear space is the direct sum of the quotient spaces $G_n = U_n(L)/U_{n-1}(L)$ $(n = 1, 2, \dots)$ and in which the operation is given by factorization of the multiplication operation $\cdot : U_m \otimes_K U_n \to U_{m+n}$ in $U(L)$. From relation (52.7) it follows that $\mathrm{Gr}\, U(L)$ is an associative and commutative K-algebra which is isomorphic to the free associative and commutative K-algebra $S(L)$ over the space L, i.e., any K-linear mapping of the space L into an arbitrary associative and commutative algebra can be extended to a unique homomorphism of $\mathrm{Gr}\, U(L)$ into this algebra.

52.2. Shirshov basis in a free Lie algebra. Let $X = \{x_1, x_2, \dots\}$ be a countable alphabet. We shall assume that the letters of this alphabet are totally ordered in a natural way: $x_i < x_{i+1}$. The expressions of the form $x_{i_1} \cdots x_{i_k}$, where $x_{i_j} \in X$, are said to be *words* in the alphabet X and the number k is called the length of the word ($k = 0$ for the empty word). For any word v in the alphabet X we denote by $|v|$ its length and denote by v^l the word $\underbrace{vv \cdots v}_{l \text{ times}}$. The set of all words in the alphabet X will be denoted by $W(X)$.

LEMMA 52.3. *Let for some nonempty words u and v in the alphabet X the equality $uv = vu$ holds. Then there exists a word w such that $u = w^k$ and $v = w^l$ for some integers l and k.*

We shall provide the proof by induction on the length of the word w. It is clear that if $|u| = |v|$, then $u = v$ and we can take u for the word w. Suppose that $|u| > |v|$. Then the equality $uv = vu$ implies $u = vu'$. Hence, $vu'v = vvu'$. Therefore, $u'v = vu'$. By the induction assumption, we have $u' = w^{k'}$ and $v = w^l$ for some word $w \in W(X)$. However, then we have $u = vu' = w^{k'+1}$ and $v = w^l$. The lemma is proved.

A total order on the alphabet X permits one to introduce by induction a total order $>$ on the set $W(X)$ by the following rules:

a) the empty word is greater that any other word,

b) if $u = x_i u'$, $v = x_j v'$, and $x_i > x_j$, then $u > v$,

c) if $u = x_i u'$ and $v = x_i v'$, then $u > v$ if and only if $u' > v'$.

The total order thus defined is called the *lexicographic order* on the set of words. It depends on the ordering of the letters in the alphabet X only.

Let us add a new letter y to the alphabet X. We shall assume that in the new alphabet we have $x_{k-1} < y < x_k$ and another letters are ordered in the same way as in X. Let us denote this new alphabet by the letter Y. For any letter x_i $(x_i < x_k)$ we define a mapping $s_{ik}\colon W(x) \to W(Y)$ that transforms any word of the form

$$u_1 x_k x_i u_2 x_k x_i u_3 \cdots u_l x_k x_i u_{l+1},$$

where no word u_j $(j = 1, 2, \dots, l+1)$ contains the subword $x_k x_i$, into the word

$$u_1 y u_2 y u_3 \cdots u_l y u_{l+1}.$$

LEMMA 52.4. *Suppose that $u, v \in W(X)$ and x_i is the smallest letter occurring in these words. Then for any $k > i$ the inequality $u \underset{X}{<} v$ holds if and only if $s_{ik}(u) \underset{Y}{<} s_{ik}(v)$.*

PROOF. Suppose that $u \underset{X}{<} v$. Then the following cases are possible.

1. v is the empty word. Then $s_{ik}(v)$ is the empty word and $s_{ik}(v) > s_{ik}(u)$.

2. $u = x_j u'$, $v = x_t v'$, $x_t > x_j$. If $x_t \neq x_k$ & $x_k \neq x_i$, then $s_{ik}(u) = x_j s_{ik}(u') \underset{Y}{<} x_t s_{ik}(v') = s_{ik}(v)$ by inductive arguments. If $x_t = x_k \vee x_j = x_k$, then $x_t > x_j$ implies that the first letter of the word $s_{ik}(v)$ is greater than the first letter of the word $s_{ik}(u)$.

3. $u = x_j u'$, $v = x_j v'$, $j \neq k$. In this case $u' < v'$ and from inductive arguments it follows that $s_{ik}(u') < s_{ik}(v')$. Since $j \neq k$, we have

$$s_{ik}(u) = x_j s_{ik}(u') < x_j s_{ik}(v') = s_{ik}(v).$$

4. $u = x_k x_t u'$, $v = x_k$. In this case it is clear that $s_{ik}(u) < s_{ik}(v)$.

5. $u = x_k x_t u'$, $v = x_k x_j v'$, $j \geq t$. If $j = i$, then $t = i$, because x_i is the smallest letter of the words u and v. In this case $v' > u'$, and from inductive arguments we have $s_{ik}(u') < s_{ik}(v')$. Therefore, $s_{ik}(u) = y s_{ik}(u') < y s_{ik}(v') = s_{ik}(v)$. If $j > i$ & $t = i$, then $s_{ik}(v) = x_k s_{ik}(x_j v') > y s_{ik}(u') = s_{ik}(u)$. Finally, if $t \neq i$ & $j \neq i$, then $x_t u' < x_j v'$, and from inductive arguments we have $s_{ik}(x_t u') < s_{ik}(x_j v')$. Therefore, $s_{ik}(u) = x_k s_{ik}(u') < x_k s_{ik}(v') = s_{ik}(v)$. The lemma is proved.

DEFINITION 52.3. A word u in an alphabet X is called *regular* if for any its nontrivial decomposition $u = u_1 u_2$ a strict inequality $u > u_2 u_1$ holds. A word u in the alphabet X is called *cyclic* if $u = w^k$ for some word w and $k > 1$.

It is said that the word v is obtained from u by a cyclic permutation if $v = u_2 u_1$ & $u = u_1 u_2$. For an arbitrary word u, let us consider the set of words that are its cyclic permutations. From Lemma 52.3 it follows that the greatest word in this set is either cyclic or regular.

LEMMA 52.5. *Let u be a word in an alphabet X and x_i the smallest letter occurring in this word. Suppose that $k > i$. Then u is a regular word in the alphabet X if and only if $v = s_{ik}(u)$ is a regular word in the alphabet $Y = X \cup \{y\}$ $(x_{k-1} < y < x_k)$.*

PROOF. Let u be a regular word. Then $u = x_t u'$, where x_t is the greatest letter in the word u and $x_t > x_i$. Suppose that $s_{ik}(u) = v_1 v_2$ is a nontrivial decomposition of the word $s_{ik}(u)$ in the alphabet Y. Then we have $u = u_1 u_2$, where $u_j = s_{ik}^{-1}(v_j)$ $(j = 1, 2)$. Since u is a regular word, we have $u_2 u_1 < u_1 u_2$ and, by Lemma 52.4, $s_{ik}(u_2 u_1) < s_{ik}(u_1) s_{ik}(u_2) = v$. However, the word u_1 begins with the letter $x_t > x_i$. Therefore, $s_{ik}(u_2 u_1) = s_{ik}(u_2) s_{ik}(u_1) = v_2 v_1$. Thus, $v = v_1 v_2 > v_2 v_1$. This proves the "only if" part of the assertion of the lemma.

Conversely, suppose that v is a regular word in the alphabet Y. Then $v = zv'$, where z is the greatest letter from the alphabet Y occurring in v. Since x_i is the smallest letter of the word u, we have $z > x_i$ and the word u cannot begin with the letter x_i. Suppose that $u = u_1 u_2$ is a decomposition of the word u for which $u_2 u_1 \geq u_1 u_2$. In this case the word u_2 also cannot begin with the letter x_i. Therefore,

$$s_{ik}(u_2 u_1) = s_{ik}(u_2) s_{ik}(u_1), \qquad s_{ik}(u_1 u_2) = s_{ik}(u_1) s_{ik}(u_2). \tag{52.8}$$

By Lemma 52.4 we find that $s_{ik}(u_2 u_1) \geq s_{ik}(u_1 u_2)$. This inequality, together with relation (52.8), contradicts the fact that the word $v = s_{ik}(u)$ is regular in the alphabet Y. The lemma is proved.

LEMMA 52.6. *Any word $u \in W(X)$ can be represented in the form*

$$u = v_1 v_2 \cdots v_k,$$

where $v_1, v_2, \ldots, v_k$ are regular words and $v_1 \leq v_2 \leq \cdots \leq v_k$.

PROOF. Let x_i be the smallest letter occurring in the notation of the word u. If $u = x_i u'$, then from inductive arguments we have $u' = v_1' v_2' \cdots v_l'$, where v_i' are regular words and $v_1' \leq v_2' \leq \cdots \leq v_l'$. Since x_i is the smallest letter of the word u and v_1' is a regular word, the inequality $x_i \geq v_1'$ can hold only if $v_1' = x_i$. This proves the assertion of the lemma for the case in which the word u begins with the smallest letter participating in it.

It remains to consider the case in which $u = u' x_k x_i u''$, where $x_k > x_i$ (x_i is the smallest letter of the word u). However, in this case the letter $w = s_{ik}(u)$ in the alphabet Y has lesser length than the word u, and it follows from the inductive arguments that there exists a decomposition

$$w = w_1 \cdots w_k,$$

where $w_1, \ldots, w_k$ are regular words in the alphabet Y and $w_1 \underset{Y}{\leqslant} w_2 \underset{Y}{\leqslant} \cdots \underset{Y}{\leqslant} w_k$. From Lemma 52.5 it follows that the words $v_j = s_{ik}^{-1}(w_j)$ are regular in the alphabet X, and Lemma 52.4 shows that in this case

$$v_1 \underset{X}{\leqslant} v_2 \underset{X}{\leqslant} \cdots \underset{X}{\leqslant} v_k.$$

Hence, a desired decomposition is obtained in this case as well. The lemma is proved.

COROLLARY. *Let v be a regular word in the alphabet X of length $|v|$ not less than two. Then there exists a nontrivial decomposition $v = v_1 v_2$, where v_1 and v_2 are regular words and $v_1 > v_2$.*

PROOF. By Lemma 52.5, $w = s_{ik}(v)$ is a regular word in the alphabet Y and its length is lesser than the length of the word v in the alphabet X, where x_i is the smallest letter in the word v and $v = v' x_k x_i v''$. If $|w|_Y > 1$, then from inductive arguments it follows that $w = w_1 w_2$, where w_1 and w_2 are regular words in the alphabet Y and $w_1 \underset{Y}{>} w_2$. In this case from Lemmas 52.4 and 52.5 it follows that $v = v_1 v_2$, where $v_j = s_{ik}(w_j)$ $(j = 1, 2)$, is a desired decomposition. The corollary is proved.

Let $\mathfrak{M}$ be the absolutely free Lie K-algebra with the countable set of free generators X. In $\mathfrak{M}$ we define a subset $M(X)$ by the following inductive rules:

a) all elements $x_i \in X$ belong to $M(X)$;

b) if $u, v \in M(X)$, then $[u, v] \in M(X)$.

The elements of the set $M(X)$ are called *commutators*.

By Corollary of Theorem 52.1, we can assume that the Lie algebra $\mathfrak{M}$ is embedded in the free associative K-algebra $H(X)$ and that the linear space of $\mathfrak{M}$ is spanned by $M(X)$.

For any element $w \in H(X)$ we denote by $l(w)$ the greatest word in $W(X)$ which occurs with a nonzero coefficient in the decomposition of the word w with respect to the basis $W(X)$.

LEMMA 52.7. *In every regular word $v \in W(X)$ we can place square brackets in such a way that for the commutator $[u] \in M(X)$ thus obtained the equality $l([u]) = u$ will hold in the algebra $H(X)$.*

THE PROOF will proceed by induction on the length of the regular word v. The base of induction is $v = x_i \in X$. In this case $[v] = x_i$. Let $|v| \geq 2$. Then from Corollary of Lemma 52.6 it follows that $v = v_1 v_2$, where v_1 and v_2 are regular words of lesser length and for those, by inductive arguments, there exist desired arrangements of square brackets $[v_1]$ and $[v_2]$. We put $[v] \rightleftharpoons [[v_1], [v_2]]$. It is clear that

$$l([v_1] \cdot [v_2]) = v_1 v_2, \qquad l([v_2] \cdot [v_1]) = v_2 v_1,$$
$$[v] = [v_1] \cdot [v_2] - [v_2] \cdot [v_1].$$

Since $v_1 v_2$ is a regular word, this implies that $v_1 v_2 > v_2 v_1$ and $l([v]) = v_1 v_2$. This proves the possibility of an inductive step. The lemma is proved.

Any such arrangement of square brackets on a regular word will be called *admissible*.

THEOREM 52.2 (Shirshov). *Let us fix for any regular word $v \in W(X)$ exactly one admissible arrangement of brackets $[u]$. Then the subset $[M]$ of commutators $[u]$ of this kind, where u ranges over all regular words, forms a basis of the absolutely free Lie algebra $\mathfrak{M}$.*

PROOF. Let $w = \sum_{i=1}^{n} \beta_i [v_i]$, where $\beta_1 \neq 0$ and $v_1 > \cdots > v_n$. Then $l(w) = v_1$. This shows that the set $[M]$ consists of linearly independent elements of the Lie algebra

$\mathfrak{M}$. On the other hand, Lemmas 52.6 and 52.7 imply that any word u is representable in the form

$$u = [v_1][v_2]\cdots[v_k] + \sum_i \beta_i u_i,$$

where $v_1 \leq \cdots \leq v_k$ are regular words, the words $u_i < u$, and u_i and u have the same length and the same composition. From indictive arguments this equality implies that all expressions of the form $[v_1] \cdot \ldots \cdot [v_t]$, where $t = 1, 2, \ldots$ and $v_1 \leq \cdots \leq v_t$ are regular words, span the linear space of the whole algebra $H(X)$. Since $H(X)$ is the universal enveloping algebra for $\mathfrak{M}$, this is possible, by Theorem 52.1, only if $[M]$ is a basis of the Lie algebra $\mathfrak{M}$. The theorem is proved.

COROLLARY. *For any nonzero Lie element $w \in \mathfrak{M}$ the word $l(w)$ is regular.*

We note that if in a word v all participating letters occur exactly once, then the word is regular provided the greatest letter is standing on the first place. For example, let $v = zx_{\sigma(1)}\cdots x_{\sigma(N)}$, where $z > x_i$ and σ is an arbitrary permutation of the set $\{1, \ldots, N\}$. Then $[v] = [[\cdots[[z, x_{\sigma(1)}], x_{\sigma(2)}], \ldots], x_{\sigma(N)}]$ is an admissible arrangement of square brackets on v. Commutators with such arrangement of brackets are called *left-normed* and are usually denoted by $[z, x_{\sigma(1)}\cdots x_{\sigma(N)}]$, having in mind left-normed arrangement of brackets. Theorem 52.2 implies that the Lie polynomial

$$\sum_{\sigma \in S_N} \beta_\sigma [z, x_{\sigma(1)} \cdots x_{\sigma(N)}] \qquad (\beta_\sigma \in K)$$

is equal to zero if and only if all β_σ's are zero.

§53. Modules over associative algebras

53.1. Basic notions. Let U be an associative K-algebra and let V be a linear space over the field K. Suppose that a bilinear operation $\times\colon U \otimes_K V \to V$ satisfies the condition

$$(a \cdot b) \times v = a \times (b \times v),$$

where $a, b \in U$ and $v \in V$. Then V is called a *left module* over the associative K-algebra U or a *(left) U-module*. A linear mapping $\varphi\colon V_1 \to V_2$ is called a *homomorphism* of left U-modules if for any $a \in U$ and $v \in V_1$ we have the equality

$$\varphi(a \times v) = a \times (\varphi(v)).$$

The notions of isomorphism, monomorphism, and endomorphism are naturally defined.

A linear subspace V' of a U-module V is called an *U-submodule* if $U \times V' \subseteq V'$. It is said that the U-module V is the *sum* of its submodules V_i $(i \in I)$ if any element $v \in V$ is representable in the form

$$v = v_{i_1} + v_{i_2} + \cdots + v_{i_k} \qquad (v_{i_j} \in V_{i_j}). \tag{53.0}$$

The module V is the *direct sum* of its submodules V_i $(i \in I)$ if the decomposition (53.0) is unique. If the set I is finite and $I = \{1, \ldots, k\}$, we write

$$V = V_1 \oplus \cdots \oplus V_k.$$

Note that if we fix a basis $\{u_j \mid j \in J\}$ in the algebra U and introduce in each U-module V a unary operation ω_j $(j \in J)$ by the rule $\omega_j(v) = u_j \times v$, then the class of all U-modules can be considered as a variety of universal algebras whose signature consists of the unary operations ω_j $(j \in J)$ and the identities defining this variety have the form

$$\omega_i(\omega_j(v)) = \sum_q c_{ij}^q \omega_q(v) \qquad (i, j \in J,\ v \in V),$$

where $c_{ij}^q \in K$ are the structure constants of the associative K-algebra U uniquely determined by the equalities

$$u_i \cdot u_j = \sum_q c_{ij}^q u_q \qquad (i, j \in J).$$

This permits one to use the universal algebra language in the theory of modules.

53.2. Injective modules. A module V over an associative K-algebra U will be called *injective* if for any U-module M and any submodule M' of it an arbitrary module homomorphism $\varphi\colon M' \to V$ can be extended to a homomorphism $\psi\colon M \to V$ $(\psi|_{M'} = \varphi)$.

Let us adjoin a unit element to the algebra U externally, as it was done in §51, and denote the algebra thus obtained by U_{id}. Then U is a subalgebra of U_{id} of codimension 1 and any element $a \in U_{\text{id}}$ can be uniquely represented in the form $a = \beta_a e + a'$, where $a' \in U$, e is the unit element of the algebra U_{id}, and the multiplication in U_{id} satisfies the equality

$$a \cdot b = \beta_a \beta_b \cdot e + \beta_a b' + \beta_b a' + a'b'.$$

Therefore, any U-module M can be transformed into an U_{id}-module by setting

$$a \times v = \beta_a \cdot v + a' \times v \qquad (a \in U_{\text{id}},\ v \in V). \tag{53.1}$$

We endow U_{id} with the structure of a left U-module with respect to the operation $u \times a = u \cdot a$, where $u \in U$ and $a \in U_{\text{id}}$. It is clear that any mapping $e \to v$, where v is an element of an arbitrary V-module v, can be extended to a linear mapping $\varkappa\colon U_{\text{id}} \to V$ for which $\varkappa(\beta_a e + a') = \beta_a \cdot v + a' \times v$. It is immediate that $\varkappa$ is a U-module homomorphism. This proves that U_{id} is a free U-module and e is its free generator.

THEOREM 53.1. *Any U-module can be embedded in an injective U-module.*

The proof is based on the following construction. Let M be an arbitrary linear space over the field K. We denote by $\operatorname{Hom}_K(U_{\text{id}}, M)$ the set of all K-linear transformations from U_{id} into M. This set possesses a natural structure of a linear space over K. We can endow it with the structure of a left U-module by setting

$$\langle u \times f, a\rangle = \langle f, au\rangle \qquad (u \in U,\ a \in U_{\text{id}},\ f \in \operatorname{Hom}_K(U_{\text{id}}, M)) \tag{53.2}$$

(here $\langle f, a\rangle$ is an element of M obtained as a result of an application of the transformation f to the element $a \in U_{\text{id}}$). Let M be a left U-module. Consider the linear transformation $i\colon M \to \operatorname{Hom}_K(U_{\text{id}}, M)$ for which

$$\langle i(m), u\rangle = u \times m \qquad (m \in M,\ u \in U_{\text{id}}). \tag{53.3}$$

It is clear from (53.1)–(53.3) that i is a U-module homomorphism. Since $\langle i(m), e\rangle = e \times m = m$, i is a monomorphism, and in order to prove the theorem it remains to show that $\mathrm{Hom}_K(U_{\mathrm{id}}, M)$ is an injective U-module. This is a consequence of the following two assertions.

LEMMA 53.1. *Let J be a left submodule of the U-module U_{id} and let $\varphi\colon J \to \mathrm{Hom}_K(U_{\mathrm{id}}, M)$ be a U-module homomorphism. Then there exists a module homomorphism $\psi\colon U_{\mathrm{id}} \to \mathrm{Hom}_K(U_{\mathrm{id}}, M)$ such that $\psi|_J = \varphi$.*

PROOF. Consider a linear mapping $g\colon J \to M$ for which $g(j) = \langle\varphi(j), e\rangle$ and extend it in an arbitrary way to a linear mapping $f\colon U_{\mathrm{id}} \to M$. Then $f \in \mathrm{Hom}_K(U_{\mathrm{id}}, M)$. Since e is a free generator of the U-module U_{id}, the mapping $e \to f$ can be extended to a U-module homomorphism $\psi\colon U_{\mathrm{id}} \to \mathrm{Hom}_K(U_{\mathrm{id}}, M)$ for which $\psi(a) = a \times f$ ($a \in U_{\mathrm{id}}$). However, in this case from (53.1) and (53.2) and from the fact that φ is a U-module homomorphism for any $j \in J$ and $a \in U_{\mathrm{id}}$ it follows that

$$\begin{aligned}\langle\psi(j) - \varphi(j), a\rangle &= \langle j \times f, a\rangle - \langle a \times \varphi(j), e\rangle \\ &= \langle f, a \times j\rangle - \langle\varphi(a \times j), e\rangle = g(a \times j) - \langle\varphi(a \times j), e\rangle.\end{aligned}$$

Hence, $\psi|_J = \varphi$ and ψ is a desired homomorphism. The lemma is proved.

LEMMA 53.2 (Baer). *A U-module V is injective if and only if any homomorphism $J \to V$ of an arbitrary submodule J of the U-module U_{id} can be extended to a U-module homomorphism $U_{\mathrm{id}} \to V$.*

PROOF. The necessity of the condition is obvious.

Let us prove its sufficiency. Let $\varphi\colon M' \to V$ be a homomorphism of U-modules and let M' be a submodule of the U-module M. Denote by the letter S the set whose elements are the pairs (L, ψ), where L is a U-submodule of M containing M and ψ is a homomorphism of the U-module L into V for which $\psi|_{M'} = \varphi$. We introduce a partial ordering on the set S for which $(L_1, \psi_1) \subseteq (L_2, \psi_2)$ if and only if $L_1 \subseteq L_2$ & $\psi_2|_{L_1} = \psi_1$. Since for any totally ordered subset $\{(L_i, \psi_i) \mid i \in I\}$ of the set S the pair (L', ψ'), where $L' = \cup_i L_i$ and $\psi'|_{L_i} = \psi_i$, is an element of the set S for which $(L', \psi') \supseteq (L_i, \psi_i)$ ($i \in I$), the set S satisfies the conditions of Zorn's lemma with respect to the partial ordering $\subseteq$ and, hence, it contains some maximal element (L, ψ).

We shall show that $L = M$. Assume the converse. Then $L \neq M$ and there exists an element m of M not belonging to L. We denote by L' the U-submodule of M consisting of all elements of the form $g + u \times m + \beta \cdot m$, where $g \in L$, $u \in U$, $\beta \in K$. Since e is a free generator of the U-module U_{id}, the mapping $e \to m$ can be extended to a homomorphism $\varkappa\colon U_{\mathrm{id}} \to M$ for which $\varkappa(a) = a \times m$. Denote by the letter J the U-submodule of U_{id} equal to $\varkappa^{-1}(L \cap \varkappa(U_{\mathrm{id}}))$. Then by the assumption imposed on the U-module V the homomorphism $\psi\varkappa\colon J \to V$ can be extended to a U-module homomorphism $\delta\colon U_{\mathrm{id}} \to V$. Define a K-linear mapping $\psi'\colon L' \to V$ by setting

$$\psi'(g+u\times m+\beta\cdot m) = \psi(g)+u\times\delta(e)+\beta\cdot\delta(e) \qquad (g \in L,\ u \in U,\ \beta \in K). \tag{53.4}$$

We verify that this mapping is well defined. To this end, it suffices to show that if $g + u \times m + \beta \cdot m = 0$ in M, then

$$\psi(g) + u \times \delta(e) + \beta \cdot \delta(e) = 0 \tag{53.5}$$

in V. Since $g = -u \times m - \beta \cdot m = 0$, we have $g \in L \cap \varkappa(U_{\mathrm{id}})$ and $-u \times e - \beta \cdot e \in J$. But then

$$\begin{aligned} -u \times \delta(e) - \beta \cdot \delta(e) &= \delta(-u \times e - \beta \cdot e) \\ &= \psi \cdot \varkappa(-u \times e - \beta \cdot e) = \psi(-u \times m - \beta \cdot m) = \psi(g), \end{aligned}$$

and this is another form of equality (53.5). From the definition (53.4) of the mapping ψ' it follows directly that $\psi' \colon L' \to V$ is a U-module homomorphism and $\psi'|_L = \psi$. However, in this case $(L, \psi) \subsetneqq (L', \psi')$, which contradicts the maximality of the element (L, ψ) in the set S.

Hence, $L = M$ and $\psi \colon M \to V$ is a desired extension of the homomorphism $\varphi \colon M' \to V$. The lemma is proved.

Lemmas 53.1 and 53.2 show that $\mathrm{Hom}_K(U_{\mathrm{id}}, M)$ is an injective U-module containing a submodule isomorphic to the U-module M. Theorem 53.1 is proved.

The proof of the next assertion follows immediately from the definition of injective modules.

PROPOSITION 53.1. *Suppose that a U-module V is the direct sum of a finite set of its U-submodules $V_1, \dots, V_k$. Then V is an injective module if and only if each direct summand V_i is an injective module.*

DEFINITION 53.1. An injective U-module P is called an *injective hull* of the U-module M if M is a submodule of P and the intersection of any nonzero U-submodule of P with M is a nonzero submodule.

THEOREM 53.2. *For any U-module M there exists an injective hull which is unique up to isomorphism.*

PROOF. *Uniqueness.* Suppose that there exist two injective U-modules P_1 and P_2 containing M as a U-submodule. Then the identity homomorphism $\mathrm{id} \colon M \to M$ can be extended to a U-module homomorphism $\varkappa \colon P_1 \to P_2$. If P_1 is an injective hull of the U-module M, then the kernel $\operatorname{Ker} \varkappa$ of the homomorphism $\varkappa$ is equal to the zero submodule, because otherwise we have $\operatorname{Ker} \varkappa \cap M \neq 0$, which contradicts the fact that $\varkappa|_M = \mathrm{id}$ is the identity mapping. Hence, $\varkappa$ is a monomorphism and the U-submodule $\varkappa(P_1)$ in P_2 is isomorphic to P_1. Since P_1 is an injective U-module, the isomorphism $\varkappa^{-1} \colon \varkappa(P_1) \to P_1$ can be extended to an epimorphism $\psi \colon P_2 \to P_1$. It is clear that $\operatorname{Ker} \psi \cap \varkappa(P_1) = 0$ and $\operatorname{Ker} \psi \neq 0$ if and only if $\varkappa(P_1) \neq P_2$. Therefore, if P_2 is an injective hull of $M = \varkappa(M)$, then $\operatorname{Ker} \psi = 0$. Hence, ψ is an epimorphism and a monomorphism. This proves that for any two injective hulls P_1 and P_2 of the U-module M there exists an isomorphism $\varkappa$ of these acting identically on M.

Existence of the injective hull. By Theorem 53.1 we can assume that M is contained in some injective U-module V. We denote by the letter S the set of all U-submodules of the module V and consider two subsets S_1 and S_2 of it. The set S_1 is formed by those and only those U-submodules L for which $L \supseteq M$ and the intersection of any nonzero submodule of L with M is a nonzero submodule. The set S_2 consists of those and only those U-submodules L for which $L \cap M = 0$. The sets S_1 and S_2 are partially ordered with respect to the natural inclusion of modules. It is clear that if $\{L_i \mid i \in I\}$ is an arbitrary subset of U-modules of S_1 totally ordered by inclusion, then the U-module

$\bigcup_{i \in I} L_i$ containing all submodules L_i is also contained in S_1. Therefore, by Zorn's lemma, there is a maximal element in the partially ordered set S_1. We denote this submodule by P_1. Similarly we find that there exists a maximal element in S_2. We denote it by P_2.

Since $P_2 \cap M = 0$, we have $P_2 \cap P_1 = 0$ (otherwise the intersection of the nonzero submodule $P_1 \cap P_2$ of P_1 with M should be a nonzero submodule). Hence, by the theorem on homomorphisms, the image of the module P_1 under the canonical homomorphism $\varkappa\colon V \to P = V/P_2$ is isomorphic to the U-module P_1. Since V is an injective module, the isomorphism $\varkappa^{-1}\colon \varkappa(P_1) \to P_1$ can be extended to a homomorphism $\psi\colon P \to V$. Let L be an arbitrary nonzero submodule of the U-module P. Then $L \cap \varkappa(M) \neq 0$, since otherwise we have $\varkappa^{-1}(L) \supsetneqq P_2$ and $\varkappa^{-1}(L) \cap M = 0$, which contradicts the maximality of P_2 in S_2. Since $\varkappa(P_1) \cap \operatorname{Ker} \psi = 0$, we conclude that the kernel of the homomorphism ψ cannot be a nonzero submodule of P. Hence, ψ is a monomorphism and the intersection of any nonzero submodule of the U-module $\psi(P)$ with M is a nonzero submodule. This shows that $\psi(P) \in S_1$. Since $\psi\varkappa(P_1) = P_1$, we have $\psi(P) \supseteq P_1$. By the maximality of P_1 in S_1 we find that $\psi(P) = P_1$. Thus, for the homomorphism $\varphi = \psi\varkappa$ we have

$$\varphi^2 = \varphi, \qquad \operatorname{Ker} \varphi = P_2, \qquad \varphi(V) = P_1.$$

Hence, any element $v \in V$ is uniquely representable in the form

$$v = \varphi(v) + (v - \varphi(v)), \qquad v - \varphi(v) \in P_2,$$

and the module V is the direct sum of its submodules P_1 and P_2. By Lemma 53.3, the module P_1 is injective. Since $P_1 \in S_1$, P_1 is an injective hull of the U-module M. The theorem is proved.

COROLLARY. *If the K-algebra U and the U-module M are finite-dimensional over the field K, then the injective hull of the U-module M is also finite-dimensional over this field.*

Indeed, from the proof of Theorems 53.1 and 53.2 it follows that the injective hull of the U-module M can be realized as a subspace of $\operatorname{Hom}_K(U_{\mathrm{id}}, M)$ and, therefore, its dimension is at most $(\dim_K U + 1)\dim_K M$.

53.3. Density theorem. By a representation of an associative K-algebra U on a linear space V we mean a homomorphism of K-algebras $\rho\colon U \to \operatorname{End}_K V$.

Each representation ρ permits us to endow V with the structure of a U-module with respect to the operation $\times\colon U \otimes_K V \to V$ defined by the formula

$$u \times v = \rho(u)v.$$

This formula shows that, conversely, any module V over an associative K-algebra U determines a representation of the associative K-algebra U on the linear space V.

A representation $\rho\colon U \to \operatorname{End}_K V$ (a U-module V) is called *faithful* if its kernel $\operatorname{Ker} \rho$ is equal to zero. By the *centralizer* of a U-module V (of a representation $\rho\colon U \to \operatorname{End}_K V$) we mean the subset of $\operatorname{End}_K V$ consisting of those and only those linear transformations c for which we have the equalities

$$(53.6) \qquad c \cdot (u \times v) = u \times (c \cdot v) \qquad (\rho(u) \cdot c = c \cdot \rho(u)),$$

where u ranges over U and v ranges over V. This subset is an associative K-subalgebra of $\operatorname{End}_K V$; it is denoted by $\operatorname{End}_U V$.

A U-module V (a representation $\rho\colon U \to \operatorname{End}_K V$) is called *irreducible* if in V there are no U-submodules, except for V and the zero submodule, and $U \times V \neq 0$ ($\rho(U) \neq 0$).

It is said that an associative K-algebra U is *primitive* if it possesses a faithful irreducible representation on some linear space V.

DEFINITION 53.2. An associative K-algebra U is called a *skew field* if

a) the algebra C has the unit element e,

b) for any nonzero element $a \in C$ there exists an element $b \in C$ such that

$$a \cdot b = b \cdot a = e$$

(the element b is called the *inverse* element for a and is usually denoted by a^{-1}).

SCHUR'S LEMMA. *The centralizer of any irreducible U-module V is a skew field.*

PROOF. From (53.6) it follows that the centralizer of the U-module V is precisely the set of all endomorphisms of the U-module V. Therefore, the identity (linear) transformation of the space V is contained in the centralizer and is the unit element of this algebra. Furthermore, for any $c \in \operatorname{End}_U V$ the kernel $\operatorname{Ker} c$ and the image $c \cdot V$ of the homomorphism c are U-submodules of V. By irreducibility of the U-module V the submodules $\operatorname{Ker} c$ and $c \cdot V$ must coincide with either V or the zero submodule. If $c \neq 0$, then there is only one possibility: $c \cdot V = V$ and $\operatorname{Ker} c = 0$. However, in this case the theorem on homomorphisms implies that $c\colon V \to V$ is an isomorphism. Therefore, the linear mapping inverse to c is also an endomorphism of the U-module V and, therefore, belongs to the centralizer. The lemma is proved.

Elements $v_1, \dots, v_k$ of an irreducible U-module V are called *linearly independent over the centralizer* $\operatorname{End}_U V$ if the equality

$$c_1 \cdot v_1 + \dots + c_k \cdot v_k = 0 \qquad (c_i \in \operatorname{End}_U V,\ i = 1, 2, \dots, k)$$

holds only if each c_i is equal to zero.

THEOREM 53.3. *Let v_i and w_i ($i = 1, \dots, k$) be arbitrary elements of an irreducible U-module V. Suppose that $v_1, \dots, v_k$ are linearly independent over the field* $\operatorname{End}_U V$. *Then in the algebra U there exists an element a for which in the module V we have the equalities*

$$a \times v_i = w_i \qquad (i = 1, 2, \dots, k).$$

PROOF. Let us turn the U-module V into a U_{id}-module by formula (53.1). For any elements $e_1, \dots, e_q \in V$ we denote by $\operatorname{Ann}\{e_1, \dots, e_q\}$ the subset of U_{id} containing those and only those elements $b \in U_{\mathrm{id}}$ for which $b \times e_i = 0$ ($i = 1, 2, \dots, q$).

LEMMA 53.3. *Let v be an element of an irreducible U-module V such that $a \times v = 0$ for any $a \in$* $\operatorname{Ann}\{e_1, \dots, e_q\}$. *Then in the centralizer* $\operatorname{End}_U V$ *there exist elements $c_1, \dots, c_q$ such that*

$$v = c_1 \cdot e_1 + \dots + c_q \cdot e_q. \tag{53.7}$$

PROOF. We denote q isomorphic copies of the injective hull P_0 of the U-module V by $P_1, \dots, P_q$ and denote their direct sum by P. Elements of the U-module P are arbitrary strings $(w_1, \dots, w_q)$ ($w_i \in P_0$), and by definition of the direct sum we have

$$a \times (w_1, \dots, w_q) = (a \times w_1, \dots, a \times w_q) \qquad (a \in U_{\mathrm{id}}).$$

Therefore, a permutation σ of the set $\{1, \dots, q\}$ induces an endomorphism c_σ of the U-module P for which

$$c_\sigma(w_1, \dots, w_q) = (w_{\sigma(1)}, \dots, w_{\sigma(q)}).$$

Denote by e_0 the element $(e_1, \dots, e_q) \in P$. Then from the assumption of the lemma it follows that the mapping

$$\varphi \colon U_{\mathrm{id}} \times e_0 \to P$$

for which $\varphi(a \times e_0) \rightleftharpoons (a \times v, 0, \dots, 0)$, is well defined and is a U_{id}-module homomorphism. Since P is an injective U-module by Proposition 53.1, this homomorphism of U-modules can be extended to an endomorphism c_0 of the U-module P. Then

$$(v, 0, \dots, 0) = c_0 \cdot e_0 = c_0 \sum_{i=0}^{q-1} c_{\sigma^i}(e_i, 0, \dots, 0),$$

where σ is a permutation for which $1 \to 2 \to \cdots \to q \to 1$. Taking the projections of both the sides of this equality onto the first direct summand P_1, we find that in the injective hull P_1 of the U-module V we have equality (53.7) for some $c_i \in \mathrm{End}_U P_1$. It remains to show that $c \cdot V \subseteq V$ for any $c \in \mathrm{End}_U P_1$. If $c \cdot V = 0$, then this inclusion is evident. If $c \cdot V \neq 0$, then the intersection with V of the nonzero submodule $c \cdot V$ of the injective hull of the U-module V must be a nonzero submodule. However, the module V (and, hence, the module $c \cdot V$) is irreducible. Therefore, $V = (c \cdot V) \cap V = c \cdot V$ and the restriction of any endomorphism of P_1 to V is an endomorphism of the U-module V. Hence, we can assume that all c_i's in equality (53.7) belong to the centralizer of the U-module V. The lemma is proved.

By the assumption of the theorem, the elements $v_1, \dots, v_k$ are linearly independent over $\mathrm{End}_U V$. Therefore, by Lemma 53.3, for any $i \in \{1, \dots, k\}$ there exists an element $a_i \in U_{\mathrm{id}}$ for which

$$a_i \times v_i = w_i' \neq 0, \qquad a_i \times v_j = 0 \qquad (j \neq i).$$

Since the U-module V is irreducible, we have $w_i = u_i \times w_i'$ for some $u_i \in U$. However, in this case $a = \sum_{i=1}^{k} u_i \cdot a_i$ is a desired element of the associative K-algebra U. The theorem is proved.

§54. Properties of enveloping algebras

54.1. Centralizers of irreducible representations of finite-dimensional Lie algebras. Let L be a Lie K-algebra and let $U(L)$ be its universal enveloping associative K-algebra. By Theorem 53.1, any representation $\rho \colon L \to (\mathrm{End}_K V)^{(-)}$ of the Lie algebra L on a linear space V can be uniquely extended to a representation $\rho \colon U(L) \to \mathrm{End}_K V$ of its universal enveloping algebra on the same linear space. Therefore, any representation of the Lie K-algebra L can be obtained from some representation of its universal enveloping algebra $U(L)$.

A representation of the Lie algebra L on a linear space V is called *irreducible* if the $U(L)$-module V corresponding to this representation is irreducible.

THEOREM 54.1. *Suppose that the ground field K is algebraically closed and that $\mathfrak{g}$ is a finite-dimensional Lie K-algebra. Then the centralizer* $\operatorname{End}_{U(\mathfrak{g})} V$ *of any irreducible representation V of the Lie algebra $\mathfrak{g}$ is isomorphic to the field K and coincides with $K \cdot 1$, where 1 is the identity operator on the space V.*

A U-module M is called *cyclic* (*strictly cyclic*) if there exists an element $v \in M$ for which $M = U_{\text{id}} \times v$ ($M = U \times v$).

LEMMA 54.1. *Let V be a linear space over an algebraically closed field K, let $c \in \operatorname{End}_K V$, and let $K[c]$ be an associative K-subalgebra of* $\operatorname{End}_K V$ *generated by c and the identity operator. Suppose that in the $K[c]$-module V there exists a chain of $K[c]$-submodules $0 = V_0 \subset V_1 \subset V_2 \subset \cdots$ with the following properties:*

a) *the quotient modules V_i/V_{i-1} are strictly cyclic* $(i = 1, 2, \dots)$;

b) *there exists a polynomial $f(t) \in K[t]$ such that for any finite-dimensional quotient module V_i/V_{i-1} we have $f(c) \times V_i \subseteq V_{i-1}$;*

c) $V = \cup_{i=0}^{\infty} V_i$.

If $V = (c - \beta \cdot 1) \times V$ & $f(\beta) \neq 0$ for some $\beta \in K$, then all the quotient modules V_i/V_{i-1} are finite-dimensional.

PROOF. Since all conditions of the lemma imposed on V are satisfied for the quotient module V/V_i $(i = 0, 1, 2, \dots)$ as well, it suffices to show that V_1 is a finite-dimensional $K[c]$-module.

We assume that the chain of modules V_i is strictly increasing. Then from properties a) and c) it follows that we can choose in V nonzero elements $v_1, v_2, \dots$ such that $v_i \in V_i$ & $v_i \notin V_{i-1}$ and any element $v \in V_q$ is representable in the form

$$v = c_1 v_1 + \cdots + c_q v_q \qquad (c_i \in K[c]). \tag{54.1}$$

Suppose that $\dim_K V_1 = \infty$. Then for any polynomial $g(t) \in K[t]$ the element $g(c) \times v_1$ is nonzero and the elements $c^i \times v_1$ $(i = 0, 1, \dots)$ form a basis of the subspace V_1.

Let $\beta \in K$. Suppose that $f(\beta) \neq 0$, where $f(t)$ is the polynomial mentioned in property b). Since by the assumption of the lemma $(c - \beta \cdot 1) \times V = V$, there exists an element $v \in V$ such that

$$(c - \beta \cdot 1) \times v = v_1, \tag{54.2}$$

We assume that $v \in V_q$ & $v \notin V_{q-1}$. Three cases are possible.

1. $q = 1$. Then v is uniquely representable in the form

$$v = \beta_0 v_1 + \beta_1 c \times v_1 + \cdots + \beta_l c^l \times v_1 \qquad (\beta_i \in K,\ \beta_l \neq 0),$$

and the element $(c - \beta \cdot 1) \times v - v_1$ has a nonzero coefficient (equal to β_l) at $c^{l+1} \times v_1$ in the decomposition with respect to the basis $c^i \times v_1$ $(i \in \mathbb{N})$ of the subspace V_1. Hence, equality (54.2) is impossible.

2. $\dim_K V_q/V_{q-1} = \infty$ & $q > 1$. Then the elements $c^i \times v_q + V_{q-1}$ $(i \in \mathbb{N})$ form a basis of the quotient space V_q/V_{q-1} and, as in the first case, we have $(c - \beta \cdot 1) \times v \notin V_{q-1}$. Since $v_1 \in V_{q-1}$, we have again that equality (54.2) is impossible.

3. $\dim_K V_q/V_{q-1} < \infty$ & $q > 1$. Then condition b) implies $f(c) \times V_q \subseteq V_{q-1}$, and, since $f(\beta) \neq 0$, we have $f(t) = (t - \beta)g(t) + \beta'$, where $0 \neq \beta' \in K$ and $g(t) \in K[t]$. Equality (54.2) implies $(c - \beta \cdot 1) \times w = 0$ in the quotient space V_q/V_{q-1}, where $w \neq 0$ is the image of the element v under the canonical homomorphism $V_q \to V_q/V_{q-1}$. Multiplying this equality by $g(c)$, we obtain

$$0 = g(c) \times (c - \beta \cdot 1) \times w = f(c) \times w - \beta' \cdot w = -\beta' \cdot w,$$

which contradicts the fact that $\beta' \neq 0$ and $w \neq 0$.

From condition c) and the three cases discussed above it follows that equality (54.2) does not hold for any $v \in V$. The contradiction thus obtained shows that $\dim_K V_1 < \infty$. The lemma is proved.

PROOF OF THEOREM 54.1. Let $e_1, \dots, e_m$ be a basis of the Lie K-algebra $\mathfrak{g}$. For any collection $s = \langle s_1, \dots, s_m \rangle \in \mathbb{N}^m$ of m positive integers we denote by e^s the element of $U_{\text{id}}(\mathfrak{g})$ equal to $e_1^{s_1} \cdots e_m^{s_m}$. Let us introduce a total order $<$ on $\mathbb{N}^m$ by setting $r = \langle r_1, \dots, r_m \rangle < s = \langle s_1, \dots, s_m \rangle$ if $|r| = r_1 + \cdots + r_m < |s| = s_1 + \cdots + s_m$ and by ordering lexicographically m-tuples with equal sums of coordinates. Let v_0 be an arbitrary nonzero element of V. Then it follows from the irreducibility of the $U(\mathfrak{g})$-module V and from Theorem 53.1 that any element $v \in V$ is representable in the form

$$v = \sum_{r \leq s} \beta_r v_r \qquad (\beta_r \in K,\ v_s = e^s v_0) \tag{54.3}$$

for some $s \in \mathbb{N}^m$, and formula (52.7) implies

$$e^r v_q = v_{r+q} + \sum_{p < r+q} \beta_p v_p \qquad (\beta_p \in K,\ p, q, r \in \mathbb{N}^m). \tag{54.4}$$

Let c be an arbitrary element of the centralizer $\operatorname{End}_{U(\mathfrak{g})} V$ and let $K[c]$ be an associative K-subalgebra generated by c and by the identity operator. Denote by V_s ($s \in \mathbb{N}^m$) a $K[c]$-submodule generated by all elements v_r, where $r < s$, and denote by V_s^+ the $K[c]$-submodule equal to $V_s + K[c] \times v_s$. It is clear that for $K[c]$-submodules V_s the conditions a) and c) of Lemma 54.1 are satisfied and that $V_s \subset V_q$ for $s \leq q$.

Let us show that condition b) of this lemma is also satisfied for these submodules. To this end, we denote by S the subset of $\mathbb{N}^m$ consisting of those and only those s for which $\dim_K V_s^+/V_s < \infty$. Then, by Lemma 51.3, either the set S is empty or in S there exists a finite subset S_0 with the following property: any element s of S is representable in the form $s = q + r$, where $q \in S_0$ and $r \in \mathbb{N}^m$. If the set S_0 is empty, then contition is satisfied. If S is nonempty, then we denote by $f(t)$ the product of all nonzero characteristic polynomials $f_q(t) = \operatorname{Det}(t \cdot 1 - c)$ of the operator c in the space V_q^+/V_q, where q ranges over S_0. Then, by the Hamilton–Cayley theorem, we have $f(c) \cdot V_q^+ \subseteq V_q$ ($q \in S_0$), and formula (54.4) implies

$$f(c) \times e_{r+q} = e^r f(c) v_q - \sum_{p < r+q} \beta_p f(c) v_p \subset f(c) \times V_{r+q}.$$

Hence, $f(c) \times V_s^+ \subseteq V_s$ for any $s \in S$, and condition b) of Lemma 54.1 is satisfied in this case as well.

Since an algebraically closed field is infinite, we can choose β in K in such a way that $f(\beta) \neq 0$ and $c - \beta \cdot 1 \neq 0$. But in this case, by Schur's lemma (see §53) we have $(c - \beta \cdot 1)V = V$ and, by Lemma 54.1, $f(c)v_0 = 0$. Since the field K is algebraically closed, we have $f(t) = (t - \beta_1) \cdots (t - \beta_n)$ for some $\beta_i \in K$. Therefore, in the chain

$$w_0 = v_0, \qquad w_i = (c - \beta_i \cdot 1) \times w_{i-1} \qquad (i = 1, \dots, n-1)$$

there is an element $w_i \neq 0$ for which $(c - \beta_{i+1} \cdot 1)w_i = 0$. But in this case we have $V = U(\mathfrak{g}) \times w_i$ and $(c - \beta_{i+1} \cdot 1) \cdot V = 0$. Hence, $c \in K \cdot 1$ for any $c \in \operatorname{End}_{U(\mathfrak{g})} V$. The theorem is proved.

54.2. Noether property of enveloping algebras. A module V over an associative K-algebra U is called *Noetherian* if any ascending chain $V_1 \subseteq V_2 \subseteq \cdots$ of U-submodules of V is stabilized in a finite number of steps, i.e., $V_{N+k} = V_N$ for some $N \in \mathbb{N}$ and any $k \in \mathbb{N}$. It is clear that any submodule of a Noetherian module is Noetherian. From the theorem on homomorphisms it follows that a quotient module of a Noetherian module is also Noetherian. Furthermore, if for a U-module V its U-submodule V' and the quotient module V/V' are Noetherian, then V is also a Noetherian U-module. This implies, by induction on the number of direct summands, that the direct sum of a finite set of Noetherian U-modules is also a Noetherian U-module.

It is said that a U-module V is *finitely generated* if

$$V = U_{\mathrm{id}} \times v_1 + \cdots + U_{\mathrm{id}} \times v_q \tag{54.5}$$

for some $v_1, \dots, v_q \in V$, i.e., any element $v \in V$ is representable in the form

$$v = \beta_1 \cdot v_1 + u_1 \times v_1 + \cdots + \beta_q \cdot v_q + u_q \times v_q \qquad (\beta_i \in K,\ u_i \in U).$$

LEMMA 54.2. *A module V over an associative K-algebra U is Noetherian if and only if any its U-submodule is finitely generated.*

PROOF. *Necessity.* Let V' be a U-submodule of a Noetherian module V. Define by induction an ascending chain of U-submodules of V' by setting $V_0 = 0$ and $V_{i+1} = V_i + U_{\mathrm{id}} \times v_{i+1}$, where $v_{i+1} \in V' \setminus V_i$. Since V is Noetherian, we must have $V' = V_N$ for some $N \in \mathbb{N}$, hence, $v_1, \dots, v_N$ are generators of the U-module V.

Sufficiency. Let $V_1 \subseteq V_2 \subseteq \cdots$ be an ascending chain of U-submodules of V. Then $V' = \cup_{i=1}^{\infty} V_i$ is a finitely generated U-submodule of V. Hence, $V' = \sum_{i=1}^{q} U_{\mathrm{id}} \times v_i$ for some $v_i \in V'$. But in this case we have $v_i \in V_N$ for a sufficiently large number N. Therefore, $V' = V_N = V_{N+k}$. The lemma is proved.

As it has already been noticed in §53, any associative K-algebra U can be turned into a left U-module by setting $a \times u = a \cdot u$ $(a, u \in U)$. An associative K-algebra U is called *left Noetherian* if it is Noetherian as a left U-module.

PROPOSITION 54.1. *Any finitely generated left module V over an associative left Noetherian K-algebra U is a Noetherian U-module.*

PROOF. Suppose that for the module V decomposition (54.5) holds. Consider a linear mapping $\varphi\colon \oplus_{i=1}^{q} U_{\mathrm{id}} \to V$ for which

$$\varphi(a_1, \dots, a_q) = a_1 \times v_1 + \cdots + a_q \times v_q \qquad (a_i \in U_{\mathrm{id}}).$$

Obviously, φ is an epimorphism of U-modules. Since U_{id} is a Noetherian U-module ($U_{\text{id}} \supset U$ & $\dim_K U_{\text{id}}/U = 1$), the direct sum of q copies of it is also Noetherian. Therefore, the U-module V is Noetherian as a quotient module of a Noetherian module. Proposition 54.1 is proved.

THEOREM 54.2. *Let U be an associative K-algebra with unit and let $\mathcal{E}$ be a commutative Noetherian K-subalgebra of U containing the unit element of the algebra U. Suppose that U is generated over $\mathcal{E}$ by elements $e_1, \dots, e_m$ for which*

$$e_i e_j = e_j e_i + \sum_{q=1}^{m} c_{ij}^q e_q \qquad (c_{ij}^q \in \mathcal{E},\ i, j = 1, \dots, m), \tag{54.6}$$

$$e_j c = c e_j \qquad (j = 1, \dots, m,\ c \in \mathcal{E}). \tag{54.7}$$

Then U is a left Noetherian K-algebra.

PROOF. Let us introduce on the set $\mathbb{N}^m$ the total order $<$ in the same way as in the proof of Theorem 54.1 and denote by e_r, for any collection $r = \langle r_1, \dots, r_m \rangle \in \mathbb{N}^m$, the element of the algebra U equal to $e_1^{r_1} \cdots e_m^{r_m}$.

From relations (54.7) it follows that any element of $\mathcal{E}$ commutes with any element of U. Therefore, the linear space of U is spanned by the elements $c e_{i_1} \cdots e_{i_t}$, where c ranges over $\mathcal{E}$, $t \in \mathbb{N}$, and $i_1, \dots, i_t \in \{1, \dots, m\}$. Using relations (54.6) for $i > j$ in the same way as it was done in the proof of Poincaré–Birkhoff–Witt theorem, we can express these elements as sums of elements of the form $c \cdot e_r$, where $c \in \mathcal{E}$ and $r \in \mathbb{N}^m$. Hence, any element $v \in U$ is representable in the form

$$v = c(s)e_s + \sum_{p<s} c(p)e_p, \tag{54.8}$$

where $s, p \in \mathbb{N}^m$ and $c(s), c(p) \in \mathcal{E}$. In particular, relations (54.6) permit one to obtain for any $r, q \in \mathbb{N}^m$ and some $d_{rq}^p \in \mathcal{E}$ the following equality in U:

$$e_r \cdot e_q = e_{r+q} + \sum_{p<r+q} d_{rq}^p e_p \qquad (p \in \mathbb{N}^m). \tag{54.9}$$

Let V be an arbitrary nonzero left U-submodule of U. We consider all possible decompositions of the form (54.8) for all elements $v \in V$ and denote by $J(V)$ the subset of $\mathcal{E}$ consisting of all nonzero coefficients $c(s)$ which occur in all these decompositions.

For any element $f \in J(V)$ denote by $S(f)$ the subset of $\mathbb{N}^m$ consisting of those and only those elements s for which there exist elements $c(p) \in \mathcal{E}$ ($p \in \mathbb{N}^m$, $p < s$) such that

$$f e_s + \sum_{p<s} c(p)e_p \in V.$$

Let us consider the subset $S_0(f)$ of $S(f)$ formed by those and only those elements $r \in S(f)$ which cannot be represented in the form $r = r' + t$, where $r' \in S_0$ and $t \in \mathbb{N}^m$ & $t \neq (0, \dots, 0)$. Since $S(f)$ is a nonempty set, Lemma 51.3 implies that $S_0(f)$ is a finite set.

LEMMA 54.3. *The set* $T(V) = \cup_{f \in J(V)} S_0(f)$ *is finite.*

PROOF. Suppose that $T(V)$ is an infinite set. Then from the proof of Lemma 51.3 it follows that in $T(V)$ we can choose a countable set of distinct elements $s_1, s_2, \dots$ such that $s_{i+1} = s_i + t_i$, where $t_i \in \mathbb{N}^m$ and $t_i \neq (0, \dots, 0)$. For any s_i we choose $f_i \in J(V)$ such that $s_i \in S_0(f_i)$. From the definition of the set $S_0(f_i)$ it follows that there exist elements $c_i(p)$, $p \in \mathbb{N}^m$, such that the elements

$$v_i = f_i e_{s_i} + \sum_{p<s_i} c_i(p) e_p$$

belong to V. Since the algebra $\dot{\mathcal{E}}$ is left Noetherian (see Proposition 54.1), for a sufficiently large n we have in $\mathcal{E}$ the equality $f_{n+1} = g_1 f_1 + \dots + g_n f_n$ for some $g_i \in \mathcal{E}$. But in this case formula (54.9) implies

$$\sum_{i=1}^{n} g_i \cdot e_{s_n - s_i} \cdot v_i = f_{n+1} \cdot e_{s_n} + \sum_{p<s_n} c(p) e_p \in V.$$

Therefore, $s_n \in S(f_{n+1})$. However, $s_{n+1} = s_n + t_n$, where $t_n \neq (0, \dots, 0)$, which contradicts the fact that $s_{n+1} \in S_0(f_{n+1})$. The contradiction thus obtained proves the lemma.

From the definition of the set $S(f)$, the condition $f \in J(V)$ implies that for any $s \in S(f)$ there exists an element $v(s, f) \in V$ such that we have a decomposition of the form

$$v(s,t) = f \cdot e_s + \sum_{p<s} c(s,f,p) \cdot e_p \qquad (p \in \mathbb{N}^m,\ c(s,f,p) \in \mathcal{E}). \tag{54.10}$$

LEMMA 54.4. *The elements* $v(s, f)$, *where* s *ranges over* $T(V)$ *and* f *ranges over* $J(V)$, *are generators of the left* U*-module* V.

PROOF. Let v be an arbitrary nonzero element of V. Then it can be represented in the form (54.8), where $c(s) \neq 0$ and $c(s) \in J(V)$ & $s \in S(c(s))$. From the definition of the set $S_0(c(s))$ it follows that in $S_0(c(s))$ there exists an element r such that $s = r + t$ $(t \in \mathbb{N}^m)$. Then from formulas (54.8) and (54.10) we conclude that the element $v' = v - e_{s-s_0} v(s_0, c(s))$ is representable in the form

$$c'(s') e_{s'} + \sum_{p<s'} c'(p) e_p \qquad (s', p \in \mathbb{N}^m,\ c' \in \mathcal{E}),$$

where $s' < s$. Since the set of elements from $\mathbb{N}^m$ that are less than s is finite, the process of deleting the leading component just indicated will be interrupted at some finite step and we shall find that $v = \sum_{r \leq s} e_t \cdot v(r, f)$, where the sum is taken over a finite set $r \in T(V)$ and $f \in J(V)$. The lemma is proved.

By Lemma 54.2, in order to prove the theorem it suffices to show that the U-module V has a finite set of generators. From Lemma 54.4 it follows that we are to search for these generators in the set $L = \{ v(r, f) \mid r \in T(V),\ f \in J(V) \}$. Consider in a left $\mathcal{E}$-module V the $\mathcal{E}$-submodule $V' = \mathcal{E}L$. Since $T(V)$ is a finite set (Lemma 54.3), formula (54.10) implies that V' is contained in an $\mathcal{E}$-module V'' generated by a finite subset $\{e_p \mid p \leq q\}$, where q is a maximal element of $T(V)$ with respect to the comparison $<$. Since V'' is a finitely generated module over the Noetherian algebra $\mathcal{E}$, V'' is a Noetherian $\mathcal{E}$-module by Proposition 54.1, and the $\mathcal{E}$-module V' has a finite set of generators by Lemma 54.2. Denote these generators by $v_1, \dots, v_k$. Since $V' = \mathcal{E}L$, we have

$$v_i = \sum_{r,f} c_i(r, f) \cdot v(r, f) \qquad (i = 1, 2, \dots, k).$$

Let us denote by L' the subset of the elements $v(r, f)$ of L which occur in all these equalities with nonzero coefficients. Then $v_i \in \mathcal{E}L'$. Hence, $V' = \mathcal{E}L' \supset L$. Therefore, $V = U \cdot L = U \cdot \mathcal{E} \cdot L' = U \cdot L'$ and L' is a desired finite set of generators of the U-module V. The theorem is proved.

COROLLARY 1. *The universal enveloping algebra $U(\mathfrak{g})$ of an arbitrary finite-dimensional Lie algebra $\mathfrak{g}$ is a left Noetherian K-algebra.*

PROOF. As it was mentioned above, any associative K-algebra U is left Noetherian if and only if the K-algebra U_{id} is left Noetherian. Therefore, it suffices to prove that U_{id} is left Noetherian. Then, choosing an arbitrary basis $e_1, \dots, e_m$ of the Lie K-algebra $\mathfrak{g}$ and putting $\mathcal{E} = K \cdot 1$, we find that for U_{id} all assumptions of Theorem 54.2 hold. The corollary is proved.

COROLLARY 2. *Any finitely generated associative and commutative K-algebra $\mathcal{E}$ is (left) Noetherian.*

PROOF. Let $e_1, \dots, e_m$ be generators of the K-algebra $\mathcal{E}$ and let $\mathfrak{g}$ be a linear subspace of $\mathcal{E}$ spanned by these elements. It is clear that $\mathfrak{g}$ is a Lie subalgebra of the Lie algebra $\mathcal{E}^{(-)}$. Then, by the Poincaré–Birkhoff–Witt theorem, $\mathcal{E}$ is a homomorphic image of the universal enveloping algebra $U(\mathfrak{g})$, which is left Noetherian by Corollary 1 of Theorem 54.2. Hence, $\mathcal{E}$ is a left Noetherian K-algebra. The corollary is proved.

HISTORICAL SURVEY

Distinguishing of classes of objects under investigation by means of identities is a well-known algebraic device widely used in mathematics when considering various mathematical structures. In this way classes of algebraic objects that have become classical have been determined: the class of groups, the class of associative algebras, and the classes of alternative and Jordan algebras. At present, in the literature on the subject, two names of classes of universal algebras given by identical relations have taken root: "a variety of algebras" (G. Birkhoff, B. Neumann) and "a primitive class of algebras" (A. I. Mal′tsev).

As early as in the fundamental work of G. Birkhoff, on universal algebras [Bi] published in 1935, it was stated that each class of universal algebras given by identical relations is closed with respect to three operations concerning algebras of this class:

a) the operation of taking full Cartesian products,

b) the operation of taking subalgebras,

c) the operation of taking homomorphic images;

moreover, it was proved by him that this property of classes is a characteristic one, i.e., any class of universal algebras closed with respect to these three operations can be given by its own identities. G. Birkhoff's statement has clarified the notion of variety; it has explicitly indicated, in universal algebra, the relationship between the axiomatic and the model approaches when investigating primitive classes of algebras, and it has fixed two modes of defining such classes:

The first mode. A variety is given explicitly by its identities.

The second mode. A variety of algebras under investigation is given by concrete algebras which generate this variety with respect to the three operations mentioned above.

The availability of these two modes, and the further comprehension of the problem of effective construction of mathematical objects, has inevitably led to the general tasks of assessing the interrelation of these two modes. In the case of the first mode, of determining a variety by means of a concrete enumerable system of identities, a question is naturally bound to arise: is it possible to generate a given variety, with respect to the three operations mentioned, by effectively given algebras, for instance, by algebras with the solvable word problem. A special case of this task is the solvability of the word problem in the free algebra of arbitrary finite rank for the variety given in this way. The second mode of defining a variety leads, sooner or later, to the question of the possibility of an effective description of identities defining a given variety and, in particular, to the question of the possibility of determining this variety by a finite number of identities.

As a rule, the second mode of defining a class of algebraic objects under investigation preceded the first one. Concrete examples of new algebras of the same type appear at an earlier stage of research. It is only at the next stage that the question arises of discovering general properties of the objects introduced that ensure some or other theorems and results concerning these objects. In particular, this leads to a description of identities of algebras and the properties of algebras determined by these identities. But it is not always possible to obtain all the defining identities of the initially given algebras; in the majority of cases only those identities appear to be sufficient which ensure a formal derivation of most of the algebraic properties close to the properties of the original objects that are of interest to the investigator. In this case a primitive class of algebras analysed is implicitly broadened and a variety of algebras given by the discovered identities is introduced. The primitive class of Jordan algebras, for which the special Jordan algebras have been the initial objects under investigation, can serve as an example of such an introduction of a variety.

It is quite obvious that the language of identities and varieties is very poor. Many interesting classes of algebras cannot be given only by identities. Moreover, usually even within a variety, properties of concrete classes of algebras, for instance, of simple algebras, cannot be derived only from identical relations. These drawbacks seem to bypass another theory which uses the whole apparatus of mathematical logic, i.e., the theory of models that goes back to 1936, when the works of A. Tarsky on the theory of sets endowed with a given system of relations appeared. A. I. Mal′tsev, who in his paper published in 1936, pointed out the possibility of good use of the apparatus of mathematical logic in algebraic research both in the theory of universal algebras and in the classical section of algebra, namely, in group theory, became the passionate propagandist of this theory's ideas in the USSR. The further introduction of concepts and methods of mathematical logic into algebraic investigations led to the foundation of the theory of algebraic systems at the junction of algebra and logic by A. I. Mal′tsev, and to the subsequent research of axiomatic classes of algebraic systems.

Nevertheless, algebraists' interest in studying identities and varieties was gradually raising. In the first place this could be accounted for by the fact that complicated problems of the theory of concrete varieties were, little by little, acquiring acceptance and were becoming more and more common to a wide circle of algebraists, and their solution was becoming matter of prestige.

In this connection it is necessary to note the great role played by the initial papers of B. Neumann concerning the theory of subvarieties of the variety of groups and the subsequent works of Australian algebraists on the subject. The results of this work were collected in the monograph by H. Neumann "Varieties of Groups" (1967) translated into Russian in 1969. This book presented a very successful selection of problems reflecting 30 years of investigation of group varieties and determining the directions of the research for the following fifteen years. The same problems were reflected in the theory of varieties of associative, Jordan, alternative, and Lie algebras.

The problem of finite basis property of identities of concrete varieties appeared to be the central problem of this theory. Examples of group varieties without a finite basis of identities were given in 1969 by A. Yu. Ol′shanskiĭ and then practically simultaneously by S. I. Adyan and Vaughan-Lee (see [Ol, Ad1, V-L2]). This provided the negative solution of the general problem of finite basis property of subvarieties of the variety of all groups. In the papers of S. I. Adyan and Vaughan-Lee the varieties were explicitly given by independent systems of identities.

A great influence on the subsequent research of varieties and identities was exerted by the paper published in 1964 by S. Oates and M. B. Powell (see [OaP]) who had established the finite basis property for identities of any subvariety of the variety of all groups generated by a finite group. Later on an analog of this result was obtained for varieties of associative rings (I. V. L′vov, 1969, see [L′v1]), of Lie rings (Yu. A. Bakhturin and A. Yu. Ol′shanskiĭ, 1975, see [BakO]), of alternative rings (I. V. L′vov, 1975, see [L′v2]), of Jordan algebras (Yu. A. Medvedev, 1979, see [Med]), and of group representations (S. M. Vovsi, 1979, see [Vov]).

An example of a finite algebra not possessing a finite basis of identities was constructed in 1974 by S. V. Polin (see [Po]). In the class of all algebras this example outlined the bounds within which analogs of the S. Oates and M. B. Powell theorem should be looked for.

A paramount role in the reorientation and attracting the attention of algebraists toward problems on identities was played by problems of Burnside type. The 40s–50s became the turning point of this reorientation when these problems were solved for the most part.

For groups, the Burnside problem concerning the local finiteness of groups of restricted exponent appeared to be such a task. In 1958 for a sufficiently large odd number n P. S. Novikov found a word algorithm in a free group given by the identity $x^n = 1$ which implied the negative solution of the Burnside problem for such groups (a brief notice on the subject was published by P. S. Novikov in 1959, see [Nov], a detailed publication, for $n \geq 4381$, by P. S. Novikov and S. I. Adyan in 1968, see [NA]; further inprovement to $n \geq 665$ can be found in the monograph of S. I. Adyan [Ad2]).

For associative algebras the paper of J. Levitzky, published in 1946, proved to be valuable. This paper contains the positive solution of A. G. Kurosh's restricted problem on the local nilpotency of algebras satisfying the identity $x^n = 0$ and on the local finite dimensionality of algebraic algebras of restricted degree (see [Levit]). Together with the paper published in 1948 by I. Kaplansky (see [Kap1]) on the structure of primitive associative algebras satisfying a nontrivial polynomial identity, the paper of J. Levitzky marked the beginning of subsequent investigations of associative algebras satisfying nontrivial polynomial identities (the so called PI algebras) and of the study of identities of associative algebras.

Out of the results of the 50s in this direction the following theorems should be mentioned primarily: published in 1957 the theorems of S. A. Amitsur on the local nilpotency of the Jacobson radical of a finitely generated PI algebra and on the coincidence of the Jacobson radical of a relatively free PI algebra with the ideal of identities of a full matrix algebra over a field (see [Am]), and also the theorem of A. I. Shirshov on the height (see [Shi2]) which provided a constructive solution of the A. G. Kurosh's restricted problem.

Also in 1957, I. Kaplansky published a note (see [Kap3] and also [Kap4]) with a number of problems which proved to be useful for the further development of the theory of PI algebras and for the subsequent investigations of identical relations of associative algebras. Out of the publications of the 70s on the subject the following papers should be noted: published in 1972, the papers of E. Formanek [Fo] and of the author [Raz3], which gave the positive answer to the problem of I. Kaplansky on the existence of a multilinear central polynomial for full matrix algebras over an arbitrary commutative ring with unit, published in the same year as theorems of A. Regev on the tensor product of PI algebras and on the exponential growth of dimensions of spaces of multilinear polynomials in n variables in any free PI algebra over a field of characteristic zero (see [Reg]), published in 1974 the author's theorem on the nilpotence of the Jacobson radical of a finitely generated PI algebra in which all the identities of a certain full matrix algebra over a field of arbitrary characteristic or all the Capelli identities of some order over a field of characteristic zero are satisfied (see [Raz6]).

At the same period of time the interest for the problem of finite basis property for identities of varieties associative algebras was not diminishing. The papers of V. N. Latyshev (see [Lat1–Lat5]) were concerned with this subject. Sufficiently general conditions for the finite basis property of identities of some variety of associative algebras and all its subvarieties (this property of varieties was called the Specht property by V. N. Latyshev) have been independently discovered in the papers of V. N. Latyshev [Lat2] and G. K. Genov [Gen]. In particular, these results imply the Specht property of the variety generated by algebras of triangular matrices of arbitrary order over a field of characteristic zero.

The 80s were marked by the summing up of more than thirty-five years' activities of algebraists concerning specification and reinforcement of the aforementioned results of J. Levitzky and S. A. Amitsur. In the papers of the author, A. R. Kemer, and A. Braun (see [Raz11, Kem1, Bra]) the nilpotency of the Jacobson radical of an arbitrary finitely generated PI algebra was proved. (The author has proved that over a field of characteristic zero the Jacobson radical of a finitely generated PI algebra is nilpotent if and only if an algebraic identity of certain degree holds in this algebra (see [Raz11]). A. R. Kemer showed that from the results of A. Regev concerning the exponential growth of any nontrivial variety of associative algebras it follows that for any variety of PI algebras an algebraic identity of a sufficiently large degree holds (see [Dn]). This provided the solution of the problem for PI algebras over fields of characteristic zero. A. Brown succeeded in reducing the problem on nilpotency of the Jacobson radical for PI algebras of arbitrary characteristic to the aforementioned result of the present author concerning nilpotency of the radical of a finitely generated PI algebra in which all the identities of some full matrix algebra over a field are satisfied.)

The weakened Burnside problem on the boundedness of orders of all finite groups with a given number of generators and with the identity $x^p = 1$ and a series of papers by A. I. Kostrikin on the Engel Lie algebras, published in the 50s, have exerted a great influence on the study of identities of Lie algebras and their representations. As early as the beginning of the 50s it was well known that for any prime number p this problem is solved positively if any $(p-1)$th Engel Lie algebra over the field with p elements is locally finite. A. I. Kostrikin's papers (see [Kos2, Kos1]) not only founded the local nilpotency of Engel Lie algebras and provided the solution of an old and well-known problem that is the weakened Burnside problem for the groups of prime period p, but also gave incentive to the further study of identical relations of Lie algebras and of representations of Lie algebras. Substantially these papers for the first time introduced the notion of an identity of an associative-Lie pair (A, L) (L is a Lie algebra and A is its associative enveloping algebra) as a formula $f(t_1, \dots, t_l) = 0$ (f is an associative polynomial in noncommuting variables $t_1, t_2, \dots, t_l$) that turns out to be a true equality in the algebra A under the substitution of any elements of the Lie algebra L instead of $t_1, \dots, t_l$. This notion organically combines the notion of identity for an associative algebra (to this end it is enough to consider the identities of associative-Lie pairs (A, L) for which L coincides with $A^{(-)}$) and the notion of identity of the Lie algebra $\mathfrak{g}$ (to this end it suffices to restrict oneself by considering the identities of the associative-Lie pair (A, L), where $L = \operatorname{ad}\mathfrak{g}$ and A is the associative subalgebra of $\operatorname{End}_K \mathfrak{g}$ generated by $\operatorname{ad}\mathfrak{g}$). In its turn, this made it possible to study identities of associative algebras and of Lie algebras from unified positions by means of associative-Lie pairs and this also predetermined the further apportionment of the theory of varieties of associative-Lie pairs and their identities as a subject of independent research.

2. The method of 2-words appeared as a result of the solution of a series of concrete problems concerning the theory of varieties of Lie algebras, of associative algebras, and of groups. The notion of 2-word appeared in the author's paper [Raz1] published in 1971 in the solution of the nonnilpotency problem for $(p-1)$th Engel Lie algebras over a field of characteristic $p \geq 5$ posed by A. I. Kostrikin in 1959. In embryonic form, the notion of $\boldsymbol{\alpha}$-function on 2-words was used by me in the paper [Raz2] to give an explicit constructive tracing of a nonnilpotent $(p-2)$th Engel variety of Lie algebras for which any proper subvariety is nilpotent and to construct a locally finite almost Cross nonsolvable variety of groups of exponent p. The construction of a nonnilpotent $(p-2)$th Engel Lie algebra and an almost Cross variety proposed in the papers [Raz1, Raz2] was natural only for the case of the 3rd Engel Lie algebra over a field of characteristic 5 and for $p > 5$ it seemed to be a purely technical method which could hardly be used in other circumstances.

The observation essential for the formation of the method of 2-words was made in 1971. It was noticed that the function α on 2-words constructed in the papers [Raz1, Raz2] can be obtained in a natural way by the two-dimensional representation of the simple three-dimensional Lie algebra. This fact valuably simplified the proof of the nonnilpotency of the $(p-2)$th Engel Lie algebra and explained in some sense the existence of the Lie algebras of this kind over fields of characteristic $p \geq 5$; it was noticed by me in the paper [Raz3] published in 1972.

This observation gave incentive to the investigation in two related directions.

First, the construction of the function α on 2-words for the two-dimensional irreducible representation of the simple three-dimensional Lie algebra possessed a generalization for the case of an arbitrary irreducible representation of any semisimple Lie algebra $\mathfrak{g}$ in a finite-dimensional linear space V. Even in the special case of $\mathfrak{g} = \mathrm{sl}(n, K)$ and $\dim_K V = n$ this construction gave a complete description of all central polynomials of the full matrix algebra over the field, M_n, and led to the solution of the existence problem for central polynomials on the algebra M_n (see [Raz3]) posed by I. Kaplansky in 1957.

Secondly, the function α on 2-words constructed for the two-dimensional representation of the simple three-dimensional Lie algebra $\mathfrak{g} = \mathrm{sl}(2, K)$ permits one to apply the ideas of the paper [Raz2] for a formal study of identical relations of this representation and of the Lie algebra $\mathfrak{g}$, omitting model considerations. This part of the work was completed at the beginning of 1973 in the paper [Raz4], where the finite basis property was proved and bases of identities for the Lie algebra $\mathrm{sl}(2, K)$, for the quaternion algebra, and for the full matrix algebra M_2 of second order over a field of characteristic zero were constructed. This gave a solution of the problem posed by S. Amitsur.

An important step for the formation of the method of 2-words was taken in 1973. It was noticed that for any irreducible module V over a finite-dimensional Lie algebra $\mathfrak{g}$ possessing an invariant symmetric bilinear form a function α on 2-words with values in an algebraically closed field K can be constructed in a natural way, and this function has the properties which permit one to determine a variety of associative-Lie pairs $\mathfrak{B}_\alpha$ starting from this function only. Furthermore, $\mathfrak{B}_\alpha \subseteq \mathrm{var}(\mathrm{End}_K V, \mathfrak{g})$, and the equality could be not attained. This construction was applied for the Lie algebra $M_n = \mathrm{gl}(n, \mathbb{C})$ and its n-dimensional irreducible representation in the space V. (In this case we have $\mathfrak{B}_\alpha = \mathrm{var}\, M_n$.) The algorithm of calculating this special function α implies: a) the value of this function on any 2-word is a polynomial in the parameter n; b) the algorithm itself (of calculating the function α) requires an extension of the signature of the associative algebra by adding a unary operation "trace". This led, on the one hand, to the setting of the problem on trace identities for the full matrix algebra M_n and to the proof of the theorem that all trace identities in M_n are equivalent to the only trace identity expressing the Hamilton–Cayley theorem, and, on the other hand, and it was much more important, it gave the setting of a problem on the $\boldsymbol{\alpha}$-function itself for the first time. Certainly, this problem was stated only in a particular case, namely, only for the function α under consideration attaining its values in the polynomial ring in one variable. However, all the principal points were present: an extension of the signature, the ring of coefficients for the $\boldsymbol{\alpha}$-function, a description of all ideals of the commutative ring to which nontrivial ideals of trace identities are corresponding, and the description of trace identities determined by these ideals. All these results were published in 1974 in the author's paper [Raz7].

The conceptual formation of the method of 2-words was basically completed by the end of 1976, and both the basic notions and the main problem of this method were stated by me in the report "The nonsolvability of the variety of groups of exponent 4" at the International Conference on Ring Theory in Belgium in 1978 (see [Raz10]).

This made it possible to estimate the results obtained before from common positions, to simplify their proofs and above all to considerably extend the previous results on the identities of algebras and their representations and to obtain new ones.

Here are the major new results which were obtained by means of the method of 2-words after 1976.

Using the $\boldsymbol{\alpha}$-function for the three-dimensional irreducible representation of the simple three-dimensional Lie algebra, I succeeded in constructing a nonsolvable group of exponent 4 (see [Raz9]), which gave the answer to the problem posed by M. Hall and G. Higman in 1950. The $\boldsymbol{\alpha}$-function on 2-words constructed for any representation of the simple three-dimensional Lie algebra $\mathfrak{g}$ was studied, which led to the solution of the problem on the Specht property of the variety of associative-Lie pairs generated by the pair $(U(\mathfrak{g}), \mathfrak{g})$, where $U(\mathfrak{g})$ is the universal enveloping algebra of the Lie algebra $\mathfrak{g}$, (see [Raz15]) and led to the discovery of finite bases of identities for any irreducible representation of the Lie algebra $\mathfrak{g}$ over an algebraically closed field of characteristic zero (see [Tri1]). The study of properties of the $\boldsymbol{\alpha}$-function α on 2-words constructed for an arbitrary irreducible representation V of a semisimple Lie algebra $\mathfrak{g}$ over an algebraically closed field of characteristic zero gave the equality $\mathfrak{B}_\alpha = \mathrm{var}(U, \mathfrak{g})$, where U is a simple enveloping algebra of the Lie algebra $\mathfrak{g}$ whose central character coincides with the one of the $\mathfrak{g}$-module V. This equality led, first, to the construction of central polynomials of a special form for almost all irreducible representations of reductive Lie algebras and, second, to the construction of an associative polynomial which permits one to restore, in a constructive way, the algebra of infinitely differentiable functions on a smooth variety in terms of the Lie algebra of all vector fields of this variety (see [Raz18, Raz20]). The classification of ideals of trace identities in the already mentioned paper [Raz7], which was given within the method of 2-words for the series of representations of the Lie algebras $\mathfrak{g}_n \rightleftharpoons \mathrm{gl}(n, \mathbb{C}) = M_n$ in n-dimensional spaces, obtained an unexpected further development. It turned out that all the ideals of trace identities discovered in this paper admit a natural interpretation via ideals of trace identities of associative matrix superalgebras M_{nk} on which we must take the supertrace for the operation "trace". This gave the finite basis property and a complete description of trace identities for all matrix superalgebras M_{nk} (see [Raz23, Raz24]).

All the aforementioned results obtained by means of the method of 2-words constitute the content of Chapters II-V of the present book. In my opinion the perspectives of obtaining new results on identities of algebras and their representations are available in each of the three sections connected with the third, the fourth, and the fifth chapters of the monograph, but the main hope in the method of 2-words and $\boldsymbol{\alpha}$-functions is likely to be linked with the searching of new successful series of irreducible representations of reductive Lie algebras and with the solution for these series of the main problem of the theory of 2-words, mentioned in §11, in the spirit of Theorems 26.1, 26.2, 27.1, and 27.2.

REFERENCES

[Ad1] S. I. Adyan, *Infinite irreducible systems of group identities*, Izv. Akad. Nauk SSSR Ser. Mat. **34** (1970), no. 4, 715–734; English transl. in Math. USSR-Izv. **4** (1970).

[Ad2] ———, *The Burnside problem and identities in groups*, "Nauka", Moscow, 1975; English transl., Springer-Verlag, Berlin and New York, 1979.

[Am] S. A. Amitsur, *A generalization of Hilbert's Nullstellensatz*, Proc. Amer. Math. Soc. **8** (1957), 649–656.

[AR] S. A. Amitsur and A. Regev, *PI-algebras and their cocharacters*, J. Algebra **78** (1982), 248–254.

[BM] S. Bachmuth and H. Y. Mochizuki, *Third Engel groups and Macdonald–Neumann conjecture*, Bull. Austral. Math. Soc. **5** (1971), 379–386.

[BMW] S. Bachmuth, H. Y. Mochizuki, and D. W. Walkup, *A nonsolvable group of exponent* 5, Bull. Amer. Math. Soc. **76** (1970), 638–640.

[Bak] Yu. A. Bakhturin, *Identities in Lie algebras*, "Nauka", Moscow, 1985; English transl., VNU Science Press, Utrecht, 1987.

[BakO] Yu. A. Bakhturin and A. Yu. Ol′shanskiĭ, *Identical relations in finite Lie rings*, Mat. Sb. **96** (1975), no. 4, 543–559; English transl. in Math. USSR-Sb. **25** (1975).

[BMa] W. E. Baxter and W. S. Martindale III, *Central closure of semiprime non-associative rings*, Comm. Algebra **7** (1979), 1103–1132.

[Bi] G. Birkhoff, *On the structure of abstract algebras*, Proc. Cambridge Philos. Soc. **31** (1935), 433–454.

[BL] G. Birkhoff and J. D. Lipson, *Heterogeneous algebras*, J. Combinatorial Theory **8** (1970), 115–133.

[BiK] R. L. Bishop and R. J. Crittenden, *Geometry of manifolds*, Pure Appl. Math., vol. XV, Academic Press, New York, 1964.

[BW1] R. E. Block and R. L. Wilson, *On filtered Lie algebras and divided power algebras*, Comm. Algebra **3** (1975), 571–589.

[BW2] ———, *The restricted simple Lie algebras are of classical or Cartan type*, Proc. Nat. Acad. Sci. U.S.A. **81** (1984), 5271–5274.

[Bou] N. Bourbaki, *Groupes et algèbres de Lie*, Chaps. 1–3, Actualités Sci. Indust., nos. 1285, 1349, Hermann, Paris, 1971, 1972; English transl., Hermann, Paris and Addison–Wesley, Reading, MA, 1975.

[Bra] A. Braun, *The radical in a finitely generated P. I. algebras*, Bull. Amer. Math. Soc. **7** (1982), 385–386.

[Car] É. Cartan, *Les groupes de transformation continues, infinis, simples*, Ann. Sci. Ècole Norm. Sup. **26** (1905), 93–161.

[Coh] P. M. Cohn, *Universal algebra*, Harper & Row, New York and London, 1965.

[CR] C. W. Curtis and I. Reiner, *Representation theory of finite groups and associative algebras*, Pure Appl. Math., vol. XI, Interscience, New York, 1962.

[Dix] J. Dixmier, *Algèbres enveloppantes*, Gauthier–Villars, Paris, 1974; English transl., North–Holland Math. Library, vol. 14, North–Holland, Amsterdam and New York, 1977.

[Dn] V. A. Andrunakievich (ed.), *The Dniester notebook*, Inst. Mat., Sibirsk. Otdel., Akad. Nauk SSSR, Novosibirsk, 1982, pp. 17, 35. (Russian)

[Dre1] V. S. Drenskiĭ, *Identities in Lie algebras*, Algebra i Logika **13** (1974), no. 3, 265–290; English transl. in Algebra and Logic **13** (1974).

[Dre2] ———, *A minimal basis for identities of a second order matrix algebra over a field of characteristic* 0, Algebra i Logika **20** (1981), no. 3, 282–290; English transl. in Algebra and Logic **20** (1981).

[Dre3] ———, *Representations of the symmetric group and varieties of linear algebras*, Mat. Sb. **115** (1981), no. 1, 98–115; English transl. in Math. USSR-Sb. **43** (1981).

[Duf] M. Duflo, *Construction of primitive ideals in an enveloping algebra*, Lie Groups and Their Representations (Proc. of the Summer School, Bolyai Janoś Math. Soc., Budapest, 1971), Halsted, New York, 1975, pp. 77–93.

[ErMO] T. S. Erickson, W. S. Martindale III, and J. M. Osborn, *Prime nonassociative algebras*, Pacific J. Math. **60** (1975), 49–63.

[Evg] M. A. Evgrafov, *Asymptotic estimates and entire functions*, 3rd ed., "Nauka", Moscow, 1979; English transl. of 1st ed., Gordon and Breach, New York, 1961.

[Fo] E. Formanek, *Central polynomials for matrix rings*, J. Algebra **23** (1972), 129–132.

[Gen] G. K. Genov, *The Spechtness of certain varieties of associative algebras over a field of zero characteristic*, C. R. Acad. Bulgare Sci. **29** (1976), no. 7, 939–941. (Russian)

[GTsYu] V. M. Glushkov, G. E. Tseĭtlin, and E. L. Yushchenko, *Algebra, languages, programming*, "Naukova Dumka", Kiev, 1974. (Russian)

[Gol] E. S. Golod, *On nil-algebras and finitely approximable p-groups*, Izv. Akad. Nauk SSSR Ser. Mat. **28** (1964), 273–276; English transl. in Amer. Math. Soc. Transl. Ser. 2 **48** (1965).

[Hal1] M. Hall, Jr., *The theory of groups*, Macmillan, New York, 1959.

[Hal2] ———, *Combinatorial theory*, Blaisdell, Waltham, MA, 1967.

[Her] I. N. Herstein, *Noncommutative rings*, Carus Math. Monographs, No. 15, Math. Assoc. America, Washington, DC, 1968.

[Higg1] P. J. Higgins, *Algebras with a scheme of operators*, Math. Nachr. **27** (1963), 115–132.

[Higg2] ———, *Lie rings satisfying the Engel condition*, Proc. Cambridge Philos. Soc. **50** (1954), 8–15.

[Higg3] ———, *Notes on categories and groupoids*, Nostrand Reinhold, London, 1971.

[Higm] G. Higman, *On a conjecture of Nagata*, Proc. Cambridge Philos. Soc. **52** (1956), 1–4.

[Il] A. V. Il′tyakov, *The Specht property of ideals of identities of certain simple nonassociative algebras*, Algebra i Logika **24** (1985), no. 3, 327–351; English transl. in Algebra and Logic **24** (1985).

[Jam] G. D. James, *The representation theory of the symmetric groups*, Lecture Notes in Math., vol. 682, Springer-Verlag, Berlin, 1978.

[Jac1] N. Jacobson, *Lie algebras*, Pure Appl. Math., vol. X, Interscience, New York, 1962.

[Jac2] ———, *Structure of rings*, Amer. Math. Soc. Colloq. Publ., vol. 37, Amer. Math. Soc., Providence, RI, 1964.

[Kap1] I. Kaplansky, *Rings with a polynomial identity*, Bull. Amer. Math. Soc. **54** (1948), 575–580.

[Kap2] ———, *An introduction to differential algebra*, Actualités Sci. Indust., No. 1251, Hermann, Paris, 1957.

[Kap3] ———, *Problems in the theory of rings*, Report of a conference on linear algebras, June 1956, Publ. 502, Nat. Acad. Sci., National Research Council, Washington, DC, 1957, pp. 1–3.

[Kap4] ———, *"Problems in the theory of rings" revisited*, Amer. Math. Monthly **77** (1970), 445–454.

[Kats] V. G. Kats [V. Kac], *Simple irreducible graded Lie algebras of finite growth*, Izv. Akad. Nauk SSSR Ser. Mat. **32** (1968), no. 6, 1323–1367; English transl. in Math. USSR-Izv. **2** (1968).

[Kem1] A. R. Kemer, *Capelli identities and nilpotency of the radical of a finitely generated PI algebra*, Dokl. Akad. Nauk SSSR **255** (1980), no. 4, 793–797; English transl. in Soviet Math. Dokl. **22** (1980).

[Kem2] ———, *Decomposition of varieties*, Algebra i Logika **20** (1981), no. 4, 395-418; English transl. in Algebra and Logic **20** (1981).

[Kem3] ———, *Varieties and Z_2-graded algebras*, Izv. Akad. Nauk SSSR Ser. Mat. **48** (1984), no. 5, 1042–1059; English transl. in Math. USSR-Izv. **25** (1985).

[Kha1] V. K. Kharchenko, *The Galois theory of semiprime rings*, Algebra i Logika **16** (1977), no. 3, 313–363; English transl. in Algebra and Logic **16** (1977).

[Kha2] ———, *On centralizers in prime rings*, Algebra i Logika **20** (1981), no. 2, 231–247; English transl. in Algebra and Logic **20** (1981).

[Klya] A. A. Klyachko, *Lie elements in a tensor algebra*, Sibirsk. Mat. Zh. **15** (1974), no. 6, 1296–1304; English transl. in Siberian Math. J. **15** (1974).

[Kos1] A. A. Kostrikin, *Lie rings satisfying the Engel condition*, Izv. Akad. Nauk SSSR Ser. Mat. **21** (1957), no. 4, 515–540; English transl. in Amer. Math. Soc. Transl. Ser. 2 **45** (1965).

[Kos2] ———, *The Burnside problem*, Izv. Akad. Nauk SSSR. Ser. Mat. **23** (1959), no. 1, 3–34; English transl. in Amer. Math. Soc. Transl. Ser. 2 **36** (1964).

[Kos3] ———, *Sandwiches in Lie algebras*, Mat. Sb. **110 (152)** (1979), no. 1, 3–12; English transl. in Math. USSR-Sb. **38** (1981).

[Kos4] ———, *Around Burnside*, "Nauka", Moscow, 1986; English transl., Springer-Verlag, Berlin, 1990.

[KoSh] A. A. Kostrikin and I. R. Shafarevich, *Graded Lie algebras of finite characteristic*, Izv. Akad. Nauk SSSR. Ser. Mat. **33** (1969), no. 2, 251–322; English transl. in Math. USSR-Izv. **3** (1969).

[Kru] R. Kruse, *Identities satisfied by a finite ring*, J. Algebra **26** (1973), 298–318.

[Kuz] E. N. Kuz′min, *On the Nagata–Higman theorem*, Proc. of the Conf. dedicated to the 60th Birthday of Academician L. Iliev, Sofia, 1975, pp. 101–107. (Russian)

[KuR] A. Kh. Kushkuleĭ and Yu. P. Razmyslov, *Varieties generated by irreducible representations of Lie algebras*, Vestnik Moskov. Univ. Ser. I Mat. Mekh. **1983**, no. 5, 4–7; English transl. in Moscow Univ. Math. Bull. **38** (1983).

[Lam] J. Lambek, *Lectures on rings and modules*, Blaisdell, Waltham, MA, 1966.

[Lan] S. Lang, *Algebra*, Addison–Wesley, Reading, MA, 1965.

[Lat1] V. N. Latyshev, *On Regev's theorem on identities in a tensor product of* PI *algebras*, Uspekhi Mat. Nauk **27** (1972), no. 4, 213–214. (Russian)

[Lat2] ———, *Partially ordered sets and nonmatrix identities of associative algebras*, Algebra i Logika **15** (1976), no. 1, 53–70; English transl. in Algebra and Logic **15** (1976).

[Lat3] ———, *Finite basis property of the identities of certain rings*, Uspekhi Mat. Nauk **32** (1977), no. 4, 259–260. (Russian)

[Lat4] ———, *The complexity of nonmatrix varieties of associative algebras.* I, II, Algebra i Logika **16** (1977), no. 2, 149–183, 184–199; English transl. in Algebra and Logic **16** (1977).

[Lat5] ———, *Nonmatrix varieties of associative algebras*, Mat. Zametki **27** (1980), no. 1, 147–156; English transl. in Math. Notes **27** (1980).

[Levin] B. Ya. Levin, *Entire functions*, Izdat. Moskov. Gos. Univ., Moscow, 1971. (Russian)

[Levit] J. Levitzki, *A problem of A. Kurosch*, Bull. Amer. Math. Soc. **52** (1946), 1033–1035.

[L′v1] I. V. L′vov, *Varieties of associative rings.* I, II, Algebra i Logika **12** (1973), no. 3, 6, 269–297, 667–688; English transl. in Algebra and Logic **12** (1973).

[L′v2] ———, *Varieties generated by finite alternative rings*, Algebra i Logika **17** (1978), no. 3, 282–286; English transl. in Algebra and Logic **17** (1978).

[L′v3] ———, *The Brown theorem on the radical of a finitely generated* PI *algebra*, Preprint No. 63, Inst. Mat., Sibirsk. Otdel., Akad. Nauk SSSR, Novosibirsk, 1984. (Russian)

[Mac] S. Maclane, *Homology*, Grundlehren Math. Wiss., Band 114, Springer-Verlag, Berlin and New York, 1967.

[Mal1] A. I. Mal′tsev, *Untersuchungen aus dem Gebiete der mathematischen Logik*, Mat. Sb **1 (43)** (1936), 323–336; English transl., Chapter I in A. I. Mal′cev, The Metamathematics of Algebraic Systems. Collected Papers: 1936-1967, Studies in Logic and Foundations of Math., vol. 66, North–Holland, Amsterdam, 1971.

[Mal2] ———, *Model correspondences*, Izv. Akad. Nauk SSSR Ser. Mat. **23** (1959), no. 3, 313–336; English transl., Chapter XI in A. I. Mal′cev, The Metamathematics of Algebraic Systems. Collected Papers: 1936-1967, Studies in Logic and Foundations of Math., vol. 66, North–Holland, Amsterdam, 1971.

[Mal3] ———, *Algebraic systems*, "Nauka", Moscow, 1970; English transl., Grundlehren Math. Wiss., Band 192, Springer-Verlag, Berlin and New York, 1973.

[Mat] J. Matzuk, *Central closure of semiprime tensor products*, Comm. Algebra **10** (1982), 262–278.

[Med] Yu. A. Medvedev, *Identities of finite Jordan Φ-algebras*, Algebra i Logika **18** (1979), no. 6, 723–748; English transl. in Algebra and Logic **18** (1979).

[Mi] S. P. Mishchenko, *On the problem of the Engel property*, Mat. Sb. **124(166)** (1984), no. 1, 56–67; English transl. in Math. USSR-Sb. **52** (1985).

[B. N] B. H. Neumann, *Identical relations in groups*, Math. Ann. **114** (1937), 506–525.

[H. N] H. Neumann, *Varieties of groups*, Ergeb. Math. Grenzgeb., Band 37, Springer-Verlag, New York, 1967.

[Nov] P. S. Novikov, *On periodic groups*, Dokl. Akad. Nauk SSSR **127** (1959), 749–752. (Russian)

[NA] P. S. Novikov and S. I. Adyan, *Infinite periodic groups.* I, II, III, Izv. Akad. Nauk SSSR Ser. Mat. **32** (1968), no. 1, 2, 3, 212–244, 251–524, 709–731; English transl. in Math. USSR-Izv. **2** (1968).

[OaP] S. Oates and M. B. Powell, *Identical relations in finite groups*, J. Algebra **1** (1964), 11–39.

[Okh] S. V. Okhitin, *Stable T-ideals and central polynomials*, Vestnik Moskov. Univ. Ser. I Mat. Mekh. **1986**, no. 3, 85–88; English transl. in Moscow Univ. Math. Bull. **41** (1986).

[Ol] A. Yu. Ol′shanskiĭ, *The finite basis problem for identities in groups*, Izv. Akad. Nauk SSSR Ser. Mat. **34** (1970), no. 2, 376–384; English transl. in Math. USSR-Izv. **4** (1970).

[PlV] B. I. Plotkin and S. M. Vovsi, *Varieties of group representations. General theory, connections, and applications*, "Zinatne", Riga, 1983. (Russian)

[Po] S. V. Polin, *Identities of finite algebras*, Sibirsk. Mat. Zh. **17** (1976), no. 6, 1356–1366; English transl. in Siberian Math. J. **17** (1976).

[Pro1] C. Procesi, *The invariant theory of* $n \times n$ *matrices*, Adv. Math. **19** (1976), 306–381.

[Pro2] ———, *Trace identities and standard diagrams*, Ring Theory, Lecture Notes in Pure and Appl. Math., vol. 51, Dekker, New York, 1979, pp. 191–218.

[Raz1] Yu. P. Razmyslov, *Lie algebras satisfying Engel conditions*, Algebra i Logika **10** (1971), no. 1, 33–44; English transl. in Algebra and Logic **10** (1971).

[Raz2] ———, *A certain example of nonsolvable almost Cross varieties of groups*, Algebra i Logika **11** (1972), no. 2, 186–205; English transl. in Algebra and Logic **11** (1972).

[Raz3] ———, *A certain problem of Kaplansky*, Izv. Akad. Nauk SSSR Ser. Mat. **37** (1973), no. 3, 483–501; English transl. in Math. USSR-Izv. **7** (1973).

[Raz4] ———, *The existence of a finite basis for the identities of the matrix algebra of order two over a field of characteristic zero*, Algebra i Logika **12** (1973), no. 1, 83–113; English transl. in Algebra and Logic **12** (1973).

[Raz5] ———, *Existence of a finite base for certain varieties of algebras*, Algebra i Logika **13** (1974), no. 6, 685–693; English transl. in Algebra and Logic **13** (1974).

[Raz6] ———, *The Jacobson radical in* PI *algebras*, Algebra i Logika **13** (1974), no. 3, 337–360; English transl. in Algebra and Logic **13** (1974).

[Raz7] ———, *Identities with trace in full matrix algebras over a field of characteristic zero*, Izv. Akad. Nauk SSSR Ser. Mat. **38** (1974), no. 4, 723–756; English transl. in Math. USSR-Izv. **8** (1974).

[Raz8] ———, *The nonsolvability of the variety of groups of period* 4, Sixth All-Union Group Theory Symposium, Abstracts of Reports, "Naukova Dumka", Kiev, 1978, p. 50. (Russian)

[Raz9] ———, *The Hall–Higman problem*, Izv. Akad. Nauk SSSR Ser. Mat. **42** (1978), no. 4, 833–847; English transl. in Math. USSR-Izv. **13** (1978).

[Raz10] ———, *The nonsolvability of the variety of groups of exponent* 4, Ring Theory, Lecture Notes in Pure and Appl. Math., vol. 51, Dekker, New York, 1979, pp. 219–231.

[Raz11] ———, *Algebras satisfying identity relations of Capelli type*, Izv. Akad. Nauk SSSR Ser. Mat. **45** (1981), no. 1, 143–166; English transl. in Math. USSR-Izv. **18** (1982).

[Raz12] ———, *On a constructive way of reconstruction of the algebra of differentiable functions from the Lie algebra of all vector fields*, Fifth All-Union Symposium on The Theory of Rings, Algebras, and Modules, Abstracts of Reports, Inst. Mat., Sibirsk. Otdel., Akad. Nauk SSSR, Novosibirsk, 1982, p. 108. (Russian)

[Raz13] ———, *Varieties of representations of finite dimensional algebras in prime algebras*, Vestnik Moskov. Univ. Ser. I Mat. Mekh. **1982**, no. 6, 31–37; English transl. in Moscow Univ. Math. Bull **13** (1982).

[Raz14] ———, *Central polynomials in irreducible representations of a Lie algebra*, Fifth All-Union Symposium on The Theory of Rings, Algebras, and Modules, Abstracts of Reports, Inst. Mat., Sibirsk. Otdel., Akad. Nauk SSSR, Novosibirsk, 1982, p. 109. (Russian)

[Raz15] ———, *The existence of a finite basis for the identities of representations of the simple three-dimensional Lie algebra over a field of characteristic zero*, Algebra, Izdat. Moskov. Gos. Univ., Moscow, 1982, pp. 139–150; English transl. in Amer. Math. Soc. Transl. Ser. 2 **140** (1988).

[Raz16] ———, *Varieties generated by simple algebras*, Seventeenth All-Union Algebra Conf., Abstracts of Reports. Part 1, Inst. Mat., Akad. Nauk BSSR, Minsk, 1983, pp. 160–161. (Russian)

[Raz17] ———, *Trace identities on the superalgebras* $M(N, K)$, Seventeenth All-Union Algebra Conf., Abstracts of Reports. Part 1, Inst. Mat., Akad. Nauk BSSR, Minsk, 1983, p. 160. (Russian)

[Raz18] ———, *Central polynomials in irreducible representations of a semisimple Lie algebra*, Mat. Sb. **122 (164)** (1983), no. 1, 97–125; English transl. in Math. USSR-Sb. **50** (1985).

[Raz19] ———, *Bases of identities of irreducible representations of the three-dimensional Lie algebra*, Issledowaniya po Algebre (Studies in Algebra), Tbilisi Univ., Tbilisi, 1985, pp. 142–159. (Russian)

[Raz20] ———, *Constructive reconstruction of a smooth affine algebraic variety from its Lie algebra of vector fields, and simple Lie algebras satisfying a standard identity of degree* 5, Vīsnik Kiïv. Unīv. Ser. Mat. Mekh. **27** (1985), 94–96. (Russian)

[Raz21] ______, *Simple Lie algebras in varieties generated by Lie algebras of Cartan type*, Eighteenth All-Union Algebra Conf., Abstracts of Reports. Part 2, Inst. Mat., Vychisl. Tsentr., Akad. Nauk Moldav. SSR, Kishinev, 1985, p. 120. (Russian)

[Raz22] ______, *Simple Lie algebras that satisfy the standard Lie identity of degree* 5, Izv. Akad. Nauk SSSR Ser. Mat. **49** (1985), no. 3, 592–634; English transl. in Math. USSR-Izv. **26** (1986).

[Raz23] ______, *Trace identities and central polynomials in matrix superalgebras* $M(n,k)$, Mat. Sb. **128 (170)** (1985), no. 2, 194–215; English transl. in Math. USSR-Sb. **56** (1987).

[Raz24] ______, *Polynomial c-dual sets and central polynomials in matrix superalgebras* $M_{n,k}$, Vestnik Moskov. Univ. Ser. I Mat. Mekh. **1986**, no. 1, 49–52; English transl. in Moscow Univ. Math. Bull **41** (1986).

[Reg] A. Regev, *Existence of identities in* $A \otimes B$, Israel J. Math. **11** (1972), 131–152.

[Rit] J. F. Ritt, *Differential algebra*, Amer. Math. Soc. Colloq. Publ., vol. 33, Amer. Math. Soc., New York, 1950.

[Rob] G. de B. Robinson, *Representation theory of the symmetric group*, Univ. of Toronto Press, Toronto, 1961.

[San] I. N. Sanov, *Solution of Burnside's problem for exponent* 4, Leningrad. Gos. Univ. Uchen. Zap. Ser. Mat. Nauk **10** (1940), 166–170. (Russian)

[Ser] J.–P. Serre, *Lie algebras and Lie groups*, Lectures given at Harvard University, 1964, Benjamin, New York, 1965.

[SiS] I. M. Singer and S. Sternberg, *The infinite groups of Lie and Cartan.* I. *The transitive groups*, J. Analyse Math. **15** (1965), 1–114.

[Shab] B. V. Shabat, *Introduction to complex analysis*, Parts I, II, 3rd ed., "Nauka", Moscow, 1985; English transl. of Part II, Amer. Math. Soc., Providence, RI, 1992.

[Shaf] I. R. Shafarevich, *Foundations of algebraic geometry*, "Nauka", Moscow, 1972; English transl., Grundlehren Math. Wiss, Band 213, Springer-Verlag, Berlin and New York, 1974; Revised ed., 1977.

[Shi1] A. I. Shirshov, *On free Lie rings*, Mat. Sb. **45 (87)** (1958), 113-122. (Russian)

[Shi2] ______, *On rings with identity relations*, Mat. Sb. **43 (85)** (1957), no. 2, 277–283; English transl. in Amer. Math. Soc. Transl. Ser. 2 **119** (1983).

[Shi3] ______, *On the bases of a free Lie algebra*, Algebra i Logika **1** (1962), 14–19. (Russian)

[Sto] V. V. Stovba, *Finite basis property of certain varieties of Lie algebras and associative algebras*, Vestnik Moskov. Univ. Ser. I Mat. Mekh. **1982**, no. 2, 54–58; English transl. in Moscow Univ. Math. Bull. **37** (1982).

[Sum] E. A. Sumenkov, *An example in Lie algebra theory*, Third All-Union Symposium on The Theory of Rings, Algebras, and Modules, Abstracts of Reports, Tartu Univ., Tartu, 1976, pp. 97–98. (Russian)

[Tar] A. Tarski, *Der Wahrheitsbegriff in der formalisierten Sprachen*, Studia Philos. **1** (1936), 261–405.

[Tri1] I. M. Trishin, *Identities of irreducible representations of the three-dimensional simple Lie algebra*, Algebra i Logika **22** (1983), no. 3, 316–342; English transl. in Algebra and Logic **22** (1983).

[Tri2] ______, *Identities of representations of the Lie triple system of a bilinear form*, Preprint No. 875-B86, deposited Vsesoyuz. Inst. Nauchn. i Tekhn. Inform., Izdat. Moskov. Gos. Univ., Moscow, 1985. (Russian)

[V-L1] M. R. Vaughan–Lee, *Uncountably many varieties of groups*, Bull. London Math. Soc. **2** (1970), 280–286.

[V-L2] ______, *Derived lengths of Burnside groups of exponent* 4, Quart. J. Math. Oxford Ser. (2) **30** (1979), 495–504.

[Vei] B. Yu. Veĭsfeĭler [Boris Weisfeiler], *Infinite dimensional filtered Lie algebras and their connection with graded Lie algebras*, Funktsional. Anal. i Prilozhen. **2** (1968), no. 1, 94–95; English transl. in Functional Anal. Appl. **2** (1968).

[Vov] S. M. Vovsi, *Locally finite varieties of representations of groups*, Izv. Vyssh. Uchebn. Zaved. Mat. **1979**, no. 6, 16–27; English transl. in Soviet Math. (Iz. VUZ) **23** (1979).

[VdW] B. L. van der Waerden, *Algebra*, Ungar, New York and Amsterdam, 1970.

[We] H. Weyl, *The classical groups. Their invariants and representations*, Princeton Univ. Press, Princeton, NJ, 1939.

[Zhe] D. P. Zhelobenko, *Compact Lie groups and their representations*, "Nauka", Moscow, 1970; English transl., Transl. Math. Monographs, vol. 40, Amer. Math. Soc., Providence, RI, 1973.

[Zyr] A. N. Zyrichev, *Locally nilpotent* $(p-2)$*th Engel nonsolvable varieties of pairs over fields of characteristic* p, Seventeenth All-Union Algebra Conf., Abstracts of Reports, Part 1, Inst. Mat., Akad. Nauk BSSR, Minsk, 1983, p. 80. (Russian)

[ZR] A. N. Zyrichev and Yu. P. Razmyslov, *Skew-symmetric elements in the group algebra of a symmetric group*, Algebra i Logika **23** (1984), no. 2, 123–135; English transl. in Algebra and Logic **23** (1984).

Recent Titles in This Series

(Continued from the front of this publication)

97 **Itiro Tamura,** Topology of foliations: An introduction, 1992
96 **A. I. Markushevich,** Introduction to the classical theory of Abelian functions, 1992
95 **Guangchang Dong,** Nonlinear partial differential equations of second order, 1991
94 **Yu. S. Il′yashenko,** Finiteness theorems for limit cycles, 1991
93 **A. T. Fomenko and A. A. Tuzhilin,** Elements of the geometry and topology of minimal surfaces in three-dimensional space, 1991
92 **E. M. Nikishin and V. N. Sorokin,** Rational approximations and orthogonality, 1991
91 **Mamoru Mimura and Hirosi Toda,** Topology of Lie groups, I and II, 1991
90 **S. L. Sobolev,** Some applications of functional analysis in mathematical physics, third edition, 1991
89 **Valeriĭ V. Kozlov and Dmitriĭ V. Treshchëv,** Billiards: A genetic introduction to the dynamics of systems with impacts, 1991
88 **A. G. Khovanskiĭ,** Fewnomials, 1991
87 **Aleksandr Robertovich Kemer,** Ideals of identities of associative algebras, 1991
86 **V. M. Kadets and M. I. Kadets,** Rearrangements of series in Banach spaces, 1991
85 **Mikio Ise and Masaru Takeuchi,** Lie groups I, II, 1991
84 **Đáo Trọng Thi and A. T. Fomenko,** Minimal surfaces, stratified multivarifolds, and the Plateau problem, 1991
83 **N. I. Portenko,** Generalized diffusion processes, 1990
82 **Yasutaka Sibuya,** Linear differential equations in the complex domain: Problems of analytic continuation, 1990
81 **I. M. Gelfand and S. G. Gindikin, Editors,** Mathematical problems of tomography, 1990
80 **Junjiro Noguchi and Takushiro Ochiai,** Geometric function theory in several complex variables, 1990
79 **N. I. Akhiezer,** Elements of the theory of elliptic functions, 1990
78 **A. V. Skorokhod,** Asymptotic methods of the theory of stochastic differential equations, 1989
77 **V. M. Filippov,** Variational principles for nonpotential operators, 1989
76 **Phillip A. Griffiths,** Introduction to algebraic curves, 1989
75 **B. S. Kashin and A. A. Saakyan,** Orthogonal series, 1989
74 **V. I. Yudovich,** The linearization method in hydrodynamical stability theory, 1989
73 **Yu. G. Reshetnyak,** Space mappings with bounded distortion, 1989
72 **A. V. Pogorelev,** Bendings of surfaces and stability of shells, 1988
71 **A. S. Markus,** Introduction to the spectral theory of polynomial operator pencils, 1988
70 **N. I. Akhiezer,** Lectures on integral transforms, 1988
69 **V. N. Saliĭ,** Lattices with unique complements, 1988
68 **A. G. Postnikov,** Introduction to analytic number theory, 1988
67 **A. G. Dragalin,** Mathematical intuitionism: Introduction to proof theory, 1988
66 **Ye Yan-Qian,** Theory of limit cycles, 1986
65 **V. M. Zolotarev,** One-dimensional stable distributions, 1986
64 **M. M. Lavrent′ev, V. G. Romanov, and S. P. Shishat·skiĭ,** Ill-posed problems of mathematical physics and analysis, 1986
63 **Yu. M. Berezanskiĭ,** Selfadjoint operators in spaces of functions of infinitely many variables, 1986
62 **S. L. Krushkal′, B. N. Apanasov, and N. A. Gusevskiĭ,** Kleinian groups and uniformization in examples and problems, 1986
61 **B. V. Shabat,** Distribution of values of holomorphic mappings, 1985
60 **B. A. Kushner,** Lectures on constructive mathematical analysis, 1984
59 **G. P. Egorychev,** Integral representation and the computation of combinatorial sums, 1984
58 **L. A. Aĭzenberg and A. P. Yuzhakov,** Integral representations and residues in multidimensional complex analysis, 1983

(See the AMS catalog for earlier titles)